Lecture Notes in Computer Science　　9261

Commenced Publication in 1973
Founding and Former Series Editors:
Gerhard Goos, Juris Hartmanis, and Jan van Leeuwen

Editorial Board

More information about this series at http://www.springer.com/series/7409

Qiming Chen · Abdelkader Hameurlain
Farouk Toumani · Roland Wagner
Hendrik Decker (Eds.)

Database and Expert Systems Applications

26th International Conference, DEXA 2015
Valencia, Spain, September 1–4, 2015
Proceedings, Part I

 Springer

Editors
Qiming Chen
Hewlett-Packard Enterprise
Sunnyvale, CA
USA

Abdelkader Hameurlain
Paul Sabatier University
Toulouse
France

Farouk Toumani
Blaise Pascal University
Aubiere
France

Roland Wagner
University of Linz
Linz
Austria

Hendrik Decker
Universidad Politécnica de Valencia
Valencia
Spain

ISSN 0302-9743   ISSN 1611-3349   (electronic)
Lecture Notes in Computer Science
ISBN 978-3-319-22848-8   ISBN 978-3-319-22849-5   (eBook)
DOI 10.1007/978-3-319-22849-5

Library of Congress Control Number: 2015946079

LNCS Sublibrary: SL3 – Information Systems and Applications, incl. Internet/Web, and HCI

Printed on acid-free paper

Springer International Publishing AG Switzerland is part of Springer Science+Business Media
(www.springer.com)

Preface

Since 1990, DEXA has established itself as a premier forum in the intersection of data management, knowledge engineering, and artificial intelligence, providing a unique opportunity for the researchers, users, practitioners, and developers from multiple disciplines to present the state-of-the-art in these fields, exchange research ideas, share industry experiences, and explore future directions, which will help proliferate interdisciplinary discovery, drive innovation, and provide commercial opportunity. The 26th International Conference on Database and Expert Systems Applications took place in Valencia, Spain, during September 1–4, 2015. We are proud to present its program in these proceedings.

Each of the main days of the conference started out with a keynote by a distinguished scientist: Shahram Ghandeharizadeh from USC Database Laboratory of University of Southern California, USA; Juan Carlos Perez Cortes from Universidad Politecnica de Valencia (UPV), Spain; and Roland Traunmüller from Johannes Kepler Universität Linz, Austria. Accompanying the main conference were eight related conferences and workshops.

We received 125 submissions for the research track and industrial tracks, which is a testament to our community's high regard for DEXA, and we thank all authors for submitting their innovative work to the conference.

The Program Committee (PC) comprised highly reputational scientists from all over the word. The paper selection process consisted of three distinct phases: the initial reviews by the PC members with each paper being evaluated by four reviewers on average, the early opinion exchanges, and the further discussion for fine-tuning of the review results. This year many submissions report competitive scientific and engineering breakthroughs; however, because of space limitations we had to reject some high-quality papers. The final program featured 40 full papers and 32 short papers.

In these volumes, we also include, as a special section, the papers accepted for the 8th International Conference on Data Management in Cloud, Grid and P2P Systems (Globe 2015), which took place in Valencia, Spain, on September 1, 2015. The Globe conference provides opportunities for academics and industry researchers to present, exchange, and discuss the latest data management research and applications in cloud, grid, and peer-to-peer systems. Globe 2015 received 13 submissions for the research track. Each paper was evaluated by three reviewers on average. The reviewing process led to the acceptance of eight full papers for presentation at the conference and inclusion in this LNCS volume. The selected papers focus mainly on MapReduce framework (e.g., load balancing, optimization), security, data privacy, query rewriting, and streaming.

As is the tradition of DEXA conferences, all accepted papers are published by Springer. Authors of selected papers presented at the conference will be invited to submit extended versions of their papers for publication in the Springer journal *Transactions on Large-Scale Data- and Knowledge-Centered Systems (TLDKS)*.

The success of DEXA 2015 is a result of collegial teamwork from many individuals. We thank Isidro Ramos, Technical University of Valencia, Spain, the honorary chair of DEXA 2015. We thank Vladimir Marik, Czech Technical University, Czech Republic, the publication chair, as well as Marcus Spies, Ludwig-Maximilians-Universität München, Germany, A. Min Tjoa, Technical University of Vienna, Austria, and Roland R. Wagner, FAW, University of Linz, Austria, the workshop chairs. Our appreciations also go to all the PC members, the external reviewers, and the session chairs. Furthermore, we especially express our deep appreciation to Gabriela Wagner, the DEXA secretary, and to the local organization team, for their outstanding work put in over many months.

Finally, we once again thank all the authors, presenters, and participants of the conference. We hope that you enjoy the proceedings.

June 2015   Abdelkader Hameurlain
                                                                         Qiming Chen
                                                                       Farouk Toumani

Organization

General Chair

Hendrik Decker   Instituto Tecnológico de Informática, Valencia, Spain
Roland R. Wagner   Johannes Kepler University Linz, Austria

Program Committee Co-chairs

Abdelkader Hameurlain   IRIT, Paul Sabatier University, Toulouse, France
Qiming Chen   HP Labs Palo Alto, CA, USA

Honorary Chair

Isidro Ramos   Technical University of Valencia, Spain

Program Committee

Abdennadher Slim	German University, Cairo, Egypt
Abramowicz Witold	The Poznan University of Economics, Poland
Afsarmanesh Hamideh	University of Amsterdam, The Netherlands
Albertoni Riccardo	Institute of Applied Mathematics and Information Technologies - Italian National Council of Research, Italy
Alfaro Eva	UPV, Spain
Anane Rachid	Coventry University, UK
Appice Annalisa	Università degli Studi di Bari, Italy
Atay Mustafa	Winston-Salem State University, USA
Bakiras Spiridon	City University of New York, USA
Bao Jie	Microsoft Research Asia, China
Bao Zhifeng	National University of Singapore, Singapore
Bellatreche Ladjel	ENSMA, France
Bennani Nadia	INSA Lyon, France
Benyoucef Morad	University of Ottawa, Canada
Berrut Catherine	Grenoble University, France
Biswas Debmalya	Swisscom, Switzerland
Bouguettaya Athman	RMIT, Australia
Boussaid Omar	University of Lyon, France
Bressan Stephane	National University of Singapore, Singapore
Bu Yingyi	University of California, Irvine, USA
Camarinha-Matos Luis M.	Universidade Nova de Lisboa + Uninova, Portugal
Catania Barbara	DISI, University of Genoa, Italy

Ceci Michelangelo	University of Bari, Italy
Chen Cindy	University of Massachusetts Lowell, USA
Chen Phoebe	La Trobe University, Australia
Chen Qiming	HP Labs Palo Alto, CA, USA
Chen Shu-Ching	Florida International University, USA
Cheng Hao	Yahoo, USA
Chevalier Max	IRIT - SIG, Université de Toulouse, France
Choi Byron	Hong Kong Baptist University, Hong Kong, SAR China
Christiansen Henning	Roskilde University, Denmark
Chun Soon Ae	City University of New York, USA
Dahl Deborah	Conversational Technologies, USA
Darmont Jérôme	Université de Lyon, ERIC Lyon 2, France
de Carvalho Andre	University of Sao Paulo, Brazil
De Virgilio Roberto	Università Roma Tre, Italy
Decker Hendrik	Instituto Tecnológico de Informática, Valencia, Spain, Spain
Deng Zhi-Hong	Peking University, China
Deufemia Vincenzo	Università degli Studi di Salerno, Italy
Dibie-Barthélemy Juliette	AgroParisTech, France
Ding Ying	Indiana University, USA
Dobbie Gill	University of Auckland, New Zealand
Dou Dejing	University of Oregon, USA
du Mouza Cedric	CNAM, France
Eder Johann	University of Klagenfurt, Austria
El-Beltagy Samhaa	Nile University, Cairo, Egypt
Embury Suzanne	The University of Manchester, UK
Endres Markus	University of Augsburg, Germany
Fazzinga Bettina	ICAR-CNR, Italy
Fegaras Leonidas	The University of Texas at Arlington, USA
Felea Victor	"Al. I. Cuza" University of Iasi, Romania
Ferilli Stefano	University of Bari, Italy
Ferrarotti Flavio	Software Competence Center Hagenberg, Austria
Ferrucci Filomena	Università di Salerno, Italy
Fomichov Vladimir	School of Business Informatics, National Research University Higher School of Economics, Moscow, Russian Federation
Frasincar Flavius	Erasmus University Rotterdam, The Netherlands
Freudenthaler Bernhard	Software Competence Center Hagenberg GmbH, Austria
Fukuda Hiroaki	Shibaura Institute of Technology, Japan
Furnell Steven	Plymouth University, UK
Gangopadhyay Aryya	University of Maryland Baltimore County, USA
Gao Yunjun	Zhejiang University, China
Garfield Joy	University of Worcester, UK
Gergatsoulis Manolis	Ionian University, Greece

Grabot Bernard	LGP-ENIT, France
Grandi Fabio	University of Bologna, Italy
Gravino Carmine	University of Salerno, Italy
Groppe Sven	Lübeck University, Germany
Grosky William	University of Michigan, USA
Grzymala-Busse Jerzy	University of Kansas, USA
Guerra Francesco	Università degli Studi Di Modena e Reggio Emilia, Italy
Guerrini Giovanna	University of Genova, Italy
Guzzo Antonella	University of Calabria, Italy
Hameurlain Abdelkader	Paul Sabatier University, France
Hamidah Ibrahim	Universiti Putra Malaysia, Malaysia
Hara Takahiro	Osaka University, Japan
Hsu Wynne	National University of Singapore, Singapore
Hua Yu	Huazhong University of Science and Technology, China
Huang Jimmy	York University, Canada, Canada
Huang Xiao-Yu	South China University of Technology, China
Huptych Michal	Czech Technical University in Prague, Czech Republic
Hwang San-Yih	National Sun Yat-Sen University, Taiwan
Härder Theo	TU Kaiserslautern, Germany
Iacob Ionut Emil	Georgia Southern University, USA
Ilarri Sergio	University of Zaragoza, Spain
Imine Abdessamad	Inria Grand Nancy, France
Ishihara Yasunori	Osaka University, Japan
Jin Peiquan	University of Science and Technology of China, China
Kao Anne	Boeing, USA
Karagiannis Dimitris	University of Vienna, Austria
Katzenbeisser Stefan	Technische Universität Darmstadt, Germany
Kim Sang-Wook	Hanyang University, Republic of Korea
Kitagawa Hiroyuki	University of Tsukuba, Japan
Kleiner Carsten	University of Applied Sciences & Arts Hannover, Germany
Koehler Henning	Massey University, New Zealand
Korpeoglu Ibrahim	Bilkent University, Turkey
Kosch Harald	University of Passau, Germany
Krátký Michal	Technical University of Ostrava, Czech Republic
Kremen Petr	Czech Technical University in Prague, Czech Republic
Küng Josef	University of Linz, Austria
Lammari Nadira	CNAM, France
Lamperti Gianfranco	University of Brescia, Italy
Laurent Anne	LIRMM, University of Montpellier 2, France
Léger Alain	FT R&D Orange Labs Rennes, France
Lhotska Lenka	Czech Technical University, Czech Republic
Liang Wenxin	Dalian University of Technology, China
Ling Tok Wang	National University of Singapore, Singapore

Link Sebastian	The University of Auckland, New Zealand
Liu Chuan-Ming	National Taipei University of Technology, Taiwan
Liu Hong-Cheu	University of South Australia, Australia
Liu Rui	HP, USA
Lloret Gazo Jorge	University of Zaragoza, Spain
Loucopoulos Peri	Harokopio University of Athens, Greece
Lu Jianguo	University of Windsor, Canada
Lumini Alessandra	University of Bologna, Italy
Ma Hui	Victoria University of Wellington, New Zealand
Ma Qiang	Kyoto University, Japan
Maag Stephane	TELECOM SudParis, France
Masciari Elio	ICAR-CNR, Università della Calabria, Italy
Medjahed Brahim	University of Michigan - Dearborn, USA
Mishra Alok	Atilim University, Ankara, Turkey
Mishra Harekrishna	Institute of Rural Management Anand, India
Misra Sanjay	University of Technology, Minna, Nigeria
Mocito Jose	MakeWise, Portugal
Moench Lars	University of Hagen, Germany
Mokadem Riad	IRIT, Paul Sabatier University, France
Moon Yang-Sae	Kangwon National University, Republic of Korea
Morvan Franck	IRIT, Paul Sabatier University, France
Munoz-Escoi Francesc	Universitat Politecnica de Valencia, Spain
Navas-Delgado Ismael	University of Málaga, Spain
Ng Wilfred	Hong Kong University of Science & Technology, Hong Kong, SAR China
Nieves Acedo Javier	University of Deusto, Spain
Oussalah Mourad	University of Nantes, France
Ozsoyoglu Gultekin	Case Western Reserve University, USA
Pallis George	University of Cyprus, Cyprus
Paprzycki Marcin	Polish Academy of Sciences, Warsaw Management Academy, Poland
Pastor Lopez Oscar	Universidad Politecnica de Valencia, Spain
Patel Dhaval	Indian Institute of Technology Roorkee, India
Pivert Olivier	Ecole Nationale Supérieure des Sciences Appliquées et de Technologie, France
Pizzuti Clara	Institute for High Performance Computing and Networking (ICAR)-National Research Council (CNR), Italy
Poncelet Pascal	LIRMM, France
Pourabbas Elaheh	National Research Council, Italy
Qin Jianbin	University of New South Wales, Australia
Rabitti Fausto	ISTI, CNR Pisa, Italy
Raibulet Claudia	Università degli Studi di Milano-Bicocca, Italy
Ramos Isidro	Technical University of Valencia, Spain
Rao Praveen	University of Missouri-Kansas City, USA
Rege Manjeet	University of St. Thomas, USA

Resende Rodolfo F.	Federal University of Minas Gerais, Brazil
Roncancio Claudia	Grenoble University/LIG, France
Ruckhaus Edna	Universidad Simon Bolivar, Venezuela
Ruffolo Massimo	ICAR-CNR, Italy
Sacco Giovanni Maria	University of Turin, Italy
Saltenis Simonas	Aalborg University, Denmark
Sansone Carlo	Università di Napoli "Federico II", Italy
Santos Grueiro Igor	Deusto University, Spain
Sanz Ismael	Universitat Jaume I, Spain
Sarda N.L.	I.I.T. Bombay, India
Savonnet Marinette	University of Burgundy, France
Scheuermann Peter	Northwestern University, USA
Schewe Klaus-Dieter	Software Competence Centre Hagenberg, Austria
Schweighofer Erich	University of Vienna, Austria
Sedes Florence	IRIT, Paul Sabatier University, Toulouse, France
Selmaoui Nazha	University of New Caledonia, New Caledonia
Siarry Patrick	Université Paris 12 (LiSSi), France
Silaghi Gheorghe Cosmin	Babes-Bolyai University of Cluj-Napoca, Romania
Skaf-Molli Hala	Nantes University, France
Sokolinsky Leonid	South Ural State University, Russian Federation
Srinivasan Bala	Monash University, Australia
Straccia Umberto	ISTI - CNR, Italy
Sunderraman Raj	Georgia State University, USA
Taniar David	Monash University, Australia
Teisseire Maguelonne	Irstea - TETIS, France
Tessaris Sergio	Free University of Bozen-Bolzano, Italy
Teste Olivier	IRIT, University of Toulouse, France
Teufel Stephanie	University of Fribourg, Switzerland
Teuhola Jukka	University of Turku, Finland
Thalheim Bernhard	Christian-Albrechts-Universität zu Kiel, Germany
Thevenin Jean-Marc	University of Toulouse 1 Capitole, France
Thoma Helmut	Thoma SW-Engineering, Basel, Switzerland
Tjoa A. Min	Vienna University of Technology, Austria
Torra Vicenc	University of Skövde, Sweden
Truta Traian Marius	Northern Kentucky University, USA
Tzouramanis Theodoros	University of the Aegean, Greece
Vaira Lucia	University of Salento, Italy
Vidyasankar Krishnamurthy	Memorial University of Newfoundland, Canada
Vieira Marco	University of Coimbra, Portugal
Wang Junhu	Griffith University, Brisbane, Australia
Wang Qing	The Australian National University, Australia
Wang Wendy Hui	Stevens Institute of Technology, USA
Weber Gerald	The University of Auckland, New Zealand
Wijsen Jef	Université de Mons, Belgium
Wu Huayu	Institute for Infocomm Research, A*STAR, Singapore
Xu Lai	Bournemouth University, UK

Yang Ming Hour	Chung Yuan Christian University, Taiwan
Yang Xiaochun	Northeastern University, China
Yao Junjie	ECNU, China
Yin Hongzhi	The University of Queensland, Australia
Yokota Haruo	Tokyo Institute of Technology, Japan
Zeng Zhigang	Huazhong University of Science and Technology, China
Zhang Xiuzhen (Jenny)	RMIT University Australia, Australia
Zhao Yanchang	RDataMining.com, Australia
Zhu Qiang	The University of Michigan, USA
Zhu Yan	Southwest Jiaotong University, China

External Reviewers

Yimin Yang	Florida International University, USA
Hsin-Yu Ha	Florida International University, USA
Samira Pouyanfar	Florida International University, USA
Anas Katib	University of Missouri-Kansas City, USA
Ermelinda Oro	ICAR-CNR - Italy
Ángel Luis Garrido	University of Zaragoza, Spain
Pasquale Salza	University of Salerno, Italy
Erald Troja	City University of New York, USA
Konstantinos Nikolopoulos	City University of New York, USA
Bin Mu	City University of New York, USA
Gang Qian	University of Central Oklahoma, USA
Alok Watve	Google, USA
Satya Motheramgari	Harman International, USA
Adrian Caciula	Georgia Southern University, USA
Weiqing Wang	The University of Queensland, Australia
Hao Wang	University of Oregon, USA
Fernando Gutierrez	University of Oregon, USA
Sabin Kafle	University of Oregon, USA
Yun Peng	Qilu University of Technology, China
Kiki Maulana	Monash University, Australia
Qiong Fang	HKUST, Hong Kong, SAR China
Valentina Indelli Pisano	University of Salerno, Italy
Meriem Laifa	University Bordj Bou Arreridj, Algeria
Jiyi Li	Kyoto University, Japan
Chenyi Zhuang	Kyoto University, Japan
Matthew Damigos	NTUA, Greece
Lefteris Kalogeros	Ionian University, Greece
Franca Debole	ISTI-CNR, Italy
Claudio Gennaro	ISTI-CNR, Italy
Fabrizio Falchi	ISTI-CNR, Italy
Sajib Mistry	RMIT, Australia
Hai Dong	RMIT, Australia

Azadeh Ghari-Neiat	RMIT, Australia
Xu Zhuang	Southwest Jiaotong University, China
Paul de Vrieze	Bournemouth University, UK
Amine Abdaoui	LIRMM, France
Mike Donald Tapi Nzali	LIRMM, France
Florence Wang	LIRMM, France
Wenxin Liang	Dalian University of Technology, China
Yosuke Watanabe	Nagoya University, Japan
Xi Fang	Tokyo Institute of Technology, Japan
Gianvito Pio	University of Bari, Italy
Cyril Labbé	University of Grenoble, France
Patrick Roocks	University of Augsburg, Germany
Jorge Bernardino	Polytechnic Institute of Coimbra, Portugal
Bruno Cabral	University of Coimbra, Portugal
Ivano Elia	University of Coimbra, Portugal
Nuno Antunes	University of Coimbra, Portugal
Souad Boukhadouma	University of Nantes, France
Nouredine Gasmallah	University of Nantes, France
Quentin Grossetti	CNAM, France
Camelia Constantin	UPMC, France
Frederic Flouvat	PPME, University of New Caledonia
Jia-Ning Luo	Ming Chuan University, Taiwan
Christos Kalyvas	University of the Aegean, Greece
Athanasios Kokkos	University of the Aegean, Greece
Eirini Molla	University of the Aegean, Greece
Stéphane Jean	LIAS/ISAE-ENSMA, France
Selma Khouri	LIAS/ISAE-ENSMA, France
Selma Bouarar	LIAS/ISAE-ENSMA, France
Julius Köpke	University of Klagenfurt, Austria
Frédéric Flouvat	PPME University of New Caledonia, New Caledonia

Organization of the Special Section Globe 2015 (8th International Conference on Data Management in Cloud, Grid and P2P Systems)

Conference Program Chairs

Abdelkader Hameurlain	IRIT, Paul Sabatier University, Toulouse, France
Farouk Toumani	LIMOS, Blaise Pascal University, Clermont-Ferrand, France

Program Committee

Fabricio B. Alves	FIOCRUZ - Fundação Oswaldo Cruz - Rio de Janeiro, Brazil
Djamal Benslimane	LIRIS, University of Lyon, France
Qiming Chen	HP Labs Palo Alto, CA, USA
Thomas Cerqueus	University College Dublin, Ireland
Frédéric Cuppens	Telecom, Bretagne, France
Tran Khanh Dang	HCMC University of Technology, Ho Chi Minh City, Vietnam
Bruno Defude	Telecom INT, Evry, France
Tasos Gounaris	Aristotle University of Thessaloniki, Greece
Maria Indrawan-Santiago	Faculty of Information Technology Monash University, Melbourne, Australia
Sergio Ilarri	University of Zaragoza, Spain
Rui Liu	HP Labs Palo Alto, CA, USA
Gildas Menier	IRISA-UBS, University of South Bretagne, France
Anirban Mondal	Xerox Research Lab, Bangladore, India
Riad Mokadem	IRIT, Paul Sabatier University, Toulouse, France
Franck Morvan	IRIT, Paul Sabatier University, Toulouse, France
Kjetil Nørvåg	Norwegian University of Science and Technology, Trondheim, Norway
Jean-Marc Pierson	IRIT, Paul Sabatier University, Toulouse, France
Claudia Roncancio	LIG, Grenoble University, France
Soror Sahri	LIPADE, Descartes Paris University, France
Florence Sedes	IRIT, Paul Sabatier University, Toulouse, France
Mário J. Gaspar da Silva	ST/INESC-ID, Lisbon, Portugal
Hala Skaf-Molli	LINA, Nantes University, France
Wolfram Wöß	FAW, University of Linz, Austria
Shaoyi Yin	IRIT, Paul Sabatier University, Toulouse, France

Keynote Talks

SQL, NoSQL, and Next Generation Data Stores (Extended Abstract)

Shahram Ghandeharizadeh

Computer Science Department
University of Southern California
Los Angeles, CA 90089, USA
shahram@usc.edu

Introduction

SQL systems have been a cornerstone of data intensive applications for decades. Examples include commercial products by vendors such as Oracle and Microsoft and open source software such as MySQL. These system provide the following common features:

- An implementation of the relational data model.
- SQL as a query language that enables a user to identify "what" data should be retrieved.
- The concept of a transaction as a sequence of SQL statements and other programming constructs (e.g., if-else, for) that execute atomically.
- Indexing mechanisms for efficient processing of SQL queries, updates, and deletes.
- A query optimizer to identify an efficient plan for processing a SQL statement.
- Concurrency control to guarantee isolated execution of multiple transactions executing simultaneously.
- Crash-recovery for durability of transactions in the presence of failures.
- Physical data independence, the ability to change the details of the data organization such as an indexing strategy or storage devices with no re-write of the application software.

With SQL systems aging beyond a quarter of century, a debate is raging on the design decisions of next generation data stores. These data stores are anticipated to scale both vertically and horizontally for simple operations [2]. A simple operation reads and writes a small amount of data. This is in contrast to complex queries or joins that may read and write a large number of records in a table. An example simple operation is to look up the profile of a member of a social networking site such as Facebook or LinkedIn and display simple analytic such as the member's number of friends. Vertical scaling is the ability to process a larger number of simple operations by increasing the resources of a single server. Example resources include a server's number of CPU cores, number of network interface cards, disk bandwidth using RAID, bandwidth provided by the disk host bust adapter, etc. Horizontal scaling is the ability to use multiple servers to process a larger number of simple operations. This is realized by distributing both data and the load of the simple operations across the available servers. Ideally, each simple operation should be processed by a single server to avoid the communication overhead of parallelism associated with coordinating multiple servers.

These new data stores may differ from a traditional SQL system in a number of ways. First, they may abandon the relational data model for an alternative such as a graph or a document (JSON) data model. Second, they may use a weaker forms of consistency such as Basically Available, Soft state, Eventually Consistent (BASE). Third, they may augment a SQL system with a cache such as memcached [4] that stores the result of operations for future look up, avoiding processing of repeated queries. Fourth, they may fundamentally change physical design of data for processing queries to consider the new emerging non-volatile memory, NVM. Table 1 shows several candidate NVMs and their comparison with today's DRAM, NAND Flash, and magnetic disk.

Table 1. Alternative data storage technologies [3].

	Memristor	FeRAM	PCM	STT-RAM	DRAM	NAND Flash	Disk
Read Time (ns)	< 10	20–40	20–70	10–30	10–50	25,000	$2\text{-}8\times10^6$
Write Time (ns)	20–30	10–65	50–500	13–95	10–50	200,000	$4\text{-}8\times10^6$
Retention	> 10 years	~10 years	< 10 years	Weeks	< 100 msec	~10 years	~10 years
Energy/bit $(pJ)^2$	0.1–3	0.01–1	2–100	0.1–1	2–4	$10\text{-}10^4$	$10^6\text{-}10^7$
3D capability	Yes	Yes	Yes	No	No	Yes	N/A

This invited keynote provides an overview of these new data stores and their design decisions. We identify the scarcity of benchmarks to substantiate their claims as an open research opportunity. Next, we present a social networking benchmark named BG [1] and its use to compare a graph data store with a document data store and a traditional SQL system. We conclude by presenting next-generation data stores that exercise the byte-addressability feature of the emerging NVM.

References

1. Barahmand, S., Ghandeharizadeh, S.: BG: a Benchmark to evaluate interactive social networking actions. In: Proceedings of 2013 CIDR, January 2013
2. Cattell, R.: Scalable SQL and NoSQL data stores. SIGMOD Rec. **39**, 12–27 (2011)
3. Fink, M.: Beyond DRAM and Flash, Part 2: New Memory Technology for the Data Deluge, HP Next (2014). http://www8.hp.com/hpnext/posts/beyond-dram-and-flash-part-2-new-memory-technology-data-deluge.vcb6vrbcfe8
4. Nishtala, R., Fugal, H., Grimm, S., Kwiatkowski, M., Lee, H., Li, H.C., McElroy, R., Paleczny, M., Peek, D., Saab, P., Stafford, D., Tung, T., Venkataramani, V.: Scaling memcache at facebook. In: NSDI (2013)

Pattern Recognition in Embedded Systems: An Overview

Juan-Carlos Perez-Cortes

Instituto Tecnologico de Informatica, Universitat Politecnica de Valencia,
Valencia, Spain
jcperez@iti.upv.es

Abstract. Many Pattern Recognition tasks must be implemented with embedded systems, since interactions with humans and with the environment are among the most frequent functions they perform. Traditionally seen as an area of Artificial Intelligence (AI) that comes after the perception process, a classical PR application takes a set of physical measures obtained by sensors or communication subsystems and generates actions using actuators or provides information through the display or an I/O port.

In this talk, some of the application domains, optimization techniques and important, tradeoffs associated to the computational issues and time constraints involved, along with open problems and solution methodologies, are presented.

Contents – Part I

Keynote Talk

Pattern Recognition in Embedded Systems: An Overview   3
   Juan-Carlos Perez-Cortes

Temporal, Spatial and High Dimensional Databases

Restricted Shortest Path in Temporal Graphs 13
   Sudip Biswas, Arnab Ganguly, and Rahul Shah

An Efficient Distributed Index for Geospatial Databases 28
   Le Hong Van and Atsuhiro Takasu

The xBR$^+$-tree: An Efficient Access Method for Points 43
   George Roumelis, Michael Vassilakopoulos, Thanasis Loukopoulos,
   Antonio Corral, and Yannis Manolopoulos

Semantic Web and Ontologies

Probabilistic Error Detecting in Numerical Linked Data 61
   Huiying Li, Yuanyuan Li, Feifei Xu, and Xinyu Zhong

From General to Specialized Domain: Analyzing Three Crucial Problems
of Biomedical Entity Disambiguation. 76
   Stefan Zwicklbauer, Christin Seifert, and Michael Granitzer

Ontology Matching with Knowledge Rules 94
   Shangpu Jiang, Daniel Lowd, and Dejing Dou

Modeling, Linked Open Data

Detection of Sequences with Anomalous Behavior in a Workflow Process . . . 111
   Marcelo G. Armentano and Analía A. Amandi

An Energy Model for Detecting Community in PPI Networks 119
   Yin Pang, Lin Bai, and Kaili Bu

Filtering Inaccurate Entity Co-references on the Linked Open Data 128
   John Cuzzola, Ebrahim Bagheri, and Jelena Jovanovic

Quality Metrics for Linked Open Data. 144
   Behshid Behkamal, Mohsen Kahani, and Ebrahim Bagheri

NoSQL, NewSQL, Data Integration

A Framework of Write Optimization on Read-Optimized Out-of-Core
Column-Store Databases   155
   Feng Yu and Wen-Chi Hou

Integrating Big Data and Relational Data with a Functional SQL-like
Query Language   170
   Carlyna Bondiombouy, Boyan Kolev, Oleksandra Levchenko,
   and Patrick Valduriez

Comparative Performance Evaluation of Relational and NoSQL Databases
for Spatial and Mobile Applications   186
   Pedro O. Santos, Mirella M. Moro, and Clodoveu A. Davis Jr.

Uncertain Data and Inconsistency Tolerance

Query Answering Explanation in Inconsistent Datalog+/−
Knowledge Bases   203
   Abdallah Arioua, Nouredine Tamani, and Madalina Croitoru

PARTY: A Mobile System for Efficiently Assessing the Probability
of Extensions in a Debate   220
   Bettina Fazzinga, Sergio Flesca, Francesco Parisi,
   and Adriana Pietramala

Uncertain Groupings: Probabilistic Combination of Grouping Data   236
   Brend Wanders, Maurice van Keulen, and Paul van der Vet

Database System Architecture

Cost-Model Oblivious Database Tuning with Reinforcement Learning   253
   Debabrota Basu, Qian Lin, Weidong Chen, Hoang Tam Vo,
   Zihong Yuan, Pierre Senellart, and Stéphane Bressan

Towards Making Database Systems PCM-Compliant   269
   Vishesh Garg, Abhimanyu Singh, and Jayant R. Haritsa

Workload-Aware Self-Tuning Histograms of String Data   285
   Nickolas Zoulis, Effrosyni Mavroudi, Anna Lykoura,
   Angelos Charalambidis, and Stasinos Konstantopoulos

Data Mining I

Data Partitioning for Fast Mining of Frequent Itemsets in Massively
Distributed Environments   303
   Saber Salah, Reza Akbarinia, and Florent Masseglia

Does Multilevel Semantic Representation Improve Text Categorization?   319
Cheng Wang, Haojin Yang, and Christoph Meinel

Parallel Canopy Clustering on GPUs   334
*Yusuke Kozawa, Fumitaka Hayashi, Toshiyuki Amagasa,
and Hiroyuki Kitagawa*

Query Processing and Optimization

Efficient Storage and Query Processing of Large String in Oracle   351
George Eadon, Eugene Inseok Chong, and Ananth Raghavan

SAM: A Sorting Approach for Optimizing Multijoin Queries   367
Yong Zeng, Amy Nan Lu, Lu Xia, Chris Xing Tian, and Y.C. Tay

GPU Acceleration of Set Similarity Joins   384
*Mateus S.H. Cruz, Yusuke Kozawa, Toshiyuki Amagasa,
and Hiroyuki Kitagawa*

Data Mining II

Parallel Eclat for Opportunistic Mining of Frequent Itemsets   401
*Junqiang Liu, Yongsheng Wu, Qingfeng Zhou, Benjamin C.M. Fung,
Fanghui Chen, and Binxiao Yu*

Sequential Data Analytics by Means of Seq-SQL Language   416
*Bartosz Bebel, Tomasz Cichowicz, Tadeusz Morzy, Filip Rytwiński,
Robert Wrembel, and Christian Koncilia*

Clustering Attributed Multi-graphs with Information Ranking   432
*Andreas Papadopoulos, Dimitrios Rafailidis, George Pallis,
and Marios D. Dikaiakos*

Indexing and Decision Support Systems

Building Space-Efficient Inverted Indexes on Low-Cardinality Dimensions . . .   449
Vasilis Spyropoulos and Yannis Kotidis

A Decision Support System for Hotel Facilities Inventory Management   460
*Giuseppe Monteleone, Raffaele Di Natale, Piero Conca,
Salvatore Michele Biondi, Antonio Rosario Intilisano, Vincenzo Catania,
and Daniela Panno*

TopCom: Index for Shortest Distance Query in Directed Graph   471
Vachik S. Dave and Mohammad Al Hasan

A Universal Distributed Indexing Scheme for Data Centers
with Tree-Like Topologies   481
  Yuang Liu, Xiaofeng Gao, and Guihai Chen

Data Mining III

Improving Diversity Performance of Association Rule Based
Recommender Systems   499
  M. Kumara Swamy and P. Krishna Reddy

A Prime Number Based Approach for Closed Frequent Itemset Mining
in Big Data   509
  Mehdi Zitouni, Reza Akbarinia, Sadok Ben Yahia,
  and Florent Masseglia

Multilingual Documents Clustering Based on Closed Concepts Mining   517
  Mohamed Chebel, Chiraz Latiri, and Eric Gaussier

Modeling, Extraction, Social Networks

Analyzing the Strength of Co-authorship Ties with Neighborhood Overlap. . .   527
  Michele A. Brandão and Mirella M. Moro

Event Extraction from Unstructured Text Data   543
  Chao Shang, Anand Panangadan, and Viktor K. Prasanna

A Cluster-Based Epidemic Model for Retweeting Trend Prediction on
Micro-blog   558
  Zhuonan Feng, Yiping Li, Li Jin, and Ling Feng

Author Index   575

Contents – Part II

Knowledge Management and Consistency

A Logic Based Approach for Restoring Consistency in P2P Deductive
Databases.. 3
    Luciano Caroprese and Ester Zumpano

Expert System with Web Interface Based on Logic of Plausible Reasoning .. 13
    Grzegorz Legień, Bartłomiej Śnieżyński, Dorota Wilk-Kołodziejczyk,
    Stanisława Kluska-Nawarecka, Edward Nawarecki,
    and Krzysztof Jaśkowiec

Extending Knowledge-Based Profile Matching in the Human Resources
Domain.. 21
    Alejandra Lorena Paoletti, Jorge Martinez-Gil,
    and Klaus-Dieter Schewe

Sensitive Business Process Modeling for Knowledge Management 36
    Mariam Ben Hassen, Mohamed Turki, and Faïez Gargouri

Mobility, Privacy and Security

Partial Order Preserving Encryption Search Trees 49
    Kyriakos Ispoglou, Christos Makris, Yannis C. Stamatiou,
    Elias C. Stavropoulos, Athanasios K. Tsakalidis, and Vasileios Iosifidis

mobiSurround: An Auditory User Interface for Geo-Service Delivery 57
    Keith Gardiner, Charlie Cullen, and James D. Carswell

A Diversity-Seeking Mobile News App Based on Difference Analysis
of News Articles... 73
    Keisuke Kiritoshi and Qiang Ma

K^{UR}-Algorithm: From Position to Trajectory Privacy Protection
in Location-Based Applications............................... 82
    Trong Nhan Phan, Josef Küng, and Tran Khanh Dang

Data Streams, Web Services

Candidate Pruning Technique for Skyline Computation Over Frequent
Update Streams.. 93
    Kamalas Udomlamlert, Takahiro Hara, and Shojiro Nishio

Mining Frequent Closed Flows Based on Approximate Support
with a Sliding Window over Packet Streams   109
   Imen Brahmi, Hanen Brahmi, and Sadok Ben Yahia

D-FOPA: A Dynamic Final Object Pruning Algorithm to Efficiently
Produce Skyline Points Over Data Streams   117
   *Stephanie Alibrandi, Sofia Bravo, Marlene Goncalves,
and Maria-Esther Vidal*

GraphEvol: A Graph Evolution Technique for Web Service Composition   134
   Alexandre Sawczuk da Silva, Hui Ma, and Mengjie Zhang

Distributed, Parallel and Cloud Databases

Can Data Integration Quality Be Enhanced on Multi-cloud Using SLA?   145
   *Daniel A.S. Carvalho, Plácido A. Souza Neto, Genoveva Vargas-Solar,
Nadia Bennani, and Chirine Ghedira*

An Efficient Gear-Shifting Power-Proportional Distributed File System   153
   Hieu Hanh Le, Satoshi Hikida, and Haruo Yokota

Highly Efficient Parallel Framework: A Divide-and-Conquer Approach   162
   Takaya Kawakatsu, Akira Kinoshita, Atsuhiro Takasu, and Jun Adachi

Ontology-Driven Data Partitioning and Recovery for Flexible Query
Answering   177
   Lena Wiese

Information Retrieval

Detecting Near-Duplicate Documents Using Sentence Level Features   195
   Jinbo Feng and Shengli Wu

A Dwell Time-Based Technique for Personalised Ranking Model   205
   Safiya Al-Sharji, Martin Beer, and Elizabeth Uruchurtu

An Evaluation of Diversification Techniques   215
   *Duong Chi Thang, Nguyen Thanh Tam, Nguyen Quoc Viet Hung,
and Karl Aberer*

XML and Semi-structured Data

TOIX: Temporal Object Indexing for XML Documents   235
   Rasha Bin-Thalab and Neamat El-Tazi

Expressing and Processing Path-Centric XML Queries.................   250
  Huayu Wu, Dongxu Shao, Ruiming Tang, Tok Wang Ling,
  and Stéphane Bressan

A Logical Framework for XML Reference Specification   258
  C. Combi, A. Masini, B. Oliboni, and M. Zorzi

XQuery Testing from XML Schema Based Random Test Cases   268
  Jesús M. Almendros-Jiménez and Antonio Becerra-Terón

Data Partitioning, Indexing

Grid-File: Towards to a Flash Efficient Multi-dimensional Index..........   285
  Athanasios Fevgas and Panayiotis Bozanis

Supporting Fluctuating Transactional Workload   295
  Ibrahima Gueye, Idrissa Sarr, Hubert Naacke, and Joseph Ndong

Indexing Multi-dimensional Data in Modular Data Centers..............   304
  Libo Gao, Yatao Zhang, Xiaofeng Gao, and Guihai Chen

Data Mining IV, Applications

A Modified Tripartite Model for Document Representation in Internet
Sociology..   323
  Mikhail Alexandrov, Vera Danilova, and Xavier Blanco

Improving Financial Time Series Prediction Through Output Classification
by a Neural Network Ensemble   331
  Felipe Giacomel, Adriano C.M. Pereira, and Renata Galante

Mining Strongly Correlated Intervals with Hypergraphs...............   339
  Hao Wang, Dejing Dou, Yue Fang, and Yongli Zhang

WWW and Databases

Diagnosis Model for an Organization Based on Social Network Analysis....   351
  Dongwook Park, Soungwoong Yoon, and Sanghoon Lee

Integration Method for Complex Queries Based on Hyponymy Relations....   359
  Daisuke Kitayama and Takuma Matsumoto

An Expertise-Based Framework for Supporting Enterprise Applications
Development   369
  Devís Bianchini, Valeria De Antonellis, and Michele Melchiori

Data Management Algorithms

A Linear Program for Holistic Matching: Assessment on Schema Matching
Benchmark.   383
    Alain Berro, Imen Megdiche, and Olivier Teste

An External Memory Algorithm for All-Pairs Regular Path Problem   399
    Nobutaka Suzuki, Kosetsu Ikeda, and Yeondae Kwon

Special Section Globe 2015 – MapReduce Framework: Load Balancing, Optimization and Classification

An Efficient Solution for Processing Skewed MapReduce Jobs.   417
    Reza Akbarinia, Miguel Liroz-Gistau, Divyakant Agrawal,
    and Patrick Valduriez

MapReduce-DBMS: An Integration Model for Big Data Management
and Optimization.   430
    Dhouha Jemal, Rim Faiz, Ahcène Boukorca, and Ladjel Bellatreche

A Travel Time Prediction Algorithm Using Rule-Based Classification
on MapReduce   440
    HyunJo Lee, Seungtae Hong, Hyung Jin Kim, and Jae-Woo Chang

Special Section Globe 2015 – Security, Data Privacy and Consistency

Numerical SQL Value Expressions Over Encrypted Cloud Databases   455
    Sushil Jajodia, Witold Litwin, and Thomas Schwarz

A Privacy-Aware Framework for Decentralized Online Social Networks   479
    Andrea De Salve, Paolo Mori, and Laura Ricci

CaLibRe: A Better Consistency-Latency Tradeoff for Quorum Based
Replication Systems   491
    Sathiya Prabhu Kumar, Sylvain Lefebvre, Raja Chiky,
    and Eric Gressier-Soudan

Special Section Globe 2015 – Query Rewriting and Streaming

QTor: A Flexible Publish/Subscribe Peer-to-Peer Organization Based
on Query Rewriting.   507
    Sébastien Dufromentel, Sylvie Cazalens, François Lesueur,
    and Philippe Lamarre

Model for Performance Analysis of Distributed Stream Processing
Applications... 520
  Filip Nalepa, Michal Batko, and Pavel Zezula

Author Index 535

Keynote Talk

Pattern Recognition in Embedded Systems: An Overview

Juan-Carlos Perez-Cortes[✉]

Instituto Tecnologico de Informatica, Universitat Politecnica de Valencia,
Valencia, Spain
jcperez@iti.upv.es

Abstract. Many Pattern Recognition tasks must be implemented with embedded systems, since interactions with humans and with the environment are among the most frequent functions they perform. Traditionally seen as an area of Artificial Intelligence (AI) that comes after the perception process, a classical PR application takes a set of physical measures obtained by sensors or communication subsystems and generates actions using actuators or provides information through the display or an I/O port.

In this talk, some of the application domains, optimization techniques and important, tradeoffs associated to the computational issues and time constraints involved, along with open problems and solution methodologies, are presented.

Keywords: Pattern recognition · Embedded systems · Sensors · Signal processing · Artificial intelligence

1 Introduction

Some embedded systems work exclusively with direct physical measurements taken from sensors and produce actions after simple computations. Other systems, at the next level of complexity, include additional inputs coming from user interaction, such as push-buttons, switches or potentiometers. And the most sophisticated embedded systems need to take complex decisions from inputs which are ambiguous, incomplete or with a high variability, and include more advanced user interactions such as finger gestures, voice or images.

Due to the computational needs and algorithmic complexity of some Pattern Recognition processes, along with the high speed and, sometimes, strict real time constraints of the tasks they deal with, the adaptation or development of Embedded Pattern Recognition Systems is an important challenge to the designers. In this talk, an overview of typical application domains and design guidelines are presented along with open problems and current trends.

2 Pattern Recognition

Pattern Recognition is a multidisciplinary area with a long history [7] and a broad range of sources, theoretical foundations, functional paradigms, techniques

© Springer International Publishing Switzerland 2015
Q. Chen et al. (Eds.): DEXA 2015, Part I, LNCS 9261, pp. 3–10, 2015.
DOI: 10.1007/978-3-319-22849-5_1

and applications, as depicted in Fig. 1. From a reductionist but practical point of view, Pattern Recognition can be characterized as a discipline that studies the potential capabilities of machines to perform one or more of the following tasks on a signal or set of signals:

- Extract non-trivial information, learning to distinguish *patterns* of interest according to some criterion.
- Take a set of actions or decisions after analyzing some *patterns* in the input.
- Output a semantic representation of the results of a cognitive task on the input *patterns*.

Where a *pattern* is an entity defined in the context of the task at hand. We can find elements as different as a human face, a DNA sequence, a heartbeat in an electrocardiogram, a spoken sentence, a fingerprint image, a handwritten word or a chromosome image.

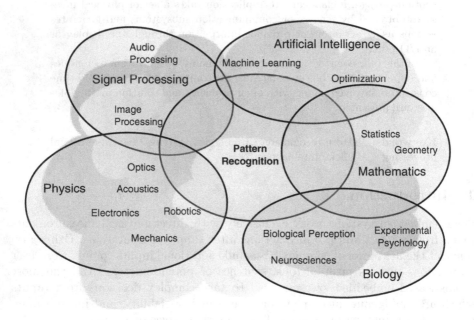

Fig. 1. Different disciplines contributing to pattern recognition

Embedded systems have traditionally used Digital Signal Processing algorithms and techniques to deal with analog or digital sensor measurements [5]. Complex Pattern Recognition applications can be seen as an evolution of those first succesful systems where the calculations were kept in the signal space, i.e. with no need to define a new representation space where probability density functions had to be estimated, to obtain an output from the sensed inputs.

The evolution describes a path where new techniques rely on the results of the previous stages to get a final result. This can be seen in Fig. 2

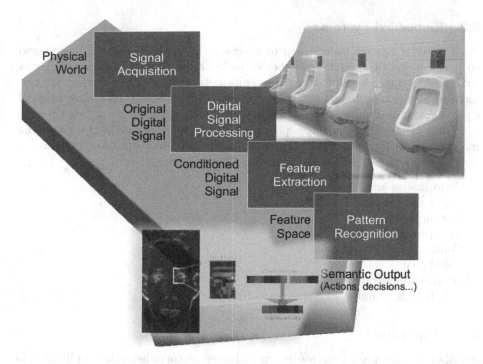

Fig. 2. Evolution path of Embedded Systems from Signal Processing to Pattern Recognition. Some systems should be kept as simple as possible since they do not need expensive hardware or sensors (the upper-right image in the figure is an example of a presence detection system where a single photocell is clearly preferable to a video camera)

2.1 Different Signals Used in Pattern Recognition Tasks

Among the signals that are typically processed by Pattern Recognition systems, sound is prevalent. It is one of the main carriers of sensorial data in humans. Therefore, a number of useful tasks have to deal with audio signals, such as: speech recognition, speaker identification, speech translation, music transcription, music retrieval, preventive maintenance, plague detection, etc.

On the other hand, vision is the richest of the human senses and a large amount of daily tasks have to do with images. Even in some cases other kind of information is first translated into images and then interpreted. Typical examples are: medical imaging, many biometric modalities, optical character recognition, behaviour analysis, industrial inspection, remote sensing, robot vision, etc.

And there are also many other kinds of input data that can be processed by Pattern Recognition algorithms to carry out practical tasks, such as: biolectrical signals, temperature, pressure, etc., signals from industrial sensors, time series or data sets from physical processes, social or economic activities, scientific measurements, etc.

2.2 Embedded Systems

An *Embedded System*, as opposed to a *General Purpose Computer*, is designed to perform one or a limited number of tasks. Examples of embedded systems with the potential to use Pattern Recognition are telephones and other communication devices, systems on board of vehicles (cars, boats, planes...), medical equipment, industrial equipment, household appliances, etc.

There are specific challenges posed by Embedded Systems in this case. The main peculiarities to be addressed when a Pattern Recognition application is developed for or adapted to an Embedded System are:

– Performance and Memory limits.
– Power dissipation.
– System size.
– Real-time requirements.
– Constrained user interface.
– Reliability.
– On-site adaptation or re-training.
– Installation, maintenance and upgrades.

2.3 Applications

Another classification criterion is the industry sector where the application is used. Typical sectors are: **Medical**, **Security**, **Manufacturing** and **Mobility**.

A relevant medical field is *medical imaging*. It sometimes deals with signals that are not strictly images, but can be processed as such, like ultrasound, X-ray or magnetic resonance. Security-related applications include voice, fingerprints, face and iris *biometrics*, among other, as well as *surveillance*. In manufacturing, *industrial inspection* and *quality control* are the most usual tasks. And, finally, in the mobility area, a lot of innovative features are being introduced in the huge markets of *vehicles*, as well as *handheld mobile devices*.

Mobility Applications. Mobile consumer and profesional devices represent one of the fastest-growing application areas where Pattern Recognition can be widely used [3] in tasks such as: biometric user verification (face, voice, fingerprint...), advanced user interfaces (voice, pen, accelerometers...), language recognition and translation, remote medical assistance, remote technical operations, etc.

The main problems and challenges associated to a Pattern Recognition application to be deployed on mobile devices are:

– Very limited performance and memory.
– Power dissipation.
– Less-than-optimal sensors.
– Error-rate and user acceptance.
– Cost.
– User support.

Manufacturing Applications. In the manufacturing industry, Pattern Recognition can be used mainly to improve productivity and quality control [8], in applications like: Biometric operator authorization and presence control, Advanced machine user interfaces (hands-free operation), Quality control (2D, 3D, range sensors, metrology...), Advanced automation (robot manipulation, transport...), Preventive Maintenance...

The main elements to be considered when a Pattern Recognition application has to be installed in an industrial setting are:

- Hostile physical environment.
- Safety regulations.
- Constrained user interfaces and operation protocols.
- Error-rate and cost/benefit tradeoff.
- Reliability.
- Support.

Security Applications. Embedded systems in security are also increasingly used. In this application area, Pattern Recognition can help to improve the security of installations and persons in many innovative ways, like: biometric access control [1], intrusion detection, behaviour analysis, vehicle control, forensic analysis, etc.

The main details that challenge and influence the development of a Pattern Recognition application related to Security are:

- Flexible physical placement.
- Legal regulations.
- User interfaces must be well designed (fatigue, attention).
- False positives/False negatives tradeoff.
- Site adaptation and re-training.
- Installation, maintenance and support.

2.4   Medical Applications

Medical equipment is a classical area of early adoption of new technology [6]. In this case, Pattern Recognition can be very useful to improve diagnosis and clinical practice using embedded systems in areas such as: interfaces (hands-free operation, precision surgery, etc.), computer aided analysis (macro or microscopic lesions, pathogens, etc.), large-scale screening and epidemiological studies, remote medical assistance and tests by non-specialized personnel, etc.

The main topics and details to be taken into account for an Embedded Pattern Recognition Medical Application are:

- Variability of conditions and subjects.
- Operator training and support.
- Safety and legal regulations.
- Specialized user interfaces and operation protocols.
- Error-rates and cost/benefit tradeoff.
- Reliability.
- Support.

3 Approaches

We cannot find a universal recipe to address the often daunting problem of implanting an intricate algorithm into a low-power, resource-limited, closed, autonomous, reliable and self-supporting system, sometimes with real-time requirements and always with a fast response expectation from the user. Many proposals exist in the literature to achieve those goals, ranging from the optimization of code for conventional processors with limited CPU power and memory, to the design of special purpose hardware:

Using Off-the-Shelf Architectures requires a detailed low level code optimization and sometimes the adaptation of some operations to specialized DSP's, GPU's and *Media Processors* [2].

Using Special Architectures like parallel interconnected processors, optical computation devices and other special designs is also possible [4].

Using *Ad-Hoc* **or Reconfigurable Hardware** to implement core PR algorithms is another option. The designs can become VLSI chips or be loaded into reconfigurable circuits like FPGA's [9].

3.1 Using Off-the-Shelf Architectures

When to use off-the-shelf architectures

Some Embedded Pattern Recognition tasks can be successfully performed by off-the-shelf hardware conveniently adapted to the task. Especially for applications where: very small size or power consumption is not needed, the computational needs are not extreme, development costs are an issue, time to market is very important, it is not a high-volume application and system cost is not critical

The main design elements and guidelines to take into account when one wants to build a system using standard components are:

- Use industrial-grade components for reliability.
- Use long-term supported subsystems with stable specifications.
- Test the sensors in the expected conditions.
- Use the simplest possible interfaces.
- Strip the software components as much as possible.
- Take into account the final combined power and cooling needs.

3.2 Using Specialized DSP's

Some applications demand dedicated Digital Signal Processors in addition to a conventional CPU. The tasks that require this kind of architecture are characterized by: small cost, size or power consumption is needed, the computational needs are more on the signal processing phases and less in the last symbolic computation stages, slightly higher development costs are acceptable, slightly higher time to market is acceptable.

The main design criteria in this case are:

- Balance the load of the CPU(s) and the DSP(s).
- Subdivide the process into Preprocessing, Feature Extraction and Recognition to get results from the DSP(s) and into the CPU(s) at an adequate rate to keep the pipeline filled.
- Adapt the necessary algorithms to the DSP(s) chosen. Use preferably Fixed Point Arithmetic.

3.3   Using Massively Parallel HW (e.g. GPUs)

Some tasks pose huge processing demands and the most economically viable option is to use massive parallelism or Graphical Processor Units. This is the case when: the application involves a high complexity or large data volumes, high speed or a large throughput is needed, the most costly parts of the algorithm can be split into many parallel small operations, higher size, power needs and system costs are acceptable.

The relevant design topics and guidelines in these cases are:

- Balance the load of the CPU(s) and the GPU(s) or massively parallel subsystems.
- Isolate the part(s) of the process that can be atomized.
- The data transfer between the main CPU and the GPU should be bounded.
- Parallelize and adapt the algorithms to the GPU. Use preferably short Arithmetic.

3.4   Using *Ad-Hoc* or Reconfigurable HW

Some projects target large-volume cost-competitive products that deserve the design of *ad-hoc* hardware or reconfigurable components such as FPGAs or ASICs. In the cases where: The application cannot meet the given requirements just by integrating standard components, the expected volumes justify a longer development work, the most costly parts of the algorithm can be re-designed for hw, a large amount of inherent parallelism or "segmentability" is present in the algorithms.

The relevant design topics here are:

- Isolate the part(s) of the process that can be re-designed for a HW implementation.
- Do not stick to the well-known sequential algorithms when thinking the redesign.
- Keep under control the data transfer between the CPU and the reconfigurable subsystem.
- Use special memory and large data buses if needed, to take advantage of the FPGAs or *ad-hoc* hardware.

Conclusions

Every day, the applications of embedded systems reach new areas and get more elaborate. This leads to many recent applications that require a contribution of PR algorithms. Some of them deal with an **Audio** Input signal. Among them, speech recognition, speech translation and audio biometrics (speaker recognition and verification). Other tasks take **Images** as their input (Computer Vision). They are increasingly popular and include advanced features in digital photography, like face detection or smile recognition, also in surveillance applications, with intruder detection, behaviour recognition, biometrics, information gathering from crowds, vehicles, etc., and other systems deal with very **Diverse Inputs** from physical devices or from the environment, as in manufacturing, automotive or aeronautic applications, where complex decision capabilities and autonomous function are demanding for increasingly sophisticated PR and AI subsystems.

In all those cases, we can conclude that Pattern Recognition applications on embedded systems are natural, already present in the market and growing very quickly. It is also evident that there are many different target systems and there are very different requirements for the large variety of applicable PR tasks. And, finally, a *very detailed analysis*, has to be performed *before* starting the design of an embedded PR system.

References

1. Aaraj, N., Ravi, S., Raghunathan, A., Jha, N.K.: Architectures for efficient face authentication in embedded systems. In: DATE 2006: Proceedings of the Conference on Design, Automation and Test in Europe, pp. 1–6. European Design and Automation Association, 3001 Leuven, Belgium, Belgium (2006)
2. Kisacanin, B.: Examples of low-level computer vision on media processors. p. 135, June 2005
3. Lara, O.D., Labrador, M.A.: A survey on human activity recognition using wearable sensors. IEEE Commun. Surv. Tutor. **15**(3), 1192–1209 (2013)
4. MacLean, W.J.: An evaluation of the suitability of FPGAs for embedded vision systems. In: CVPR 2005: Proceedings of the 2005 IEEE Computer Society Conference on Computer Vision and Pattern Recognition (CVPR 2005) - Workshops, p. 131. IEEE Computer Society, Washington (2005)
5. Marwedel, P.: Embedded System Design: Embedded Systems Foundations of Cyber-Physical Systems. Embedded Systems. Springer, Netherlands (2010). https://books.google.es/books?id=EXboa4sXlRsC
6. Meyer-Baese, A., Schmid, V.: Pattern Recognition and Signal Analysis in Medical Imaging. Elsevier Science, New York (2014). https://books.google.es/books?id=dPSAAgAAQBAJ
7. Rosenfeld, A., Wechsler, H.: Pattern recognition: historical perspective and futuredirections. Int. J. Imaging Syst. Technol. **11**(2), 101–116 (2000). http://dx.doi.org/10.1002/1098-1098(2000)11:2⟨101::AID-IMA1⟩3.0.CO;2-J
8. Sanz, J.: Advances in Machine Vision. Springer Series in Perception Engineering. Springer, New York (2012). https://books.google.es/books?id=lNTTBwAAQBAJ
9. Yu, F., Gregory, D.: Optical pattern recognition: architectures and techniques. Proc. IEEE **84**(5), 733–752 (1996)

Temporal, Spatial and High Dimensional Databases

Restricted Shortest Path in Temporal Graphs

Sudip Biswas, Arnab Ganguly[✉], and Rahul Shah

School of Electrical Engineering and Computer Science, Louisiana State University,
Baton Rouge, USA
{sbiswa7,agangu4}@lsu.edu, rahul@csc.lsu.edu

Abstract. The *restricted shortest path* (RSP) problem on directed networks is a well-studied problem, and has a large number of applications such as in *Quality of Service* routing. The problem is known to be *NP-hard*. In certain cases, however, the network is not static i.e., edge parameters vary over time. In light of this, we extend the RSP problem for general networks to that for *temporal networks*. We present several exact algorithms for this problem, one of which uses heuristics, and is similar to the A^* algorithm. We experimentally evaluate these algorithms by simulating them on both existing temporal networks, and synthetic ones. Furthermore, based on one of the pseudo-polynomial exact algorithms, we derive a fully polynomial time approximation scheme.

1 Introduction

The restricted shortest path (RSP) problem on directed networks is an important problem both from theoretic and application point of view. The practical importance can be mainly attributed to its enormous application in *Quality of Service* (QoS) routing [12], where the objective is to find a path that satisfies various QoS attributes. Other applications include resource planning, airline industry [3], railway networks [16], and etc. One can visualize this problem as follows: a traveller wishes to travel from one place to another in the fastest possible way, but only has a limited amount of money. Therefore, she has to choose a path that provides the fastest travel time, but maintains her budget constraints.

In a large number of applications, however, the network is not static i.e., the edge attributes are dependent on time. An RSP in these networks has to maintain the edge constraints while taking the *temporal* information into account. In continuation with our example of a traveller above, this can be visualized as follows. Suppose, the traveller books multiple flights to reach her destination. Then, the departure of any flight from an intermediate city must be after the flight at which she arrived at this city.

1.1 Related Work

The RSP problem is *NP-hard* even for acyclic graphs [5]. The best known exact algorithms (in terms of complexity) are based on dynamic programming [8].

This research is funded in part by National Science Foundation (NSF) Grants CCF 1218904 and CCF 1527435.

Using these (pseudo-polynomial) algorithms, the best known fully-polynomial time approximation scheme (FPTAS) has complexity $O(mn(\log \log n + 1/\epsilon))$ for general graphs [13] and $O(mn/\epsilon)$ for acyclic graphs [4]. Here, n and m are the number of vertices and edges respectively. Other techniques (see [2,10,11,17] and references therein), both exact and approximate, are based on label setting approach, Lagrangian relaxation, K-shortest paths approach, pre-processing schemes, heuristics, and A^* search.

A related problem is that of finding RSP in time varying graphs, where the length and the constraint of an edge depends on a particular time-stamp [1]. In temporal networks, on the other hand, every edge e has a start time, and and an arrival time, say t, which is also the arrival time of a path ending with e. Only the edges starting after (or, at) t can be taken to extend this path. Clearly, the RSP problem on temporal networks is at least as hard as that in static networks.

To the best of our knowledge, RSP problem on temporal graphs has not been previously studied. However, other path problems and connectivity problems in temporal networks have received significant attention. An important result was that of Kempe et al. [9], who showed that classical Menger's theorem is violated for temporal networks, and the problem of computing number of source-destination disjoint paths becomes NP-complete. Later, Mertzios et al. [14] showed that an alternative formulation of Menger's theorem holds for all temporal graphs. Recently, Wu et al. [18] showed that under reasonable preprocessing of the graph, single-source shortest paths can be found in $O(n + m)$ time. Travelling salesman problem on temporal networks has also been studied [15].

1.2   Notations and Problem Formulation

Let $\mathcal{G} = (\mathcal{V}, \mathcal{E})$ be a directed finite temporal graph on n vertices and m edges. Though, the correctness of our algorithms does not depend on n and m, we assume $m \geq n$. Let $\delta^+(v)$ (resp. $\delta^-(v)$) be the set of outgoing edges (resp. incoming edges) of a vertex v, and let $\Delta_{in} = \max\{| \delta^-(v) | \| v \in \mathcal{V}\}$. Each edge $e \in \mathcal{E}$ has a source vertex $src(e)$, a target vertex $tgt(e)$, a length $len(e)$, a penalty (i.e., constraint) $pen(e)$, a start time $start(e)$, and an arrival time $end(e)$. We assume that length, penalty, start time and arrival time are all positive and integral. The transit time of an edge is the difference $end(e) - start(e)$.

For any non-empty subset $\mathcal{E}'$ of $\mathcal{E}$, let ith smallest edges of $\mathcal{E}'$ denote those edge in $\mathcal{E}'$ having the ith smallest $end(e)$ value. The graph can be initially preprocessed, such that for any edge $e \in \delta^-(v)$, in $O(1)$ time we can determine its rank among all edges in $\delta^-(v)$; this requires an additional $O(m)$ space.

Throughout this paper, s and g denote the source and destination vertices, and $[t_\alpha, t_\beta]$ denotes a time interval. An s-g *temporal path* w.r.t $[t_\alpha, t_\beta]$ is a path, say $\pi = \langle e_1, e_2, \ldots e_k \rangle$, such that $start(e_1) \geq t_\alpha$, $start(e_{i+1}) \geq end(e_i)$, $1 \leq i \leq k - 1$, and $end(e_k) \leq t_\beta$, where s $= src(e_1)$ and g $= tgt(e_k)$. The start time (resp. arrival time) of π is $start(e_1)$ (resp. $end(e_k)$), and is denoted by $start(\pi)$ (resp. $end(\pi)$). An s-g *restricted temporal path* (RTP), say π, w.r.t $[t_\alpha, t_\beta]$ and a positive integer P, is an s-g temporal path w.r.t $[t_\alpha, t_\beta]$ that has penalty at most P i.e., $pen(\pi) = \sum_{e \in \pi} pen(e) \leq$ P. An s-g *restricted shortest temporal path* (RSTP) is an s-g RTP π that minimizes the length $len(\pi) = \sum_{e \in \pi} len(e)$. Throughout this paper, P is a positive integral upper-bound on the allowed penalty of an s-g temporal path, and unless stated otherwise, the terms RTP and RSTP are used w.r.t $[t_\alpha, t_\beta]$ and P. We denote by B_U (resp. B_L), an upper-bound (resp. lower-bound) on L_{opt}, the length of an s-g RSTP.

Problem 1. (Restricted Shortest Temporal Path). *Let $\mathcal{G}$ be a temporal graph where each edge has a length, a penalty, a start time, and an arrival time.*
Input: *Two vertices s and g, a maximum penalty P, and a time interval $[t_\alpha, t_\beta]$.*
Output: *An s-g restricted shortest temporal path with respect to $[t_\alpha, t_\beta]$ and P.*

1.3   Minimum Penalty Temporal Path

In order to find an RSTP from s to g w.r.t $[t_\alpha, t_\beta]$ and P, we need to first determine whether there exists any RTP for the same parameters. If we are provided with an algorithm (see Lemma 1 below) that computes the minimum penalty temporal path (MPTP) from s to g w.r.t $[t_\alpha, t_\beta]$, then we can easily figure out whether there exists an RTP, or not. The actual algorithm is a straight-forward adaptation of Dijkstra's algorithm. We omit the algorithm description, and the proof of Lemma 1 due to space constraints.

Lemma 1. *There exists an algorithm that can find the minimum penalty s-g temporal path with respect to $[t_\alpha, t_\beta]$, or reports that no such path exists. The complexity is $O(m \log m + m \Delta_{in})$.*

1.4   Organization of the Paper

The rest of the paper is organized as follows. In Sect. 2, we present three pseudo-polynomial exact algorithms. In Sect. 3, we present an A^* algorithm. In Sect. 4, we present an FPTAS. Simulation results of the exact algorithms are presented in Sect. 5. Finally, we conclude the paper in Sect. 6.

2   Exact Algorithms Using Dynamic Programming

In this section, we present several (pseudo-polynomial) algorithms based on dynamic programming. In Sect. 2.1, we first present two algorithms whose complexities are independent of the query time span i.e., t_α and t_β. In Sect. 2.2, we present an algorithm whose complexity depends on the query time span.

2.1   Query Time Span Independent Algorithms

Our approach is to find the minimum penalty by which one can reach a vertex v from s, using a temporal path of length ℓ that arrives at v at time t, where $t \in [t_\alpha, t_\beta]$. We incrementally check this for every $\ell \in [0, B_U]$, and for every vertex v. Once we reach g via a path having penalty at most P, we terminate. Following are important properties of RSTP and MPTP.

Algorithm 1.

```
 1: procedure DYNAMICLENGTH(s, g, P, B_U, t_α, t_β, len, pen, start, end)
 2:     open(0) ← {s}; P(s, 0, t_α) ← 0; E(s, 0, t_α) ← ∅; add t_α to T(s, 0);
 3:     for (ℓ ← 0; ℓ ≤ B_U; ℓ ← ℓ + 1) do
 4:         if (g ∈ open(ℓ)) then return ℓ;
 5:         end if
 6:         for (every vertex u ∈ open(ℓ) and every edge e⁺ = (u, v) ∈ δ⁺(u)) do
 7:             let ℓ⁺ = ℓ + len(e⁺), t⁺ = end(e⁺);
 8:             if (ℓ⁺ ≤ B_U and t⁺ ≤ t_β) then
 9:                 for (every t in T(u, ℓ)) do
10:                     let p⁺ = P(u, ℓ, t) + pen(e⁺);
11:                     if (start(e⁺) ≥ t and p⁺ ≤ P) then
12:                         if (v ∉ open(ℓ⁺)) then add v to open(ℓ⁺);
13:                         end if
14:                         if (t⁺ ∉ T(v, ℓ⁺)) then P(v, ℓ⁺, t⁺) ← p⁺; E(v, ℓ⁺, t⁺) ←
                           ⟨e⁺, t⟩; add t⁺ to T(v, ℓ⁺);
15:                         else if (p⁺ < P(v, ℓ⁺, t⁺)) then
16:                             P(v, ℓ⁺, t⁺) ← p⁺; E(v, ℓ⁺, t⁺) ← ⟨e⁺, t⟩;
17:                         end if
18:                     end if
19:                 end for
20:             end if
21:         end for
22:     end for
23:     return ∞;
24: end procedure
```

Observation 1. *For a vertex v, let π_1 and π_2 be two s to v temporal paths having same length and arriving at v at the same time. If $pen(\pi_1) < pen(\pi_2)$, and an s-g RSTP contains π_2, then π_2 can be replaced by π_1 in the RSTP.*

Observation 2. *For any vertex $v \neq s$, let $\pi = \langle e_1, e_2, \ldots, e_k \rangle$ be an s to v minimum penalty temporal path that arrives at $v = tgt(e_k)$ at time $end(e_k)$. Then, π is a minimum penalty temporal path that arrives at $tgt(e_{k-1})$ at time $end(e_{k-1}) \leq start(e_k)$, and has length $len(\pi) - len(e_k)$.*

Correctness and Complexity. Algorithm 1 presents a detailed implementation of the above observations. Here, $open(\ell)$ is the set of vertices v, such that an RTP from s to v having length ℓ has been discovered. Denote by $P(v, \ell, t)$, the minimum penalty among all s to v RTPs that has length ℓ and arrives at v at time t. Also, $E(v, \ell, t)$ maintains the tuple $\langle e, t' \rangle$, where $e = (u, v)$ is the incoming edge of v on the path defined by $P(v, \ell, t)$, and t' is the arrival time of this path at u. Further, $T(v, \ell)$ is the set of distinct arrival times t, such that an s-v RTP, having length ℓ, and arriving at v at time t has been discovered.

The correctness follows from Observations 1, 2, and 3.

Observation 3. *The length of an RSTP from s to g w.r.t timespan $[t_α, t_β]$ and penalty P is $\min\{\ell \mid P(g, \ell, t) \leq P \text{ and } t \in [t_α, t_β]\}$.*

We maintain each $\mathcal{T}(v, \ell)$ as a bit-array of length at most $|\delta^-(v)|$, where the ith index in the array corresponds to the ith smallest edges in $\delta^-(v)$. Recall that the graph has been preprocessed so that in $O(1)$ time, we can find the rank of any edge $e \in \delta^-(v)$ among all edges in $\delta^-(v)$.If the ith bit is set to 1, then an s-v RTP that uses an ith smallest edge in $\delta^-(v)$ has been found, otherwise not. Likewise, for any edge $e = (u, v)$, the entry $P(v, \ell, t)$, where $t = end(e)$, can also be accessed (and updated) in constant time.

Recall that we assume that there is an RTP w.r.t P and $[t_\alpha, t_\beta]$. Therefore, line 3 is executed exactly L_{opt} times, and for each iteration, line 6 is executed at most m times. Since, the size of $\mathcal{T}(u, \ell)$ is at most $|\delta^-(u)|$, the time required by Algorithm 1 can be bounded by $O(m\Delta_{in}\mathsf{L}_{opt})$.

For every $v \in \mathcal{V}$, $\ell \in [0, \mathsf{B}_\mathsf{U}]$, and $t \in \{end(e) \mid e \in \delta^-(v)\}$, the data-structures $P(v, \ell, t)$ requires $O(m\mathsf{B}_\mathsf{U})$ space. Likewise, maintaining all $E(v, \ell, t)$ and $\mathcal{T}(v, \ell)$ structures requires $O(m\mathsf{B}_\mathsf{U})$ space. Maintaining the structure to find out the rank of any edge e in $\delta^-(v)$ requires $O(m)$ space. Therefore, total space required is bounded by $O(m\mathsf{B}_\mathsf{U})$.

Constructing the Path. The RSTP can be constructed easily using the information in $E(\cdot, \cdot, \cdot)$ as follows. Let the s-g RSTP be $\langle v_1, e_1, v_2, \ldots, e_{k-1}, v_k \rangle$, where $v_1 = $ s, $v_k = $ g, and for $1 \le i < k$, $v_i = src(e_i)$ and $v_{i+1} = tgt(e_i)$. Furthermore, let the length (resp. arrival time) of the s-v_i subpath be ℓ_i (resp. t_i). Then, $t_1 = t_\alpha$, and $\ell_1 = 0$. For $1 < i \le k$, $t_i = end(e_{i-1})$, $\ell_i = \ell_{i-1} + len(e_{i-1})$, where $\ell_k = \mathsf{L}_{opt}$. Clearly, $\varnothing \in E(v_1, 0, t_\alpha)$, and for every $1 < i \le k$, $\langle e_{i-1}, t_{i-1} \rangle \in E(v_i, \ell_i, t_i)$. Using this information, we can easily construct the s-g RSTP, say π^*, in $O(|\pi^*|)$ time, where $|\pi^*| < n$ denotes the number of edges in π^*.

An alternate algorithm is based on the following somewhat orthogonal approach to Algorithm 1. Constrained by space, we provide a sketch of the algorithm, and refer the reader to [8] for a penalty bounded algorithm for the RSP problem.

We maintain $L(v, p, t)$ i.e., the length of the minimum length s to v temporal path reaching v at time t with penalty p. For every $p \in [0, \mathsf{P}]$, $v \in V$, and $t \in \{end(e) \mid e \in \delta^-(v)\} \cap [t_\alpha, t_\beta]$, we maintain and update $L(v, p, t)$ according to Observations 4 and 5. Finally, we return the optimal length based on Observation 6. The correctness claim is immediate. As in the case of Algorithm 1, the path can also be constructed by maintaining additional information. Space and time complexity of $O(m\mathsf{P})$ and $O(m\Delta_{in}\mathsf{P})$ respectively can be derived based on the arguments in complexity analysis of Algorithm 1.

Observation 4. *For a vertex v, let π_1 and π_2 be two s to v temporal paths having same penalty and arriving at v at the same time. If $len(\pi_1) < len(\pi_2)$, then any RSTP from s to g does not contain the path π_2.*

Observation 5. *For any vertex $v \neq $ s, let $\pi = \langle e_1, e_2, \ldots, e_k \rangle$ be an s to v minimum length temporal path that arrives at $v = tgt(e_k)$ at time $end(e_k)$. Then, π is a minimum length temporal path that arrives at $tgt(e_{k-1})$ at time $end(e_{k-1}) \le start(e_k)$, and has penalty $pen(\pi) - pen(e_k)$.*

Observation 6. *The length of an RSTP from s to g w.r.t $[t_\alpha, t_\beta]$ and penalty P is $\min\{L(g, p, t) \mid p \in [0, \mathsf{P}] \text{ and } t \in [t_\alpha, t_\beta]\}$.*

Algorithm 2.

1: **procedure** TIMESPAN($s, g, P, t_\alpha, t_\beta, len, pen, start, end$)
2:     $L(v, p, t) \leftarrow \infty$ for every $v \in \mathcal{V}, 1 \le p \le P$ and $t \in [t_\alpha, t_\beta]$; $E(s, 0, t_\alpha) \leftarrow \varnothing$;
3:     let $\mathcal{E}'$ be the set of edges e in $\mathcal{E}$, such that $start(e) \ge t_\alpha$ and $end(e) \le t_\beta$;
4:     sort the edges e of $\mathcal{E}'$ in ascending order based on $end(e)$;
5:     **for** (every edge $e = (u, v) \in \mathcal{E}'$) **do**
6:         **for** (every entry $L(u, p', t')$ of u with $t' \le start(e)$) **do**
7:             let $\ell = L(u, p', t') + len(e), p = p' + pen(e)$;
8:             **if** ($p \le P$ and $\ell < L(v, p, end(e))$) **then**
9:                 $L(v, p, end(e)) \leftarrow \ell$, $E(v, p, end(e)) \leftarrow \langle e, t' \rangle$;
10:            **end if**
11:        **end for**
12:    **end for**
13:    **return** the minimum value in $\{L(g, p, t) \mid t_\alpha \le t \le t_\beta \text{ and } 0 \le p \le P\}$;
14: **end procedure**

By Lemma 1, verifying the existence of an RTP is achieved in $O(m \log m + m\Delta_{in})$ time. Following the discussions in this section, we arrive at the following theorem.

Theorem 1. *By maintaining data-structures occupying $O(mB_U)$ and $O(mP)$ space, we can find* s-g *RSTP w.r.t $[t_\alpha, t_\beta]$ and* P *in $O(m \log m + m\Delta_{in}L_{opt})$ and $O(m \log m + m\Delta_{in}P)$ time respectively.*

2.2   Query Time Span Dependent Algorithm

In the previous section, we introduced two exact algorithms with complexities $O(m \log m + m\Delta_{in}P)$ and $O(m \log m + m\Delta_{in}L_{opt})$. Larger values of Δ_{in} and L_{opt} can be a bottleneck for these algorithms, resulting in inefficient query time. In this section, we tackle those scenarios and propose an $O(mP(t_\beta - t_\alpha))$ time algorithm. Following is yet another useful property of RSTP.

Observation 7. *For a vertex v, let π_1 and π_2 be two s to v temporal paths having equal length and penalty. If $end(\pi_1) < end(\pi_2)$ and an* s-g *RSTP contains π_2, then π_2 can be replaced by π_1 in the RSTP.*

Correctness and Complexity. We use the greedy properties of Observations 4 and 7 at every intermediate vertex v while searching an s-g RSTP. This way the number of useful paths reaching v at a time-stamp t is bounded by P. Let $L(v, p, t)$ be the length of the shortest s to v temporal path reaching v at time t with penalty p. The idea is to store $L(v, p, t)$ for every value of $p \in [0, P]$, and update it based on Observation 4. By maintaining additional information, as described in the Algorithm 1, the RSTP can be constructed.

Correctness follows from Observations 4, 5, 6, and 7.

The total size of $L(u, \cdot, \cdot)$ can be bounded by $O(P(t_\beta - t_\alpha))$. Therefore, total space required over all vertices is bounded by $O(nP(t_\beta - t_\alpha))$. Since we only consider the edges with start time and arrival time falling within $[t_\alpha, t_\beta]$, we can sort the edges in step 4 by using bucket sort with $(t_\beta - t_\alpha)$ buckets. This step can then be performed in $O(m + (t_\beta - t_\alpha))$ time. The size of $\mathcal{E}'$ is at most m; thus steps 5–12 require time $O(m\mathsf{P}(t_\beta - t_\alpha))$. Finally, returning the optimal length takes $O(\mathsf{P}(t_\beta - t_\alpha))$ time, and we have the following theorem.

Theorem 2. *By maintaining data-structures occupying $O(n\mathsf{P}(t_\beta - t_\alpha))$ space, we can find* s-g *RSTP w.r.t* $[t_\alpha, t_\beta]$ *and* P *in* $O(m\mathsf{P}(t_\beta - t_\alpha))$ *time.*

3 A^* Algorithm

We start with a few notations. Let π^* be an s-g RSTP. For a vertex v on π^*, let $g_{len}(\pi^*, v)$ be the length of the s-v subpath of π^*, and $h_{len}(\pi^*, v)$ be the length of the v-g subpath of π^*. Likewise, we define $g_{pen}(\pi^*, v)$ and $h_{pen}(\pi^*, v)$. Clearly, $g_{len}(\pi^*, v) + h_{len}(\pi^*, v) = len(\pi^*) = \mathsf{L}_{opt}$ and $g_{pen}(\pi^*, v) + h_{pen}(\pi^*, v) = pen(\pi^*)$. Further, if a vertex v' is not on π^*, then $g_{len}(\pi^*, v') = g_{pen}(\pi^*, v') = \infty$ and $h_{len}(\pi^*, v') = h_{pen}(\pi^*, v') = \infty$. For a vertex w (not necessarily on an s-g RSTP),

- $h_{len}(w)$ is an underestimate of the length of the w-g subpath (if any) of π^* i.e., $h_{len}(w) \le h_{len}(\pi^*, w)$. Likewise, we define $h_{pen}(w)$. Since, every edge has positive length and penalty, we let $h_{len}(w) \ge 0$ and $h_{pen}(w) \ge 0$.
- $open(w)$ is a set of *non-dominated* tuples $x = \langle \ell, p, t, y, e \rangle$, where ℓ and p are the length and penalty of an s-w path, e is the incoming edge of w on this path, and y is the tuple which led to the formation of x. If $w = $ s then $t = t_\alpha$, otherwise $t = end(e)$. For any vertex v, a tuple $x = \langle \ell, p, t, y, e \rangle$ in $open(v)$ is *said to be non-dominated* if there is no other $x' = \langle \ell', p', t', y', e' \rangle$ in $open(v)$ such that $t' \le t$, $\ell' \le \ell$, and $p' \le p$. For $y = \langle \ell', p', t', y', e' \rangle \in open(u)$, and an edge $e = (u, w)$, we modify $open(w)$ from y as follows (see lines 9–22 in Algorithm 3). Let $x = \langle \ell' + len(e), p' + pen(e), end(e), y, e \rangle$. If $p' + pen(e) + h_{pen}(w) \le \mathsf{P}$, $t' \le start(e)$, $end(e) \le t_\beta$, and there is no previously removed tuple from $open(w)$ that dominates x, merge x with the tuples at $open(w)$, keeping only non-dominated ones in $open(w)$. Initially, $open(\mathsf{s})$ contains $\langle 0, 0, t_\alpha, \varnothing, \varnothing \rangle$, and every other $open(\cdot)$ is empty.
- $closed(w)$ is a set of tuples that have been selected (and removed) from $open(w)$. Initially, every $closed(\cdot)$ is empty. We select a tuple from $open(w)$ as follows (see lines 5–8 in Algorithm 3). If $open(v)$ is not empty, then $openVertices$ contains the vertex v keyed by $\ell_v + h_{len}(v)$, where ℓ_v is the length of the minimum length tuple in $open(v)$. We first remove the vertex w from $openVertices$ with the minimum key, and then remove the minimum length tuple from $open(w)$.

Constructing the Path. The previous tuple y and incoming edge e is maintained in each tuple x so that we can efficiently construct the s-g RSTP π^*. Let, $\pi^* = \langle v_1, e_1, v_2, \ldots, e_{k-1}, v_k \rangle$, where $v_1 = $ s and $v_k = $ g. For each vertex v_i, $1 \le i \le k$, there is a tuple $x_i = \langle \ell_i, p_i, t_i, x_{i-1}, e_{i-1} \rangle$ in $closed(v_i)$. Here, $\ell_1 = p_1 = 0$, $t_1 = t_\alpha$, and $x_0 = e_0 = \varnothing$. For each x_i, by using the incoming edge e_{i-1} and the previous tuple x_{i-1}, we can construct π^*.

Algorithm 3.

1: **procedure** $A^*(\mathsf{s}, \mathsf{g}, \mathsf{t}_\alpha, \mathsf{t}_\beta, \mathsf{P}, len, pen, start, end, h_{len}, h_{pen})$
2:     $closed(v), open(v) \leftarrow \varnothing$, for all $v \in \mathcal{V}$;
3:     add $\langle 0, 0, \mathsf{t}_\alpha, \varnothing, \varnothing \rangle$ to $open(\mathsf{s})$; add s to $openVertices$ keyed by $h_{len}(\mathsf{s})$;
4:     **while** ($openVertices$ is not empty) **do**
5:         remove the vertex u from $openVertices$ with minimum key, with ties broken arbitrarily, but always in favour of vertex g; remove the minimum length tuple $x^* = \langle \ell, p, t, x, e \rangle$ from $open(u)$; add x^* to $closed(u)$;
6:         if $open(u)$ is not empty, then add u to $openVertices$ keyed by $\ell_{min} + h_{len}(u)$, where $\ell_{min} = \min\{\ell' \mid \langle \ell', p', t', x', e' \rangle \in open(u)\}$;
7:         **if** ($u = \mathsf{g}$) **then** construct and return the s-g path having length ℓ and penalty p, which arrives at g at $end(e)$;
8:         **end if**
9:         **for** (each edge $e^+ = (u, v) \in \delta^+(u)$) **do**
10:             let $p^+ = p + pen(e^+)$, $\ell^+ = \ell + len(e^+)$; $x' = \langle \ell^+, p^+, end(e^+), x^*, e^+ \rangle$;
11:             **if** ($p^+ + h_{pen}(v) \leq \mathsf{P}$, $t \leq start(e^+)$ and $end(e^+) \leq \mathsf{t}_\beta$) **then**
12:                 **if** (there is no tuple in $closed(v)$ that dominates x') **then**
13:                     **if** (there is no tuple in $open(v)$) **then** add x' to $open(v)$;
14:                         add v to $openVertices$ keyed by $\ell^+ + h_{len}(v)$;
15:                     **else if** (no triplet in $open(v)$ dominates x') **then**
16:                         remove the triplets from $open(v)$ that are dominated by x'; add x' to $open(v)$; update the key of v in $openVertices$ with $\ell^+_{min} + h_{len}(v)$, where $\ell^+_{min} = \min\{\ell' \mid \langle \ell', p', t', x', e' \rangle \in open(v)\}$;
17:                     **end if**
18:                 **end if**
19:             **end if**
20:         **end for**
21:     **end while**
22:     **return** $\varnothing$;
23: **end procedure**

Remark 1. *We can also prune paths using the transit time information. For any vertex u, let $e = (u, v) \in \delta^+(u)$, and $h_{transit}(v)$ be an underestimate of the minimum total transit time of v to g sub-path of π^*. If $end(e) + h_{transit}(v) > \mathsf{t}_\beta$, then we can ignore the edge e for a tuple selected from $open(u)$. If the transit time of an edge is same as its length, then $h_{transit}$ is same as h_{len}.*

3.1   Correctness and Complexity

In Algorithm 3, by taking temporal information into account, we extend the A^* algorithm of Li et al. [11] for the RSP problem. According to the definition of h_{len} and h_{pen}, both are *admissible* i.e., for any vertex v and any s-g RSTP π^*, we have $h_{len}(v) \leq h_{len}(\pi^*, v)$ and $h_{pen}(v) \leq h_{pen}(\pi^*, v)$. In Lemma 2, we show that given admissible heuristics, Algorithm 3 finds an RSTP w.r.t P and $[\mathsf{t}_\alpha, \mathsf{t}_\beta]$. Due to space constraint, we omit the proofs in this section, which are based on the A^* algorithm for the shortest path problem [7].

Lemma 2. *If h_{len} and h_{pen} are both admissible, then Algorithm 3 finds an* s-g *RSTP w.r.t $[t_\alpha, t_\beta]$ and* P.

Consistency Assumption. For any two vertices u and v, denote by $L(u,v)$, the length of a minimum length path (not necessarily temporal or restricted) from u to v. The consistency assumption is

$$h_{len}(u) \leq L(u,v) + h_{len}(v)$$

The following theorem summarizes our main result in this section.

Theorem 3. *Suppose h_{len} is consistent and h_{pen} is admissible. Then, Algorithm 3 finds an* s-g *RSTP in time $O(m\bar{L}\Delta_{in} + m\bar{L}\log(n\bar{L}\Delta_{in}))$, where $\bar{L} = \mathsf{L}_{opt} + \max_{e \in \mathcal{E}}\{len(e)\}$.*

3.2   Obtaining Estimates

For any two vertices w and w', denote by $L(w, w')$, (resp. $P(w, w')$) the length (resp. penalty) of a minimum length (resp. penalty) path from w to w'. For every vertex v, $h_{len}(v) = L(v, \mathsf{g})$, and $h_{pen}(v) = P(v, \mathsf{g})$. Clearly, both $L(v, \mathsf{g})$ and $P(v, \mathsf{g})$ are admissible. Since, $L(u, v) + h_{len}(v) - h_{len}(u) = L(u, v) + L(v, \mathsf{g}) - L(u, \mathsf{g}) \geq 0$, h_{len} (and likewise, h_{pen}) is consistent. Observe that $L(v, \mathsf{g})$ (resp. $P(v, \mathsf{g})$) in a graph is same as $L(\mathsf{g}, v)$ (resp. $P(\mathsf{g}, v)$ in the same graph, but with edge directions reversed. Therefore, we can obtain $L(v, \mathsf{g})$ and $P(v, \mathsf{g})$ in $O(m + n\log n)$ time by first reversing the edges of the graph, and then running Dijkstra's algorithm twice, once for edge lengths and then for edge penalties.

Note that the minimum length (resp. penalty) of a v-g temporal path w.r.t $[t_\alpha, t_\beta]$ also serves as an admissible $h_{len}(v)$ (resp. $h_{pen}(v)$) of any v-g RSTP. These again can be obtained using suitably modified forms (with the complexity remaining unchanged) of the algorithm in Lemma 1. Also note that the estimates using minimum length and minimum penalty temporal paths is at least as tight an estimate as those using Dijkstra's algorithm.

Since for both these heuristic measures, we need to find shortest (temporal) paths from every vertex v to g, the time required to find the underestimates can be very expensive. To this end, we adapt the landmark based approach in [6] for the shortest path problem to that for the RSTP problem.

Landmark Based Approach. A set of landmarks, denoted by $\mathcal{L}$, is a subset of the set of the vertices of the graph i.e., $\mathcal{L} \subseteq \mathcal{V}$. We show how to compute $h_{len}(v)$ for any vertex v; $h_{pen}(v)$ can be computed similarly. Let $L(v', v'', t_1, t_2)$ be the minimum length of a temporal path from v' to v'' that lies in the interval $[t_1, t_2]$. For any $w \in \mathcal{V}$, let $t_{min}^{w+} = \min\{start(e) \mid e \in \delta^+(w)\}$, $t_{max}^{w+} = \max\{start(e) \mid e \in \delta^+(w)\}$, and $t_{max}^{w-} = \max\{end(e) \mid e \in \delta^-(w)\}$. For $u \in \mathcal{L}$ and $v \in \mathcal{V}$, we have

$$L(v, \mathsf{g}) \geq L(v, u) - L(\mathsf{g}, u)$$
$$L(v, \mathsf{g}, t_{min}^{v+}, t_\beta) \geq L(v, u, t_{min}^{v+}, t_{max}^{u-}) - L(\mathsf{g}, u, t_\beta, t_{max}^{u-})$$

Algorithm 4.

1: **procedure** TEST($s, g, P, B_L, B_U, t_\alpha, t_\beta, \epsilon, len, pen, start, end$)
2:     $\lambda \leftarrow \epsilon \times B_L/n$;
3:     $len'(e) \leftarrow \lfloor len(e)/\lambda \rfloor + 1$, for every edge $e \in \mathcal{E}$;
4:     $B \leftarrow \lfloor B_U/\lambda \rfloor + n$;
5:     **return** DYNAMICLENGTH($s, g, P, B, t_\alpha, t_\beta, len', pen, start, end$);
6: **end procedure**
7:
8: **procedure** APPX($s, g, P, t_\alpha, t_\beta, \epsilon, len, pen, start$)
9:     compute B_L and B_U such that $B_U < nB_L$;
10:     **while** ($B_U > 2B_L$) **do**
11:         $B \leftarrow \sqrt{B_U \cdot B_L}$;
12:         **if** (TEST($s, g, P, B, B, t_\alpha, t_\beta, 1, len, pen, start, end$) $= \infty$) **then** $B_L \leftarrow B$
13:         **else** $B_U \leftarrow B$
14:         **end if**
15:     **end while**
16:     **return** TEST($s, g, P, B_L, 2B_L, t_\alpha, t_\beta, \epsilon, len, pen, start, end$);
17: **end procedure**

Note that both $L(v, g)$ and $L(v, g, t_{min}^{v+}, t_\beta)$ can be used as $h_{len}(v)$. For every $u \in \mathcal{L}$, and every $v \in \mathcal{V}$, we store the following: (i) $L(v, u)$, (ii) $L(v, u, t_{min}^{v+}, t_{max}^{u-})$, and (iii) $L(v, u, t_{max}^{v+}, t_{max}^{u-})$. Total space is bounded by $O(n|\mathcal{L}|)$.

If $t_\beta \leq t_{max}^{g+}$, then $L(g, u, t_\beta, t_{max}^{u-}) \leq L(g, u, t_{max}^{g+}, t_{max}^{u-})$; otherwise, there is no g-u temporal path in the time interval $[t_\beta, t_{max}^{u-}]$. Let $\mathcal{L}' \subseteq \mathcal{L}$ be the set of vertices in the former case. For any vertex v, we obtain $h_{len}(v)$ as shown below.

$$h_{len}(v) = \max\{0, h'_{len}(v), \max_{u \in \mathcal{L}}\{L(v, u) - L(g, u)\}\}, \text{ where}$$
$$h'_{len}(v) = \max_{u \in \mathcal{L}'}\{L(v, u, t_{min}^{v+}, t_{max}^{u-}) - L(g, u, t_{max}^{g+}, t_{max}^{u-})\}.$$

4 Approximation Algorithm

In this section, we present an FPTAS for the RSTP problem, which is based on the FPTAS for the RSP problem [4,13]. The idea is to scale down the edge lengths, and then use Algorithm 1 on this scaled down instance. Due to lack of space, we omit the proofs in this section, which are similar to that in [4,13].

Let $\mathbb{I}$ be an instance of the RSTP problem. Consider the *TEST* procedure in Algorithm 4, which scales down both the edge lengths, and an appropriate upper-bound B_U by λ, and then calls Algorithm 1. Denote this scaled down instance by $\mathbb{I}_\lambda$. For the instance $\mathbb{I}_\lambda$, if there exists an s-g RTP, say π, having length at most the scaled down upper-bound B, then Algorithm 1 returns $len(\pi)$; otherwise, it returns ∞. The following lemma proves important in obtaining the FPTAS.

Lemma 3. *Suppose TEST procedure in Algorithm 4 returns the length of a path π_f. Then, $len(\pi_f) < L_{opt} + \epsilon \cdot B_L$*

Clearly, the complexity of the $TEST$ procedure can be bounded by $O(nm\Delta_{in}(1 + \frac{B_U}{\epsilon \cdot B_L}))$. Further, if B_U (and hence, L_{opt}) is within a constant factor of B_L, then $TEST$ is an FPTAS for the RSTP problem. The objective, therefore, is to obtain B_L and B_U such that B_U is within a constant factor of B_L. To this end, as described in the following lemma, we first find B_L and B_U such that $B_U < nB_L$.

Lemma 4. *An upper bound B_U and a lower bound B_L of the length of a RSTP such that $B_U < nB_L$ can be found in $O(m \log^2 m + m\Delta_{in} \log m)$ time.*

Finally, if both B_U and B_L are taken in the logarithmic scale, then by using Lemmas 3 and 4, we arrive at the following theorem.

Theorem 4. *APPX procedure in Algorithm 4 is an FPTAS for Problem 1 having complexity $O(m\Delta_{in} \log m + nm\Delta_{in}(\log \log n + 1/\epsilon))$.*

5   Experimental Evaluation

Since, there is no previously known work on finding RSP in temporal graphs, we present a comparative analysis of our proposed exact algorithms by varying different parameters. For simplicity of notation, we denote the algorithms having complexity $O(m \log m + m\Delta_{in}L_{opt})$, $O(m \log m + m\Delta_{in}P)$ and $O(mP(t_\beta - t_\alpha))$ as *DLen*, *DPen* and *DTime* respectively. For Algorithm 3, we use *DijkA**, *TempA** and *ALT* to denote the Dijkstra estimates, temporal estimates and landmark based estimates respectively.

5.1   Settings and Dataset

All experiments are performed using the JAVA programming language on a 2.6 Ghz Intel dual-core E5300 processor and 6 Gb RAM machine running LINUX.

We use the flight dataset (*flight*) located at Bureau of Transportation Statistics (http://www.transtats.bts.gov/). The origin (resp. destination) airport ID has been taken as edge source (resp. edge target). The departure (resp. arrival) time has been taken as the start (resp. end) time of an edge. We take the transit time of an edge as its length, and the distance traversed as the edge penalty.

We also use the well-known temporal datasets located at the Koblenz Large Network Collection (http://konect.uni-koblenz.de/). Specifically, we choose the Enron communication network database (*enron*), YouTube social network database (*youtube*), DBLP co-author index database (*dblp*), and the Wiktionary English authorship database (*wiktionary*). For each dataset, the first two columns are taken as the source and target of an edge, the third column is edge length, and the fourth column is the start time of an edge. The arrival time of an edge is assumed to be its start time plus its length. The penalty of an edge is a randomly assigned integer in the interval [1, 100].

Table 1 presents some statistics of the datasets. Here, **Disksize** and **Load-time** is the space occupied by the dataset on the disk, and the time required to load the graph onto the memory. The latter includes the time required to create

Table 1. Dataset statistics

Dataset	Disksize	Loadtime	n	m	δ_{avg}	Δ_{in}	$\mathcal{T}$
flight	11.2 Mb	3247 ms	6084	456964	75	29172	2642
enron	25.9 Mb	4693 ms	87273	1148072	13	6166	220364
youtube	257.1 Mb	32994 ms	3223589	9375374	2	61556	203
dblp	518.7 Mb	62880 ms	1314050	18986618	14	2630	72
wiktionary	215.9 Mb	33957 ms	2133892	8998641	4	34536	8459765

the reverse graph so that we can compute minimum length (penalty) heuristics, or minimum temporal length (penalty) heuristics, as discussed in Sect. 3.2. The average degree has been denoted by δ_{avg}, and $\mathcal{T}$ is the number of distinct arrival times. In every dataset barring the *flight* dataset, every edge has unit length (or, transit time); therefore, $\mathcal{T}$ is also the set of distinct start times.

5.2   Algorithms on *flight* dataset

In this set of experiments, we use the flight dataset to evaluate the effect of varying query time span and that of varying P on running time. For Algorithm 1, the upper-bound B_U on the length of the RSTP, was taken to be length of the minimum penalty temporal path from s to g. Each experiment was run for 100 randomly selected s-g pairs, and the average time was taken.

Varying Query Time Span. In Fig. 1(a), we show the average running times for time spans $t_\beta - t_\alpha = 80, 100, 120, 140, 160, 180, 200$ and fixed penalty restriction P $= 150$. In terms of performance, Algorithm $DTime$ shows a linear progression with increasing time span. Running time of Algorithms $DLen$ and $DPen$ stay stable, which supports their theoretical time bounds. Also, as expected from the theoretical bounds, the running time of $DTime$ is lower than both $DLen$ and $DPen$ for lower time span values. Running time for the heuristic estimate based algorithms are significantly lower.

Varying Query Penalty. In Fig. 1(b), we use query penalties P $= 160, 180, 200,$ $220, 240, 260, 280$ and fixed time span $t_\beta - t_\alpha = 60$. Again only Algorithm $DTime$ is affected and shows significant increase in running time, while Algorithms $DLen$ and $DPen$ remain stable. Running time for the heuristic estimate based algorithms are negligible compared to the other exact algorithms.

The running time of $DLen$ can be attributed to its theoretical complexity; however, $DPen$ algorithm shows an anomaly, as it should linearly scale with P. A possible explanation is that for most values of $p \in [1, \mathsf{P}]$, $open(p)$ is empty or contains very few vertices. Therefore, the running time becomes negligible when compared to that of $DTime$, in which every edge with arrival time t is scanned once for every $t \in [t_\alpha, t_\beta]$ and every $p \in [0, \mathsf{P}]$.

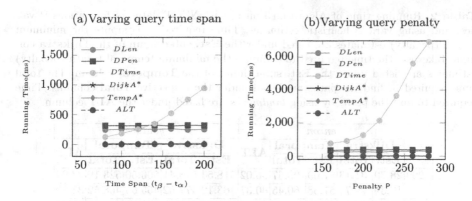

Fig. 1. Running time for varying query time span and penalty restriction.

5.3 A^* Algorithm on KONECT Dataset

The space occupied by the algorithms in Sect. 2 can be very high depending on n, Δ_{in}, P, B_U, and $(t_\beta - t_\alpha)$. For e.g., the *dblp* dataset has 1314050 vertices, and average degree is 14. Therefore, the space occupied by only $P(\cdot,\cdot,\cdot)$ structure in Algorithm 1, even for $B_U = 10$ can be as bad as 5.5 Gb. Additional space required by other data structures, and the graph itself, makes these algorithms impractical for large datasets. In fact, we observed that these algorithms suffer from heap size overflow exception for every KONECT dataset. The strength of the A^* algorithm lies in the fact that it prunes those paths which can never lead to an s-g RSTP not only based on $[t_\alpha, t_\beta]$, but also on the dominance criteria among tuples, and the penalty underestimate $h_{pen}(v)$. Since, in these data sets, the edge transit time is also its length, we also prune paths based on Remark 1.

As expected using a tighter heuristic of minimum temporal lengths (penalty), the path finding time (i.e., **Total - Est.** time) is smaller than that using minimum length (penalty) estimates. However, note that in Table 2, the bulk of time in computing the s-g RSTP (or determining that there is none) is due to computing the estimates. Therefore, the objective is to reduce this estimate computation time at the expense of the path computing time. This is achieved using estimates based on landmarks, which along with the corresponding values of $L(\cdot,\cdot)$, $L(\cdot,\cdot,\cdot,\cdot)$, $P(\cdot,\cdot)$, and $P(\cdot,\cdot,\cdot,\cdot)$ is precomputed and stored. Therefore, we only need to compute $h_{len}(v)$ and $h_{pen}(v)$ based on these stored values, as discussed in Sect. 3.2, whenever a tuple is added to $open(v)$ for the first time. It can be seen that in most cases, the landmark based approach is much faster.

We chose the landmarks randomly, and the size of the landmark set $\mathcal{L}$ is fixed at 10. Furthermore, $t_\alpha = \min\{start(e) \mid e \in \delta^+(s)\}$ and $t_\beta = \max\{end(e) \mid e \in \delta^-(g)\}$. Each experiment was repeated for 100 s-g pairs, and then the average was taken. We observed that on increasing the size of $\mathcal{L}$, the running time increases, and in fact in few cases, even for an empty $\mathcal{L}$ (i.e., with no heuristic information), Algorithm 3 performs better than using minimum (temporal) length and penalty

Table 2. Running time of Algorithm 3 on the KONECT datasets for various P values, and using various heuristic estimates. Time required to compute the minimum length (penalty) estimates are listed under the **Est.** sub-column of the **Dijkstra** column. Likewise, the time required to compute the minimum temporal length (penalty) estimates are listed under the **Est.** sub-column of the **Temporal** column. The total time required to find the paths are listed under the respective **Total** columns. Time required to find the paths by using *landmarks* are listed under the **ALT** column.

P	enron					youtube				
	Dijkstra		Temporal		ALT	Dijkstra		Temporal		ALT
	Est.	Total	Est.	Total		Est.	Total	Est.	Total	
200	28.79	31.13	87.09	90.67	56.02	518.54	528.41	506.50	523.19	23.99
400	19.39	29.70	57.58	60.45	60.37	463.12	477.62	455.69	468.52	35.87
600	27.22	49.54	56.85	64.03	44.63	458.16	468.80	403.73	416.09	13.72
800	23.16	23.62	52.60	52.93	49.01	345.66	355.79	374.92	388.84	15.17
1000	26.70	60.85	63.59	76.08	43.24	481.94	494.27	388.85	402.06	15.77
Avg	25.06	38.97	63.55	68.84	50.66	453.49	464.99	425.94	439.74	20.91

P	dblp					wiktionary				
	Dijkstra		Temporal		ALT	Dijkstra		Temporal		ALT
	Est.	Total	Est.	Total		Est.	Total	Est.	Total	
200	3111.96	3128.84	2052.05	2057.07	72.47	47.66	55.31	28.04	35.33	11.91
400	3572.06	4120.01	3063.62	3097.18	2057.39	67.71	102.2	88.98	120.26	190.13
600	3374.16	4094.12	1994.92	2016.98	1515.96	40.39	48.19	26.27	35.51	13.22
800	3735.48	4205.83	1945.71	2089.19	1737.14	65.54	98.05	84.91	115.92	43.30
1000	3358.02	4552.57	1898.25	1908.88	1727.06	53.05	61.61	32.41	40.91	12.06
Avg	3430.34	4020.28	2190.91	2233.86	1422.01	54.87	73.08	52.13	69.59	54.13

estimates. Also for the given datasets, we observed that the choice of t_α and t_β does not affect the running time of Algorithm 3 to a great extent.

6   Conclusion

In this paper, we extend the restricted shortest path problem in static networks to that for temporal networks. We present various exact algorithms, discuss their complexities, and experimentally evaluate them. Furthermore, using one of these exact algorithms, we show how to obtain an FPTAS. For the A^*-like algorithm, we show how to obtain various underestimates, one of which uses landmark based techniques. However, the selection of the landmarks has been done randomly. As a future effort, we would like to extend this by adapting "more intelligent" landmark selection strategies. Also, we have not discussed lower bounds of the complexities of the problem. It would be interesting to study whether algorithms with better complexities exist for the RSTP problem.

References

1. Cai, X., Kloks, T., Wong, C.K.: Time-varying shortest path problems with constraints. Networks **29**(3), 141–150 (1997)
2. Carlyle, W.M., Royset, J.O., Kevin Wood, R.: Lagrangian relaxation and enumeration for solving constrained shortest-path problems. Networks **52**(4), 256–270 (2008)
3. Day, P.R., Ryan, D.M.: Flight attendant rostering for short-haul airline operations. Oper. Res. **45**(5), 649–661 (1997)
4. Ergun, F., Sinha, R., Zhang, L.: An improved FPTAS for restricted shortest path. Inf. Process. Lett. **83**(5), 287–291 (2002)
5. Garey, M.R., Johnson, D.S.: Computers and Intractability: A Guide to the Theory of NP-Completeness. W.H. Freeman and Co., New York (1979)
6. Goldberg, A.V., Harrelson, C.: Computing the shortest path: a search meets graph theory. In: Proceedings of the Sixteenth Annual ACM-SIAM Symposium on Discrete Algorithms, SODA 2005, pp. 156–165. Society for Industrial and Applied Mathematics, Philadelphia (2005)
7. Hart, P., Nilsson, N., Raphael, B.: A formal basis for the heuristic determination of minimum cost paths. IEEE Trans. Syst. Sci. Cybern. **4**(2), 100–107 (1968)
8. Hassin, R.: Approximation schemes for the restricted shortest path problem. Math. Oper. Res. **17**(1), 36–42 (1992)
9. Kempe, D., Kleinberg, J.M., Demers, A.J.: Spatial gossip and resource location protocols. In: Proceedings on 33rd Annual ACM Symposium on Theory of Computing, Heraklion, Crete, Greece, 6–8 July 2001, pp. 163–172 (2001)
10. Korkmaz, T., Krunz, M.: Multi-constrained optimal path selection. In: Proceedings IEEE INFOCOM 2001, The Conference on Computer Communications, Twentieth Annual Joint Conference of the IEEE Computer and Communications Societies, Twenty Years into the Communications Odyssey, Anchorage, Alaska, USA, 22–26 April 2001, pp. 834–843 (2001)
11. Li, Y., Harms, J., Holte, R.: Fast exact multiconstraint shortest path algorithms. In: IEEE International Conference on Communications, ICC 2007, June 2007, pp. 123–130 (2007)
12. Lorenz, D., Orda, A.: QoS routing in networks with uncertain parameters. IEEE/ACM Trans. Netw. **6**(6), 768–778 (1998)
13. Lorenz, D.H., Raz, D.: A simple efficient approximation scheme for the restricted shortest path problem. Oper. Res. Lett. **28**(5), 213–219 (2001)
14. Mertzios, G.B., Michail, O., Chatzigiannakis, I., Spirakis, P.G.: Temporal network optimization subject to connectivity constraints. In: Fomin, F.V., Freivalds, R., Kwiatkowska, M., Peleg, D. (eds.) ICALP 2013, Part II. LNCS, vol. 7966, pp. 657–668. Springer, Heidelberg (2013)
15. Michail, O., Spirakis, P.G.: Traveling salesman problems in temporal graphs. In: Csuhaj-Varjú, E., Dietzfelbinger, M., Ésik, Z. (eds.) MFCS 2014, Part II. LNCS, vol. 8635, pp. 553–564. Springer, Heidelberg (2014)
16. Nachtigall, K.: Time depending shortest-path problems with applications to railway networks. Eur. J. Oper. Res. **83**(1), 154–166 (1995)
17. Van Mieghem, P., Kuipers, F.: Concepts of exact QoS routing algorithms. IEEE/ACM Trans. Netw. **12**(5), 851–864 (2004)
18. Wu, H., Cheng, J., Huang, S., Ke, Y., Lu, Y., Xu, Y.: Path problems in temporal graphs. Proc. VLDB Endow. **7**(9), 721–732 (2014)

An Efficient Distributed Index for Geospatial Databases

Le Hong Van[1]([⊠]) and Atsuhiro Takasu[1,2]

[1] SOKENDAI (The Graduate University for Advanced Studies),
Kanagawa, Japan
{l-van,takasu}@nii.ac.jp
[2] National Institute of Informatics, Tokyo, Japan

Abstract. The recent and rapid growth of GPS-enabled devices has resulted in an explosion of spatial data. There are three main challenges for managing and querying such data: the massive volume of data, the need for a high insertion throughput and enabling real-time spatial queries. Although key–value store databases handle large-scale data effectively, they are not equipped with effective functions for supporting spatial data. To solve this problem, we propose an efficient spatial index structure based on HBase, a standard key–value store database. We first use Geohash as the rowkey in HBase to sustain high insert throughput. We present a novel data structure, the binary Geohash rectangle-partition tree, that partitions data into subrectangles, then add these subrectangles into an R-Tree to support spatial queries. Our experiments demonstrate high scalability and an improved performance with spatial queries, when compared with a state-of-the-art system. They also show a good real-time query-processing capability, with response times of less than one second.

Keywords: HBase · GeoHash · R-Tree · Spatial index · Real-time

1 Introduction

With the rapid growth of spatial data, support for high-performance spatial queries about large-scale data is becoming essential in both scientific research and daily life. For example, intelligent transportation systems such as advanced traffic management systems are exploiting the massive volume of real-time sensor data from probe cars and GPS-enabled devices to provide efficient traffic balancing and trajectory recommendation services. Commuters can take advantage of these systems by sending data continuously about their current location and then receiving relevant analyses of their real-time traffic situation.

The main challenges for these systems are the massive volume of data, the need for a high insertion throughput and support for real-time spatial queries. For instance, traffic management systems in busy cities such as Tokyo have to support a large number of location updates per minute from probe cars such as

© Springer International Publishing Switzerland 2015
Q. Chen et al. (Eds.): DEXA 2015, Part I, LNCS 9261, pp. 28–42, 2015.
DOI: 10.1007/978-3-319-22849-5_3

taxis and buses and from collaborating private cars and GPS-enabled devices. At the same time, they have to analyze the data and respond in real time.

To meet these requirements, systems need database management systems (DBMS) that have good scalability while guaranteeing satisfactory performance with real-time spatial queries. Although distributed-computing technologies such as Hadoop MapReduce [8] are reasonable choices for handling large volumes of data, with some frameworks based on MapReduce such as SpatialHadoop [9] supporting spatial queries, they have high latency in comparison with a real-time system. Key–value store databases such as HBase, with their scalability, fault tolerance, availability and random real-time read/write capability, have shown promise. However, they do not have native support for spatial queries.

Some methods [13, 16] use Geohash [2], a linearization technique that transforms two-dimensional spatial data into a one-dimensional hashcode, to handle spatial data in HBase. However, they do not address edge cases and the Z-order limitations of Geohash, which can lead to incorrect results and can decrease the performance on spatial queries. In this paper, we tackle the limitations of Geohash with spatial data by proposing an efficient distributed spatial index (Fig. 1), which uses Geohash and the R-Tree [12] with HBase. We use Geohash as the key for key–value pairs for HBase, then further utilize the R-Tree, a multi-dimensional index structure for geographical data, to generate correct results and improve query performance. To bridge Geohash and the R-Tree, we propose a novel data structure, the binary Geohash rectangle-partition tree (BGRP Tree). By using the BGRP Tree, we partition a Geohash range into nonoverlapping subrectangles, then insert these subrectangles into the R-Tree.

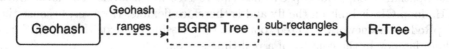

Fig. 1. Two-tier index of Geohash and the R-Tree using the BGRP Tree to bridge the tiers

Using Geohash as the key for HBase enables geospatial databases to achieve a high insert throughput because Geohash encoding is not computation intensive. The R-Tree index helps improve the performance and the accuracy of spatial queries. Our experimental results indicate that querying that can take advantage of the Geohash and R-Tree index outperformed both querying that used only Geohash on HBase and querying in a MapReduce-based system. We also observed that our proposed index could process queries in real time, with response times of less than one second.

This paper is organized as follows. Section 2 briefly surveys related work. The design of our proposed index and the new data structure, the BGRP Tree, are presented in Sects. 3 and 4, respectively. Experimental evaluation is discussed in Sect. 5, followed by our conclusions and plans for future work.

2   Related Work

Managing and querying spatial data can be processed efficiently using traditional relational DBMS such as Oracle Spatial and PostGIS [4]. However, these systems cannot meet the major requirements of a geospatial system, namely, scaling and analyzing a large volume of data, because of insufficient scalability.

When considering scalable data-processing systems, systems based on MapReduce [8] such as Hadoop [19] have dominated. Hadoop provides not only scalability to petabytes of data but is also fault tolerant and highly available. However, Hadoop is not aimed primarily at supporting the processing and analysis of spatial data. There are some extensions, such as SpatialHadoop [9] and Hadoop-GIS [5], which do support high-performance spatial queries based on the MapReduce framework.

SpatialHadoop is an extension of Hadoop that is designed with a view to handling massive volumes of spatial data. It comprises four components: a high-level spatial language, spatial operations, MapReduce and storage. The key idea in SpatialHadoop is a two-tier index (global and local) that uses an R-Tree to index spatial data inside the Hadoop distributed file system (HDFS). When compared with the original Hadoop, R-Tree indexing in SpatialHadoop performs better with spatial data. However, because the R-Tree is a balanced search tree, rebalancing is necessary after a number of data insertions. With continuous insertion from spatial data sources such as sensor data, this rebalancing can be computationally expensive.

Hadoop-GIS is another scalable and high-performance spatial data processing system, which is integrated into Hive [18] to support declarative spatial queries. Like SpatialHadoop, Hadoop-GIS has two R-Tree index levels. The local index in Hadoop-GIS is built on the fly and is therefore flexible. However, because the MapReduce framework is suited to a batch-processing context, Hadoop-GIS has a high latency in the context of a real-time system.

Spatial support has been also extended to NoSQL-based solutions. For instance, MD-HBase [15] is a scalable multidimensional database for location-aware services, which layers a multidimensional index over the key–value store. It uses linearization techniques such as Z-ordering [14] to transform multidimensional data into one-dimensional form, then stores the data in HBase. It also layers multidimensional index structures, such as K-d trees [6] and quad-trees [10], on top of the key–value stores. For our work, we adopt a different approach by using Geohash as a linearization technique and the R-tree as a multidimensional index. The common prefix in Geohash offers a better way to find nearest neighbors and the R-Tree can accelerate a nearest-neighbor search.

Lee et al. [13] and Pal et al. [16] also adopt Geohash to handle spatial data in HBase. In [13], spatial queries with the Geohash index performed better than spatial queries with the R-Tree index. However, they did not present solutions for the limitations of Geohash.

3   Distributed Spatial Index

3.1   Basis of Distributed Spatial Index

Apache HBase [11] is a distributed scalable database that takes advantage of the HDFS. Because HBase stores data as a sorted list of key–value pairs, it is also a key–value store database. Within a table, data are stored according to rows that are uniquely identified by their rowkeys, which therefore play an important role when searching or scanning. Rowkey design is one of the most important aspects of HBase schema.

One of the advantages of HBase is its auto-sharding capability, which means it can dynamically split tables when the quantity of data becomes excessive. The basic unit of scalability and load balancing in HBase is called a region. Because HBase keeps rowkeys in lexicographical order, each region stores a range of rowkeys between a start rowkey and an end rowkey.

Data stored in HBase are accessed by a single rowkey. However, spatial data are represented by two coordinates (longitude and latitude), which are equally important in defining a location. Geohash, which was invented by Gustavo Niemeyer in 2008 [2], provides a solution to transform longitude/latitude coordinates into unique geocodes.

Geohash has some notable properties. It is a hierarchical spatial data structure that subdivides space into grid-shaped buckets. Points that share the same prefix will be close to each other geographically, and all Geohashes with the same prefix are included in the rectangle for that prefix. The shorter a Geohash is, the larger will be its rectangle. Geohash has a Z-order traversal of rectangles in each resolution (Fig. 2). To best take advantage of Geohash, we store the spatial data in HBase where the rowkeys are Geohashes. In this way, points that are close geographically are stored close together.

Despite its advantages, Geohash has some limitations.

Edge-Case Problem. Locations on the opposite sides of the 180-degree meridian are close to each other geographically but have no common prefix in Geohash codes. Points close together at the North and South poles may also have very different Geohashes.

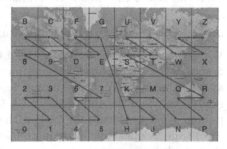

Z-Order Problem. This problem is related to Z-order traversal. Some points having a common Geohash prefix are not always close to each other geographically. For instance, in Fig. 2, locations in rectan-

Fig. 2. Z-order traversal in Geohash

gles whose Geohash codes are *7* and *8* will be stored close together, but are not close geographically. In contrast, nearby locations such as in rectangles *7* and *e* have widely separated Geohash codes.

3.2   Index Design

Although using Geohash as the rowkey in HBase helps the system support high insert throughput, using Geohash alone does not guarantee efficient spatial-query processing. For instance, nearby points often share the same Geohash prefix, enabling us to find the k nearest neighbors (kNN) of a query point using a prefix filter. However, the edge-case problem would lead to insufficient or incorrect results when the query point has a longitude near 180 or -180 degrees. The Z-order problem can cause redundant scanning. For example, in Fig. 2, when the system wants to scan all points ranging from rectangles with Geohash *1* to *4*, it will scan rectangles *1*, *4* and the unrelated rectangles *2* and *3*.

To obtain accurate query results and prune unnecessary scanning, we use the R-Tree as a secondary index tier. The R-tree stores only rectangles, rather than points. In this way, we can avoid frequent rebalancing of the tree, which is a time-consuming process, even if new points arrive continuously. In our method, we first partition regions into rectangles using the *longest common prefix* (LCP). Because there is overlap between these rectangles, we use the BGRP Tree (described in Sect. 4) for further partitioning of rectangles into subrectangles until there is no overlap between them. Finally, all subrectangles are inserted into the R-Tree. When processing spatial queries, we find the rectangles in the R-Tree that may contain query results before scanning. We then scan only the found rectangles, thereby pruning the scanning on unrelated regions.

As described in Sect. 3.1, the data are split into regions dynamically. HBase performs a finer-grained partition for the denser places, i.e., those that contain more points in a small area. The R-tree is constructed based on these HBase partitions and the proposed index can therefore manage any data skew by this adaptive partitioning.

All the steps for R-Tree creation are processed in the background, removing the need for R-Tree creation overhead when processing queries. Because the R-Tree is stored in memory and the number of nodes in the tree is not large, the R-Tree search overhead is also very small. We evaluate this overhead empirically in Sect. 5.

3.3   LCP-Based Region Partition

We partition regions by using the LCP of the Geohash codes. Algorithm 1 describes the algorithm for region partition. We first find the LCP of the start and end rowkeys. We then obtain the character next to the LCP from the start rowkey (c_1) and the end rowkey (c_2). Geohash uses base32 [1] to encode spatial points. We concatenate the LCP with each character from c_1 to c_2 in base32 to create the Geohash of a new rectangle. By using this algorithm, we can ensure that the list of rectangles will cover all points in the region.

Table 1 shows an example of region partitioning. In this example, we have three regions in the Geohash range from *ww4durf3yp21* to *wx1g1zbnwxnv*. The result of region partitioning is a list of rectangles that cover all points in the three regions. Figure 3a shows the results of the region partitioning on a map. In this figure, we can assume that the areas involving rectangles with longer Geohashes, i.e., *ww5e* and *ww5f*, have higher point density than other areas.

Algorithm 1. Region Partitioning

[1] $lcp \longleftarrow findLongestCommonPrefix(startRowKey, endRowKey)$
[2] $lengthOfLcp \longleftarrow length(Lcp)$
[3] $c_1 \longleftarrow startRowKey[lengthOfLcp]$
[4] $c_2 \longleftarrow endRowKey[lengthOfLcp]$
[5] /* Loop on list of symbols of base32   */
[6] for $c \leftarrow c_1$ to c_1 do
[7] $addToResult(lcp + c)$

Table 1. An example of region partitioning

Region	Start Rowkey	End Rowkey	LCP	Rectangles
Region 1	**ww**4durf3yp21	**ww**5eh56ebb98	ww	ww4, ww5
Region 2	**ww5**eh56ebb98	**ww5**ffn87crgs	ww5	ww5e, ww5f
Region 3	ww5ffn87crgs	wx1g1zbnwxnv	w	ww, wx

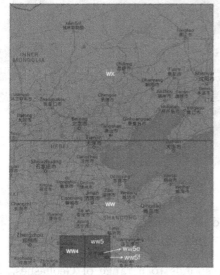

(a) Rectangles generated from the example in Table 1

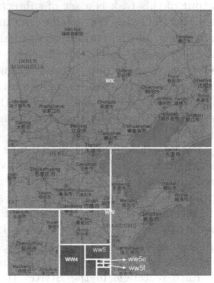

(b) Rectangles after using the BGRP Tree for partitioning

Fig. 3. Partition using the BGRP tree

4   The BGRP Tree

In Fig. 3a, some rectangles overlap, such as *ww*, *ww5* and *ww5e*. Inserting these rectangles directly into the R-Tree would lead to redundancy in the query results. For instance, when searching the R-Tree for a rectangle containing a point with prefix ww5e, the results would be ww, ww5 and ww5e.

The BGRP Tree is a data structure for representing Geohash rectangles. A BGRP Tree satisfies the following properties.

- The level of a node is the length of the Geohashes.
- All nodes (rectangles) are within the Geohash range.
- All nodes have between 1 and 32 children.
- There is no overlap between nodes.
- The leaves cover the whole surface of the original rectangles.
- Tree construction does not depend on the order of rectangle insertion.

4.1   The BGRP Task

A BGRP task is necessary in many insertion cases, as explained in Sect. 3. The input for this task is two rectangles, with the bigger rectangle containing the smaller one. Both could be presented by a Geohash or a range of Geohashes. The output of this task is a set of subrectangles that includes the smaller rectangle and covers all the surface of the bigger one. Our BGRP method recursively subdivides a rectangle into two subrectangles until one of the two subrectangles matches the smaller input rectangle. This method is similar to the binary space-partitioning method [17].

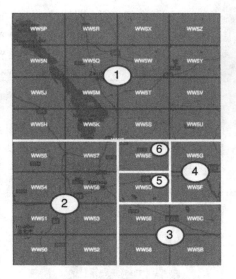

Fig. 4. The BGRP task

Figure 4 describes the BGRP task for the inputs $ww5$ and $ww5e$. First, we divide $ww5$ into two parts, a subrectangle sr_1 with the Geohash from $ww50$ to $ww5g$ and a subrectangle sr_2 with the Geohash from $ww5h$ to $ww5z$. The rectangle sr_2 is inserted as is, whereas sr_1 is split further into two subrectangles. The process is recursively applied until one of the subrectangles is equal to $ww5e$. As a result, six subrectangles are obtained (Fig. 4).

Each rectangle in the BGRP Tree corresponds to a range of Geohashes. For instance, the Geohash range of sr_2 is from $ww5h$ to $ww5z$, whereas $ww5$ corresponds to a range from $ww50$ to $ww5z$. For a rectangle whose Geohash range length is r, the task will require log_2r binary partition steps and will result in a set of $log_2r + 1$ subrectangles.

4.2   BGRP Tree Searching

The search algorithm descends the tree from the root, recursively searching subtrees for the highest-level node that contains the input rectangle, as shown in Algorithm 2. Like search algorithms in R-Trees, if a node does not contain the query point, we do not search inside that node. If there are n nodes in the tree, the search requires $O(log(n))$ time.

Algorithm 2. BGRP Tree Search

```
search(node, rect)
      /* this function calls initially with node = root        */
      input  : node, rect
      output: highest-level node that contains input rectangle
[1]   if node = rect then
[2]       return node
[3]   if node is leaf then
[4]       return node
[5]   else
[6]       for child ∈ children of node do
[7]           if child = rect then
[8]               return child
[9]           else if child contains rect then
[10]              return search(child, rect)
```

4.3   BGRP Tree Insertion

Insertion is the most complicated aspect of the BGRP Tree. Figure 5 shows the insertion of all rectangles in Fig. 3a into a BGRP Tree. The resulting tree is shown in Fig. 3b.

Algorithm 3 describes the insertion algorithm for the BGRP Tree. If the tree has been initialized, the first step is to search for the highest-level node containing the insert rectangle (as described in Sect. 4.2). If no node is found, we create a new root by obtaining the LCP between the Geohash of the root and the rectangle. For instance, in Fig. 5, when inserting $ww5$, the LCP between

---
Algorithm 3. BGRP Tree Insertion

```
[1]  if no root then
[2]      root ⟵ rect
[3]      return
[4]  containNode ⟵ search(root, rect)
[5]  if containNode is null then
[6]      oldRoot ⟵ root
[7]      root ⟵ getLongestCommonPrefix(oldRoot, rect)
[8]      insertWithoutPartition(root, oldRoot)
[9]      insertWithoutPartition(root, rect)
[10] else
[11]     if containNode = rect then
[12]         if containNode is not leaf and is not yet fully partitioned then
[13]             binaryNodePartition(containNode)
[14]     else
[15]         if containNode is leaf then
[16]             binaryLeafPartition(containNode, rect)
[17]         else
[18]             insertWithoutPartition(containNode, rect)
```

---

the Geohash of the current root ($ww4$) and $ww5$ is ww. We therefore create the new root ww, then insert $ww5$ into the new root without partitioning.

There are two types of insertion in the BGRP Tree, namely insertion with a binary partition and insertion without partition. Insertion without partition is applied when inserting into an existing nonleaf node. We create a binary partition when inserting into a leaf, as shown for the insertion of $ww5e$ and $ww5f$ in Fig. 5. Binary partitioning is also applied when the insert rectangle exists in the tree as a nonleaf node that is not yet fully partitioned. Insertion of ww in Fig. 5 is an example of this case. Before being inserted, ww was a nonleaf node with only two children, $ww4$ and $ww5$, which do not cover the entire surface of ww. Therefore, we need to partition ww to ensure that the surface of ww is completely covered in leaves.

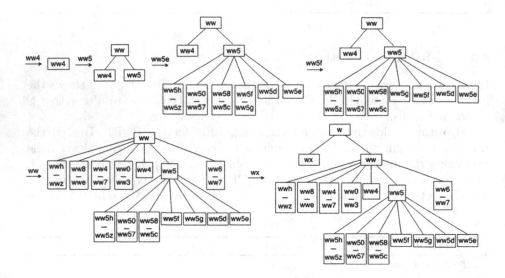

Fig. 5. BGRP tree insertion

Binary partitioning is different for leaf and nonleaf nodes. For a nonleaf node, the existing children of the node need to be considered and we therefore partition the nonleaf node until all its children are contained in the results.

5   Experimental Evaluation

5.1   Experimental Setup

HBase. We built a cluster comprising one HMaster, 60 Region Servers and three Zookeeper Quorums. Each node had a virtual core, 8 GB memory and a 64 GB hard drive. The operating system for the nodes was CentOS 7.0 (64-bit).

We used Apache HBase 0.98.7 with Apache Hadoop 2.4.1 and Zookeeper 3.4.6. Replication was set to two.

SpatialHadoop. To evaluate the performance of our method, we compared it with SpatialHadoop. For this, we used a cluster with one master and 64 slaves. Each node had one virtual core, 8GB memory and a 64GB hard drive with CentOS 7.0 (64-bit). Replication was also set to two. We installed SpatialHadoop v2.1, which shipped with Apache Hadoop 1.2.1.

5.2   Dataset

We used two real-world datasets, namely T-Drive [20,21] and OpenStreetMap (OSM) [3].

T-Drive. T-Drive was generated by 30,000 taxis in Beijing over a period of three months. The total number of records in this dataset is 17,762,390. It requires about 700 MB, and was split into 16 regions for the HBase cluster.

OSM. OSM is an open dataset that was built by contributions from a large number of community users. It contains spatial objects presented in many forms such as points, lines and polygons. We used the dataset of nodes for the whole world. It is a 64 GB dataset that includes 1,722,186,877 records. By using Snappy compression, we inserted the OSM dataset into 251 regions on the HBase cluster.

5.3   Comparison Method

To evaluate the performance of spatial queries based on our proposed index, we executed kNN queries using the two datasets, T-Drive and OSM. We evaluated three kNN query-processing methods using HBase and compared them with SpatialHadoop as a baseline method. The three methods are described below.

HBase with Geohash. We used Geohash as the rowkey for storing data in HBase. When processing a kNN query, the client first sends a scan request with a prefix filter to the region servers, obtains the scan results, then calculates kNN results for the received data. The prefix filter enables scanning of only those points that share the prefix with the query point.

HBase Parallel with Geohash. HBase supports efficient computational parallelism by using "coprocessors". The idea of HBase coprocessors was inspired by Google's BigTable [7] coprocessors. By using coprocessors, we could apply a divide-and-conquer approach for the kNN query. In parallel kNN, the kNN is processed inside each of the regions. In this way, instead of returning the unprocessed scan results, each region returns the kNN for its region to the client.

HBase Parallel with Geohash and R-Tree. With our proposed index design, the client searches the R-Tree for the rectangle that contains the query point before scanning. If the number of neighbors inside that rectangle is insufficient, the client continues the scan inside the neighbors of the rectangle until sufficient results are found.

SpatialHadoop. We also experimented with kNN queries using SpatialHadoop for comparison with our method.

We chose two groups of points for the experiments, namely, a group of high-density points and a group of low-density points. High-density points are the points in crowded areas, which will contain many points. Conversely, low-density points are the points in uncongested areas, which contain few points. Because HBase can handle data skew via its auto-sharding capability, we visualized the start rowkeys and end rowkeys for all regions on the map, then randomly chose points in high-density and low-density areas. In low-density areas, when k is large, a query may not find enough neighbors in its early stages. In such cases, we continue by scanning an increasingly large area (by decreasing the length of the prefix in the prefix filter) until sufficient results are retrieved. The query-processing time is measured by calculating the elapsed time between when the system starts scanning data and when all results are returned.

5.4   Results and Discussion

Insert Efficiency. We evaluated the insert performance for both the T-Drive and OSM datasets. Table 2 shows the insertion-process efficiency in terms of the number of inserted points per second. It is calculated by dividing the number of inserts by the elapsed time. This experiment involved insertion from a single client. Evaluation of insertion from multiple clients is proposed for future work. As shown in the table, our index structure can process about 80,000 inserts per second. Note that, unlike other systems with a tree-structured index, no extra tree balancing is needed even for heavy inserts. Because creating rectangles to insert into the R-Tree is based on the start and end rowkeys of regions, recreating the R-Tree is only necessary when a region exceeds a predefined maximum size and is partitioned into two regions.

kNN Queries for the T-Drive Dataset. Figure 6 shows the performance of kNN queries for the T-Drive dataset. We see that parallel kNN using Geohash and R-Tree outperforms all other kNN query methods. In this experiment, note that our index design operates about 60 to 90 times faster than SpatialHadoop. This is because HBase does not require the startup, cleanup, shuffling and sorting tasks of MapReduce. Another reason is that we store kNN procedures in every region server beforehand, thereby needing only to invoke that procedure locally on each server. In contrast, MapReduce has to send the procedure to slave servers for every query, thereby taking more time for network communication.

Table 2. Insert throughput

Dataset	Number of inserts	Elapsed time (sec)	Throughput (inserts/sec)
T-Drive	17,762,390	399	44,517
OSM	1,722,186,877	21,268	80,975

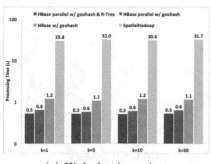

(a) High-density points

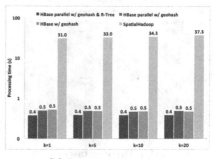

(b) Low-density points

Fig. 6. Performance of kNN queries for the T-Drive dataset

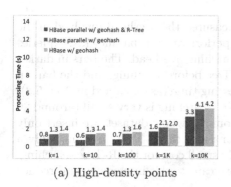

(a) High-density points

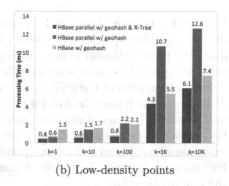

(b) Low-density points

Fig. 7. Performance of kNN queries for the OSM dataset

kNN Query for the OSM Dataset. As Fig. 7 shows, when processing the larger dataset, a kNN query with the proposed two-tier index also outperforms other query methods. For parallel kNN query with Geohash, I/O bottlenecks may occur because regions that do not contain neighbors of the query point still send responses to the client (with no useful results). The T-Drive data spread across only 16 regions, giving a less-than-critical I/O load. However, for the larger OSM dataset, with its number of regions increased 15-fold, the client I/O load is much heavier. For our proposed method, where we search the R-Tree before scanning to limit the regions requiring processing, we can reduce this I/O load.

Unlike the experiments using the T-Drive dataset, experiments with the OSM dataset differed in performance, depending on whether high-density or low-density points were being processed. For low-density points and larger values of k, the queries could not find sufficient neighbors during the initial search. Therefore, queries had to search again over a larger area, which led to higher latency. By using the R-Tree to search rectangles near the query point, the queries could reduce the scanning of unrelated areas, thereby achieving an improved performance.

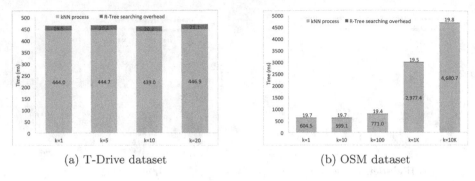

(a) T-Drive dataset   (b) OSM dataset

Fig. 8. R-Tree searching overhead (Color figure online)

R-Tree Searching Overhead. We also measured the overhead involved in the R-Tree searching step. Figure 8 shows the performance of parallel kNN queries using the proposed index with an R-Tree searching overhead. The bars in darker color represent the cost of searching the R-Tree before scanning, and the bars in lighter color represent the kNN query processing time (as explained in Sect. 5.3). As shown in the figure, the cost for the R-Tree searching is very small (around 20 milliseconds). Instead of inserting points from the whole dataset, we insert only big rectangles representing dataset partitions into the R-Tree, implying only a small number of nodes and a correspondingly small cost for the R-Tree searching. Furthermore, because the R-Tree is small, we can store it in memory and reduce the reading-time latency.

6 Conclusion

In this paper, we proposed a two-tier index for spatial data in a key–value store database using Geohash and an R-tree structure. We also defined the BGRP Tree data structure to bridge the two tiers. Using our proposed index, HBase can support spatial queries efficiently. Experimental results for two real-world datasets demonstrated the performance improvement for spatial queries when using the two-tier index with Geohash and an R-Tree. The results also showed that the R-Tree searching overhead is very small and the response time for queries would meet the requirements of real-time systems.

In future work, we plan to consider other complex spatial queries such as range queries, spatial joins and convex hulls. For some applications such as route planners based on current traffic data, a spatiotemporal database plays an important role in the data analysis. Spatiotemporal indexing for a key–value store database is one of our ongoing projects. We are also working on stream data, such as data being collected continuously from sensors. Stream data pose many new challenges, such as processing data when its spatial relationships are changing continuously and maintaining consistency while guaranteeing high read/write throughput and effective query processing.

Acknowledgments. This work was partly supported by the research promotion program for national-level challenges Research and development for the realization of next-generation IT platforms? by MEXT, Japan and the Strategic Innovation Promotion Program of the Japanese Cabinet Office.

References

1. Base32. http://en.wikipedia.org/wiki/Base32
2. GeoHash. http://geohash.org
3. Openstreetmap. http://www.openstreetmap.org
4. PostGIS. http://postgis.net
5. Aji, A., Wang, F., Vo, H., Lee, R., Liu, Q., Zhang, X., Saltz, J.: Hadoop GIS: a high performance spatial data warehousing system over mapreduce. Proc. VLDB Endow. **6**(11), 1009–1020 (2013)
6. Bentley, J.L.: Multidimensional binary search trees used for associative searching. Commun. ACM **18**(9), 509–517 (1975)
7. Chang, F., Dean, J., Ghemawat, S., Hsieh, W.C., Wallach, D.A., Burrows, M., Chandra, T., Fikes, A., Gruber, R.E.: Bigtable: a distributed storage system for structured data. ACM Trans. Comput. Syst. (TOCS) **26**(2), 4 (2008)
8. Dean, J., Ghemawat, S.: Mapreduce: simplified data processing on large clusters. Commun. ACM **51**(1), 107–113 (2008)
9. Eldawy, A., Mokbel, M.F.: A demonstration of spatialhadoop: an efficient mapreduce framework for spatial data. Proc. VLDB Endow. **6**(12), 1230–1233 (2013)
10. Finkel, R.A., Bentley, J.L.: Quad trees a data structure for retrieval on composite keys. Acta informatica **4**(1), 1–9 (1974)
11. George, L.: HBase: The Definitive Guide. O'Reilly Media Inc., Sebastopol (2011)
12. Guttman, A.: R-Trees: A Dynamic Index Structure for Spatial Searching, vol. 14. ACM, New York (1984)
13. Kisung, L., Ganti, R.K., Srivatsa, M., Liu, L.: Efficient spatial query processing for big data. Framework **7**(11), 4–12 (2014)
14. Morton, G.M.: A Computer Oriented Geodetic Data Base and a New Technique in File Sequencing. International Business Machines Company, New York (1966)
15. Nishimura, S., Das, S., Agrawal, D., Abbadi, A.E.: MD-HBase: a scalable multi-dimensional data infrastructure for location aware services. In: 2011 12th IEEE International Conference on Mobile Data Management (MDM), vol. 1, pp. 7–16. IEEE (2011)
16. Pal, S., Das, I., Majumder, S., Gupta, A.K., Bhattacharya, I.: Embedding an extra layer of data compression scheme for efficient management of big-data. In: Mandal, J.K., Satapathy, S.C., Sanyal, M.K., Sarkar, P.P., Mukhopadhyay, A. (eds.) Information Systems Design and Intelligent Applications, pp. 699–708. Springer, India (2015)
17. Schumacker, R.A., Brand, B., Gilliland, M.G., Sharp, W.H.: Study for applying computer-generated images to visual simulation. Technical report, DTIC Document (1969)
18. Thusoo, A., Sarma, J.S., Jain, N., Shao, Z., Chakka, P., Anthony, S., Liu, H., Wyckoff, P., Murthy, R.: Hive: a warehousing solution over a map-reduce framework. Proc. VLDB Endow. **2**(2), 1626–1629 (2009)
19. White, T.: Hadoop: The Definitive Guide. O'Reilly Media Inc., Sebastopol (2012)

20. Yuan, J., Zheng, Y., Xie, X., Sun, G.: Driving with knowledge from the physical world. In: Proceedings of the 17th ACM SIGKDD International Conference on Knowledge Discovery and Data Mining, pp. 316–324. ACM (2011)
21. Yuan, J., Zheng, Y., Zhang, C., Xie, W., Xie, X., Sun, G., Huang, Y.: T-drive: driving directions based on taxi trajectories. In: Proceedings of the 18th SIGSPATIAL International Conference on Advances in Geographic Information Systems, pp. 99–108. ACM (2010)

The xBR+-tree:
An Efficient Access Method for Points

George Roumelis[1], Michael Vassilakopoulos[2]($\boxtimes$), Thanasis Loukopoulos[3],
Antonio Corral[4], and Yannis Manolopoulos[1]

[1] Department of Informatics, Aristotle University of Thessaloniki,
Thessaloniki, Greece
{groumeli,manolopog}@csd.auth.gr
[2] Department of Electrical and Computer Engineering, University of Thessaly,
Volos, Greece
mvasilako@inf.uth.gr
[3] Department of Computer Science and Biomedical Informatics,
University of Thessaly, Lamia, Greece
luke@dib.uth.gr
[4] Department of Informatics, University of Almeria, Almeria, Spain
acorral@ual.es

Abstract. Spatial indexes, such as those based on Quadtree, are important in spatial databases for efficient execution of queries involving spatial constraints. In this paper, we present improvements of the xBR-tree
(a member of the Quadtree family) with modified internal node structure and tree building process, called xBR+-tree. We highlight the differences of the algorithms for processing single dataset queries between the
xBR and xBR+-trees and we demonstrate performance results (I/O efficiency and execution time) of extensive experimentation (based on real
and synthetic datasets) on tree building process and processing of single
dataset queries, using the two structures. These results show that the two
trees are comparable, regarding their building performance, however, the
xBR+-tree is an overall winner, regarding spatial query processing.

Keywords: Spatial access methods · Quadtrees · xBR-trees · Query
processing.

1 Introduction

Hierarchical index structures are useful because of their ability to focus on the
interesting subsets of the data [8]. This focusing results in an efficient representation and improved execution times on query processing and is thus particularly useful for performing spatial operations [13]. Important advantages of

G. Roumelis, M. Vassilakopoulos, T. Loukopoulos, A. Corral and Y. Manolopoulos—
Work funded by the GENCENG project (SYNERGASIA 2011 action, supported
by the European Regional Development Fund and Greek National Funds); project
number 11SYN_8_1213.

© Springer International Publishing Switzerland 2015
Q. Chen et al. (Eds.): DEXA 2015, Part I, LNCS 9261, pp. 43–58, 2015.
DOI: 10.1007/978-3-319-22849-5_4

these structures are their conceptual clarity and their great capability for query processing. The Quadtree is a well known hierarchical index structure, which has been applied successfully on Geographical Information Systems (GISs), image processing, spatial information analysis, computer graphics, digital databases, etc. [11,12]. It was introduced in the early 1970s [2], and it is based on the principle of recursive decomposition of space, and have become an important access method for spatial data [3].

The External Balanced Regular (xBR)-tree [15] belongs to the family of Quadtrees and it has been shown to be competitive to the R*-tree [1] for spatial queries involving a single dataset [10]. In this paper, we present an improved version of the xBR-tree [15], called xBR$^+$-tree, which is also a secondary memory structure that belongs to the Quadtree family. The xBR$^+$-tree improves the xBR-tree node structure and tree building process. The node structure of the xBR$^+$-tree stores information related to the quadrant subregions that contain point data. On the contrary, information related to the position of the quadrants is not stored explicitly in xBR$^+$-tree, but is computed when needed. These, make query processing more efficient.

Apart from the presentation of the xBR$^+$-tree, other contributions of this paper are the conclusions arising from an extensive experimental comparison (based on real and synthetic datasets) of xBR and xBR$^+$-trees, regarding I/O performance and execution time for

Table 1. List abbreviations.

Abbreviation	Term
PLQ	Point Location Query
WQ	Window Query
DRQ	Distance Range Query
K-NNQ	K-Nearest Neighbor Query
CK-NNQ	Constrained K-Nearest Neighbor Query
K-DJQ	K-Distance Join Query
MBR	Minimum Bounding Rectangle
DBR	Data Bounding Rectangle
Address	Directional digits that specify position and size of a node
REG	Size and position of Quadrant
qside	Size of Quadrant
NAclN	North America Cultural Landmarks Dataset
NAppN	North America Populated Places Dataset
NArrN	North America Rail Roads Centers Dataset
NArrND	North America Rail Roads MBR Coordinates Dataset
NArdN	North America Roads Centers Dataset
NArdND	North America Roads MBR Coordinates Dataset
XXXKCN	XXX thousands of Clustered Normalized Points

- Tree building,
- Point Location Queries (*PLQs*),
- Window Queries (*WQs*) and Distance Range Queries (*DRQs*),
- K-Nearest Neighbor Queries (*K-NNQs*) and Constrained K-Nearest Neighbor Queries (*CK-NNQs*).

To improve readabilty, in Table 1 we present the above and other abbreviations used in this paper.

This paper is organized as follows. In Sect. 2 we review Related Work on Quadtrees and comparable access methods, regarding query processing and provide the motivation for this paper. In Sect. 3, we review the xBR-tree and present the improvements of the xBR$^+$-tree, paying special attention to the node structure and the tree building process. In Sect. 4, we present the algorithms for processing single dataset spatial queries over xBR and xBR$^+$-trees. In Sect. 5, we present representative results of the extensive experimentation performed, using real and synthetic datasets, for comparing the performance of the two Quadtree-based structures. Finally, in Sect. 6 we provide the conclusions arising from our work and discuss related future work directions.

2   Related Work and Motivation

A Quadtree is a class of hierarchical data structures whose common property is that they are based on the principle of *recursive decomposition of space* and it contains two types of nodes: non-leaf (internal) nodes and leaf (external) nodes. It is most often used to partition a 2d space by recursively subdividing it into four quadrants or regions: NW (North West), NE (North East), SW (South West) and SE (South East). According to [14], types of Quadtrees can be classified by following three principles: (1) the type of data that they are used to represent (points, regions, curves, surfaces and volumes), (2) the principle guiding the decomposition process, and (3) the resolution (variable or not).

In order to represent Quadtrees, there are two major approaches: *pointer-based Quadtree* and *pointerless Quadtree*. In general, the *pointer-based Quadtree* representation is one of the most natural ways to represent a Quadtree structure. In this method, every node of the Quadtree will be represented as a record with pointers to its four sons. Sometimes, in order to achieve special operations, an extra pointer from the node to its father could also be included. This representation should be taken into account when considering space requirements for recording the pointers and internal nodes. The xBR-tree belongs to the category of *pointer-based Quadtree*. On the other hand, the *pointerless representation of a Quadtree* defines each node of the tree as a unique locational code [12]. By using the regular subdivision of space, it is possible to compute the locational code of each node in the tree. The linear Quadtree is an example of pointerless Quadtree. We refer the reader to [11–13, 16] for further details.

Regarding to the performance comparison of spatial query algorithms using the most cited spatial access methods (R-trees and Quadtrees), several previous research efforts have been published.

- In [5] a qualitative comparative study is performed taking into account three popular spatial indexes (R*-tree, R$^+$-tree and PMR Quadtree), in large line segment databases. The main conclusion reached was that the R$^+$-tree and PMR Quadtree have the best performance when the operations involve search, since they result in a disjoint decomposition of space.
- In [6], various R-tree variants (R-tree, R*-tree and R$^+$-tree) and the PMR Quadtree have been compared for the traditional overlap spatial join operation. They showed that the R$^+$-tree and PMR Quadtree outperform the R-tree and R*-tree using 2d spatial data for overlap join, since they are spatial data structures based on a disjoint decomposition of space.
- In [7], the authors have compared the performance of the R*-tree and the Quadtree for evaluating the *K-NN* and the *K* Distance Join (*K-DJ*) query operations and the index construction methods (dynamic insertion and bulk-loading algorithm). It was shown that the query processing performance of R*-tree is significantly affected by the index construction methods, while the Quadtree is relatively less affected by the index construction method. The regular and disjoint partitioning method used by the Quadtree has an inherent structural advantage over the R*-tree in performing *KNN* and distance join queries. Finally, the Quadtree-based index structure can be a better choice than the widely used R*-tree for studied spatial queries when indices are constructed dynamically.
- In [10], the performance of R*-trees and xBR-trees is compared for the most usual spatial queries, like *PLQs*, *WQs*, *DRQs*, *K-NNQs* and *CK-NNQs*. The conclusions arising from this comparison show that the two indexes are competitive. The xBR-tree is more compact and it is built faster than the R*-tree. The performance of the xBR-tree is higher for *PLQs*, *DRQs* and *WQs*, while the R*-tree is slightly better for *K-NNQs* and needs less disk access for *CK-NNQs*.

Finally, xBR-trees have been presented in [15] and results related to the analysis of their performance have been presented in [4]. Using xBR-trees for processing *PLQs*, *WQs* or *DRQs* is rather straightforward [10], due to the organization of the xBR-tree. However, algorithms for processing *K-NNQs* and *CK-NNQs* by using these trees have only recently been developed and tested with real datasets [9,10], with excellent performance. Therefore, the main objective of this paper is to improve the xBR-tree, obtaining the xBR$^+$-tree, and compare its performance against the performance of the xBR-tree, considering the most representative spatial queries where a single index is involved.

3   The xBR-tree Family

In this section we describe the xBR-tree ([10,15]) and illustrate the extension of it, the xBR$^+$-tree, which is the primary contribution of this paper. When a characteristic is common for both trees, we will refer to the *xBR-tree family*. Otherwise, we will refer explicitly to the tree type that has this characteristic.

For 2d the hierarchical decomposition of space in the xBR-tree family is that of Quadtrees (the space is recursively subdivided in 4 equal subquadrants). The space indexed by a member of the xBR-tree family is a *square*. The nodes of members of the xBR-tree family are disk pages of two kinds: *leaves*, which store the actual multidimensional data themselves and *internal nodes*, which provide a multiway indexing mechanism.

3.1 Internal Nodes

As described in [10], *internal* nodes of xBR-trees contain entries of the form (*Shape, Address, REG, Pointer*). An *Address* is used to determine the region of a child node and is accompanied by the *Pointer* to this child. Since *Addresses* are of variable size, the number of entries fitting in each node is not predefined. Apparently, the space occupied by all entries within a node must not exceed the size of this node. The maximum size of an *Address* is only limited by the node size and in practice it never reaches this limit. *Shape* is a flag that determines if the region of the child is a complete or non-complete square (the area remaining, after one or more splits; explained later in this subsection). This field will be used widely in queries. Finally, *REG* stores the coordinates of the region referenced by *Address*. We measured the execution time for queries and we found that it is more expensive if we do not save this field, but calculate its value when needed.

Each *Address* represents a subquadrant which has been produced by Quadtree-like hierarchical subdivision of the current space. It consists of a number of directional digits that make up this subdivision. The NW, NE, SW and SE subquadrants of a quadrant are distinguished by the directional digits 0, 1, 2 and 3, respectively. For example, the *Address* 1 represents the NE quadrant of the current space, while the *Address* 10 the NW subquadrant of the NE quadrant of the current space. The address of the left child is * (has zero digits), since the region of the left child is the whole space minus the region of the right child.

However, the region of a child is, in general, the subquadrant of the related *Address* minus a number of smaller subquadrants. The region of this child is the subquadrant determined by the *Address* in its entry, minus the subquadrants corresponding to the next entries of the internal node (the entries in an internal node are saved sequentially, in preorder traversal of the Quadtree that corresponds to the internal node). For example, in Fig. 1 an internal node (a root) that points to 2 internal nodes that point to 7 leaves is depicted. The region of the root is the original space, which is assumed to have a quadrangular shape. The region of the right (left) child is the NW quadrant of the original space (the whole space minus the region of the NW quadrant - a non complete square), depicted by the union of the black regions of the leaves of this child. The * symbol is used to denote the end of a variable size address. The *Address* of the right child is 0*, since the region of this child is the NW quadrant of the original space. The *Address* of the left child is * (has zero directional digits), since the region of the left child is the whole space minus the region of the right child. Each of these *Addresses* is expressed relatively to the minimal quadrant that covers the internal node (each *Address* determines a subquadrant of this minimal

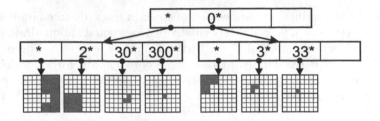

Fig. 1. An XBR$^+$-tree with two levels of internal nodes.

quadrant). For example, in Fig. 1, the *Address* 2* is the SW subquadrant of the whole space (the minimal quadrant that covers the left right child of the root). During a search, or an insertion of a data element with specified coordinates, the appropriate leaf and its region is determined by descending the tree from the root. Although, for the sake of presentation, Fig. 1 depicts a tree with nodes having up to four children, note that nodes are disk pages and they are likely to have a significant number of children (xBR-trees are multiway trees).

Internal node entries in xBR$^+$-trees contain entries of the form (*Shape, qside, DBR, Pointer*). The fields *Shape* and *Pointer* play the same role as in xBR-trees, *DBR* (Data Bounding Rectangle) stores the coordinates of the subregion that contains point data (at least one point must reside on each side of the *DBR*), while *qside* is the side length of the quadrant that corresponds to the child-node pointed by *Pointer*. In xBR$^+$-trees, the *Address* and the corresponding *REG* of the region of this child node are not explicitly stored. However, only *REG* may be needed by the query processing algorithms and it can be easily calculated using *qside* and *DBR* (in most steps of the query processing algorithms, the use of the values of *qside* and *DBR* is enough and further calculations are avoided). Note that the fields *qside* and *DBR* are represented using *double* precision floating point numbers (like *REG*) which have a fixed size defined by the implementation of the programming language. By avoiding to store the variable-sized field *Address* in xBR$^+$-trees, processing of internal nodes is simplified, since their capacity is fixed. Moreover, the use of *DBR* makes processing of several queries more efficient, since it signifies the subarea of the child node that actually contains data, which is (in general) different to and smaller than the region of this child node, leading to higher selectivity of the paths that have to be followed downwards when descending the tree and deciding the parts of the tree that may contain (part of) the query answer.

In summary, the basic structural differences between the xBR-tree and the xBR$^+$-tree are:

- Internal node entries of xBR$^+$-trees do not contain the *Address* and *REG* fields, but contain the *DBR* and *qside* fields.
- The *Address* and *REG* fields are calculated only when needed.
- Since variable fields are not stored in internal node entries of xBR$^+$-trees, internal nodes have a fixed capacity.

3.2   Leaf Nodes

External nodes (leaves) of members of the xBR-tree family simply contain the data elements and have a predetermined capacity C. When C is exceeded, due to an insertion, the leaf is partitioned according to hierarchical (Quadtree like) decomposition, until each of the resulting two regions (the one region corresponds to the most populated subquadrant and the other region to the rest of the node area) contains data elements with cardinalities $\leq xC$, $0.5 < x < 1$. The choice of x affects the number of necessary subdivisions of an overflowed node and the size of addresses that result from a node split. A value closer to 0.5, in general, results in more subdivisions and larger addresses, since it is more difficult to partition the region of the leaf in subregions with almost equal numbers of elements. Of course, such a choice provides a better guarantee for the space occupancy of leaves. We used $x = 0.75$, which leads to a good compromise between size of addresses and leaf occupancy. Splitting of a leaf creates a new entry that must be hosted by an internal node of the parent level. This can cause backtracking to the upper levels of the tree and may even cause an increase of its height. Note also that in the xBR-tree family, data elements are stored in leaves in X-order. This ordering permits us to use the *plane sweep* technique (when appropriate) during processing of the data elements of a leaf, in the process of answering certain query types.

3.3   Splitting of Internal Nodes

When an internal node of a member of the xBR-tree family overflows, it is split in two. The goal of this split is to achieve the best possible balance between the space use in the two nodes. The split in the xBR-tree family is either based on existing quadrants or in ancestors of existing quadrants. First, a Quadtree is built that has as nodes the quadrants specified in the internal node [15]. This tree is used for determining the best possible split of the internal node in two nodes that have almost equal number of bits, as proposed in [15], or entries (a simpler and equally effective criterion, according to experimentation).

3.4   Tree Building

Building of a member of the xBR-tree family consists in repetitive insertion of the data elements in the tree by descending the tree for the root, seeking for the appropriate leaf to host each new element. The leaf found may be split, which may cause further splits of internal nodes in the path up to the root. In case of the xBR$^+$-tree, the possibility to avoid creating a new node (leaf, or internal node), but to merge it with another node is considered. The nodes examined for merging are either direct descendants, or direct ancestors of the new node, when we consider the Quadtree that corresponds to the entries of the parent internal node of the new node and candidate merging-nodes.

4   Query Processing Algorithms on the xBR-tree Family

In this section, extending material that appears in [9, 10], we present algorithms for processing *PLQs*, *WQs*, *DRQs*, *K-NNQs* and *CK-NNQs* on the xBR-tree family, highlighting the differences between the xBR and xBR$^+$-trees.

PLQs can be processed in a top-down manner on the xBR-tree family. During a *PLQ* for a point with specified coordinates, the appropriate leaf and its region is determined by descending the tree from the root. Initially, the region under consideration is the whole space (the region of the root). As noted in Subsect. 3.1, the entries in an internal node are saved in preorder traversal of the Quadtree that corresponds to the internal node and are examined in reverse sequential order (this means that we examine first a subregion, before examining an enclosing region of this subregion and in this way we avoid to examine multiple times overlapping regions). So first we examine the last node of the Quadtree. If its subquadrant specified by the *Address* field of the entry, in case of the xBR-tree, or its *DBR*, in case of the xBR$^+$-tree, does not contain the query point, we continue with the next entry in reverse sequential order. The first subquadrant/*DBR* that hosts the query point determines the smallest region that hosts this point. Then we follow the *Pointer* field to the related child at the next lower level, until we reach the leaf level. This way, we reach the unique leaf that may contain the query point. Note that, it is possible the query point to fall within a subquadrant without falling inside the *DBR* of the corresponding a child node. This means that, in case of an unsuccessful search, the search is likely to stop at a middle level of the xBR$^+$-tree, while it will always continue until the leaf level of the xBR-tree.

Processing of *WQs* follows the same strategy to *PLQs*, regarding the way we examine regions/entries of an internal node. The decision about whether we are at a entry with a region likely to contain points inside the query window is the answer to the question: do the subquadrant/*DBR* of the current entry (specified by the *Address*/*DBR* field of the entry) and the query window intersect? If yes, then we follow the pointer to the related child at the next lower level. We repeat until we have examined all entries of the internal node, or until the query window is completely inscribed inside the region (subquadrant) of the entry that we examine (because none of the other, not examined, regions of the tree overlaps with this region). Note that, the use of *DBRs* in xBR$^+$-trees eliminates the possibility to visit a subtree storing data outside the query window.

DRQ follows the same strategy as *WQ*. At first, the querying circle is replaced from its MBR (the calculations are faster in this way) and if the answer about the intersection of the subquadrant/*DBR* of the current entry and the query MBR is positive, then we follow the pointer to the related child at the next lower level. If we reach a leaf with a region that intersects the query MBR, we select the points that are inside the query circle.

For *K-NNQs*, the algorithmic approach for both trees is similar. The search algorithm traverses recursively the tree in a DF (Depth-First) manner (in each node, entries are examined according to *mindist* from the query point and recursively child nodes are visited), or in a BF (Best-First) manner (among all nodes visited so far, entries are examined, and respective children are visited, according to *mindist* from the query point). Both algorithmic approaches utilize a *max K-heap* that stores the K points found so far with the shortest distances to the query point. In the case of the DF (BF) algorithm, an additional local to every internal node (global) *min heap* is kept, storing node entries according to their *mindist* from the query point, calculated using the *Address* field, in case of the xBR-tree, or the DBR field, in case of the xBR$^+$-tree. When the leaves are reached, the set of entries is sorted in ascending order on X-axis, next this set is split into two subsets, taking as reference the *query point*. Then, the algorithm scans both subsets (from left to right and from right to left) while the distance on X-axis is smaller than the distance value of the *Kth-NN* that has been found so far, inserting those points in *max K-heap*. The *CK-NNQ* algorithm inserts entries in *min heaps*/points in the *max heap*, in case *mindist* of entries/distance of points is larger than a distance threshold. More details about these algorithms for the case of xBR-trees appear in [9,10].

In summary, the basic algorithmic differences for spatial query processing between the xBR-tree and the xBR$^+$-tree are:

- In *PLQs*, due to the use of *DBRs* an unsuccessful search is likely to stop without descending the whole height of the tree.
- In *WQs*, due to the use of *DBRs* we avoid visiting subtrees that either do not store any data, or store data outside the query window.
- In *DRQs*, due to the use of *DBRs* we avoid visiting subtrees that either do not store any data, or store data outside the circumscribed square of the query circle.
- In *K-NNQs*/*CK-NNQs*, the precedence of *min heap* entries follows *mindist* of *DBRs* from the query point which gives better estimates of the actual distances of the data from the query point.

5   Experimentation

We designed and run a large set of experiments to compare xBR and xBR$^+$-trees and not R-tree variants ([10] documents comparable performance of xBR and R*-trees). We used 6 real spatial datasets of North America, representing cultural landmarks (NAclN with 9203 points) and populated places (NAppN with 24493 points), roads (NArdN with 569120 line-segments) and rail-roads (NArrN with 191637 line-segments). To create sets of 2d points, we have transformed the MBRs of line-segments from NArdN and NArrN into points by taking the center of each MBR (i.e. |NArdN| = 569120 points, |NArrN| = 191637 points). Moreover, in order to get the double amount of points from NArrN and NArdN, we chose the two points with *min* and *max* coordinates of the MBR of each line-segment (i.e. |NArdND| = 1138240 points, |NArrND| = 383274 points). The data of these 6 files were normalized in the range $[0,1]^2$. We have also created synthetic clustered datasets of 125000, 250000, 500000 and 1000000 points, with 125 clusters in each dataset (uniformly distributed in the range $[0,1]^2$), where for a set having N points, $N/125$ points were gathered around the center of each cluster, according to Gaussian distribution. The experiments were run on a Linux machine, with Intel core duo 2×2.8 GHz processor and 4 GB of RAM. We run experiments for tree building, counting tree characteristics and creation time and experiments for *PLQs*, *WQs*, *K-NNQs* and *CK-NNQs*, counting disk-page accesses (I/O) and total execution time.

In the first experiment, we built the xBR and xBR⁺-trees. We stored point coordinates as *double numbers*[1] and constructed each tree for the following node (page) sizes: 512 B, 1 KB, 2 KB, 4 KB, 8 KB and 16 KB. Results for the construction characteristics indicate that, the xBR⁺-tree in comparison to the xBR-tree has larger height (by 1, or more rarely by 2 levels) in 1/3 of the cases and creates more (54/60) or equal (6/60) internal nodes, uses less space in the most cases (46/60) (i.e. it is more compact) and the creation time varies, in some cases (22/60) is faster having maximum relative difference 18.75 % while in some cases (34/60) is slower having worst relative difference −27.18 %. Thus, the xBR⁺-tree is slightly higher and·has more internal nodes but needs less space in disk, which means that it has (in all cases) a smaller number of leafs (this is due to the use of *merging leafs or internal nodes* that improves searching efficiency). Results for the construction characteristics (Table 2) are depicted only for one node size for each dataset, due to the limited space (other results were analogous).

Table 2. Tree construction characteristics.

Dataset	Node size	Tree height		Tree size (KBytes)		Creation time (secs)	
		xBR	xBR⁺	xBR	xBR⁺	xBR	xBR⁺
NAclN	512 B	4	5	412	408	0.16	0.13
NAppN	1 KB	4	4	1034	1002	0.31	0.35
NArrN	2 KB	4	4	7596	7252	3.19	3.33
NArrND	4 KB	3	3	14932	14140	8.60	8.81
NArdN	8 KB	3	3	22160	20668	20.12	19.40
NArdND	16 KB	3	3	44272	41280	71.30	62.29
125KCN	2 KB	3	4	4582	4632	1.97	2.04
250KCN	4 KB	3	3	9112	9100	5.36	5.13
500KCN	8 KB	3	3	18168	18104	17.22	15.64
1000KCN	16 KB	3	3	36224	36000	60.32	52.53

For each dataset, we created rectangular query windows (and their inscribed circles) for studying *WQs* (*DRQs*) by splitting the whole space into 2^4, 2^6, $\cdots$, 2^{16} windows, in a row-order mapping manner. The centroids of these windows were also used as query inputs for all the other queries (*PLQs*, *DRQs*,

---

[1] Note that we used double numbers for coordinates, instead of float numbers used in [10], to be able to represent large number of points in the normalized square space. Due to this change of representation, results for specific datasets that appear in [10] for the xBR-tree differ slightly from the respective results in this paper.

K-NNQs and *CK-NNQs*). Especially, for *K-NNQs* and *CK-NNQs* we used the following set of K values: 1, 10, 100 and 1000. Since the number of experiments performed was vast, we show only representative results, since results were analogous for each query category.

For *PLQs* we executed two sets of experiments. In the first set we used as query input the original datasets and in the second one we used as query input the centroids of the query windows. The results showed that the xBR$^+$-tree needs the same number of disk accesses as its height for every query point, if that point exists in the dataset. In the other case, searching for not existing points in the dataset, the disk accesses may be less than the tree-height. The xBR-tree, in both cases, needs always the same number of disk accesses as its height. Nevertheless, *PLQ* execution time was faster in xBR$^+$-tree in almost all of the experiments. Results for the largest dataset (NArdND) are shown in Figs. 2 and 3. Especially in Fig. 2 the first set of *PLQs* experiments are shown, where the query points were the 1,138,240 original (existing) points of this dataset. In Fig. 3 the second set of *PLQs* experiments are shown, where the query points were the centroids of the 2^{16} query windows. The results for the other datasets were analogous, and always in favor of the xBR$^+$-tree.

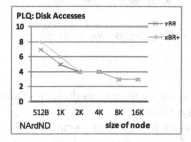 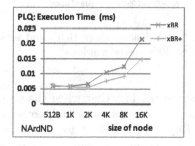

Fig. 2. Total disk accesses (left) and execution time (right) vs. node size for *PLQs* (NArdND). The query points were 1138240 points existing in the dataset.

We noticed that the xBR-tree needs a number of disk accesses equal to its tree height, while the xBR$^+$-tree needs at most this number of accesses, especially for the case of non existing query points. This finding can easily be explained from the analysis of the algorithms discussed above. This is due to the structural difference of the two trees. Internal nodes of xBR$^+$-trees contain information about *DBRs* that only include data points. In this way, if the dataset has empty regions, the xBR$^+$-tree does not store *DBRs* for these, depending on the data distribution in the space of the node. However, the xBR-tree saves in internal nodes information about the regions in which the space is split, regardless of whether they contain data points. It is important to mention that the number of disk accesses becomes lower as the size of node increases, while execution time increases (for both trees). This may be surprising at first, but can be explained

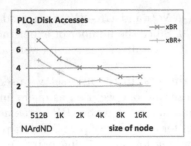
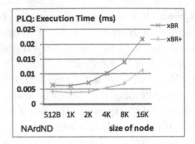

Fig. 3. Total disk accesses (left) and execution time (right) vs. node size for *PLQs* (NArdND). The query points were 16384 points non-existing in the dataset.

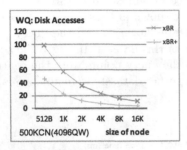
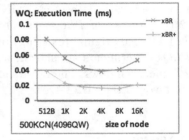

Fig. 4. Total disk accesses (left) and execution time (right) vs. node size for *WQs* (500KCN, 4096 query windows).

considering the fact that when the size of nodes increases, so does the time for main memory calculations and, consequently, the execution time.

In Fig. 4, for the *WQ*, we depict the results for the third synthetic dataset (500KCN), as one representative example. It is shown that the xBR$^+$-tree needed fewer accesses than xBR-tree, to find the population of 4096 windows with which we scanned the whole space occupied from the 500 K clustered data points of this dataset. As the size of node increases the relative I/O difference between the two trees becomes smaller. In both trees, a logarithmic dependence of the number of disk accesses to the size of the node appears. Note the reduction of the difference from the smallest node size (512 B) to the largest size (16 KB). This is due to the reduction of tree height as the size of node increases.

In the right part of Fig. 4, for the execution time, it is shown that the xBR$^+$-tree is faster for all sizes of nodes. The explanation for this fact is again related to the structural difference between two trees. Two additional remarks can be made. First, in a more detailed observation we can see that the relative difference of performance between the two trees in execution time is a little smaller than in disk accesses. This may be explained by the fact that even though xBR$^+$-tree stores *DBRs* that improve time performance, it occasionally needs to spend time to calculate the coordinates of the region pointed by the *Address* of the subquadrant. This fact does not affect the number of disk accesses. The second remark is that the number of disk accesses appears to have monotonic dependence to the size of nodes. The execution time does not have the same characteristic. Both trees seem to exhibit an inflection point in the dependence of the execution time to the size of nodes. This point of optimal execution time performance appears when the size of nodes is equal to 4 KB. This optimization in execution time is stopped for size of nodes larger than 8 KB. This behavior holds for the experiments of all datasets and all query windows.

For *DRQs* (1024 query circles, inscribed into the respective rectangular windows, δ value equal to $1/(2\sqrt{1024}) = 1/64$ of the space side length), the xBR$^+$-tree needed less disk accesses and was faster than the xBR-tree, in all cases and for all datasets (Fig. 5). The performance of the xBR$^+$-tree for all datasets and for all sets of query circles (2^4, 2^6, 2^8, 2^{10}, 2^{12} and 2^{14}) for every size of node (512 B, 1 KB, 2 KB, 4 KB, 8 KB and 16 KB) is 60-0 in all cases for the disk accesses and (59-1, 57-3, 60-0, 60-0, 60-0 and 60-0) for the execution time, where a pair like 57-3 signifies that the xBR$^+$-tree wins 57 times and the xBR-tree wins 3 times, in whole set of experiments. The lower improvement in execution time can be explained just as in the previous paragraph.

For the *K-NNQ*, the xBR-tree showed similar behavior to the *WQ*. The xBR$^+$-tree needed fewer disk accesses for finding the nearest neighbors than the xBR-tree. Furthermore, the difference became larger when the size of node (for the same dataset) increased. Regarding the execution time, the xBR$^+$-

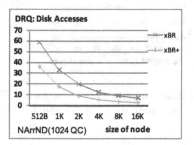
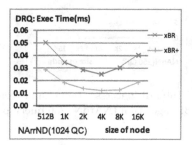

Fig. 5. Total disk accesses (left) and execution time (right) vs. node size for *DRQs* (NArrND, 1024 query circles).

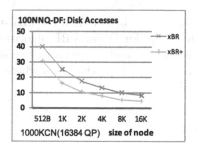
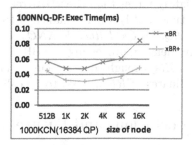

Fig. 6. Total disk accesses (left) and execution time (right) vs. node size for *KNNQs* using DF algorithm (1000KCN, 16384 query points, searching for $K = 100$).

tree showed improved performance, in relation to its I/O difference from the xBR-tree. In Fig. 6, we show results for $K = 100$ and the large synthetic dataset (1000KCN) using algorithms DF. At this point, the worse time performance of both trees, for larger node sizes (where the I/O cost is smaller) must be noted. This is due to the fact that as the node size increases, the trees become very wide and very short. In this case, a node holds many elements to be processed and branching during tree descend plays a smaller role in restricting the search space. This leads us to the conclusion that the increase of the node size leads to many more calculations in main memory, canceling the benefit of reducing I/O. For all experiments performed, the xBR+-tree shows better performance than the xBR-tree in disk accesses and execution time for both DF and BF algorithms.

Finally, for the *CK-NNQs* we noticed that the xBR+-tree was improved for both performance categories of our study. In Fig. 7, we present the results of *CK-NNQs* for the largest real dataset (NArdN) for all (256 query points, setting $K = 100$ and δ value equal to $1/(2\sqrt{256}) = 1/32$ of the space side length) using the BF algorithm. The behavior is similar to *K-NNQs*.

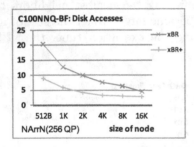
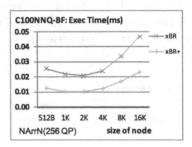

Fig. 7. Total disk accesses (left) and execution time (right) vs. node size for *CKNNQs* using BF algorithm (NArdN, 256 query points, searching for $K = 100$ and $\delta = 1/32$).

In summary, the experimental comparison showed that

- The xBR+-tree needs less space and is built in a similar time, while its height is slightly larger than the xBR-tree.
- The time and I/O performance of the xBR+-tree is better than the xBR-tree for *PLQs*, *WQs*, *DRQs*, *K-NNQs* and *CK-NNQs* for DF and BF algorithms.
- Main memory processing of xBR+-trees is simpler and faster thanks to storing *DBRs*, though some execution efficiency is spent for the cases where the coordinates of *REGs* need to be calculated.

6   Conclusions and Future Work

In [10], the xBR-tree was compared to the R*-tree and, as future work, possible variations of the xBR-tree that might improve its performance were proposed. In this paper, elaborating on these variations, we developed an improved version of the xBR-tree, called xBR$^+$-tree, and compared it experimentally to the xBR-tree. Moreover, we presented the differences of the algorithms for processing PLQs, WQs, DRQs, KNNs and CKNNs in xBR$^+$-trees.

The presented extensive experimental comparison, based on real and synthetic data, showed that the xBR$^+$-tree is a global winner in I/O and execution time, considering the most representative spatial queries that involve a single index, while building of the two trees has comparable efficiency. More specifically, I/O is improved (on the average) by 23 % in PLQs for non-existing points, by 45.4 % in WQs, by 45.4 % in DRQs, by 33.4 % in 100NNs-DF, by 42 % in 100NNs-BF, by 49.6 % in C100NNs-DF, by 55 % in C100NNs-BF and execution time is improved (on the average) by 37.4 % in PLQs for non-existing points, by 42.2 % in WQs, by 41.7 % in DRQs, by 24.2 % in 100NNs-DF, by 40.1 % in 100NNs-BF, by 40.5 % in C100NNs-DF, by 48.9 % in C100NNs-BF. In fact, in 98 % of the cases, the xBR$^+$-tree excels in I/O and time performance by at least 5 %.

Due to its improved building process, the xBR$^+$-tree is smaller and taller than the xBR-tree, while the use of the *DBR* field in xBR$^+$-trees represents populated regions more accurately, gives better estimates of the actual distances of the data from the query point, improves pruning of subtress and simplifies main memory processing.

Future work might include studying how the higher performance of the xBR$^+$-tree is achieved, in terms of Euclidean distance and X-axis distance calculations saved and heap operations performed. This insight might permit further optimizations of the new structure. Moreover, a detailed relative performance study of the xBR$^+$-tree against the R*-tree and/or R$^+$-tree for single dataset and multi-dataset queries is a target in process.

References

1. Beckmann, N., Kriegel, H.P., Schneider, R., Seeger, B.: The R*-tree: an efficient and robust access method for points and rectangles. In: SIGMOD Conference, pp. 322–331 (1990)
2. Finkel, R., Bentley, J.: Quad trees: a data structure for retrieval on composite keys. Acta Informatica **4**, 1–9 (1974)
3. Gaede, V., Gunther, O.: Multidimensional access methods. ACM Comput. Surv. **30**(2), 170–231 (1998)
4. Gorawski, M., Bugdol, M.: Cost Model for XBR-tree. In: Kozielski, S., Wrembel, R. (eds.) New Trends in Data Warehousing and Data Analysis. Springer, USA (2009)
5. Hoel, E.G., Samet, H.: A qualitative comparison study of data structures for large line segment databases. In: SIGMOD Conference, pp. 205–214 (1992)
6. Hoel, E.G., Samet, H.: Benchmarking spatial join operations with spatial output. In: VLDB Conference, pp. 606–618 (1995)
7. Kim, Y.J., Patel, J.: Performance comparison of the R*-tree and the quadtree for kNN and distance join queries. IEEE Trans. Knowl. Data Eng. **22**(7), 1014–1027 (2010)

8. Manolopoulos, Y., Nanopoulos, A., Papadopoulos, A., Theodoridis, Y.: R-Trees: Theory and Applications. Springer, London (2006)
9. Roumelis, G., Vassilakopoulos, M., Corral, A.: Algorithms for processing Nearest Neighbor Queries using xBR-trees. In: Panhellenic Conference on Informatics, pp. 51–55 (2011)
10. Roumelis, G., Vassilakopoulos, M., Corral, A.: Performance comparison of xBR-trees and R*-trees for single dataset spatial queries. In: Eder, J., Bielikova, M., Tjoa, A.M. (eds.) ADBIS 2011. LNCS, vol. 6909, pp. 228–242. Springer, Heidelberg (2011)
11. Samet, H.: The Design and Analysis of Spatial Data Structures. Addison-Wesley, Boston (1990)
12. Samet, H.: Applications of Spatial Data Structures: Computer Graphics, Image Processing, and GIS. Addison-Wesley, Boston (1990)
13. Samet, H.: Foundations of Multidimensional and Metric Data Structures. Morgan Kaufmann Publishers, San Francisco (2006)
14. Samet, H.: The quadtree and related hierarchical data structure. Comput. Surv. **16**(2), 187–260 (1984)
15. Vassilakopoulos, M., Manolopoulos, Y.: External balanced regular (x-BR) trees: new structures for very large spatial databases. In: Panhellenic Conference on Informatics, pp. 324–333 (2000)
16. Yin, X., Düntsch, I., Gediga, G.: Quadtree representation and compression of spatial data. In: Peters, J.F., Skowron, A., Chan, C.-C., Grzymala-Busse, J.W., Ziarko, W.P. (eds.) Transactions on Rough Sets XIII. LNCS, vol. 6499, pp. 207–239. Springer, Heidelberg (2011)

Semantic Web and Ontologies

Probabilistic Error Detecting in Numerical Linked Data

Huiying Li$^{(\boxtimes)}$, Yuanyuan Li, Feifei Xu, and Xinyu Zhong

School of Computer Science and Engineering, Southeast University,
Nanjing 210096, P.R.China
{huiyingli,220141547,220141529,xyzhong}@seu.edu.cn

Abstract. Linked Open Data (LOD) has become a vast repository with
billions of triples available in thousands of datasets. One of the most
pressing challenges to Linked Data management is detecting errors in
numerical data. This study proposes a novel probabilistic framework that
enables the detection of inconsistencies in numerical attributes including not only integer, float or double values but also date values of liked
data. We develop an automatic method to detect error between multi
attributes which can not be detected only considering single attribute.
Evaluations are performed on four DBpedia versions from 3.7 to 2014
which are a central hub dataset of LOD cloud. Results show that our
approach reaches 96 % precision when testing on DBpedia 2014 with
threshold $\alpha = 0.9$. We also compare the percentage distribution of errors
between different DBpedia versions and analyze two basic classes of
causes that lead to errors. Efficiency evaluation results confirm the scalability of our approach to large Linked Data repositories.

1 Introduction

Linked Open Data aims to publish structured data to enable such data to be
interlinked and thus enhance their utility. It shares information that can be read
automatically by computers and allows data from different sources to be connected and queried [1]. LOD comprises of an unprecedented volume of structured
datasets currently amounting to 50 billion facts represented as Resource Description Framework (RDF) triples on the web.

The large amounts of Linked Data bring many opportunities in integrating,
querying, and reusing these datasets, but the Linked Data quality is still a crucial
problem to block its reuse. A large number of high-quality datasets exist in
Linked Data, particularly in the life-sciences domain. These datasets are carefully
constructed and published on the web. However, many low-quality datasets also
exist which is extracted from unstructured and semi-structured information or
is the result of some crowd-sourced process [2].

DBpedia [3] is a representative dataset extracted from a crowd-sourced web
site, Wikipedia. Given its approach of heuristic information extraction from
a crowd-sourced web site, DBpedia contains various kinds of errors. These
errors may be attributed to two reasons: one is the input entered manually in

Q. Chen et al. (Eds.): DEXA 2015, Part I, LNCS 9261, pp. 61–75, 2015.
DOI: 10.1007/978-3-319-22849-5_5

Wikipedia, and the other is the problems during parsing. The following example presents some snippets and errors in DBpedia.

Example. (1) A snippet of DBpedia dataset is shown in Fig. 1(a)[1], which lists two instances with its type, name, date of birth and date of death. An error can be found in instance *dbres : Roberto_Penna*, i.e., his date of birth is later than the date of death. A person or an athlete's date of death is normally later than the date of birth.

```
<dbres:Dhuri> <rdf:type> <dbonto:City>.
<dbres:Dhuri> <foaf:name> "Dhuri".
<dbres:Dhuri> <dbonto:populationRural> "190674".
<dbres:Dhuri> <dbonto:populationUrban> "49406".

<dbres:Jaipur> <rdf:type> <dbonto:City>.
<dbres:Jaipur> <foaf:name> "Jaipur".
<dbres:Jaipur> <dbonto:populationRural> "3164767".
<dbres:Jaipur> <dbonto:populationUrban> "3499204".
<dbres:Jaipur> <dbonto:populationTotal> "3073350".
```

```
<dbres:Roberto_Penna> <rdf:type> <dbonto:Athlete>.
<dbres:Roberto_Penna> <foaf:name> "Roberto Penna".
<dbres:Roberto_Penna> <dbonto:birthDate> "1986-04-19".
<dbres:Roberto_Penna> <dbonto:deathDate> "1910-07-05".

<dbres:Allan_Dwan> <rdf:type> <dbonto:Person>.
<dbres:Allan_Dwan> <foaf:name> "Allan Dwan".
<dbres:Allan_Dwan> <dbonto:birthDate> "1885-04-03".
<dbres:Allan_Dwan> <dbonto:deathDate> "1981-12-28".
```

```
<dbres:Callan,_County_Kilkenny> <rdf:type> <dbonto:Town>.
<dbres:Callan,_County_Kilkenny> <foaf:name> "Callan".
<dbres:Callan,_County_Kilkenny> <dbonto:populationRural> "281".
<dbres:Callan,_County_Kilkenny> <dbonto:populationUrban> "1489".
<dbres:Callan,_County_Kilkenny> <dbonto:populationTotal> "1771".
```

(a) An example of *birthDate* and *deathDate* attributes

(b) An example of *population* attributes

Fig. 1. RDF snippet Example

(2) The snippet in Fig. 1(b) lists three instances with its type, name, population of rural, population of urban and total population. Rural population should be less than urban population in a city usually. The data of city *dbres : Dhuri* is abnormal. Obviously, the total population should be larger than rural or urban population in a city or town, and the sum of rural and urban populations should be equal to the total population. However, for instance, the total populatin of *dbres : Jaipur* is larger than rural and urban populations. Meanwhile, the sum of rural population and urban populations of *dbres : Callan,_County_Kilkenny* is not equal to the total population.

To detect these kinds of errors, we meet two challenges. The first challenge is that detecting errors in numerical attributes has to consider the relation between attributes. The approach of detecting errors in one numerical attribute does not meet the requirements. For athlete Roberto Penna, the error cannot be found if we detect only in attribute *dbonto : birthDate* or in attribute *dbonto : deathDate*. The second challenge is to employ an unsupervised error detection method that does not use domain knowledge.

To address these challenges, we consider the larger than, less than, and equal to relation between two attributes and the equal to relation among three or more attributes. We propose a probabilistic method to predicate these relations between numerical attributes of a class. An error is identified when the attribute

---

[1] The namespace *dbres* denotes http://dbpedia.org/resource, *dbonto* denotes http://dbpedia.org/ontology, *rdf* denotes http://www.w3.org/1999/02/22-rdf-syntax-ns#, *foaf* denotes http://xmlns.com/foaf/0.1/.

values of an instance in dataset are inconsistent with the prediction. As a test bed for evaluating our approach, we use four versions of DBpedia datasets published in LOD. We test the effectiveness and efficiency of our probabilistic error detection approach. We also analyze the error changes between different versions of DBpedia and the causes that lead to these errors.

In this study, we propose a novel probabilistic framework that enables to detect inconsistencies in numerical data. Our main contributions can be summed up as follows:

– We propose a novel probabilistic framework based on Bayes theorem for detecting errors in numerical data.
– We present an automatic method for error detection by predicting larger than, less than, and equal to relations between attributes.
– We test our approach on four versions of DBpedia and identify two basic classes of error causes. The evaluation of our implementation demonstrates its feasibility, with a precision of 96 %. The running time confirms the scalability of our approach to large Linked Data repositories.

The remainder of this article is organized as follows. Section 2 introduces the related work. Section 3 proposes the probabilistic error detection in numerical attribute approach. Section 4 presents the steps to implement our approach on DBpedia. Section 5 details the experimental results of our approach. Section 6 concludes the study.

2   Related Work

Some works have focused on detecting errors in numerical databases. [6] proposed a method for computing reasonable repairs of inconsistent numerical databases, for a restricted but expressive class of aggregate constraints. [7] provided an axiomatic definition of inconsistency management policies (IMPs), it empowered user to deal with his data and his mission needs. IMP allowed end users brings both their application specific knowledge as well as their own personal risk to bear when resolving inconsistencies. [8] proposed numeric functional dependencies (NFDs) to detect errors in numeric data, NFD is a class of dependencies that allow users to specify arithmetic relationships among numeric attributes. Data inconsistencies (numeric or not) can be detected in a uniform logic framework by using NFDs as data quality rules.

[9] used games with a purpose to evaluate DBpedia and spot inconcistencies. They reported that in 4, 051 statements used in the game, 265 inconsistencies have been detected by users, 121 out of which were actually inconsistencies. [10] focused on the enrichment of the DBpedia ontology by statistical methods, inconsistencies are detected during the process of the Wikipedia data extraction based on the enriched ontology. LiQuate [12] is a data and link validation tool to identify redundancies, incompleteness and inconsistencies, it maked use of a probabilistic rule-based system to infer the links that solve the identified quality problems with a certain degree of uncertainty. [13] analyzed the common errors encountered in

Linked Data and classified them according to the extent to which they are likely to be amenable to a specific form of crowdsourcing. [2] presented a methodology for assessing the quality of linked data resources, which comprised of a manual and a semi-automatic process. It identified 17 data quality problem types and 58 users assessed a total of 521 resources and pointed that 11.93 % of the evaluated DBpedia triples have some quality issues.

Recently, some works have been studied to detect error in numerical linked data. [15] studied the applications of data mining techniques in early detection of numerical typing errors by human operators through a quantitative analysis of multichannel electroencephalogram (EEG) recordings. [16] examined the possible usage of different outlier detection methods, i.e., Interquantile Range (IQR), Kernel Density Estimation (KDE), and dispersion estimators, for identifying wrong numerical statements in DBpedia. Combined with grouping by single type preprocessing, it achieved a precision of 87 % on subset of DBpedia. [17] is a subsequent work of [16].

Compared with early works, our approach is distinct in three aspects:

- We focus on detecting error in numerical attributes including not only integer, float or double values but also date values in liked data.
- It is an automatic process to detect errors by a probabilistic method.
- We detect errors between multi attributes which can not be detected by only applying outlier detection methods on single attribute. The evaluation results on four versions of DBpedia show that our approach is effective and efficient.

3   Probabilistic Error Detection in Numerical Attributes

We propose a probabilistic framework for detecting error in numerical attributes, which catches errors in numerical data by predicting arithmetic relations between numerical attributes. For an instance in dataset, we predict the probability that its relation between the numerical attributes belongs to a particular classification by Bayesian classification. An error is detected when the attribute statements of the instance are inconsistent with the prediction.

3.1   Probabilistic Model

We consider the class type of an instance as the evidence. Let $T = (t)$ be an instance that is represented by its class type t. Let R be some hypothesis, such as the relation between the numerical attributes of the instance. The probability $P(R|T)$ can be calculated based on the basis of Bayes theorem as follows:

$$P(R|T) = \frac{P(T|R)P(R)}{P(T)} \tag{1}$$

$P(R|T)$ is the posterior probability of R conditioned on T. For example, suppose that the instance is described by class *City*, and R is the hypothesis that the value of numerical attribute *populationUrban* of a city is larger than the value of numerical attribute *populationRural*, then $P(R|T)$ reflects the probability that the attribute *populationUrban* value of the instance T is larger than its attribute *populationRural* value given that we know that the class type of the instance is *City*.

Similarly, $P(T|R)$ is the posterior probability of T conditioned on R. That is, it is the probability that an instance is typed as *City* given that we know that its *populationUrban* value is larger than the *populationRural* value.

$P(R)$ is the prior probability of R. For our example, this is the probability that the *populationUrban* value of any given instance is larger than its *populationRural* value, regardless of its class type.

$P(T)$ is the prior probability of T. Using our example, it is the probability that an instance is typed as *City*. Given that $P(T)$ can be considered as a constant, denoted as γ, we have

$$P(R|T) = \frac{1}{\gamma} P(T|R)P(R) \tag{2}$$

Let D be a training set of instances. The prior probability of R is estimated by

$$P(R) = \frac{|R_D|}{|D|} \tag{3}$$

where $|R_D|$ is the number of training instances that satisfies hypothesis R. That is, it is the number of instances with *populationUrban* value larger than *populationRural* value. $|D|$ is the number of training instances.

The posterior probability $P(T|R)$ is estimated by

$$P(R) = \frac{|R_{tD}|}{|R_D|} \tag{4}$$

where $|R_{tD}|$ is the number of training instances which is typed as class t and satisfies hypothesis R. That is, it is the number of instances with type class *City* and *populationUrban* value larger than *populationRural* value.

3.2   Probabilistic Error Detection

To detect errors in numerical attributes, we consider two kinds of relations between attributes. One is the relation between two attributes: the larger than relation, less than relation, and equal to relation are considered. The other is the relation among three or more attributes: the equal to relation is considered.

The larger and less than relations help to detect the errors in two numerical attributes that one's value should be larger or less than the other's value. For example, the attribute *populationUrban* value should be larger than the attribute *populationRural* value for instances with *City* type, whereas the attribute *populationUrban* value should be less than the attribute *populationRural* value for instances with *Town* type. We consider not only ordinary numerical data (integer, float or double values) but also date values. The larger and less than relation means later and earlier than relation for date values.

The equal to relation helps detect the errors in two or more numerical attributes that one attribute's value or the sum of several attribute's value is equal to another's value. For example, the sum of attribute *populationRural* value and attribute *populationUrban* value should be equal to attribute *populationTotal* value.

The process to detect errors in numerical attributes using **Bayesian** classifier works as follows:

1. We collect all instances with numerical attributes as a training set D. Each instance is represented by its class type t, $T = (t)$.
2. Given two numerical attributes A_1, A_2, three classifications, $R_<$, $R_>$, and $R_=$ exist. Let $R_<$ represents the hypothesis that the value of A_1 should be less than the value of A_2, $R_>$ represents the hypothesis that the value of A_1 should be larger than the value of A_2, $R_=$ represents the hypothesis that the value of A_1 should be equal to the value of A_2. Given an instance, T, the classifier will predict that T belongs to the classification having the highest posterior probability, conditioned on T.

   $P(R_<|T)$, $P(R_>|T)$ and $P(R_=|T)$ are evaluated to predict the classification of T. The classifier predicts that T's value of A_1 should be less than the value of A_2 if $P(R_<|T)$ is the maximum. Similarly, T's value of A_1 should be larger than the value of A_2 if $P(R_>|T)$ is the maximum and T's value of A_1 should be equal to the value of A_2 if $P(R_=|T)$ is the maximum.
3. After the classification of T is predicted, an error in numerical attributes A_1 and A_2 can be detected if T's value of attributes A_1 and A_2 is inconsistent with the prediction. Given an instance, $T = City$. The classifier predicts that its value of attribute *populationUrban* should be larger than the value of *populationRural*. An error is reported if we find an instance typed as *City* and its value of attribute *populationUrban* is not larger than the value of attribute *populationRural* in dataset.

   To avoid over correction, we set a threshold α. Suppose that $P(R_<|T)$, $P(R_>|T)$ and $P(R_=|T)$ are evaluated, the classifier predicts that T belongs to the classification of $R_>$ because $P(R_>|T)$ is the maximum. We also require that the error is detected if it satisfies

$$\frac{P(T|R_>)}{P(T|R_<) + P(T|R_>) + P(T|R_=)} \geq \alpha \tag{5}$$

   We detect errors when the attribute statement of an instance in dataset are inconsistent with the prediction and the ratio of $P(T|R_>)$ to $P(T|R_<) + P(T|R_>) + P(T|R_=)$ is larger than threshold α. Intuitively, an error is verified only when a percentage is satisfied to avoid over correction.
4. Given three or more numerical attributes $A_1, A_2,...,A_m$, m classifications R_1, $R_2,...,R_m$ exist. Let R_i represents the hypothesis that attribute A_i's value equals to the sum of the other attribute A_j's ($1 \leq j \leq m, j \neq i$) value. Given an instance, T, the classifier will predict that T belongs to the classification having the highest posterior probability, conditioned on T. The classifier predicts that T's value of A_i should be equal to the sum of all other attributes' value if $P(R_i|T)$ is the maximum.

5. After the classification of T is predicted, a threshold α is given to avoid over correction. Suppose that $P(R_i|T)$ is the maximum and

$$\frac{P(T|R_i)}{\sum_{k=1}^{m}(P(T|R_k))} \geq \alpha \tag{6}$$

We report errors if an instance is typed as t and its value of attribute A_i is not equal to the sum of values of the other attributes in dataset.

The snippet in Fig. 1(b) is considered. Given attributes *dbonto* : *populationRural* and *dbonto* : *populationUrban*, and suppose that 10 instances are typed as *City* among 100 training instances, 8 instances with rural population is less than urban population and 2 instances with rural population is larger than urban population. Considering that $P(R_<|T)$ is the maximum among $P(R_<|T) = \frac{8}{100}$, $P(R_>|T) = \frac{2}{100}$ and $P(R_=|T) = \frac{0}{100}$, the instance typed as *City* is predicted that its rural population should be less than urban population. To avoid over correction, let threshold α set to 0.8. Given that $\frac{P(T|R_<)}{P(T|R_<)+P(T|R_>)+P(T|R_=)} = \frac{8}{10} \geq \alpha$, the abnormal of instance *dbres* : *Dhuri* city is detected because its values of attributes *dbonto* : *populationRural* and *dbonto* : *populationUrban* are inconsistent with the prediction. The error detection in three or more attributes is in the same way.

4 Implementation

We introduce the implementation of our approach in this section. The approach is applied to DBpedia, which is the central hub of LOD Cloud. DBpedia contains extracted data from Wikipedia. We concentrate on the numerical attributes in infobox subset.

4.1 Numerical Attribute Selecting

Considering that DBpedia deals with various kinds of information, given as classes, instances, and relationships, this study focuses on the detection of errors in numerical attributes using probabilistic error detection.

In its current version 2014, DBpedia contains near 2000 numerical attributes. In according with the probabilistic error detection process in Sect. 3.2, two or more numerical attributes are chosen to verify the relation hypothesis between these attributes. The classification of instance is predicated, and errors are detected. A problem affecting the efficiency our approach is that the combination number of attributes is too large to deal with.

To solve this problem, we consider selecting only the attributes with common word. Given that we tend to predicate the relation between attributes, we find that the attributes with common a word have a high opportunity to have some relation. For example, attributes *dbonto* : *birthDate* and *dbonto* : *deathDate* share a common word *Date*, i.e., they have a relation that one's death date is later (larger) than birth date. Attributes *dbonto* : *populationRural*, *dbonto* :

populationUrban, dbonto : *populationTotal* share a common word *population*. They have a relation that the sum of rural and urban populations is equal to total population.

To deduce the combination number, we list the attributes with a common word in a set. Only the combination of attributes in a set is considered. Suppose that n numerical attributes exist, set $S = \{S_1, S_2, ..., S_k\}$ contains k attribute sets, and each set $S_i = \{A_1, A_2, ...A_{i_m}\}$ contains attributes sharing a common word. The combination with two or more attributes is composed only from one set of S. During the process of selecting two attributes, the combination number is $\sum_{i=1}^{k}(C_{i_m}^2)$, which is less than the combination number C_n^2 that combines two attributes from all attributes. The combination numbers of three or more attributes are deduced in the same way.

4.2　Data Preprocessing

After numerical attributes are selected, we extract instances with their types and selected attributes from dataset in data preprocessing. We focus on two problems in data preprocessing.

One problem is to determine the class type of an instance, because it is represented by its class type in the error detection process. Three situations occur about the class type of an instance:

- For an instance with only one type, we use this class as the type of this instance.
- For an instance with more than one type, we use the class from DBpedia ontology that represents leave of the class hierarchy.
- For an instance with no type, we use class $owl : Thing^2$ as its type because everything is an instance of $owl : Thing$ in OWL.

The other problem is about extracting values of the selected attribute. Given its approach of heuristic information extraction from a crowd-sourced web site, DBpedia contains various kinds of inconsistency. Some instances in DBpedia have more than one value for a given attribute, whereas some instances have no value for a given attribute. To deal with this problem, we ignore the instance that has no value or has more than one value for selected attributes.

5　Experimental Study

In this section, we test our approach on four versions of real dataset DBpedia, i.e., DBpedia 3.7, 3.8, 3.9, and 2014. We apply the approach to detect errors in DBpedia. The effectiveness and efficiency of our probabilistic error detection approach are evaluated.

Section 5.1 describes the three versions of datasets considered in the experiments. Section 5.2 provides the effectiveness of these datasets. Section 5.3

---

[2] The namespace *owl* denotes http://www.w3.org/2002/07/owl.

presents the efficiency evaluation results of these datasets. Section 5.4 analyze the error causes. We use Jena toolkit (jena.sourceforge.net) to manage RDF data. The experiments are developed within the Eclipse environment and on two 64 bit quad Core ThinkStations with 3.10 MHz and 16 GB RAM (of which 14 GB was assigned to Java virtual machine).

5.1   Datasets

We choose the real-world RDF data DBpedia from LOD cloud to evaluate our approach. DBpedia is extracted based on the basis of hand-generated mappings of Wikipedia infoboxes or templates to a DBpedia ontology. We concentrate on the numerical attributes in infobox subset. For comparison, we consider the four versions of DBpedia Datasets, i.e., DBpedia Dataset 3.7, 3.8, 3.9, and 2014.

The newest version, DBpedia 2014, is based on updated Wikipedia dump dating from April 2014 to May 2014. This version describes 4.58 million things, out of which 4.22 million are classified in a consistent ontology, including $1,445,000$ persons, $735,000$ places, $411,000$ creative works, $241,000$ organizations, $251,000$ species, and $6,000$ diseases.

Table 1. Statistics on different versions of DBpedia

Version	# of C	# of P	# of I	# of T
3.7	319	893	9,637,079	26,988,054
3.8	359	975	13,225,166	33,742,024
3.9	529	1,406	15,894,067	41,804,710
2014	683	1,716	28,031,852	61,481,483

Table 1 shows the following statistics on different versions of DBpedia: version, number of classes (denoted by C), number of numerical properties (denoted by P), number of instances (denoted by I), and number of triples (denoted by T).

For these datasets, the number of triples increases when a new version is developed. The newest release, DBpedia 2014, includes approximately 61 million triples. The numbers of classes and numerical properties also increase with the updated version of DBpedia dataset.

5.2   Effectivity Evaluation Results

We apply our approach to different versions of DBpedia datasets and detect many errors in our quality evaluations. For example, we find that the *dbonto* : *lowerAge* value of a school should be less than its *dbonto* : *upperAge* value. An error in school *dbonto* : *The_Heritage_Private_School* is detected because its *dbonto* : *lowerAge* value is 21 and its *dbonto* : *upperAge* value is 18 in DBpedia 3.9. Also, we find that the *dbonto* : *birthDate* value of a writer should be earlier (less) than her *dbonto* : *deathDate* value. An error in writer *dbonto* : *Elizabeth_Thornton* is detected because her *dbonto* : *birthDate* value is 1940-01-24 and her *dbonto* : *deathDate* value is also 1940-01-24 in DBpedia 3.9. We find that the sum of *dbonto* : *areaWater* value and *dbonto* : *areaLand* value of a village should be equal to its *dbonto* : *areaTotal* value. An error in village *dbonto* : *Moodus, _Connecticut* is detected because its *dbonto* : *areaWater* value is 103599.52441344, its *dbonto* : *areaLand* value is 7510965.5199744 and its *dbonto* : *areaTotal* value is also 7510965.5199744 in DBpedia 3.9.

Table 2. Detected error numbers when $\alpha = 0.9$

Version	# of E_2	# of E_3	# of E_4
DBpedia 3.7	4,858	60	1
DBpedia 3.8	6,395	77	2
DBpedia 3.9	7,499	90	2
DBpedia 2014	10,411	87	0

Table 2 shows the error numbers we detect in different versions of DBpedia when threshold α is 0.9, E_2 denotes the errors detected between two attributes (include the larger than relation, less than relation, and equal to relation error detection), E_3 denotes the errors detected among three attributes (include only equal to relation error detection), E_4 denotes the errors detected among four attributes (includes just equal to relation error detection).

We find that the detected error number decreases with the increase in attributes number in Table 2. The number of detected errors between two attributes is approximately $10,000$, whereas the error number among three attributes is less than 100 in DBpedia 2014. One reason is that errors between two attributes include three kinds of relation, i.e., the larger than, less than, and equal to relation, whereas errors among three or four attributes include only equal to relation. The other reason is that considerable predicates exist for the relation between two attributes. On the contrary, the error number among four attributes is very small for the four versions of DBpedia datasets.

Table 3. Detected error numbers when $\alpha = 0.8$

version	# of E_2	# of E_3	# of E_4
DBpedia 3.7	15,070	64	1
DBpedia 3.8	19,123	84	2
DBpedia 3.9	17,515	93	2
DBpedia 2014	21,467	107	0

Tables 3 and 4 show the error numbers we detect when threshold α is 0.8 and 0.7, respectively. Threshold α obviously affects the detected error number.

Considering the error number in DBpedia 2014 for example, the error number between two attributes is approximately $10,000$ when α sets to 0.9, whereas it increases to more than $35,000$ when α sets to 0.7. Given that threshold α is set to control the over correction, a small α means we need minimal evidence to verify an error, and a large α means we need considerable evidence to verify an error. Considerable errors can be detected when α is small.

Table 4. Detected error numbers when $\alpha = 0.7$

Version	# of E_2	# of E_3	# of E_4
DBpedia 3.7	19,099	88	1
DBpedia 3.8	23,531	108	2
DBpedia 3.9	25,448	125	2
DBpedia 2014	35,958	118	0

Error number also increases with the update of DBpedia dataset. With a new version of DBpedia released, the number of detected errors always increases. One reason is that the new release always leads to an increase in the instance number and the numerical attributes. These new instances and attributes add the opportunity to lead to new errors. The other reason is that most errors in old version are not corrected in new version.

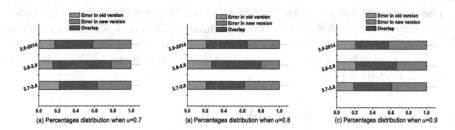

Fig. 2. Percentage distribution of errors

Figure 2 lists the error comparison between different versions of DBpedia. The labels in y-coordinate denote the two compared versions. The numbers in x-coordinate denote the percentage distribution. For two compared versions, the error number detected in the old version is n_{old}, and the error number detected in new version is n_{new}. The common error number detected in both old and new versions is n_{com}. The yellow bar denotes the percentage $\frac{n_{old}}{n_{old}+n_{new}-n_{com}}$, the blue bar denotes the percentage $\frac{n_{com}}{n_{old}+n_{new}-n_{com}}$, and the green bar denotes the percentage $\frac{n_{new}}{n_{old}+n_{new}-n_{com}}$.

We use precision to evaluate the results of error detection. The rightness of the detected errors is difficult to check because of their large number. We check

Table 5. Precision of error detection

Version	$\alpha = 0.7$	$\alpha = 0.8$	$\alpha = 0.9$
DBpedia 3.7	0.91	0.91	0.92
DBpedia 3.8	0.90	0.88	0.92
DBpedia 3.9	0.89	0.85	0.94
DBpedia 2014	0.93	0.89	0.96

the rightness of prediction instead. For every prediction, we verify whether it is right. If a prediction is right, then the errors detected by this prediction are considered right. If a prediction is wrong, then the errors detected by this prediction are considered wrong.

Table 5 lists the precisions of different DBpedia versions with different threshold values. The precisions are decreased with the version update when $\alpha = 0.7$ and $\alpha = 0.8$. The reason is that a wrong predication leads to considerable wrong results with the version update. The predication about attributes *dbonto* : *activeYearsStartYear* and *dbonto* : *draftYear* values of *dbonto* : *AmericanFootballPlayer*, leads to 433 wrong results in DBpedia 3.7, 730 wrong results in DBpedia 3.8, and 815 wrong results in DBpedia 3.9. Given that the total number of detected errors is not increased significantly with the version update, these wrong results decrease the precision. The precision for DBpedia 2014 is not decreased because the total number of detected errors is increased largely.

The precisions are the lowest when $\alpha = 0.8$ because the number of wrong predications when $\alpha = 0.8$ is more than the number when $\alpha = 0.7$. The precisions are the highest when $\alpha = 0.9$. We recommend that setting α to 0.9 if a high precision is needed in error detection and setting α to 0.7 if considerable detected errors are needed.

5.3   Efficiency Evaluation Results

We conduct performance evaluations to show the efficiency of our approach. The running time of our approach is the sum of the dataset preprocessing time and the probabilistic error detection time. It takes about 30 mins to 50 mins to preprocess different versions of DBpedia dataset. We need to conduct numerical attribute collection, word segmentation, and data extraction, which naturally consume a long time. The data preprocess is needed only one time, and the process of error detection is very fast after data are preprocessed once.

Figure 3(a) lists the probabilistic error detection times for different versions of DBpedia with different α values. The running time is the sum of error detecting time among two, three and four attributes. Approximately, 10 min to 30 min are required to detect error. Figure 3(a) shows that threshold α has an insignificant effect to the running time. There is little difference in running time when α is different. The important factor that affects the running time is the triple

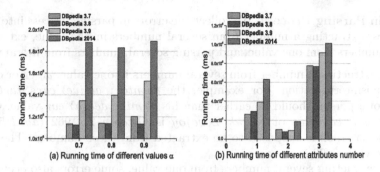

Fig. 3. Running time

number in dataset. DBpedia 2014 consumes the longest running time because it contains about 61 million triples, i.e., it takes approximately 30 min to detect all errors. This result confirms the scalability of our approach to large Linked Data repositories.

To analyze the most time consuming step in error detection, we list the running times to detect error among two, three and four attributes separately. Figure 3(b) lists the running times for different versions of DBpedia with $\alpha = 0.8$. Numbers 2 to 4 in the x-coordinate denotes the detecting time among two to four attributes.

5.4   Error Analysis

Based on the results obtained in our error detection, we further examine common patterns in the errors we find to identify their causes. Errors have two basic classes, i.e., the errors that exist in Wikipedia, and the errors that occur while parsing the data from Wikipedia to DBpedia.

Errors in Wikipedia. We collect the errors in Wikipedia into two categories: error value in Wikipedia and inconsistent semantics in Wikipedia.

– Error value in Wikipedia means that the data are already wrong at the source in Wikipedia. For example, the page about *Callan, _County_Kilkenny* gives the urban population of $1,489$ and the rural population of 281, but the town population, $1,771$, is not the sum of urban and rural populations. Such an error is difficult to quantify because it does not seem to follow a specific pattern.
– Inconsistent semantics in Wikipedia means that some infobox keys in Wikipedia are used with inconsistent semantics. For example, the value of property *Years of service* for *Merrill Sanford* is four years, which means the total years of service. The value for *Alexander Mackenzie* is 1866–1874, which means the service start and end years. The value for *Beverly Robertson* is 1849–1861 (*USA*) 1861–1865 (*CSA*), which means the service start and end years in different agencies. Such inconsistent semantics leads to numerous of errors.

Errors in Parsing Process. We collect the errors in parsing process into three categories: extracting a number from several numbers in one value, extracting several numbers from one value, and parsing several numbers from literal value.

- When extracting a number from several numbers in one value, an error comes from misinterpretation. For example, the *dbonto* : *activeYearsStartYear* value of a person should be earlier than his *dbonto* : *deathYear* value. Given that the *Died* value of *Frank Lovejoy* is *October* 2, 1962 (*aged* 50) in Wikipedia. 1962 is supposed to be extracted from these numbers. The error occurs when 0002 is extracted.
- When extracting several numbers from one value, some errors also occur. For example, the service end year of a military person should be later than his service start year. While there is an error for instance *dbres* : *Beverly_Robertson*, his service end year and start year are both 1849. The *Years of service* value of *Beverly Robertson* is 1849–1861 (*USA*) 1861–1865 (*CSA*) in Wikipedia. There are many numbers in one value denote the start and end service year in different agency. Errors occur when parsing such a complicated value.
- When parsing several numbers from a literal value, there are also some errors occur. For example, an error for instance *dbres* : *Alan_Mruvka* happens, his values of *dbonto* : *activeYearsEndYear* and *dbonto* : *activeYearsStartYear* are both 1984. The *years active* value of *Alan Mruvka* is 1984 − *present* in Wikipedia. An error occurs when parsing the literal value such as *present*.

6   Conclusions

To the best of our knowledge, this study is the first proposed approach to detect errors automatically in numerical data in consideration of the relation between attributes. We propose a novel probabilistic framework that enables to detect inconsistencies in numerical data. We consider more than one numerical attributes at the same time. A probabilistic method is proposed to predicate the larger than, less than, and equal to relations between attributes of a class. Error is identified when the attribute values of an instance in dataset are inconsistent with the prediction. The evaluation of our implementation demonstrates its feasibility with a precision of 96 %. The running time confirms the scalability of our approach to large Linked Data repositories. With this study, we not only detect errors in numerical data of DBpedia but also analyze the error causes, which help to improve the quality in future versions. We also aim to correct errors automatically and fill in absent values in our future work.

Acknowledgments. The work is supported by the Natural Science Foundation of Jiangsu Province under Grant BK20140643, the National Natural Science Foundation of China under Grant 61170165 and the 863 program under Grant 2015AA015406.

References

1. Bizer, C., Heath, T., Berners-Lee, T.: Linked data - the story so far. IJSWIS **5**(3), 1–22 (2009)

2. Zaveri, A., Kontokostas, D., Sherif, M.A., Bühmann, L., Morsey, M., Auer, S., Lehmann, J.: User-driven quality evaluation of DBpedia. In: 9th International Conference on Semantic Systems (I-SEMANTICS 2013), pp. 97–104 (2013)
3. Auer, S., Bizer, C., Kobilarov, G., Lehmann, J., Cyganiak, R., Ives, Z.G.: DBpedia: a nucleus for a web of open data. In: Aberer, K., Choi, K.-S., Noy, N., Allemang, D., Lee, K.-I., Nixon, L.J.B., Golbeck, J., Mika, P., Maynard, D., Mizoguchi, R., Schreiber, G., Cudré-Mauroux, P. (eds.) ASWC 2007 and ISWC 2007. LNCS, vol. 4825, pp. 722–735. Springer, Heidelberg (2007)
4. Millard, I.C., Glaser, H., Salvadores, M., Shadbolt, N.: Consuming multiple linked data. In: COLD 2010 - Workshop at the 9th International Semantic Web Conference (2010)
5. Nickel, M., Tresp, V., Kriegel, H.P.: Factorizing YAGO: scalable machine learning for linked data. In: Proceedings of the 21st Annual Conference on World Wide Web (WWW 2012), pp. 271–280 (2012)
6. Flesca, S., Furfaro, F., Parisi, F.: Querying and repairing inconsistent numerical databases. ACM Trans. Database Syst. **35**(2), 14 (2010)
7. Martinez, M.V., Parisi, F., Pugliese, A., Simari, G.I., Subrahmanian, V.S.: Policy-based inconsistency management in relational databases. Int. J. Approx. Reason. **55**(2), 501–528 (2014)
8. Fan, G., Fan, W., Geerts, F.: Detecting errors in numeric attributes. In: Li, F., Li, G., Hwang, S., Yao, B., Zhang, Z. (eds.) WAIM 2014. LNCS, vol. 8485, pp. 125–137. Springer, Heidelberg (2014)
9. Waitelonis, J., Ludwig, N., Knuth, M., Sack, H.: Who knows? evaluating linked data heuristics with a quiz that cleans up DBpedia. Interact. Technol. Smart Edu. **8**(4), 236–248 (2011)
10. Töpper, G., Knuth, M., Sack, H.: DBpedia ontology enrichment for inconsistency detection. In: The 8th International Conference on Semantic Systems, pp. 33–40 (2012)
11. Hogan, A., Umbrich, J., Harth, A., Cyganiak, R., Polleres, A., Decker, S.: An empirical survey of linked data conformance. J. Web Semant. **14**, 14–44 (2012)
12. Ruckhaus, E., Baldizán, O., Vidal, M.-E.: Analyzing linked data quality with LiQuate. In: Demey, Y.T., Panetto, H. (eds.) OTM 2013 Workshops 2013. LNCS, vol. 8186, pp. 629–638. Springer, Heidelberg (2013)
13. Acosta, M., Zaveri, A., Simperl, E., Kontokostas, D., Auer, S., Lehmann, J.: Crowdsourcing linked data quality assessment. In: Alani, H., Kagal, L., Fokoue, A., Groth, P., Biemann, C., Parreira, J.X., Aroyo, L., Noy, N., Welty, C., Janowicz, K. (eds.) ISWC 2013, Part II. LNCS, vol. 8219, pp. 260–276. Springer, Heidelberg (2013)
14. Kontokostas, D., Zaveri, A., Auer, S., Lehmann, J.: TripleCheckMate: a tool for crowdsourcing the quality assessment of linked data. In: Klinov, P., Mouromtsev, D. (eds.) KESW 2013. CCIS, vol. 394, pp. 265–272. Springer, Heidelberg (2013)
15. Wang, S., Lin, C.J., Wu, C., Chaovalitwongse, W.: Early detection of numerical typing errors using data mining techniques. IEEE Trans. Syst. Man Cybern. Part A: Syst. Hum. **41**(6), 1199–1212 (2011)
16. Wienand, D., Paulheim, H.: Detecting incorrect numerical data in DBpedia. In: Presutti, V., d'Amato, C., Gandon, F., d'Aquin, M., Staab, S., Tordai, A. (eds.) ESWC 2014. LNCS, vol. 8465, pp. 504–518. Springer, Heidelberg (2014)
17. Fleischhacker, D., Paulheim, H., Bryl, V., Völker, J., Bizer, C.: Detecting errors in numerical linked data using cross-checked outlier detection. In: Mika, P., Tudorache, T., Bernstein, A., Welty, C., Knoblock, C., Vrandečić, D., Groth, P., Noy, N., Janowicz, K., Goble, C. (eds.) ISWC 2014, Part I. LNCS, vol. 8796, pp. 357–372. Springer, Heidelberg (2014)

From General to Specialized Domain: Analyzing Three Crucial Problems of Biomedical Entity Disambiguation

Stefan Zwicklbauer[✉], Christin Seifert, and Michael Granitzer

University of Passau, 94032 Passau, Germany
{stefan.zwicklbauer,christin.seifert,michael.granitzer}@uni-passau.de

Abstract. Entity disambiguation is the task of mapping ambiguous terms in natural-language text to its entities in a knowledge base. Most disambiguation systems focus on general purpose knowledge bases like DBpedia but leave out the question how those results generalize to more specialized domains. This is very important in the context of Linked Open Data, which forms an enormous resource for disambiguation. We implement a ranking-based (Learning To Rank) disambiguation system and provide a systematic evaluation of biomedical entity disambiguation with respect to three crucial and well-known properties of specialized disambiguation systems. These are (i) entity context, i.e. the way entities are described, (ii) user data, i.e. quantity and quality of externally disambiguated entities, and (iii) quantity and heterogeneity of entities to disambiguate, i.e. the number and size of different domains in a knowledge base. Our results show that (i) the choice of entity context that is used to attain the best disambiguation results strongly depends on the amount of available user data, (ii) disambiguation results with large-scale and heterogeneous knowledge bases strongly depend on the entity context, (iii) disambiguation results are robust against a moderate amount of noise in user data and (iv) some results can be significantly improved with a federated disambiguation approach that uses different entity contexts. Our results indicate that disambiguation systems must be carefully adapted when expanding their knowledge bases with special domain entities.

Keywords: Entity disambiguation · Learning to rank · Linked data · Semantic web

1 Introduction

Semantically structured information like Linked Data exhibits huge potential for improving unstructured information management processes in different domains like the Web, enterprises or research. In particular, textual information can be linked to concepts found in the Linked Open Data (LOD) cloud to improve retrieval, storage and analysis of large document repositories. Entity disambiguation algorithms establish such links by identifying the correct semantic meaning

© Springer International Publishing Switzerland 2015
Q. Chen et al. (Eds.): DEXA 2015, Part I, LNCS 9261, pp. 76–93, 2015.
DOI: 10.1007/978-3-319-22849-5_6

from a set of candidate meanings, referred to as the knowledge base (KB), to a selected text fragment, also called surface form. For instance, given a sentence with surface form "Ford", an entity disambiguation algorithm determines whether the surface form refers to the actor (Harrison Ford), the 38th President of the United States (Gerald Ford), the organization (Ford Motor Company) or the place (Ford Island) [24].

Entity disambiguation has been studied extensively in the past 10 years. Most prior work focus on disambiguating entities of general KBs like Wikipedia and other encyclopedias [9,12,15,18,20]. Recent work takes on LOD data sets as KB, but still focuses on generic entities like cities, persons etc. [18,20]. However, its results do not hold true for disambiguating entities of more specialized domains. When taking specialized entities from the LOD cloud, disambiguation is more difficult due to special domain characteristics. For instance, LOD data sets that contain biomedical entities often lack appropriate entity descriptions (e.g. genes) or provide domain-specifity (e.g. UniProt focuses on genes only). Overall, a systematic evaluation of specialized entity disambiguation w.r.t special domain properties with entities of the LOD cloud is missing.

In our work we first identify the following three crucial special domain properties:

1. entity context, i.e. the way how entities are described
2. user data, i.e. quantity and quality of externally disambiguated entities
3. quantity and heterogeneity of entities to disambiguate, i.e. the number and size of different domains in a KB.

Further, to evaluate these special domain properties, we focus on the biomedical domain which is extensively represented by several large data sets in the LOD cloud. Biomedical entity disambiguation is a challenging task due to a considerable extent of ambiguity and thus has attained much attention in research in the last decade [24]. While many biomedical disambiguation algorithms apply common String matching approaches, we combine well-established disambiguation features in a ranking approach (Learning to Rank) to perform an in-depth evaluation of our special domain properties.

Overall, our **contributions** are the following:

- We provide a systematic evaluation of biomedical entity disambiguation with respect to entity context, user data as well as quantity and heterogeneity of entities.
- We show that the choice of entity context that is used to attain the best disambiguation results strongly depends on the amount of available user data.
- We show that entity contexts strongly affect disambiguation results with large-scale and heterogeneous KBs.
- We show that results are robust against a moderate amount of noise in user data.
- We show that by using a federated approach with different entity contexts some results can be improved significantly (Mean Reciprocal Rank as well as robustness against large-scale and heterogeneous KBs).

The remainder of the paper is structured as follows: In Sect. 2 we identify and model the evaluated special domain properties. Section 3 describes the implementation of our disambiguation system. Section 4 analyzes the biomedical data set CALBC which is used in our evaluation. Section 5 presents experiments in form of an in-depth evaluation. In Sect. 6 we review related work. Finally, we conclude our paper in Sect. 7.

2 Problem Statement and Modeling

First, we identify the properties entity context, user data and quantity and heterogeneity of entities which resemble core properties for specialized disambiguation systems. Second, we introduce how we model these properties in the context of our work.

2.1 Identifying Important Properties of a Specialized Disambiguation System

Problem 1: Disambiguating domain-specific entities demands a specialized disambiguation system that covers the entire range of entities belonging to the respective domain. The creation of such a system includes the choice of a data set that describes all entities as effectively as possible. Generally, an entity can be defined intensionally, i.e. through a description, or extensionally, i.e. through instances and usage [13]. Intensional definitions can be understood as a thesaurus or logical representation of an entity, as it is provided by LOD repositories. Extensional definitions resemble information on the usage context of an entity, as it is provided by entity-annotated documents. Many disambiguation systems apply LOD repositories on general knowledge (e.g. DBpedia) due to its rich feature set (e.g. descriptions, relations). LOD repositories comprising special-domain entities regularly lack such features [24]. For instance, entities like "FV3-049L" in the UniProt KB lack extensive disambiguation-relevant descriptions or relations.

This raises the question of how disambiguation with intensional and extensional entity descriptions performs in specialized domain. Additionally, the question remains to which extent federated disambiguation with both entity contexts improves the results. We refer to both extensional and intensional entity descriptions, as **entity context**.

Problem 2: Extensionally constructed KBs contain information about the entities usage context in terms of entity-annotated documents. These textual documents contain words or phrases that were linked to their entities either manually by users or automatically by disambiguation systems. In specialized domains the quantity and quality of available annotated documents is generally very limited.

The question remains to which extent quantity and quality of annotated documents influence disambiguation with different entity contexts on specialized domains. We denote words or phrases and their mapping to entity identifiers as **user data**.

Problem 3: Several general-domain disambiguation systems use DBpedia as KB
due to its wide-ranging and high quality entities. DBpedia also comprises a broad
range of popular entities from several specialized domains (e.g. Influenza) but
lacks very specific entities [19] (e.g. IIV3-011L gene). However, the LOD cloud
offers several data sets comprising entities belonging to a specific subdomain.
For instance, the UniProt KB contains genes/proteins only and therefore also
contains very unpopular and rare occurring entities. To cover all entities of a
specialized domain, we collect the entities of several LOD data sets. This may
lead to (extremely) large and heterogeneous KBs.

The question remains how quantity and heterogeneity of entities affect disambiguation accuracy in specialized domains. In the following we refer to this
property as the **quantity and heterogeneity of entities.**

2.2   Modeling the Properties in Context of a Biomedical Disambiguation System

After identifying important properties of a specialized disambiguation system,
we focus on the biomedical domain which is perfectly suitable for our analysis.
In the following we specify and model the properties (i) entity context, (ii) user
data and (iii) quantity and heterogeneity of entities in context of our work.

Modeling Entity Context. Entities are described either extensionally or intensionally. We model these entity context forms as an *entity-centric* (intensional
entity representation) or *document-centric* KB (extensional entity representation) which comprise disambiguation-relevant entity information extracted by
the original data sets. Figure 1 illustrates our model. The edge between extensional data and entity-centric KB depicts the usage of user data in the entity-
centric KB (e.g. surface forms, synonyms).

Formally, we define an *entity-centric* KB as

$$KB_{ent} = \{e_0, ..., e_n | e_i \in E, n \in \mathbb{N}\} \tag{1}$$

The set of all entities available in KB_{ent} is denoted as
E, with e_i being a single
entity. All entities $e_i \in KB_{ent}$
have a unique primary key ID
which combines the name of
the knowledge source as well as
its identifier in the knowledge
source. Additionally, a variable number of fields k contain

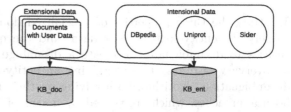

Fig. 1. Modeling entity-centric and document-centric KBs.

domain-independent attributes, e.g., description, and domain-dependent information, e.g., the sequence length of genes. Formally we denote such an entity as
$e_i = (ID, Field_1, ..., Field_k)$.

A *document-centric KB* is defined as

$$KB_{\text{doc}} = \{d_0, ..., d_n | d_i \in D, n \in \mathbb{N}\} \tag{2}$$

An entry d_i in a document-centric KB consists of the title, the content, both representing a text string, and a set of annotations $\{(t_i, \Omega_i)\}$. An annotation contains a surface form t and a set Ω with entity identifiers. These entity identifiers are referred by the respective surface form t. In the following, we denote an entry in a document-centric KB as $d_i = (Title, Content, \{(t_1, \Omega_1) ... (t_k, \Omega_k)\})$.

Modeling User Data. In our work the set of all user annotations in natural-language documents is called user data. A user annotation consists of a textual representation t, the surface form, and an entity set Ω, which is referred by surface form t. Example 3 shows an annotation of surface form "H1N1", with the *id* denoting an entity's LOD resource:

...WHO declared $<$ e id="UMLS:C1615607:T005:diso $>$ H1N1 $<$ /e $>$ influenza...
$$\tag{3}$$

As depicted in Fig. 1 user data is stored in both, entity-centric and document-centric KBs. In our work we assume that user data is readily available and provided by the underlying data set (cf. Sect. 4).

Modeling Large-scale and Heterogeneous KBs. Basically, increasing the heterogeneity within a KB is caused by adding entities from other domains. Hence, we distinguish between an *intra-specific* domain extension and an *inter-specific* domain extension. An intra-specific domain extension describes a KB enrichment with entities or documents from the same domain. In our case we add entities and documents from the biomedical domain (e.g. adding a gene database). In contrast a KB enrichment with documents or entities from other domains (e.g. DBpedia) describes an inter-specific domain extension.

3   Approach

To study the three properties of specialized domain disambiguation systems, namely the entity-context, user data, and the quantity and heterogeneity of entities to disambiguate, we create a disambiguation system. Figure 2 shows an overview of our system containing an entity-centric and document-centric disambiguation algorithm, both relying on their respective KB. The results of both approaches, which are ranked by means of Learning to Rank (LTR), are combined in a federated disambiguation approach.

In the following section we first describe the methods for disambiguation with an underlying entity-centric and document-centric KB. Second, we describe the LTR feature set in our algorithms. Finally, we describe our federated disambiguation approach.

3.1 Entity-Centric and Document-Centric Disambiguation

Our entity-centric and document-centric disambiguation algorithms can be described as
ranking-based approaches for disambiguating
entities e_i. Given a knowledge base KB that
contains all available entity candidates, a sur-
face form t as well as its context words c_t^λ (λ
denotes the number of words in front of and
after surface form t), we return a ranked list
R of entities in descending score order, i.e.

$$R = rank(KB, t, c_t^\lambda) \qquad (4)$$

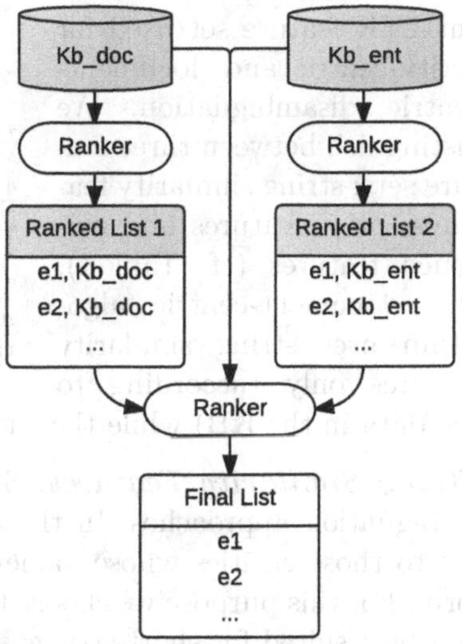

Our **entity-centric disambiguation**
approach uses a linear combination of a
weighted feature set F_{ent} to compute a score
$S_{e_i}^{ent}$ for each entity:

$$S_{e_i}^{ent} = w^\mathsf{T} f(e_i, t, c_t^\lambda) \qquad (5)$$

Variable w denotes the weight vector for our
feature set and function $f(e_i, t, c_t^\lambda)$ returns a
vector containing the feature values of entity

Fig. 2. Disambiguation system

e_i with reference to surface form t and its context c_t^λ. The disambiguation result
R consists of the Top-N scored entities.

Our **document-centric disambiguation** algorithm is similar to a K-Nearest-
Neighbor classification using majority voting. First, we obtain a predefined number τ of relevant documents using the ranking function as defined in Eq. 5 with
another feature set. A relevant document should contain similar content as given
by surface form t and surrounding context c_t^λ. The second step encompasses the
classification step. We compute the score $S_{e_i}^{doc}$ for all referenced entities K in our
queried document set T_τ:

$$S_{e_i}^{doc} = \sum_j^{T_\tau} p(e_i|d_j) \qquad (6)$$

Probability $p(e_i|d_j)$ denotes the probability of entity e_i occurring in document
d_j (with reference to all documents in KB_{doc}). To determine the probabilities
of entities occurring in documents we apply a modified Partially Labeled Latent
Dirichlet Allocation approach (PLDA) [14], which is similar to the approach of
mining evidence for entity disambiguation [10]. Due to space constraints we refer
the reader to the referenced papers for details. Again, the result list R consists
of the Top-N scored entities. The quality of the results strongly depends on
the number of annotated entities in the document set. Generally, when using a
document-centric KB, user data must be available.

3.2   Feature Choice

In the following we describe our LTR feature set used for entity-centric and document-centric disambiguation. We distinguish between three feature sets: string similarity features, prior features and evidence features (cf. Table 1). Our document-centric algorithm uses string similarity features only (according to

Table 1. Overview of LTR features

Nr.	Feature
1	Jaro-Winkler distance between surface form and entity names
2	TF-IDF weight of surface form w.r.t all entity names
3	TF-IDF weight of surface form w.r.t all entity descriptions
4	TF-IDF weight of context w.r.t all entity names
5	TF-IDF weight of context w.r.t all entity descriptions
6	BM-25 weight of surface form w.r.t all entity descriptions
7	BM-25 weight of context w.r.t all entity descriptions
8	Prior: Occurrences of an entity
9	Sense prior: Entity occurrences with a specific surface form
10	Co-occurrences: Entity-entity alignment
11	Term evidences: Entity-term alignment

the data in the KB) while the entity-centric approach applies all.

String Similarity Features: String similarity features are used in both disambiguation approaches. In the entity-centric approach we restrict our result list to those entities whose names or synonyms do not match with the surface form. For this purpose we choose the Jaro-Winkler distance [6] which is designed and best suited for short strings such as person names. Other features compute the similarity between the surface form and the entity names/synonyms as well as the entity description. Additionally, we determine the similarity between the context words and the entity names/synonyms as well as the entity description. We apply the Vector Space Model with TF-IDF weights and the Okapi BM25 model (cf. Table 1 features 2–7) for similarity computation. This similarity feature set attains the best results in our evaluation, but our approach leaves the option of choosing other metrics open.

In the document-centric approach we use the Vector Space Model (TF-IDF) and Okapi BM25 model to search for documents with similar content as given by the surface form and context words (feature 3, 5–7). TF-IDF and BM-25 weights of surface forms and surrounding context are computed w.r.t to the documents title and content. We omit the Jaro-Winkler distance as filter due querying documents instead of relevant entities. An in-depth explanation of these models is provided by [11].

Prior Features: Generally, some entities (i.e. Influenza) occur more frequent than others (i.e. IIV3-011L gene) in documents. Thus, these popular entities provide a higher probability to reoccur in other documents. In our work the *Prior* $p(e_i)$ describes the a-priori probability that an entity occurs and was initially proposed by Philip Resnik [16]. A logarithm is used for this feature to damp high values. The *Sense Prior* $p(e_i|t)$ estimates the probability of seeing an entity with a given surface form [12]. All probabilities are computed by analyzing available user data.

Evidence Features: The *Co-occurrence* feature Co_{e_i} considers context words of surface form t as potential surface forms. Basically, we assume that surface form t's real referent entity provides a higher probability to co-occur with potential but not yet disambiguated entities located in the surrounding context. First, we assume the context words c_t^λ of our surface form t to be surface forms of other entities. Hence, we compare the context words c_t^λ with all existing surface forms provided by available user data. If a context word c_j matches with one of these surface forms, we use this surface form's referent entity e_k and compute the probability of our entity candidate e_i occurring with e_k. For instance, the context word "influenza" of surface form t has already been used as surface form to address the entity "H1N1" in a document. Thus, "H1N1" constitutes a potential entity for our context word and we compute the probability of our entity e_i co-occurring with "H1N1":

$$Co_{e_i} = \sum_{c_j \in c_t^\lambda} \log(1 + \underset{e_k \in f(c_j)}{argmax}\ p(e_k|e_i)p(e_k|c_j)) \tag{7}$$

We investigate all context words c_t^λ to compute the feature score. Function $f(c_j)$ delivers a set of entities that have been annotated in combination with the possible "surface form" c_j in other documents. Further, $p(e_k|e_i)$ describes the probability of entity e_k co-occurring with our entity candidate e_i. Additionally, we take the sense prior $p(e_k|c_j)$ into account to estimate the probability of surface form c_j describing entity e_k. The logarithm is applied to attain slightly better result values.

Similar to the feature above, the *Term Evidence* feature considers probabilities of context words co-occurring with an entity candidate. For instance, the context word "disease" is an indicator of entity "Influenza" being correct. The term $p(c_j|e_i)$ denotes the probability of context word $c_j \in c_t^\lambda$ co-occurring with entity e_i. Overall, we sum up the probabilities of all context words: $\sum_j^W p(c_j|e_i)$, with $W = |c_t^\lambda|$.

To determine the entity-entity and entity-term distributions we again apply the PLDA approach [10, 14].

3.3 Federated Entity Disambiguation

In the following we present a federated entity disambiguation approach that uses both entity contexts. If an entity-centric or document-centric KB does not provide entity-relevant information, which is more likely in a specialized domain, a federated approach may still retrieve correct disambiguation results. Basically, we rerank disambiguated entities located in the result lists R_l^{ent} and R_l^{doc} of our entity-centric and document-centric disambiguation algorithms by means of LTR which serves as supervised ensemble ranker. The variables *ent* and *doc* denote the type of the KB and parameter l denotes the length of the respective approach's result list.

Overall, we compute a new score $S_{e_i}^{com}$ for every entity located in R_l^{ent} and R_l^{doc} and create a new result list. Therefore we first define an entity set M that contains all disambiguated entities of R_l^{ent} and R_l^{doc}. Further, we compute the final score $S_{e_i}^{com}$:

$$S_{e_i}^{com} = w^\mathsf{T} f(e_i), \text{with } e_i \in M \tag{8}$$

Similar to Eq. 5, variable w denotes the weight vector of our feature set F_{com} and function $f(e_i)$ returns a vector containing the feature values of entity e_i.

Our first two features represent the entity scores $S_{e_i}^{ent}, S_{e_i}^{doc}$ attained with our entity-centric and document-centric disambiguation approaches (cf. Eqs. 5 and 6). Our third feature describes the entity score attained with the combined feature set of entity-centric and document-centric disambiguation. More specific, we compute the linear combination of the weighted feature set comprising the entity-centric feature set F_{ent} and the document-centric classification feature (used in Eq. 6). The weights of the corresponding weight vector to compute this feature score are learned in a preprocessing step. Our last two features describe the probability of the entity-centric or document-centric approach retrieving a correct result given the biomedical subdomain of entity e_i. An entity may belong to one of five subdomains as given by our corpus (cf. Sect. 4). We compute the probabilities by analyzing the results of our approaches.

Overall, we use the top 50 entities of the entity-centric and document-centric algorithms as input entities to provide a good entity repertory for the federated approach.

4   Data Set

To evaluate our properties we have chosen the CALBC (Collaborative Annotation of a Large Biomedical Corpus) data set, a biomedical domain specific KB representing a very large, community-wide shared, silver standard text corpus annotated with biomedical entity references [8]. Overall, we applied the CALBC due to the following reasons:

- In contrast to gold standard corpora like the BioCreative (II) corpora[1], CALBC provides a huge set of annotations which perfectly suit for our evaluation purpose in terms of quantity (24,447 annotations in Biocreative II versus ≈120M annotations in CALBC). It is noted that despite some annotations might be erroneous the corpus most likely serves as predictive surrogate for a gold standard corpora [8].
- It already represents a document-centric KB comprising biomedical documents annotated with biomedical entities, which mostly can be linked to the LOD cloud.

Basically, the data set is released in 3 differently sized corpora: small, big and pilot. For our evaluations we use the small (CALBCSmall, 174.999 documents) and the big (CALBCBig, 714.282 documents) corpus, which mainly differ in the number of available documents. All CALBC documents cover Medline abstracts of the "Immunology" domain, a reasonably broad topic within the biomedical domain. Overall, the set of annotated entities amounts to ≈500.000 distinct biomedical entities overall, compared to ≈100.000 biomedical entities covered by DBpedia [19]. These referenced entities are categorized into four main classes

---

[1] http://www.biocreative.org/news/biocreative-ii/.

(subdomains) namely, Protein and Genes, Chemicals, Diseases and Disorders as well as Living Beings. All these entities are separated in different namespaces. Due to resources from some of the namespaces are not publicly available we restricted the data set to the available data sets, namely using the name-spaces UMLS[2], Uniprot[3], Disease (is a subset of UMLS), EntrezGene[4] and Chemlist[5].

Despite this restriction we still cover the majority of the annotated entities ($\approx$90 %) in the corpus. With these entities constituting our sample space, we are able to generate an entity-centric knowledge base by gathering information from LOD repositories. For each user annotation we are able to create a

Table 2. Data set statistics

	CALBCSmall	CALBCBig
Documents	174.999	714.282
Surface forms	2.548.900	10.304.172
Unique surface forms	50.725	101.439
Annotated entities	37.309.221	96.526.575
Unique entities	453.352	308.644
Namespaces	14	16

link of the respective RDF resource. To create a KB entry we extract labels, available synonyms, descriptions and functional information. All LOD data sets also provide its own specific properties (e.g. taxonomies) which may be used to enrich the KBs but cannot be exploited across all entities. Several domain- and repository-specific information are stored in our KB but are not used by our disambiguation system so far. Table 2 depicts important statistics of our data sets.

In CALBC, surface forms are linked to 9 entities on average due to a comprehensive classification system. Thus, we accept several valid entities per surface form.

5   Evaluation

Our approaches are implemented in Java with all queries being executed with Apache Lucene 4.8[6]. For the LTR algorithm we chose Sofia-ml[7], a machine learning framework providing algorithms for massive data sets [7]. These algorithm are mainly embedded in our publicly available disambiguation system *DoSeR*[8] (**D**isambiguation **o**f **Se**mantic **R**esources) which is being developed continuously.

First, we investigate the influence of the entity context onto disambiguation accuracy as well as how different scales of user data affect the results (Sect. 5.2). Second, we evaluate how entity context and user data influences the accuracy with large-scale and heterogeneous KBs (Sect. 5.3). Third, we analyze how disambiguation results evolve after adding different degrees of erroneous user data

---

[2] http://www.nlm.nih.gov/research/umls/.

[3] http://www.uniprot.org.

[4] http://www.ncbi.nlm.nih.gov/gene.

[5] http://www.cas.org/content/regulated-chemicals.

[6] http://lucene.apache.org/.

[7] http://code.google.com/p/sofia-ml/.

[8] http://purl.org/eexcess/components/research/doser.

(Sect. 5.4). The small data set is used for all evaluations and the big data set serves for scalability experiments. We note that our intention was not to compare our approach with other approaches: most publicly available biomedical entity annotators do not return a ranked list (e.g. NCBO annotator[9]), which is a key factor in our evaluation. Instead, the major focus in our work lies on evaluating special domain properties.

We report a set of comprehensive and established measures, comprising mean reciprocal rank (MRR), recall and mean average precision (MAP), which are averaged over 5-fold cross validation runs. The reciprocal rank is the multiplicative inverse of the rank of the first correct result in the result. Average precision denotes the average of all precision @n values of a single disambiguation task. A precision @n value is computed at every correct hit n in the result set [11].

5.1 Basic Parameter Settings

Due to an enormous amount of analyzed parameter combinations, we will only present the most important ones. The context length affects the number of words in both directions, before and after the corresponding surface form. We use a context length of 50 words due to more words worsen the results in all experiments. By using Lucene's TF-IDF score, it must be noted that Lucene's default TF-IDF score also takes internal parameters like term boosting and coordination factor into account. Entity-centric disambiguation always uses fuzzy queries to query the entity mentions and term queries to query the surrounding context. Fuzzy queries match terms with a max. edit distance of 2. Document-centric disambiguation always uses term queries for entity mentions and context queries. When using the document-centric KB, we choose $\tau = 1500$, with τ denoting the amount of documents used for classification. Our result list is trimmed to 10 entities per query to provide a good relation between recall and precision.

5.2 Entity Context and User Data

In this experiment we investigate the influence and effects of the entity context (entity-centric vs. document-centric KB) onto disambiguation accuracy. Furthermore, we use different scales of user data and investigate its effect on the results. We performed the evaluations with different fractions of user data whereby 100 % states that all available annotations are used. For all fractions all models were reconstructed accordingly.

Table 3 shows an overview of the results attained by different algorithm combinations with various user data fractions. For a better estimation we can say that 1 % of user data corresponds to 1 annotation per entity on average. We compare entity-centric disambiguation (EC), document-centric disambiguation (DC) and the federated disambiguation approach while user data must be available for document-centric and federated disambiguation. Figure 3 shows the MRR and recall of our approaches. To maintain clarity we omit MAP values in

---

[9] http://bioportal.bioontology.org/annotator.

Table 3. Disambiguation accuracy (MRR, Recall and MAP) of entity-centric, document-centric and federated disambiguation with various amount of user data.

	MRR					Recall					MAP				
UserData in %	100	20	1	0.1	0	100	20	1	0.1	0	100	20	1	0.1	0
EC	88.0	85.5	70.2	44.7	**36.7**	**76.7**	**74.2**	56.2	29.9	**25.3**	70.7	68.1	**50.9**	28.4	**25.7**
DC	75.5	75.6	71.9	57.1	-	71.7	71.8	**58.9**	**42.2**	-	59.5	59.5	47.8	**33.7**	-
Federated	**92.7**	**92.3**	**73.9**	**58.5**	-	71.8	71.6	58.0	37.3	-	**70.9**	**68.5**	50.8	27.9	-

this graph. We note that the plot's x-axis starts at 0.1 % due to its logarithmic scale to improve visualization and its necessity of user data for document-centric and federated disambiguation.

Assuming that a high amount of user data is available (all annotations in CALBC), entity-centric disambiguation attains a high MRR (88.0 %) and recall (76.7 %) and significantly outperforms the document-centric approach in all measures. Analyzing the results of the federated approach shows a (significant) increase of the MRR of 4 % in contrast to the entity-centric approach considering all available user data. A MRR of ≈93 % shows a high level of reliability in terms of ranking a correct entity on top. In contrast, the high recall values (76 %) provided by the entity-centric approach are not transfered. Instead, the federated approach attains similar results as provided by the document-centric approach (71 %). We assume that optimizing our LTR weights w.r.t recall and using additional features may overcome this deficit. The MAP values of the federated approach are slightly decreased compared to the entity-centric approach. Map values of 70 % are decent regarding the number of correct results per surface form (depending on the use case).

Analyzing Fig. 3 shows that the amount of user data strongly influences MRR and recall of entity-centric and document-centric disambiguation. While the entity-centric approach significantly outperforms the document-centric approach if enough user data is available, we note reverse results if the amount of user data (significantly) decreases. The less user data available, the higher the advance of the document-centric approach. This is explicable by the increasing dependency of

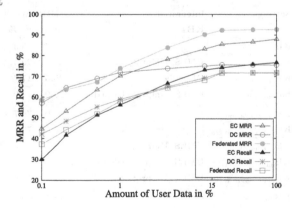

Fig. 3. Results of entity-centric, document-centric and federated disambiguation with various amount of user data.

the entity-centric approach on KB quality and availability of exploitable features across entities.

In summary, we state that neither entity-centric nor document-centric disambiguation attains the best results with all configurations. The choice of entity context that is used to attain the best results strongly depends on the amount of user data. Additionally, the federated approach attains an excellent MRR if enough user data is available.

5.3   Knowledge Base Size and Heterogeneity

In the following we analyze how entity context and user data influence the results when the size and/or heterogeneity of the KBs is increased. We extend our entity-centric KBs KB_{ent} and $KB_{ent/ua/sb}$ with additional entities. KB_{ent} denotes an entity-centric KB without user data information and $KB_{ent/ua/sb}$ denotes the enrichment of the entity-centric KB with user data information (ua) of CALBC-Small (s), CALBCBig (b) or both (sb). The set of additional entities comprises all entities belonging to UMLS, Uniprot and/or DBpedia. The document-centric KB is enriched with the CALBCBig data set (intra-specific) and/or Wikipedia pages (inter-specific).

Table 4 shows an overview of the results before and after extending the KBs. The column *Change* takes the average change of the measures MRR, recall and MAP in %. The entity-centric approach attains worse results after increasing the amount of entities when no user data is available. Additionally, increasing the domain heterogeneity by adding DBpedia entities significantly worsens the results with a decrease of 33 percent (with DBpedia only), respectively 40 percent (with DBpedia, UMLS and Uniprot) on average. An entity-centric disam-

Table 4. Results after increasing our KB with various corpora in the biomedical domain

Settings	Integrated KBs	MRR in %	Recall in %	MAP in %	#Ent/#Docs	Change in %
$KB_{ent,\ intra}$	-	36.7	25.3	25.7	265.532	-
$KB_{ent,\ intra}$	UMLS, Uniprot	30.9	20.4	19.5	32.407.960	−19.3
$KB_{ent,\ inter}$	DBpedia	25.6	17.7	18.3	4.643.509	−33.2
$KB_{ent,\ inter}$	UMLS, Uniprot, DBpedia	22.9	14.0	15.4	36.785.937	−40.4
$KB_{ent/ua/s,\ intra}$	-	88.0	76.7	70.7	265.532	-
$KB_{ent/ua/sb,\ intra}$	-	90.5	79.2	73.2	265.532	+3.1
$KB_{ent/ua/s,\ intra}$	UMLS, Uniprot	78.0	66.6	60.9	32.407.960	−9.9
$KB_{ent/ua/s,\ inter}$	UMLS, Uniprot, DBpedia	60.3	55.9	50.1	36.785.937	−29.3
$KB_{ent/ua/sb,\ inter}$	UMLS, Uniprot, DBpedia	62.7	58.0	52.4	36.785.937	−26.4
$KB_{doc,\ intra}$	-	75.5	71.7	59.5	174.999	-
$KB_{doc,\ intra}$	CALBCBig	76.0	72.2	60.1	889.282	+0.1
$KB_{doc,\ inter}$	CALBCBig, Wiki	67.3	65.0	50.8	4.267.259	−11.4
$KB_{federated,\ intra}$	-	92.7	71.8	70.9	440.531	-
$KB_{federated,\ intra}$	CALBCBig, UMLS, Uniprot	81.9	65.9	61.5	33.297.242	−11.1
$KB_{federated,\ inter}$	CALBCBig, UMLS, Uniprot, DBpedia, Wiki	75.7	60.1	51.6	37.675.219	−20.4

biguation that applies features derived from annotated documents significantly improves the robustness against an increase of entities and heterogeneity by one third. The usage of additionally mined features from CALBCBig increases these results by 3 %. The accuracy drop by about 30 % with a KB containing DBpedia, UMLS and Uniprot remains constant. All in all, disambiguation with an entity-centric KB is not robust against large-scale and heterogeneous KBs due to our feature set still does not provide enough evidence to overcome these limitations. It is an open question whether there exist features that suppress these negative effects.

When using a document-centric KB, the results do not suffer when adding more documents with biomedical content. This can be explained with the document increase does not influence the classification (cf. Sect. 3.1). Instead, the retrieval step has a wider range of documents to choose for the classification step. Selecting other documents has no negative effect on the documents' spectrum of annotated entities. An inter-specific domain extension with CALBC and Wikipedia documents causes a decrease of the disambiguation results (11 %). However, document-centric disambiguation is more robust against an inter-specific domain extension than entity-centric disambiguation.

The federated approach mitigates the accuracy decrease, compared to the entity-centric approach. With the document-centric approach being robust against the document count, the accuracy decrease after increasing heterogeneity and entity/document count stays small.

In summary, disambiguating biomedical entities from the LOD cloud with a document-centric approach is more robust against large-scale and heterogeneous KBs than entity-centric disambiguation. The results recommend to use a federated approach to yield the advantages of both approaches (result and robustness against large-scale KBs).

5.4 Noisy User Data

Available user data may contain errors caused by missing knowledge, validation etc. While the original CALBC may contain erroneous annotations due to constituting a silver-standard corpus we investigate how additional noise in its annotations influence results attained with the entity-centric, document-centric and federated disambiguation approach. We compare a user model created from the original annotations (as given by CALBC) with user models with different degrees of additional annotation errors. Prior research has already investigated the influence of noisy user data on LTR models, but the effects on disambiguation results are unknown. We modified available CALBC annotations and recreated our KBs as well as LTR models. Therefore we selected an annotation to be wrong with probability p. Instead of exchanging this entity annotation with a randomly selected entity annotation, we simulated user behavior by choosing a wrong entity from the result list of a conducted disambiguation task (entity-centric disambiguation) on the annotation's surface form. Choosing a wrong entity at the top of the result list should be more likely than choosing an entity from the end. We modeled this event with a Gaussian distributed random variable $X \sim \mathcal{N}(1, 10)$ which yields positive values only. We exchanged the correct annotation with the wrong result that was selected by the random variable. We modified the CALBC annotations with varying degree of noise. Figure 4 shows the evaluation results from 0 % additional noise (as given by CALBC) to 100 % noise (all annotations are wrong) attained with an entity-centric, document-centric and federated approach. In the following, we focus on the results with a 25 % noise rate. The MRR of the entity-centric and document-centric approach provides a slight decrease of 10 % with a noise rate of 25 %. The federated approach tops the single approaches as long as the noise stays below 33 %. In all approaches, the recall decrease is about 5 % with 25 % noise. Basically, the recall values stay high as long as the noise rate does not exceed 66 %. It another story for the MAP values which continuously decrease almost linearly from 0 to 100 % noise with the entity-centric and federated approach. However, a decrease of up to 12 % with 25 % noise in all approaches shows that the MAP results are influenced by noisy user data.

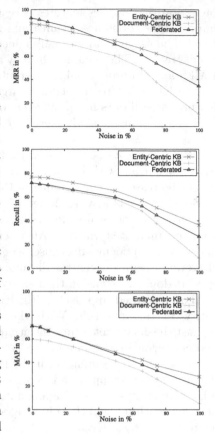

Fig. 4. Influence of noise in user data on disambiguation results.

In summary, all approaches are robust against little noise in the user data. Assuming that the amount of erroneously annotated data is about one third or less, we note that all disambiguation approaches are robust and still provide fairly satisfying results.

6 Related Work

One of the first works to disambiguate general knowledge entities (e.g. Wikipedia) defines a similarity measure to compute the cosine similarity between the text around the surface form and the referent entity candidates' Wikipedia page [1]. At the same time Cucerzan et al. introduced topical coherence for entity disambiguation [2]. The authors use the referent entity candidate and other entities within the same context to compute topical coherence by analyzing the overlap of categories and incoming links in Wikipedia. Several works use topical coherence and context similarity to improve disambiguation [9,15,18]. All these works exploit various Wikipedia features (e.g. categories). There are some more generic approaches that can be easily applied to other KBs. Subsequent work incorporate more information to improve entity context similarity comparison by exploring query expansion [3]. Another work proposes a generative entity-mention model which, similar to our work, consists of an underlying entity popularity model, entity name model and entity context model [4]. Some other works propose generic, generative topic-models for entity disambiguation which exploit context compatibility and topic coherence [5,17]. Almost all works address algorithm improvements but do not investigate the requirements to adapt the results to other domains.

Biomedical entity disambiguation has also attained much attention in research in the last decade [24]. For instance, Wang et al. classify relations between entities for biomedical entity disambiguation [21]. Biomedical entities can also be disambiguated with the help of species disambiguation. Wang et al. [22] apply language parsers for species disambiguation and attain promising results. Zwicklbauer et al. [23] compared document-centric and entity-centric KBs with a search-based algorithm. The authors report very strong results with document-centric KBs in the biomedical domain.

In terms of entity context, several works use intensional entity descriptions provided by high-quality KBs like DBpedia [12,18,20], which are similar to our entity-centric approach. Some other works store Wikipedia documents as a whole in a KB to describe entities [17], but exploit that Wikipedia articles describe one specific entity. In contrast, the authors of [5] use a document-centric KB containing arbitrary entity-annotated documents. This generative approach jointly models context compatibility, topic coherence and its correlation, while our algorithm constitutes a retrieval-based approach.

7   Conclusion and Future Work

We provide a systematic evaluation of biomedical entity disambiguation with respect to three major properties of specialized domain disambiguation systems, namely the entity context, user data and the quantity and heterogeneity of entities to disambiguate. Our evaluation reveals that the choice of entity context that is used to attain the best disambiguation results strongly depends on the amount of available user data. In this context, we indicate that the performance decrease with large-scale and heterogeneous KBs strongly depends on the underlying entity context. Additionally, we show that disambiguation results are robust against a moderate amount of noise in user data. Finally, we suggest to use a federated approach of different entity contexts to improve the reciprocal rank and to increase the robustness against large-scale and heterogeneous KBs.

In summary, we state that disambiguation systems must be carefully adapted when expanding their KBs with special domain entities. An analysis of the underlying data set is strongly required to spot the potential problem areas and integrate the appropriate approaches. In this context our future work includes the design of a model that automatically analyzes the underlying KB and chooses the best disambiguation settings.

Acknowledgments. The presented work was developed within the EEXCESS project funded by the European Union Seventh Framework Programme FP7/2007-2013 under grant agreement number 600601.

References

1. Bunescu, R., Pasca, M.: Using encyclopedic knowledge for named entity disambiguation. In: Proceedings of the 11th Conference of the European Chapter of the Association for Computational Linguistics (EACL-06), Trento, Italy, pp. 9–16 (2006)
2. Cucerzan, S.: Large-scale named entity disambiguation based on Wikipedia data. In: Proceedings of the 2007 Joint Conference on EMNLP and CoNLL, pp. 708–716. Association for Computational Linguistics, Prague, June 2007
3. Gottipati, S., Jiang, J.: Linking entities to a knowledge base with query expansion. In: Proceedings of the Conference on Empirical Methods in Natural Language Processing, EMNLP 2011, pp. 804–813. ACL, Stroudsburg (2011)
4. Han, X., Sun, L.: A generative entity-mention model for linking entities with knowledge base. In: Proceedings of the 49th Annual Meeting of the Association for Computational Linguistics, HLT 2011, pp. 945–954. ACL, Stroudsburg (2011)
5. Han, X., Sun, L.: An entity-topic model for entity linking. In: Proceedings of the 2012 Joint Conference on Empirical Methods in Natural Language Processing and Computational Natural Language Learning, EMNLP-CoNLL 2012, pp. 105–115. ACL, Stroudsburg (2012)
6. Jaro, M.A.: Advances in record-linkage methodology as applied to matching the 1985 census of Tampa, Florida. J. Am. Stat. Assoc. **84**(406), 414–420 (1989)
7. Joachims, T.: Optimizing search engines using clickthrough data. In: Proceedings of the Eighth ACM SIGKDD International Conference on Knowledge Discovery and Data Mining, KDD 2002, pp. 133–142. ACM, New York (2002)
8. Kafkas, S., Lewin, I., Milward, D., van Mulligen, E., Kors, J., Hahn, U., Rebholz-Schuhmann, D.: Calbc: releasing the final corpora. In: Proceedings of the Eight International Conference on Language Resources and Evaluation (LREC 2012), Istanbul, Turkey, May 2012
9. Kataria, S.S., Kumar, K.S., Rastogi, R.R., Sen, P., Sengamedu, S.H.: Entity disambiguation with hierarchical topic models. In: Proceedings of the 17th ACM SIGKDD International Conference on Knowledge Discovery and Data Mining, KDD 2011, pp. 1037–1045. ACM, New York (2011)
10. Li, Y., Wang, C., Han, F., Han, J., Roth, D., Yan, X.: Mining evidences for named entity disambiguation. In: Proceedings of the 19th ACM SIGKDD International Conference on Knowledge Discovery and Data Mining, KDD 2013, pp. 1070–1078. ACM, New York (2013)
11. Manning, C.D., Raghavan, P., Schütze, H.: Introduction to Information Retrieval. Cambridge University Press, New York (2008)
12. Mendes, P.N., Jakob, M., García-Silva, A., Bizer, C.: Dbpedia spotlight: shedding light on the web of documents. In: Proceedings of the 7th International Conference on Semantic Systems, I-Semantics 2011, pp. 1–8. ACM, New York (2011)
13. Ogden, C., Richards, I.A.: The Meaning of Meaning: A Study of the Influence of Language Upon Thought and of the Science of Symbolism, 8th edn. Harcourt Brace Jovanovich, New York (1923)

14. Ramage, D., Manning, C.D., Dumais, S.: Partially labeled topic models for interpretable text mining. In: Proceedings of the 17th ACM SIGKDD International Conference on Knowledge Discovery and Data Mining, KDD 2011, pp. 457–465. ACM, New York (2011)
15. Ratinov, L., Roth, D., Downey, D., Anderson, M.: Local and global algorithms for disambiguation to wikipedia. In: Proceedings of the Annual Meeting of the Association of Computational Linguistics (2011)
16. Resnik, P.: Using information content to evaluate semantic similarity in a taxonomy. In: Proceedings of the 14th International Joint Conference on Artificial Intelligence IJCAI 1995, vol. 1, pp. 448–453. Morgan Kaufmann Publishers Inc., San Francisco (1995)
17. Sen, P.: Collective context-aware topic models for entity disambiguation. In: Proceedings of the 21st International Conference on World Wide Web, WWW 2012, pp. 729–738. ACM, New York (2012)
18. Shen, W., Wang, J., Luo, P., Wang, M.: Linden: linking named entities with knowledge base via semantic knowledge. In: Proceedings of the 21st International Conference on World Wide Web, WWW 2012, pp. 449–458. ACM, New York (2012)
19. Tian, L., Zhang, W., Bikakis, A., Wang, H., Yu, Y., Ni, Y., Cao, F.: Medetect: a lod-based system for collective entity annotation in biomedicine. In: IEEE/WIC/ACM International Joint Conferences on Web Intelligence (WI) and Intelligent Agent Technologies (IAT), 2013, vol. 1, pp. 233–240. IEEE (2013)
20. Usbeck, R., Ngonga Ngomo, A.-C., Röder, M., Gerber, D., Coelho, S.A., Auer, S., Both, A.: AGDISTIS - Graph-Based disambiguation of named entities using linked data. In: Mika, P., Tudorache, T., Bernstein, A., Welty, C., Knoblock, C., Vrandečić, D., Groth, P., Noy, N., Janowicz, K., Goble, C. (eds.) ISWC 2014, Part I. LNCS, vol. 8796, pp. 457–471. Springer, Heidelberg (2014)
21. Wang, X., Tsujii, J., Ananiadou, S.: Classifying relations for biomedical named entity disambiguation. In: Proceedings of the 2009 Conference on Empirical Methods in Natural Language Processing, EMNLP 2009, vol. 3, pp. 1513–1522. ACL, Stroudsburg (2009)
22. Wang, X., Tsujii, J., Ananiadou, S.: Disambiguating the species of biomedical named entities using natural language parsers. Bioinformatics 26(5), 661–667 (2010)
23. Zwicklbauer, S., Seifert, C., Granitzer, M.: Do we need entity-centric knowledge bases for entity disambiguation? In: Proceedings of the 13th International Conference on Knowledge Management and Knowledge Technologies, i-Know 2013, pp. 4:1–4:8. ACM, New York (2013)
24. Zwicklbauer, S., Seifert, C., Granitzer, M.: Linking Biomedical Data to the Cloud. In: Holzinger, A., Röcker, C., Ziefle, M. (eds.) Smart Health. LNCS, vol. 8700, pp. 209–235. Springer, Heidelberg (2015)

Ontology Matching with Knowledge Rules

Shangpu Jiang[✉], Daniel Lowd, and Dejing Dou

Department of Computer and Information Science, University of Oregon,
Eugene, USA
{shangpu,lowd,dou}@cs.uoregon.edu

Abstract. Ontology matching is the process of automatically determining the semantic equivalences between the concepts of two ontologies. Most ontology matching algorithms are based on two types of strategies: terminology-based strategies, which align concepts based on their names or descriptions, and structure-based strategies, which exploit concept hierarchies to find the alignment. In many domains, there is additional information about the relationships of concepts represented in various ways, such as Bayesian networks, decision trees, and association rules. We propose to use the similarities between these relationships to find more accurate alignments. We accomplish this by defining soft constraints that prefer alignments where corresponding concepts have the same local relationships encoded as *knowledge rules*. We use a probabilistic framework to integrate this new *knowledge-based* strategy with standard terminology-based and structure-based strategies. Furthermore, our method is particularly effective in identifying correspondences between complex concepts. Our method achieves better F-score than the state-of-the-art on three ontology matching domains.

1 Introduction

Ontology matching is the process of aligning two semantically related ontologies. Traditionally, this task is performed by human experts from the domain of the ontologies. Since the task is tedious and error prone, especially in large ontologies, there has been substantial work on developing automated or semi-automated ontology matching systems [18]. While some automated matching systems make use of data instances, in this paper we focus on the *schema-level* ontology matching task, in which no data instance is used.

Previous automatic ontology matching systems mainly use two classes of strategies. *Terminology-based* strategies discover corresponding concepts with similar names or descriptions. *Structure-based* strategies discover corresponding groups of concepts with similar hierarchies. In many cases, additional information about the relationships among the concepts is available through domain models, such as Bayesian networks, decision trees, and association rules. A domain model can be represented as a collection of *knowledge rules*, each of which denotes a semantic relationship among several concepts. These relationships may be complex, uncertain, and rely on imprecise numeric values. In this

© Springer International Publishing Switzerland 2015
Q. Chen et al. (Eds.): DEXA 2015, Part I, LNCS 9261, pp. 94–108, 2015.
DOI: 10.1007/978-3-319-22849-5_7

paper, we introduce a new *knowledge-based strategy* which uses the structure of these knowledge rules as (soft) constraints on the alignment.

As a motivating example, consider two ontologies in the basketball game domain. One ontology has datatype properties `height,weight,center,forward` and `guard` for players, while the other ontology has the corresponding datatype properties `h,w`, and `position`. Terminology-based strategies may not identify these correspondences. However, if we know that a large value of `height` implies `center` is true in the first ontology, and the same relationship holds for `h` and `position = Center` in the second ontology, then we tend to believe that `height` maps to `h` and `center` maps to `position = Center`.

We use Markov logic networks (MLNs) [4] as a probabilistic language to combine the knowledge-based strategy with other strategies, in a formalism similar to that of [12]. In particular, we encode the knowledge-based strategy with weighted formulas that increase the probability of alignments where corresponding concepts have isomorphic relationships. We use an MLN inference engine to find the most likely alignment. We name our method Knowledge-Aware Ontology Matching (KAOM).

Our approach is also capable of identifying *complex correspondences*, an extremely difficult task in ontology matching. A complex correspondence is a correspondence between a simple concept and a complex concept (e.g., `grad_student` maps to the union of `PhD` and `Masters`). This can be achieved by constructing a set of *complex concepts* (e.g., unions of concepts) in each ontology, subsequently generating candidate complex correspondences, and using multiple strategies – including the knowledge-based strategy – to find the correct ones.

The contributions of this work are as follows:

- We show how to represent common types of domain models as knowledge rules, and how to use these knowledge rules to obtain more accurate alignments. Our approach is especially effective in identifying the correspondences of numerical or nominal datatype properties. By incorporating complex concepts, our approach is also capable of discovering complex correspondences, which is a very difficult scenario in the ontology matching task.
- We evaluate the effectiveness of KAOM in three domains with different types of knowledge rules, and show that our approach not only outperforms the state-of-the-art approaches for ontology matching in one-to-one matching, but also discovers complex correspondences successfully.

The paper is organized as follows. In Sect. 2, we define ontology matching and review previous work. In Sect. 3, we introduce the concept of "knowledge rules" with a definition and examples. In Sect. 4, we present the knowledge-based strategy. In Sect. 5, we show how to incorporate complex concepts in our method. In Sect. 6, we formalize our method with Markov logic networks. We present experimental results in Sect. 7 and conclude in Sect. 8.

2   Ontology Matching

We begin by formally defining ontology matching.

Definition 1 (Ontology Matching [5]). *Given two ontologies O_1 and O_2, a correspondence is a 3-tuple $\langle e_1, e_2, r \rangle$ where e_1 and e_2 are entities of the first and second ontologies respectively, and r is a semantic relation such as equivalence ($\equiv$) and subsumptions ($\sqsubseteq$ or $\sqsupseteq$). An alignment is a set of correspondences. Ontology matching is the task or process of identifying the correct semantic alignment between the two ontologies. In most cases, ontology matching focuses on equivalence relationships only.*

Most existing schema-level ontology matching systems use two types of strategies: terminology-based and structure-based. Terminology-based strategies are based on terminological similarity, such as string-based or linguistic similarity measures. Structure-based strategies are based on the assumption that two matching ontologies should have similar local or global structures, where the structure is represented by subsumption relationships of classes and properties, and domains and ranges of properties. Advanced ontology matching systems often combine the two types of strategies [1,10,11,14]. See [18] for a survey of ontology matching systems and algorithms.

Recently, a probabilistic framework based on Markov logic was proposed to combine multiple strategies [12]. In particular, it encodes multiple strategies and heuristics into hard and soft constraints, and finds the best matching by minimizing the weighted number of violated constraints. The constraints include string similarity, the cardinality constraints which enforce that each concept matches at most one concept, the coherence constraints which prevent inconsistency induced by the matching, and the stability constraints which penalize dissimilar local subsumption relationships.

Definition 2 (Complex Correspondences). *A complex concept is a composition (e.g., unions, complements) of one or more simple concepts. In OWL^1, there are several constructors for creating complex classes and properties (see the top part of Table 1 for an incomplete list of constructors). A complex correspondence is an equivalence relation between a simple class or property and a complex class or property in two ontologies [17].*

Previous work has taken several different approaches to find complex correspondences (i.e., complex matching). [2] constructs candidates for complex correspondences using operators for primitive classes, such as string concatenation or arithmetic operations on numbers. [17] summarizes four patterns for building up complex correspondences based on linguistic and structural features given a candidate one-to-one alignment: Class by Attribute Type (CAT), Class by Inverse Attribute Type (CIAT), Class by Attribute Value (CAV), and Property Chain pattern (PC). Finally, when aligned or overlapping data is available, inductive logic programming (ILP) techniques can be used as well [6,15].

---

[1] http://www.w3.org/TR/owl2-primer/.

Many ontology matching systems make use of data instances to some extent (e.g., [2,3,6,15]). However, in this paper, we focus on the case where data are not available or data sharing is not preferred because of communication cost or privacy concerns.

3   Representation of Domain Knowledge

In the AI community, knowledge is typically represented in *formal languages*, among which ontology-based languages are the most widely used forms. The Web Ontology Language (OWL) is the W3C standard ontology language that describes the classes and properties of objects in a specific domain. OWL and many other ontology languages are based on variations of description logics.

In ontology languages such as OWL, knowledge is represented as logic *axioms*. These axioms describe properties of classes or relations (e.g., a relation is functional, symmetric, or antisymmetric, etc.), or a relationship of several entities (e.g., the relation 'grandfather' is the composition of the two relations 'father' and 'parent').

The choice of using description logic as the foundation of the Semantic Web ontology languages is largely due to the trade-off between expressivity and reasoning efficiency. In tasks such as ontology matching, reasoning does not need to be instant, so we can afford to consider other forms of knowledge outside of a specific ontology language or description logic.

Definition 3 (Knowledge Rule). *A knowledge rule is a sentence $R(a, b, \ldots ; \theta)$ in a formal language which consists of a relation R, a set of entities (i.e., classes, attributes or relations) $\{a, b, \ldots\}$, and (optionally) a set of parameters θ. A knowledge rule carries logical or probabilistic semantics representing the relationship among these entities. The specific semantics depend on R.*

Many domain models and other types of knowledge can be represented as sets of knowledge rules, each rule describing the relationship of a small number of entities. The semantics of each relationship R can typically be expressed with a formal language. Table 1 shows some examples of the symbols used in formal languages such as description logic, along with their associated semantics.

We illustrate a few forms of knowledge rules with the following examples. For each rule, we provide a description in English, a logical representation, and an encoding as a knowledge rule with a particular semantic relationship, R_i. We define a new relationship in each example, but, in a large domain model, most relationships would be appear many times in different rules.

Example 1. The submission deadline precedes the camera ready deadline:

$$\texttt{paperDueOn} \prec \texttt{manuscriptDueOn}$$

This is represented as $R_1(\texttt{paperDueOn}, \texttt{manuscriptDueOn})$ with $R_1(a, b) : a \prec b$.

Table 1. Syntax and semantics of DL symbols (top), DL axioms (middle), and other knowledge rules used in the examples of the paper (bottom)

Syntax	Semantics
$\top$	$\mathcal{D}$
$\bot$	$\emptyset$
$C \sqcap D$	$C^{\mathcal{I}} \cap D^{\mathcal{I}}$
$C \sqcup D$	$C^{\mathcal{I}} \cup D^{\mathcal{I}}$
$\neg C$	$\mathcal{D} \backslash C^{\mathcal{I}}$
$\forall R.C$	$\{x \in \mathcal{D} \mid \forall y((x,y) \in R^{\mathcal{I}} \to y \in C^{\mathcal{I}})\}$
$\exists R.C$	$\{x \in \mathcal{D} \mid \exists y((x,y) \in R^{\mathcal{I}} \wedge y \in C^{\mathcal{I}})\}$
$R \circ S$	$\{(x,y) \mid \exists z((x,z) \in R^{\mathcal{I}} \wedge (z,y) \in S^{\mathcal{I}})\}$
R^{-}	$\{(x,y) \mid (y,x) \in R^{\mathcal{I}}\}$
$R \upharpoonright C$	$\{(x,y) \in R^{\mathcal{I}} \mid x \in C^{\mathcal{I}}\}$
$R \downharpoonright C$	$\{(x,y) \in R^{\mathcal{I}} \mid y \in C^{\mathcal{I}}\}$
$C \sqsubseteq D$	$C^{\mathcal{I}} \subseteq D^{\mathcal{I}}$
$C \sqsubseteq \neg D$	$C^{\mathcal{I}} \cap D^{\mathcal{I}} = \emptyset$
$R \prec S$	$y < y'$ for $\forall (x,y) \in R^{\mathcal{I}} \wedge (x,y') \in S^{\mathcal{I}}$
$C \Rightarrow D$	$\Pr(D^{\mathcal{I}} \mid C^{\mathcal{I}})$ is close to 1

Example 2. A basketball player taller than 81 inches and heavier than 245 pounds is likely to be a center:

$$\mathtt{h} > 81 \wedge \mathtt{w} > 245 \Rightarrow \mathtt{pos} = \mathtt{Center}$$

This rule can be viewed as a branch of a *decision tree* or an *association rule*. It can be represented as $R_2(\mathtt{h}, \mathtt{w}, \mathtt{pos=Center}, [81, 245])$, with $R_2(a,b,c,\theta) : a > \theta_1 \wedge b > \theta_2 \Rightarrow c$.

Example 3. A smoker's friend is likely to be a smoker as well:

$$\mathtt{Smokes}(x) \wedge \mathtt{Friend}(x,y) \Rightarrow \mathtt{Smokes}(y)$$

Relational rules such as this one describe relationships of attributes across multiple tables, as opposed to propositional data mining rules that are restricted to a single table. This rule can be represented as $R_3(\mathtt{Smoke}, \mathtt{Friend})$ with $R_3(a,b) : a(x) \wedge b(x,y) \Rightarrow a(y)$.

For the remainder of this paper, we will assume that the knowledge in both domains is represented as knowledge rules, as described in this section.

4   Our New Knowledge-Based Strategy

We propose a new strategy for ontology matching that uses the similarity of knowledge rules in the two ontologies. It is inspired by the structure-based strategy in many ontology matching algorithms (e.g., [11,12]). It naturally extends the subsumption relationship of entities in structure-based strategies to other types of relationships.

We use Markov logic to combine the knowledge-based strategy with other strategies. In particular, each strategy is represented as a set of *soft constraints*, each of which assigns a score to the alignments satisfying it, and the alignment with the highest total score is chosen as the best alignment. We now describe the soft constraints encoding the knowledge-based strategy. Our complete Markov logic-based approach, including the soft constraints required for the other strategies, will be described in Sect. 6.

For each relation R_k that appears in both domains, we introduce a set of soft constraints so that the alignments that preserve these relationships are preferred to those that do not:

$$+w_k \qquad R_k(a,b) \wedge \neg R_k(a',b') \Rightarrow a \not\equiv a' \vee b \not\equiv b'$$
$$+w'_k \qquad R_k(a,b) \wedge R_k(a',b') \Rightarrow a \equiv a' \wedge b \equiv b'$$
$$\forall a,b \in O_1, a',b' \in O_2$$

These formulas assume R_k is a binary relation, but they trivially generalize to any arity, e.g., $R_k(a,b,c,d,e,\ldots)$. Note that separate constraints are created for each possible tuple of constants from the respective domains. The numbers preceding the constraints (w_k and w'_k) are the *weights*. A larger weight represents a stronger constraint, since alignments are ranked based on the total weights of the constraints they satisfy. A missing weight means the constraint is a hard constraint which must be satisfied.

Example 4. A reviewer of a paper cannot be the paper's author. In the cmt[2] ontology we have $R_4(\text{writePaper}, \text{readPaper})$ and in the confOf ontology we have $R_4(\text{write}, \text{reviews})$ where $R_4(a,b) : a \sqsubseteq \neg b$ is the disjoint relationship of properties. Applying the constraint formulas defined above, we increase the score of all alignments containing the two correct correspondences: writePaper $\equiv$ writes and readPaper $\equiv$ reviews.

Rules involving continuous numerical attributes often include parameters (e.g., thresholds in Example 2) that do not match between different ontologies. In order to apply the knowledge-based strategy to numerical attributes, we make the assumption that corresponding numerical attributes roughly have a *positive linear* transformation. This assumption is often true in real applications, for instance, when an imperial measure of height matches to a metric measure of height. We propose two methods to handle numerical attributes.

The first method is to compute a *distance measure* (e.g., Kullback-Leibler divergence) between the distributions of the corresponding attributes in a candidate alignment. Although the two distributions describe different attributes, the distance can be computed by assuming a linear transformation between the

---

[2] Throughout the paper, we will use ontologies in the conference domain (cmt, confOf, conference, edas, ekaw) and the NBA domain (nba − os, yahoo) in our examples. The characteristics of these ontologies will be further described in Sect. 7.

two attributes. The coefficients of the mapping relation can be roughly estimated using the ranges of attribute values appearing in the knowledge rules (see Example 5 below).

Specifically, if the distance between rules $R(\mathtt{a}, \mathtt{b}, \ldots, \theta)$ and $R(\mathtt{a'}, \mathtt{b'}, \ldots, \theta')$ is d, then we add the constraint:

$$a \equiv a' \wedge b \equiv b' \wedge c \equiv c'$$

with a weight of $\max(d_0 - d, 0)$ for a given threshold d_0.

Example 5. In the $\mathtt{nba} - \mathtt{os}$ ontology, we have conditional rules converted from a decision tree, such as

$$\mathtt{h} > 81 \wedge \mathtt{w} > 245 \Rightarrow \mathtt{Center}$$

Similarly, in the $\mathtt{nbayahoo}$ ontology, we have

$$\mathtt{h'} > 2.06 \wedge \mathtt{w'} > 112.5 \Rightarrow \mathtt{Center'}$$

Here the knowledge rules represent the conditional distributions of multiple entities. We define the distance between the two conditional distributions as

$$d(\mathtt{h}, \mathtt{w}, \mathtt{Center}; \mathtt{h'}, \mathtt{w'}, \mathtt{Center'}) = \mathbb{E}_{p(\mathtt{h}, \mathtt{w})} d(p(\mathtt{Center}|\mathtt{h}, \mathtt{w}) \| p(\mathtt{Center'}|\mathtt{h'}, \mathtt{w'}))$$

where $\mathbb{E}(\cdot)$ is expectation and $d(p\|p')$ is a distance measure. Because $\mathtt{Center}$ and $\mathtt{Center'}$ are binary attributes, we simply use $|p - p'|$ as the distance measure. For numerical attributes, we can use the difference of two distribution histograms as the distance measure. We assume the attribute correspondences ($\mathtt{h}$ and $\mathtt{h'}$, $\mathtt{w}$ and $\mathtt{w'}$) are linear mappings, and the linear relation can be roughly estimated (e.g., by simply matching the minimum and maximum numbers in these rules). When computing the expectation over $\mathtt{h}$ and $\mathtt{w}$, we apply the linear mapping to generate corresponding values of $\mathtt{h'}$ and $\mathtt{w'}$, e.g., $\mathtt{h'} = 0.025\mathtt{h}, \mathtt{w'} = 0.45\mathtt{w}$. The distribution of the conditional attributes $p(\mathtt{h}, \mathtt{w})$ can be roughly estimated as independent and uniform over the ranges of the attributes.

The second method for handling continuous attributes is to discretize them, reducing the continuous attribute problem to the discrete problem described earlier. For example, suppose each continuous attribute x is replaced with a discrete attribute x^d, indicating the quartile of x rather than its original value. Then we have $R_5(\mathtt{h}^d, \mathtt{w}^d, \mathtt{Center})$ and $R_5(\mathtt{h'}^d, \mathtt{w'}^d, \mathtt{Center'})$ with relation $R_5(a, b, c) : a = 4 \wedge b = 4 \Rightarrow c$, and the discrete value of 4 indicates that both a and b are in the top quartile. Other discretization methods are also possible, as long as the discretization is done the same way in both domains.

Our method does not rely on the forms of knowledge rules, nor does it rely on the algorithms used to learn these rules. As long as similar techniques or tools are used on both sides of ontologies, we would always be able to find interesting knowledge-based similarities between the two ontologies.

5   Finding Complex Correspondences

Our approach can also find complex correspondences, which contain complex concepts in either or both of the ontologies. We add the complex concepts into consideration and treat them the same way as simple concepts, and then we jointly solve all the simple and complex correspondences by considering terminology, structure, and knowledge-based strategies in a single probabilistic formulation.

First, because complex concepts are recursively defined and potentially infinite, we need to select a finite subset of complex concepts and use them to generate the candidate correspondences. We will only include the complex concepts occurring in the ontology axioms or in the knowledge rules.

Second, we need to define a string similarity measure for each type of complex correspondence. For example, [17] requires two conditions for a Class by Attribute Type (CAT) matching pattern $O_1 : a \equiv O_2 : \exists p.b$ (e.g., $a =$ Accepted_Paper, $p =$ hasDecision, $b =$ Acceptance): a and b are terminologically similar, and the domain of p (Paper in the example) is a superclass of a. We can therefore define the string similarity of a and $\exists p.b$ to be the string similarity of a and b which coincides with the first condition, and the second condition is encoded in the structure stability constraints. The string similarity measure of many other types of correspondences can be defined similarly based on the heuristic method in [17]. If there does not exist a straight-forward way to define the string similarity for a certain type of complex correspondences, we can simply set it to 0 and rely on other strategies to identify such correspondences.

Lastly, we need constraints for the correspondence of two complex concepts. The corresponding component concepts and same constructor always implies the corresponding complex concepts, while in the other direction, it is a soft constraint.

$$\text{cons}_k(a, b) \equiv \text{cons}_k(a', b') \Leftarrow a \equiv a' \wedge b \equiv b'$$
$$+w_k^c \qquad \text{cons}_k(a, b) \equiv \text{cons}_k(a', b') \Rightarrow a \equiv a' \wedge b \equiv b'$$

where cons_k are different constructors for complex concepts, e.g., union, $\exists p.b$.

Some complex correspondences are almost impossible to be identified with traditional strategies. With the knowledge-based strategy, it becomes possible.

Example 6. A reviewer of a paper cannot be the paper's author. In the cmt ontology we have

$$\text{writePaper} \sqsubseteq \neg\text{readPaper}$$

and in the conference ontology we have

$$\text{contributes} \mid \text{Reviewed\_contribution} \sqsubseteq \neg(\text{contributes} \circ \text{reviews})$$

We first build two complex concepts contributes ⌊ Reviewed_contribution and contributes ∘ reviews. With $R_4(a, b) = a \sqsubseteq \neg b$ (disjoint properties), the score function would favor the correspondences writePaper ≡ contributes ⌞ Reviewed_contribution

readPaper ≡ contributes ∘ reviews

6   Knowledge Aware Ontology Matching

In this section, we present our approach, Knowledge Aware Ontology Matching (KAOM). KAOM uses Markov logic networks (MLNs) to solve the ontology matching task. The MLN formulation is similar to [12] but incorporates the knowledge-based matching strategy and treatment of complex correspondences.

An MLN [4] is a set of weighted formulas in first-order logic. Given a set of constants for individuals in a domain, an MLN induces a probability distribution over Herbrand interpretations or "possible worlds". In the ontology matching problem, we represent a correspondence in first-order logic using a binary relation, $\mathtt{match}(\mathtt{a1}, \mathtt{a2})$, which is true if concept $\mathtt{a1}$ from the first ontology is semantically equivalent to concept $\mathtt{a2}$ from the second ontology (e.g., $\mathtt{match}(\mathtt{writePaper}, \mathtt{writes})$ means $\mathtt{writePaper} \equiv \mathtt{writes}$). Each possible world therefore corresponds to an alignment of the two ontologies. We want to find the most probable possible world, which is the configuration that maximizes the sum of weights of satisfied formulas.

We define three components of the MLN of the ontology matching problem: *constants*, *evidence* and *formulas*. The logical constants are the entities in both ontologies, including the simple named ones and the complex ones. The evidence includes the complete set of OWL-supported relationships (e.g., subsumptions and disjointness) among all concepts in each ontology, and rules represented as first-order atomic predicates as described in the Sect. 3. We use an OWL reasoner to create the complete set of OWL axioms.

For the formulas, we begin with a set of formulas adapted from [12]:

1. *A-priori similarity* is the string similarity between all pairs of concepts:

$$s_{a,a'} \quad \mathtt{match}(a, a')$$

   where $s_{a,a'}$ is the string similarity between a and a', which also serves as the weight of the formula. We use the Levenshtein measure [9] for simple correspondences. This atomic formula increases the probability of matching pairs of concepts with similar strings, all other things being equal.

2. *Cardinality constraints* enforce one-to-one simple (or complex) correspondences:

$$\mathtt{match}(a, a') \wedge \mathtt{match}(a, a'') \Rightarrow a' = a''$$

3. *Coherence constraints* enforce consistency of subclass relationships:

$$\mathtt{match}(a, a') \wedge \mathtt{match}(b, b') \wedge a \sqsubseteq b \Rightarrow a' \sqsubseteq \neg b'$$

4. *Stability constraints* enforce consistency of the subclass relationships between the two ontologies. They can be viewed as a special case of the knowledge-based constraints we introduce below.

Knowledge-Based Constraints. We now describe how we incorporate knowledge-based constraints into the MLN formulation through new formulas relating knowledge rules to matchings. The *stability* constraints in [12] consider three sub-class relationships, including a is a subclass of b (`subclass`), and a is a subclass or superclass of the domain or range of a property b (`domainsub, rangesub`). We extend the relationships (knowledge rule patterns) to sub-property, disjoint properties, and user-defined relations such as ordering of dates, and non-deterministic relationships such as correlation and anti-correlation:

$$-w_k \qquad R_k(a, b, ...) \wedge \neg R_k(a', b', ...) \Rightarrow \mathtt{match}(a, a') \wedge \mathtt{match}(b, b') \wedge ..., k = 1, ..., m \tag{1}$$

where m is the number of knowledge rule patterns. User-defined relations include those derived from decision trees, association rules, expert systems, and other knowledge sources outside the ontology.

Besides the stability constraints, we introduce a new group of *similarity* constraints that encourage knowledge rules with the same pattern to have corresponding concepts.

$$+w'_k \qquad R_k(a, b, ...) \wedge R_k(a', b', ...) \Rightarrow \mathtt{match}(a, a') \wedge \mathtt{match}(b, b') \wedge ..., k = 1, ..., m \tag{2}$$

For numerical rules, we instead use MLN formulas:

$$d_0 - d \quad \mathtt{match}(a, a') \wedge \mathtt{match}(b, b') \wedge ..., k = 1, ..., m \tag{3}$$

where d is a distance measure of the two rules $R_k(a, b, ...)$ and $R'_k(a', b', ...)$ and d_0 is a threshold determining whether the rules are similar or not.

To handle complex correspondences, we add complex concepts that occur in knowledge rules as constants of the MLN, and add knowledge rules that contain these new complex concepts. We define the string similarity and enforce type constraints between simple and complex concepts, as described in Sect. 5. For complex to complex correspondences, the string similarity measure is zero, but we have constraints

$$w_k^c \qquad \begin{aligned} \mathtt{match}(a, a') \wedge \mathtt{match}(b, b') \wedge ... \Rightarrow \mathtt{match}(c, c') \\ \mathtt{match}(a, a') \wedge \mathtt{match}(b, b') \wedge ... \Leftarrow \mathtt{match}(c, c') \end{aligned}$$

where $c = \mathrm{cons}_k(a, b, ...), c' = \mathrm{cons}_k(a', b', ...)$ for each constructor cons_k.

7 Experiments

We test our KAOM approach on three domains: NBA, census, and conference. The sizes of the ontologies of these domains are listed in Table 2. These domains contain very different forms of ontologies and knowledge rules, so we can examine the generality and robustness of our approach.

We use Pellet [19] for logical inference of the ontological axioms and The-Beast[3] [16] and RockIt[4] [13] for Markov logic inference. We ran all experiments

---

[3] http://code.google.com/p/thebeast/.
[4] https://code.google.com/p/rockit/. We use RockIt for the census domain because TheBeast is not able to handle the large number of rules in that domain.

Table 2. Number of classes, object properties, data properties and nominal values of each ontology used in the experiments.

Domain	Ontology	# classes	# object props	# data props	# values
NBA	nba-os	3	3	20	3
	yahoo	4	4	21	7
Census	adult	1	0	15	101
	income	1	0	12	97
OntoFarm	cmt	36	50	10	0
	confOf	38	13	25	0
	conference	60	46	18	6
	edas	103	30	20	0
	ekaw	78	33	0	0

on a machine with 24 Intel Xeon E5-2640 cores @2500 MHz and 8GB memory. We compare our system (KAOM) with three others: KAOM without the knowledge-based strategy (MLOM), CODI [7] (a new version of [12], which is essentially a different implementation of MLOM), and logmap2 [8], a top performing system in OAEI 2014[5].

We manually specify the weights of the Markov logic formlas in KAOM and MLOM. The weights of stability constraints for subclass relationships are set with values same as the ones used in [12], i.e., the weight for subclass is -0.5, and those for sub-domain and range are -0.25. In KAOM, we also set the weights for different types of similarity rules based on our assessment of their relative importance and kept these weights fixed during the experiments.

7.1   NBA

The NBA domain is a simple setting that we use to demonstrate the effectiveness of our approach. We collected data from the NBA official website and the Yahoo NBA website. For each ontology, we used the WinMine toolkit[6] to learn a decision tree for each attribute using the other attributes as inputs.

For each pair of conditional distributions based on decision tree with up to three attributes, we calculate their similarity based on the distance measure described in Example 5. We use the Markov logic formula (3) with the threshold $d_0 = 0.2$. To make the task more challenging, we did not use any name similarity measures. Our method successfully identified the correspondence of all the numerical and nominal attributes, including height, weight and positions (center, forward and guard) of players. In contrast, without a name similarity measure, no other method can solve the matching problem at all.

---

[5] http://oaei.ontologymatching.org/2014/.

[6] http://research.microsoft.com/en-us/um/people/dmax/WinMine/Tooldoc.htm.

7.2   Census

We consider two census datasets and their ontologies from UC Irvine data repository[7]. Both datasets represent census data but are sampled and post-processed differently. These two census ontologies are flat with a single concept but many datatype properties and nominal values. For this domain, we use association rules as the knowledge. We first discretize each numerical attribute into five intervals, and then generate association rules for each ontology using the Apriori algorithm with a minimum confidence of 0.9 and minimum support of 0.001. For example, one generated rule is:

```
age='(-inf-25.5]' education='11th' hours-per-week='(-inf-35.5]'
  ==> adjusted-gross-income='<=50K' conf:(1)
```

This is represented as

$$R_6(\mathtt{age}^d, \mathtt{11th}, \mathtt{hours\text{-}per\text{-}week}^d, \mathtt{adjusted\text{-}gross\text{-}income}^d)$$

where x^d refers to the discretized value of x, split into one fifth percentile intervals, and $R_6(a, b, c, d) : a = 1 \land b \land c = 1 \Rightarrow d = 1$. For scalability reasons, we consider up to three concepts in a knowledge rule, i.e., association rules with up to three attributes. We set the weight of knowledge similarity constraints for the association rules to 0.25.

In the Markov logic formulation in [12], only the correspondences with apriori similarity measure larger than a threshold τ are added as evidence. In the experiments, we set τ with different values from 0.50 to 0.90. When τ is large, we deliberately discard the string similarity information for some correspondences. MLOM for this task is an extension of [12] by adding correspondences of *nominal values* and their dependencies with the related attributes. The results are shown in Fig. 1. We can see that KAOM always gets better recall and F1, with only a slight degradation in precision. This means our approach fully leverages the knowledge rule information and thus does not rely too much on the names of the concepts to determine the matching. For example, when τ is 0.70, KAOM finds 6 out of 8 correspondences of values of `adult : workclass` and `income : class_of_worker`, while MLOM finds none. The other two systems were not designed for nominal value correspondences. CODI only finds 7 and logmap2 only finds 3 attribute correspondences, while KAOM and MLOM find all the 12 attribute correspondences.

7.3   OntoFarm

In order to show how our system can use manually created expert knowledge bases, we use OntoFarm, a standard ontology matching benchmark for an academic conference domain as the third domain in our experiments. As part of OAEI, it has been widely used in the evaluation of ontology matching systems.

---

[7] https://archive.ics.uci.edu/ml/datasets.html.

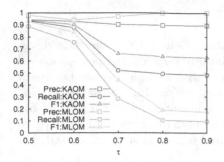
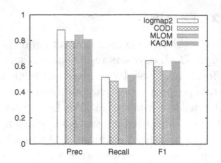

Fig. 1. Precision, recall and F1 on the census domain as a function of the string similarity threshold τ.

Fig. 2. Precision, recall and F1 on the OntoFarm domain with only the one-to-one correspondences.

The process of manually knowledge rule creation is time consuming, so we only used 5 of the OntoFarm ontologies (`cmt`, `conference`, `confOf`, `edas`, `ekaw`). Using their knowledge of computer science conferences and the structure of just one ontology, two individuals listed a number of rules (e.g., Example 1). We then translated these rules into each of the five ontologies. Thus, the same knowledge was added to each of the ontologies, but its representation depended on the specific ontology. For some ontologies, some of the rules were not representable with the concepts in them and thus had to be omitted. This manually constructed knowledge base was developed before running any experiments and kept fixed throughout our experiments. Among the 5 ontologies, we have 10 pairs of matching tasks in total. We set τ to 0.70, and the weight for the knowledge similarity constraints to 1.0.

We first compare the four methods to the reference one-to-one alignment from the benchmark (Fig. 2). KAOM achieves similar precision and F1, and better recall than other systems. It was able to identify correspondences in which the concept names are very different, for instance, `cmt : readPaper` $\equiv$ `confOf : reviews`. Note that the similarity constraints work in concert with other constraints. For instance, in Example 4, since disjointness is a symmetric knowledge rule, domain and range constraints could be helpful to identify whether `cmt : writePaper` should match to `confOf : writes` or `confOf : reviews`.

To evaluate our approach on complex correspondences, we extended the reference alignment with hand-labeled complex correspondences (Fig. 3). MLOM does not perform well in this task because the complex correspondences require a good similarity measure to become candidates (such as the linguistic features in [17]). KAOM, however, uses the structure of the rules to find many complex correspondences without relying on complex similarity measures. For this task we also tried learning the weights of the formulas[8] (KAOM-learn). For each of the 10 pairs of ontologies, we used the rest 9 pairs as training data. KAOM-learn performs slightly better than KAOM.

---

[8] We used MIRA implemented in TheBeast for weight learning.

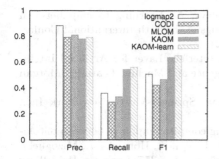
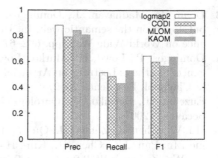

Fig. 3. Precision, recall and F1 on the OntoFarm domain with the complex correspondences.

Fig. 4. Precision-recall curve on the OntoFarm domain with the complex correspondences.

With the hand-picked or automatically learned weights, KAOM produces a single most-likely alignment. However, we can further tune KAOM to produce alignments with higher recall or higher precision. We accomplish this by adding the MLN formula $\mathtt{match}(a, a')$ with weight w. When w is positive, alignments with more matches are more likely, and when w is negative, alignments with fewer matches are more likely (all other things being equal). We adjusted this weight to produce the precision-recall curve shown in Fig. 4. KAOM dominates CODI and provides much higher recall values than logmap2, although logmap2's best precision remains slightly above KAOM's.

8   Conclusion

We proposed a new ontology matching algorithm KAOM. The key component of KAOM is the knowledge-based strategy, which is based on the intuition that ontologies about the same domain should contain similar knowledge rules, in spite of the different terminologies they use. KAOM is also capable of discovering complex correspondences, by treating complex concepts the same way as simple ones. We encode the knowledge-based strategy and other strategies in Markov logic and find the best alignment with its inference tools. Experiments on the datasets and ontologies from three different domains show that our method effectively uses knowledge rules of different forms to outperform several state-of-the-art ontology matching methods.

Acknowledgement. This research is funded by NSF grant IIS-1118050.

References

1. Cotterell, M.E., Medina, T.: A Markov model for ontology alignment. CoRR (2013)
2. Dhamankar, R., Lee, Y., Doan, A., Halevy, A., Domingos, P.: iMAP: Discovering complex semantic matches between database schemas. In: Proceedings of the 2004 ACM SIGMOD International Conference on Management of Data, pp. 383–394 (2004)

3. Doan, A., Madhavan, J., Domingos, P., Halevy, A.: Learning to map between ontologies on the semantic web. In: Proceedings of the 11th International Conference on World Wide Web, pp. 662–673 (2002)
4. Domingos, P., Lowd, D.: Markov Logic: An Interface Layer for Artificial Intelligence. Synthesis Lectures on Artificial Intelligence and Machine Learning. Morgan & Claypool, Pasadena (2009)
5. Euzenat, J., Shvaiko, P.: Ontology Matching. Springer-Verlag New York Inc., Secaucus (2007)
6. Hu, W., Chen, J., Zhang, H., Qu, Y.: Learning complex mappings between ontologies. In: Pan, J.Z., Chen, H., Kim, H.-G., Li, J., Wu, Z., Horrocks, I., Mizoguchi, R., Wu, Z. (eds.) JIST 2011. LNCS, vol. 7185, pp. 350–357. Springer, Heidelberg (2012)
7. Huber, J., Sztyler, T., Noessner, J., Meilicke, C.: CODI: combinatorial optimization for data integration-results for OAEI 2011. In: Ontology Matching, p. 134 (2011)
8. Jiménez-Ruiz, E., Grau, B.C., Zhou, Y.: LogMap 2.0: towards logic-based, scalable and interactive ontology matching. In: Proceedings of the 4th International Workshop on Semantic Web Applications and Tools for the Life Sciences, SWAT4LS 2011, pp. 45–46 (2012)
9. Levenshtein, V.: Binary codes capable of correcting deletions, insertions and reversals. Sov. Phys. Dokl. 10, 707–710 (1966)
10. Mao, M., Peng, Y., Spring, M.: An adaptive ontology mapping approach with neural network based constraint satisfaction. Web Semant. 8(1), 14–25 (2010)
11. Melnik, S., Garcia-Molina, H., Rahm, E.: Similarity flooding: a versatile graph matching algorithm. In: Proceedings of Eighteenth International Conference on Data Engineering (2002)
12. Niepert, M., Meilicke, C., Stuckenschmidt, H.: A probabilistic-logical framework for ontology matching. In: Fox, M., Poole, D. (eds.) Proceedings of the 24th AAAI Conference on Artificial Intelligence, July 2010, pp. 1413–1418 (2010)
13. Noessner, J., Niepert, M., Stuckenschmidt, H.: RockIt: exploiting parallelism and symmetry for MAP inference in statistical relational models. In: Proceedings of the Twenty-Seventh AAAI Conference on Artificial Intelligence (2013)
14. Noy, N.F., Musen, M.A.: PROMPT: algorithm and tool for automated ontology merging and alignment. In: Proceedings of the Seventeenth National Conference on Artificial Intelligence and Twelfth Conference on Innovative Applications of Artificial Intelligence, pp. 450–455 (2000)
15. Qin, H., Dou, D., LePendu, P.: Discovering executable semantic mappings between ontologies. In: Meersman, R., Tari, Z. (eds.) OTM 2007, Part I. LNCS, vol. 4803, pp. 832–849. Springer, Heidelberg (2007)
16. Riedel, S.: Improving the accuracy and efficiency of MAP inference for Markov logic. In: Proceedings of the Proceedings of the 24th Conference on Uncertainty in Artificial Intelligence (UAI-08), pp. 468–475 (2008)
17. Ritze, D., Meilicke, C., Svb-Zamazal, O., Stuckenschmidt, H.: A pattern-based ontology matching approach for detecting complex correspondences. In: Ontology Matching (OM-2009), vol. 551 (2008)
18. Shvaiko, P., Euzenat, J.: Ontology matching: State of the art and future challenges. IEEE Trans. Knowl. Data Eng. PP(99) (2011)
19. Sirin, E., Parsia, B., Grau, B.C., Kalyanpur, A., Katz, Y.: Pellet: a practical OWL-DL reasoner. Web Seman. 5(2), 51–53 (2007)

Modeling, Linked Open Data

Detection of Sequences with Anomalous Behavior in a Workflow Process

Marcelo G. Armentano[✉] and Analía A. Amandi

ISISTAN Research Institute (CONICET-UNICEN), Campus Universitario,
Paraje Arroyo Seco, 7000 Tandil, Argentina
{marcelo.armentano,analia.amandi}@isistan.unicen.edu.ar

Abstract. A workflow process consists of an organized and repeatable pattern of activities that are necessary to complete a task, within the dynamics of an organization. The automatic recognition of deviations from the expected behavior within the workflow of an organization is crucial to provide assistance to new employees to accomplish his/her tasks. In this article, we propose a two-fold approach to this problem. First, taking the process logs as an input, we automatically build a statistical model that captures regularities in the activities carried out by the employees. Second, this model is used to track the activities performed by the employees to detect deviations from the expected behavior, according to the normal workflow of the organization. An experimental evaluation with five processes logs, with different levels of noise, was conducted to determine the validity of our approach.

Keywords: Process mining · Outliers detection

1 Introduction

Every organization has a workflow dynamics that is usually not explicitly documented. All these skills and experience that is hold on the employees characterizes the "know how" of any organization and is called Organizational Memory. This workflow dynamics can be lost when employees with some specific knowledge leave the organization, a problem that is known as *corporate amnesia* [1]. For this reason it is desirable to have an explicitly representation of the workflow dynamics that can be transmitted to new employees. Automatically learning a model of the dynamics of an organization can be beneficial since the behavior of new employees can be compared with the normal behavior to detect deviations from it. In other words as a new employee performs a set of tasks, it would be desirable to non-intrusively detect, as early as possible, any deviation from expected behavior.

The increasing use of technology to align the business processes of an organization towards the same goal has made a strong trend towards process-oriented information systems, which have a whole infrastructure to support such business processes. Despite all the advantages obtained from the use of these systems,

© Springer International Publishing Switzerland 2015
Q. Chen et al. (Eds.): DEXA 2015, Part I, LNCS 9261, pp. 111–118, 2015.
DOI: 10.1007/978-3-319-22849-5_8

the continuing growth and dynamics of organizations make business processes grow in number and complexity, resulting in an increasing difficulty to support its design and its rapid adaptation to changes. To help overcome these problems, a research area known as *workflow mining* [2] has emerged. Taking data from the results of the execution of processes, which are derived from the execution logs, workflow mining is based on applying different data mining techniques to obtain additional knowledge such as building a new model of a given process in order to compare it to the original process, detecting deviations from the original process or improving the process definition itself.

In this context, the automatic recognition of the sequences of tasks that do not belong to the normal behavior according to the workflow of a give process is crucial to determine deviations from the expected behavior that might lead to an ineffective operation of the enterprise. An early recognition of this kind of outlier sequences can prevent deviations of the employees behavior from the expected behavior by providing personalized assistance. The main difference between the process mining approach and the approach proposed in this article is that we do not focus on obtaining an explicit design of the underlying workflow. Instead, we seek to obtain a model of the tasks involved in the underlying workflow in order to detect behaviors that do not fit the expected flow of activities. With this information available, a system will be able to provide personalized assistance in the execution of those tasks.

This article is organized as follows. Section 2 presents some background and related work on workflow mining and outlier detection in workflows. Section 3 presents the proposed approach to model workflow processes from activities logs and to detect sequences with abnormal behavior. Section 4 describes the experiments performed to validate our approach. Finally, in Sect. 5 we present our conclusions.

2   Related Work

Cook and Wolf investigated the mining of processes in the context of software engineering processes. In [3] three methods for process discovery are described: one using neural networks, one using an algorithmic approach and the third using a Markovian approach, concluding that the latter two methods are more effective. Wen et al [4] presented an algorithm based on two types of events that indicate the beginning and completion of tasks. Together with the causality information obtained from the activities log, they derive relationships between tasks which are then used to create a Petri net modeling the underlying process. Two disadvantages of this approach are that it is not probabilistic and that is not robust to the presence of noise in the training data. Two disadvantages of this approach are that it is not probabilistic and that it is not robust to the presence of noise in the training data.

Regarding the detection of outliers in workflow logs, Ghionna et al. [5] presented an algorithm that discovers a set of outliers, based on the computation of behavioral patterns over process logs and a clustering approach. The basic idea is to associate pattern clusters with trace clusters, and to detect as outliers those traces that do not associate to any pattern cluster or that belong to clusters whose size is smaller than the average cluster size. The research presented in [6] proposes an algorithm which makes use of the workflow's executed frequency, the concept of distance-based outlier detection, empirical rules and Method of Exhaustion to mine three types of workflow outliers, including less-occurring workflow outliers of each process, less-occurring workflow outliers of all processes and never-occurring workflow outliers. More recently, Bouarfa and Dankelman [7] derive a workflow consensus without prior knowledge using logs extracted from a clinical environment by aligning multiple sequences. This model is used to detect outliers during surgery, that is, deviations from the consensus.

In the following Section, we present our approach to obtaining a probabilistic model from execution logs. This model is then used to detect sequences with abnormal behavior.

3   Proposed Approach

3.1   Learning a Workflow Model from Execution Logs

Variable Order Markov (VOM) models arose as an alternative to fixed order Markov models to capture longer regularities while avoiding the size explosion caused by increasing the order of the model. In contrast to the Markov chain models, where each random variable in a sequence with a Markov property depends on a fixed number of random variables, in VOM models this number of conditioning random variables may vary based on the specific observed realization, known as *context*. These models consider that in realistic settings, there are certain realizations of states (represented by contexts) in which some past states are independent from the future states conducting to a great reduction in the number of model parameters.

Algorithms for learning VOM models over a finite alphabet Σ attempt to learn a subclass of Probabilistic Finite-state Automata (PFA) called Probabilistic Suffix Automata (PSA) which can model sequential data of considerable complexity. Formally, a PSA can be described as a 5-tuple $(Q, \Sigma, \tau, \gamma, \pi)$, where Q is a finite set of states, Σ is the task universe, $\tau : Q \times \Sigma \rightarrow Q$ is the transition function, $\gamma : Q \times \Sigma \rightarrow [0, 1]$ is the next task probability function, where for each $q \in Q$, $\sum_{\sigma \in \Sigma} \gamma(q, \sigma) = 1$, $\pi : Q \rightarrow [0, 1]$ is the initial probability distribution over the starting states, with $\sum_{\sigma \in \Sigma} \pi(q) = 1$.

A PFA is a PSA if the following property holds. Each state in a PSA M is labeled by a sequence of tasks with finite length in Σ^* and the set of sequences S labeling the states is suffix free. Σ is the domain task universe, that is the finite set of tasks that the employee can perform in the domain. A set of sequences S is said to be suffix free if $\forall s \in S$, $Suffix^*(s) \cap S = \{s\}$, where $Suffix^*(s) = \{s_i, \cdots, s_l | 1 \leq i \leq l\}$ is the set of all possible suffixes of s, including the empty sequence e. For every two states q_1 and $q_2 \in Q$ and for every task $\sigma \in \Sigma$, if $\tau(q_1, \sigma) = q_2$ and q_1 is labeled by a sequence s_1, then q_2 is labeled by a sequence s_2 that is a suffix of $s_1 \cdot \sigma$.

In contrast to m-order Markov models, which attempt to estimate conditional distributions of the form $Pr(\sigma|s)$, with $s \in \Sigma^N$ and $\sigma \in \Sigma$, VOM algorithms learn such conditional distributions where context lengths $|s|$ vary in response to the available statistics in the training data. Thus, PSA models provide the means for capturing both large and small order Markov dependencies based on the observed data. In [8] an algorithm for learning such models in an incremental way is proposed.

Learning a workflow model from activities logs has the main advantage that we do not need any additional information about the domain being modeled more than the tasks that can be performed in the domain. We will be able to learn regularities in the employees' behavior just by analyzing the trace examples observed in the logs. In the following section we describe how we use PSA's models to detect sequences of abnormal behavior in execution logs.

3.2   Detecting Sequences with Abnormal Behavior

In order to make the recognition process robust to the execution of noisy tasks, that are tasks that might not correspond to abnormal behavior, we use an *exponential moving average* on the prediction probability $\gamma(s, \sigma)$ at each step in the PSA. An exponential moving average (EMA) [9] is a statistic for monitoring a process that averages the data using weights that decrease as time passes. The weighting for each step decreases exponentially, giving much more importance to recent observations while still not discarding older observations entirely. By the choice of a weighting factor $0 \leq \lambda \leq 1$, the EMA control procedure can be made sensitive to a small or gradual drift in the process. Alternatively, λ may be expressed in terms of N time periods, where $\lambda = \frac{2}{N+1}$.

EMA_t expresses the value of the EMA at any time period t. EMA_1 is set to the a priori probability of the first observed task σ. EMA_t at time periods $t \geq 2$ is computed as $EMA_t = \lambda\gamma_{PSA_i}(s, \sigma) + (1-\lambda)EMA_{t-1}$. The parameter λ determines the rate at which *older* probabilities enter into the calculation of the EMA statistic. A value of $\lambda = 1$ implies that only the most recent measurement influences the EMA. Thus, a large value of λ gives more weight to recent probabilities and less weight to older probabilities; a small value of λ gives more weight to older probabilities. The value of λ is usually set between 0.2 and 0.3 [9] although this choice is somewhat arbitrary and should be determined experimentally.

In our approach, we consider that a given sequence corresponds to an abnormal behavior if at least one activity in the log sequence is predicted by the PSA model with a probability value (after applying the smoothing technique) lower than a certain threshold. This is a very strong assumption that favors the detection of outliers over normal sequences.

Table 1. Statistics for the processes used in the experiment

	Process 1	Process 2	Process 3	Process 4	Process 5
Number of AND patterns	0	2	2	1	1
Number of XOR patterns	3	1	1	2	1
Number of loops	4	2	0	1	1
Maximum number of AND branches	0	3	2	3	2
Maximum number of XOR branches	3	3	2	3	2
Total number of activities	18	18	12	18	11

Table 2. Statistics for the testing logs generated for the experiment

Sequence length	Process 1	Process 2	Process 3	Process 4	Process 5
Min.	7	14	11	6	6
Max.	293	66	11	92	33
Mean	43.78	23.16	11	26.18	10.36
Median	32	21	11	23	8
Standard deviation	37.28	8.79	0	14.98	5.34

4   Experimental Evaluation

Process logs corresponding to real-world business process are hardly available. Companies owning real process logs are often reluctant to make public their data and release only partial log files. For these reasons, we have tested our technique with artificially generated process logs, as in [10,11]. The main benefit of using simulation is that properties such as noise can be controlled.

Different tools exist for generating artificial event logs. From the evaluation presented in [12], we selected PLG [13], a framework that enables the generation of random business processes according to some specific user-defined parameters. We defined five different workflows using PLG. Table 1 shows some statistics for the five process used in the experiment.

Next, for each workflow, we use PLG to generate a maximum of 500 traces and filtered out duplicate traces. Finally, we trained five different VOM models to be used in the experiments.

We evaluated the effectiveness of our approach with various input logs, containing different percentage of noise in the sequences. Using PLG, we generated three different logs for each process. Each log contained a maximum of 500 execution traces introducing 5 %, 10 % and 20 % of noise in each sequence, representing abnormal behavior according to the workflow process. Each log was then completed with a maximum of 500 execution traces with no noise, representing normal behavior sequences[1]. Table 2 shows some statistics of the resulting logs.

---

[1] The resulting dataset is available online at: http://marcelo.armentano.isistan.unicen. edu.ar/datasets.

4.1 Selection of the Smoothing Constant λ

First, we tested different values for the smoothing constant, ranging from 0.1 to 1.0 with intervals of 0.1 for the five processes. In order to select the best λ value for each process, we computed the Matthews correlation coefficient (MCC). The MCC is in essence a correlation coefficient between the observed and predicted binary classifications and ranges between -1 and $+1$, where a coefficient of $+1$ represents a perfect prediction, 0 no better than random prediction and -1 indicates total disagreement between prediction and observation.

For processes 1, 3 and 4 we obtained better MMC with a value of $\lambda = 0.3$. For process 2, on the other hand, better results were obtained with a value of $\lambda = 0.1$, while for process 5 better results were obtained with a value of $\lambda = 0.5$. Notice that process 2 is the most complex workflow used in the experiment. For this reason, a lower smoothing constant works better for this model, since we need to consider a longer history to better predict the probability of the next performed action. Process 5, on the other hand was the most simple worflow, with few alternative paths and containing only a simple loop. The majority of variations in the log for this model corresponded to cycling in the loop in one of the alternative branches, so it was very easy to "compress" the sequences in the model and therefore only considering the last three observed actions was enough to predict the probability of the next performed action.

4.2 Recognition of Abnormal Behavior Sequences

Next, we performed the classification of each sequence in each of the three logs (with 5 %, 10 % and 20 % of noise). Figure 1 shows the results obtained for precision (a), recall (b), false positive rate (c), and false negative rate (d).

As expected, we can observe that the introduction of more noise in the sequences enables a better distinction of outliers from normal sequences. For processes 3, 4 and 5, we obtained optimal precision, that is, all the outliers detected were certainly abnormal sequences, and no normal sequences were deemed as outliers. For process 1, on the other hand, precision was over 90 % for all noise levels, while for process 2, precision varied from 66.4 % for the log containing sequences with 5 % noise to 71.9 % for the log containing sequences with 20 % noise. In the case of process 2, 38.3 % of the sequences recognized as abnormal corresponded in fact normal sequences.

Regarding recall, in the dataset with 5 % of noise, our approach detected 80.2 % of the outliers for process 1, 75 % for process 2, 85.3 % for process 3, 74.4 % for process 4 and 93.8 % for process 5. These values are increased to 98.5 %, 97.2 %, 95,4 %, 96.6 % and 98.2 % respectively in the dataset containing 20 % of noise.

The false negative rate refers to the percentage of outliers that were not detected by our approach. In the worst case, we failed to detect 25.6 % of the abnormal sequences for process 4 in the dataset containing 5 % of noise. When we increase the percentage of noise introduced to the sequences in the processes logs, we miss only 4.6 % of abnormal sequences in the worst case.

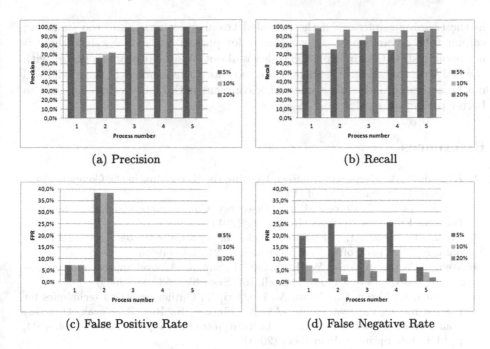

(a) Precision   (b) Recall

(c) False Positive Rate   (d) False Negative Rate

Fig. 1. Outlier detection performance

5   Conclusions

In this article, we addressed the problem of automatically building a statistical model from different processes logs generated within the workflow of an organization. This model is then used to detect sequences of tasks that do not correspond to the expected behavior according to the workflow of the organization. This can be useful to detect employees that do not follow the normal workflow in order to assist them with the completion of the expected tasks.

We evaluated the proposed approach in a simulated scenario with five different processes workflows, when introducing different amount of noise in the sequences of the workflow log. We conclude that our approach is useful to detect abnormal behavior in the tasks performed, with an average precision of 91.8 % when the executed sequences contain 5 % of noise, 92.7 % for 10 % of noise, and 93.4 % for 20 % of noise. Regarding recall, our approach was able to detect on average 81.8 % of the sequences containing abnormal behavior with 5 % of noise, 90.3 % of the sequences with 10 % of noise, and 97.2 % of the sequences with 20 % of noise.

One limitation of our approach is that expressiveness of the model built is reduced since we are not reconstructing the underlying workflow from the logs. Nevertheless the graphical representation of the VOM model in the form of a PSA can give us an idea of contexts of tasks that determine the different transitions in the model. VOM models are also sensitive to the number of actions in the training sequences: if the provided training sequences are too short, we cannot take advantage of the variable order part of our models, since there is not enough statistical information to learn about. A similar problem occurs with the EMA computation. If sequences are too short, the memory of the model will not be activated and using a smoothing constant with higher values can lead to better prediction results.

References

1. Kransdorff, A.: Corporate Amnesia: Keeping the Know-How in the Company. Butterworth Heinemann, Oxford (1998)
2. van der Aalst, W.: Process Mining: Discovery, Conformance and Enhancement of Business Processes. Springer, Heidelberg (2011)
3. Cook, J.E., Wolf, A.L.: Discovering models of software processes from event-based data. ACM Trans. Softw. Eng. Methodol. **7**, 215–249 (1998)
4. Wen, L., Wang, J., Aalst, W.M., Huang, B., Sun, J.: A novel approach for process mining based on event types. J. Intell. Inf. Syst. **32**, 163–190 (2009)
5. Ghionna, L., Greco, G., Guzzo, A., Pontieri, L.: Outlier detection techniques for process mining applications. In: An, A., Matwin, S., Ras, Z., Slezak, D. (eds.) Foundations of Intelligent Systems. Lecture Notes in Computer Science, vol. 4994, pp. 150–159. Springer, Heidelberg (2008)
6. Chuang, Y.-C., Hsu, P.Y., Wang, M.T., Chen, S.-C.: A Frequency-Based Algorithm for Workflow Outlier Mining. In: Kim, T.-H., Lee, Y.-H., Kang, B.-H., Slezak, D. (eds.) FGIT 2010. LNCS, vol. 6485, pp. 191–207. Springer, Heidelberg (2010)
7. Bouarfa, L., Dankelman, J.: Workflow mining and outlier detection from clinical activity logs. J. Biomed. Inform. **45**, 1185–1190 (2012)
8. Armentano, M., Amandi, A.: Modeling sequences of user actions for statistical goal recognition. User Model. User-Adap. Interact. **22**, 281–311 (2012)
9. Hunter, J.S.: The exponentially weighted moving average. J. Qual. Technol. **18**, 203–209 (1986)
10. Weijters, A.J.M.M., van der Aalst, W.M.P.: Rediscovering workflow models from event-based data using little thumb. Integr. Comput.-Aided Eng. **10**, 151–162 (2003)
11. Claes, J., Poels, G.: Merging computer log files for process mining: an artificial immune system technique. In: Daniel, F., Barkaoui, K., Dustdar, S. (eds.) BPM Workshops 2011, Part I. LNBIP, vol. 99, pp. 99–110. Springer, Heidelberg (2012)
12. Toon Jouck, B.D.: Generating artificial event logs to compare process discovery techniques. In: Proceedings of the 4th International Symposium on Data-driven Process Discovery and Analysis (SIMPDA 2014), vol. 1293. CEUR Workshop Proceedings, Milan, Italy (2014)
13. Burattin, A., Sperduti, A.: PLG: a framework for the generation of business process models and their execution logs. In: Muehlen, M., Su, J. (eds.) BPM 2010 Workshops. LNBIP, vol. 66, pp. 214–219. Springer, Heidelberg (2011)

An Energy Model for Detecting Community in PPI Networks

Yin Pang[1], Lin Bai[2](✉), and Kaili Bu[3]

[1] Beijing Institute of Tracking and Telecommunication Technology, Beijing, China
myown-1227@163.com
[2] School of Computer, Electronics and Information, Guangxi University,
Nanning, China
bailinbit@163.com
[3] School of Computer, Nanjing Institute of Technology, Nanjing, China
bkl_1021@163.com

Abstract. Community detection problem has been studied for years, but no agreement on the common definition of community has been reached till now. One topology is found to be of different types, unipartite or bipartite/multipartite. The previous literatures mostly proposed only one type of community. As the definition of community is understood differently, the grouping results are always applicable to some specified networks. If the definition changes, the grouping result will no longer be "good". In the present paper, it's found that some vertices "must be" in the same community, while some maybe in or not in the same community. To do the community detection on mixed protein-protein interaction (PPI) networks, an energy model with two steps is proposed. First, vertices that "must be" in the same community are grouped together by properties; second, overlapping vertices are found by functions. In the energy model, "positive energy" is defined by attraction generated between two vertices, and "negative energy" by the attraction, which weakens the attraction between the corresponding two vertices, resulting from other vertices. Energy between two vertices is the sum of their positive and negative energy. Computing the energy of each community, the community structure can be found when maximum value of the energy sum of these communities is obtained. The model is utilized to find community structure in PPI networks. The results show that the energy model is applicable to unipartite, bipartite or mixed PPI networks. Vertices with similar property/roles in the same community and overlapping vertices are found for the network.

1 Introduction

Protein-protein interactions (PPIs) play an important role in biological regulatory pathways and metabolic processes. A complete and reliable protein interaction map can provide us with an opportunity to understand the basic biological processes within a cell. Revealing community or modular structures in biological network can help us see how cells function [2]. For example, PPI networks

© Springer International Publishing Switzerland 2015
Q. Chen et al. (Eds.): DEXA 2015, Part I, LNCS 9261, pp. 119–127, 2015.
DOI: 10.1007/978-3-319-22849-5_9

were utilized for predicting gene functions, functional pathways and protein complexes [3–6]. It has been proved that PPI networks are mixed [7] (unipartite and bipartite/multipartite) and overlapped [8]. Some vertices in PPI networks must be in the same community, so communities with vertices which must be in the same community are found especially useful. Many existed methods regarded their methods as better ones compared with other methods in finding communities from different viewpoints. However, community detection seems to be a chicken-and-egg problem [1], as it is aimed at finding the actual communities which we do not know; as the real communities are unknown, it's hard to estimate the quality of a community detection method proposed. Depending on the properties of the network and the quality of measurement, one algorithm adopted may be "better" than others. In other words, algorithms, which were claimed to have good community results, were based on different understandings of community. No agreement on the common definition of community has been reached till now. As it's difficult to define community, the current algorithms are generally applicable to specified networks. But if the definition of community changes, the community grouping results will no longer be "good".

In a PPI network, vertices are connected if interaction exists between them. The fact that there are interactions between these vertices does not mean that they are similar, but certain functions do exist between them. Take a chemical action as an example, Na and Cl have an interaction to become $NaCl$, which does not indicate that Na and Cl have similar property, but Na and Cl with an interaction has a certain function.

Some vertices have properties of more than one community. In this paper, vertices are grouped into communities with one or more same kinds of properties in the first step. Then, the vertex is add into the community with the vertex's property as an overlapping vertex and into the community to get a certain function in the second step. We define the two steps as "finding vertices must be in the same community" and "finding overlapping vertices". Thus, we propose an energy model, including the above two steps, to find communities in PPI networks.

Different from methods mentioned above, we propose an energy model to measure the relationship between any two vertices to find community in this paper. The energy of each vertex pair is computed, considering attraction not only between these two vertices but also from other vertices in the whole network. The energy model is as follows: the network is considered as an undirected graph with n vertices and m edges. For each vertex pair, an attraction is supposed to exist between them. The attraction between these two vertices generates positive energy. While attraction from other vertices will weaken the attraction between these two vertices, namely generating negative energy. Vertices with larger energy tend to be in the same community.

In the current methods, the network type, unipartite or bipartite, needs to be known in advance to find its community structure. However, in many cases, researchers have no pre-knowledge of the network type. For example, when we get a network which consists of interpersonal relationships at schools, the type of network may not be sure. If the links are between students, the network will be a unipartite network; while if there are links between teachers and students, the network will be a bipartite one. Our energy model can detect communities in a mixed network, where some vertices are organized as unipartite parts while some vertices as bipartite parts. Mixed networks exist extensively in PPI network. Based on the performance evaluation, we know that the energy model can group vertices into communities with practical meaning.

Our approach proposes an energy model as a theoretical basis, followed by two steps: first, "grouping vertices must be in the same community by property"; second, "finding overlapping vertices by functions". The total energy of the network, which is the sum of energy inside and outside communities, is constant, based on the principle of the conservation of energy. Energy here is viewed as a property of vertex. Vertices inside the same community should have more energy than those among communities.

2   An Energy Model

2.1   Energy Between Vertices

There are two kinds of energy in the energy model: "positive energy" generated by the attraction between two vertices; "negative energy" produced by the attraction from other vertices, which weakens the attraction between these two vertices. Vertex pair with larger energy tends to be in the same community, so the community structure is found when energy inside communities is maximized and outside is minimized. The energy model is independent of the type of network, thus is suitable for PPI network, mixed by unipartite and bipartite parts.

The network can be viewed as an undirected graph. In the graph, we define the adjacency matrix A, a symmetric matrix with element A_{ij}. If there is an edge joining vertices i and j, $A_{ij} = 1$; otherwise, $A_{ij} = 0$. In energy model, a_i is the label of vertex i, denoting the ith vector of A. Assuming k_i the degree of vertex i,

$$k_i = |a_i|^2, K = \sum_{i=1}^{n} a_i = (k_1, k_2, ..., k_n)$$

where $|a_i|$ is the module of a_i.

The positive energy is defined as pe_{ij}, generated by the direct attraction from vertex j to vertices i as

$$pe_{ij} = a_i \cdot a_j = |a_i||a_j| \cos \theta = \sqrt{k_i} \sqrt{k_j} \cos \theta \tag{1}$$

$\theta \in [0, \pi]$ is the angle between vectors a_i and a_j. pe_{ij} decreases with the increasing of θ. The total positive energy of the network is

$$Tpe = \sum_{i,j=1}^{n} pe_{ij} = \sum_{i,j=1}^{n} a_i \cdot a_j = K \cdot K \tag{2}$$

As a result, the probability of the positive energy between vertices i and j is

$$pep_{ij} = \frac{pe_{ij}}{Tpe} \tag{3}$$

When vertices i and j attract each other, they are also attracted by other vertices in the graph at the same time. The attraction on i and j from other vertices in the graph will weaken the attraction and reduce the positive energy between vertices i and j. Relative to the positive energy from vertex j to i, a vertex k in the graph, which attracts vertex i, produces negative energy. The negative energy of vertex i is

$$ne_i = -\sum_{k=1}^{n} pe_{ik} = -\sum_{k=1}^{n} a_i \cdot a_k = -a_i \cdot K \tag{4}$$

The total negative energy is

$$Tne = \sum_{i=1}^{n} ne_i. \tag{5}$$

As the total negative energy of j is ne_j, the probability of the negative energy of other vertices applied by vertex j is

$$nep_j = \frac{ne_j}{Tne}.$$

ne_{ij} is defined as the negative energy of vertex i applied by vertex j through all vertices of the network.

$$ne_{ij} = ne_i nep_j = a_i \cdot K \frac{a_j \cdot K}{K \cdot K} \tag{6}$$

Obviously, for two random vertices i and j, $ne_{ij} = ne_{ji}$. The energy E_{ij} between vertices i and j is a sum of positive and negative energy, E_{ij} is normalized as

$$E_{ij} = \frac{1}{K \cdot K}(pe_{ij} + ne_{ij}) = [\frac{a_i \cdot a_j}{K \cdot K} - (\frac{a_i \cdot K}{K \cdot K})(\frac{a_j \cdot K}{K \cdot K})] \tag{7}$$

The energy model is as follows,

$$\begin{aligned} Q &= \sum_{k=1}^{c} \sum_{i,j \in c_k} E_{ij} \\ &= \sum_{k=1}^{c} \sum_{i,j \in c_k} [\frac{a_i \cdot a_j}{K \cdot K} - (\frac{a_i \cdot K}{K \cdot K})(\frac{a_j \cdot K}{K \cdot K})] \\ &= \sum_{k=1}^{c} \frac{d_k \cdot d_k}{K \cdot K} - (\frac{d_k \cdot K}{K \cdot K})^2 \end{aligned} \tag{8}$$

where Q is the energy within a community, $d_k = \sum_{i \in c_k} a_i$, c is the number of communities, k is the kth community. When Q is maximized, the community structure is obtained. Equation (8) is a function that divides the network into groups, where larger values indicate stronger community structure. The goal to find the largest energy inside communities is changed to find $Max(Q)$.

In the graph, E is the energy matrix. It should be noticed that the total energy of the network is $E = \sum_{i,j} E_{ij} = 0$, indicating that, in a graph the, the positive energy is equal to the negative energy. In a real complex network, vertices and edges change diversely. According to the principle of energy conservation, energy cannot be created or destroyed, but only changed or transferred. Therefore, in the graph, E is constant regardless of external variation. The energy of vertices within the same community is larger while between communities is smaller. The energy sum of vertices in the same community is non-negative.

Supposing $X = (x_1, x_2, ..., x_n)^T$ is a random vertex, it can be proved that $X^T E X \geq 0$, thus the energy matrix E is a positive semi-definite matrix. So far, it's known that E is positive semi-definite, the sum of elements in any row is 0. As a result, E, only considered as a singularity matrix, is the same as a Laplace matrix. The energy matrix E is considered as a Laplace matrix because of its good qualities, and it is widely used in clustering problem. Correspondingly, our task changes to a spectral clustering problem.

2.2   Grouping Vertices into Community

Group Vertices Which Must Be in the Same Community. An index s with n elements is defined. $s_{ik} = 1$ whe+n vertex i is in community k; otherwise, $s_{ik} = 0$, s_k is the kth vector of S, where $S^T S = n$ and each vertex is in one community. Q can be written as,

$$Q = \sum_{k=1}^{c} \sum_{i,j=1}^{n} s_{ik} s_{jk} E_{ij} = Tr(S^T E S)$$

where, matrix P is the eigenvector matrix of E, $D_{ii} = \lambda_i$, a diagonal matrix, is the ith eigenvalue of E, and p_i is the eigenvector of λ_i. therefore,

$$Q = Tr(S^T P^T D P S) = \sum_{j=1}^{n} \sum_{k=1}^{c} (\sqrt{\lambda_i} p_j s_k)^2. \tag{9}$$

Let $y_{ij} = \sqrt{\lambda_i} p_j s_k$, y_i is the row vector of matrix Y,

$$Q = \sum_{k=1}^{c} |\sum_{i \in c_k} y_i|^2. \tag{10}$$

Let $Q_k = |\sum_{i \in c_k} y_i|^2$.
   Let $Y_k = \sum_{i \in c_k} y_i$, and then Q is as follows:

$$Q = \sum_{k=1}^{c} |Y_k|^2 \tag{11}$$

Move vertex i from community l to community k, and then

$$\Delta Q = |Y_k + y_i|^2 - |Y_l - y_i|^2. \tag{12}$$

Find Overlapping Vertices. If the energy inside communities does not change, the community structure is stable. For a vertex $i \in c_l$, if $s_{ik}s_{jk} = 0, j \in c_k$, add vertex i to c_k without deleting it from c_l, thus the gain of Q is

$$\Delta Q_{ov} = (|Y_k + y_i|^2 - |Y_k|^2)$$

when $\Delta Q_{ov} \to 0$, vertex i is an overlapping vertex of c_k and c_l.

$$Max(Q_{ov}) = Max(Q) + \Delta Q_{ov} \to Max(Q).$$

Take Fig. 2 as an example,

$$\begin{aligned}
\Delta Q_{ov} &= |y_1 + y_2 + y_3 + y_4|^2 - |y_1 + y_2 + y_3|^2 \\
&+ |y_4 + y_5 + y_6 + y_7|^2 - |y_5 + y_6 + y_7|^2 \\
&= 0
\end{aligned} \tag{13}$$

3 Performance Evaluation

The community structure of some classic network is known, like Karate club network [11], Southern women network [9], Dolphin social network [10], Scotland network [12], etc., most of which are social networks. These networks have obvious characteristic of network type, unipartite or bipartite. A protein network is a mixed one. Most literatures, suitable for networks mentioned in their studies, are not suitable for a protein network. As the real community structure of protein network is unknown, the normalized mutual information is no longer useful, although former literatures utilized the normalized mutual information to compare communities.

In this paper, similarity index is utilized to evaluate the performance of methods. Communities are vertices with similar properties/roles, indicating that vertices in the same community are similar. Two famous similarity index [13], common neighbors and LHN index [14], is adopted to do the evaluation. $P - Score$ is defined to evaluate the communities as follows, $P - Score = \frac{\sum_{k=1}^{c} \sum_{i,j \in c_k} Sim_{ij}}{c \sum_{i,j=1}^{n} Sim_{ij}}$, where $Sim_{ij} = \Gamma_i \bigcap \Gamma_j$ in Common neighbor index, and $Sim_{ij} = \Gamma_i \bigcap \Gamma_j / (k_i k_j)$ in LHN index, Γ_i is the number of neighbors of vertex i. To test the effectiveness of the energy model on detecting protein communities, it is compare with BGLL [15], GCE [16] based on the classic method CFind [17], and Link community detection (called Link for short in the following sections) [1]. BGLL is the fastest method based on Modularity model for finding non-overlapping communities; CFind is the most famous method to detect communities by finding "cliques" of the network, while GCE is faster and more accurate than CFind; Link, grouping communities by links, is one of the most effective method for community detection.

Figure 1(a) is a part of the Uetz PPI network. For each protein, we first find proteins with similar roles. At the first step, there are no edges between vertices with the same color. So this network is considered as a bipartite network this step. Purple vertices in the purple dotted circle are equivalent with each other. However, we cannot decide whether these vertices are similar to the pink vertex

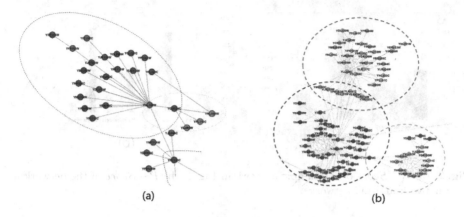

(a)   (b)

Fig. 1. (a) A part of Uetz PPI network proposed by P. Uetz, L. Giot, G. Cagney, T.A. Mansfield, et al. (b) A part of LC PPI network proposed by T. Reguly, A. Breitkreutz, L. Boucher, et al. [18]. Vertices in the same color are vertices "must be" in the same communities; dotted circles group vertices into overlapping communities (Color figure online).

"YNL189W", so they are grouped into different communities at this step, like the case in Fig. 1(b). At the second step, overlapping vertices are found out. The energy of network does not change, when the pink vertex "YNL189W" is added to the orange community as an overlapping vertex. Similarly, "YNL189W" is grouped into the purple community, and the purple vertex "YGR024C" is grouped into the orange community at step 2. "YKL130C", connected with other parts of the PPI network, is connected with other vertices, and it is overlapped with another community not shown here.

In Fig. 1(b), the network is grouped into three big communities (purple, green and light blue) and two small communities (orange and blue) at the first step. Each big community has a unipartite part with some one degree leafs. Take the purlple community as an example, vertices of the unipartite part are equivalent to each other, which should be grouped into the same community. Leafs in the purple community, connected to the unipartite part, have large energy with all the vertices in the unipartite part, like the case in Fig. 1(b). However, this time it's sure that the leafs have the similar property/role with the vertices they are connected to, because they have large energy with all the vertices of unipartite part. At the second step, some green vertices are grouped into the purple community, for they play similar roles as the leafs in purple vertices. Moreover, orange vertices are grouped into light blue community and blue vertices are grouped into purple community at the same time. It shows that step 2 not only finds overlapping vertices, but also unites small communities with large communities. The comparison among the energy model, BGLL, GCE and Link of the sample network in Fig. 2, and the Uetz network with 789 proteins and 644 interactions, and the LC network with 1528 proteins and 2835 interactions are shown in Fig. 2 respectively.

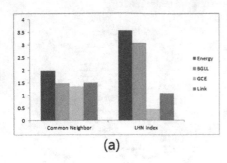
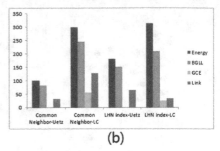

(a)   (b)

Fig. 2. (a) $P - Score$ of the sample network in Fig. 1. (b) $P - Score$ of the network in Uetz network and LC network.

As shown in Fig. 2(a) and (b), the energy model gets the best result in the sample nework, the Uetz network and the LC network. It should be noted that GCE in Uetz network does not find community structures, because there is no "clique" in the network. A "clique" is a complete subgraph of the network. If we regard two proteins with an interaction as a "clique", the GCE model becomes a modularity model. If we set a "clique" with more than two vertices, Uetz is too sparse to have "cliques" in it.

In summary, modularity can not find overlapping proteins or recognize bipartite parts in PPI networks, GCE can not find communities when the network is sparse and no "cliques" is in it. Although Link method has proved that it can obtain good result concerning networks with large number of overlapping vertices, it dose not perform pretty well in a sparse and mixed PPI network. The energy model performs well in finding community structure in mixed PPI networks, it can recognize unipartite parts and bipartite parts of the network, and find the overlapping proteins accurately at the same time.

References

1. Ahn, Y.Y., Bagrow, J.P., Lehmann, S.: Link communities reveal multiscale complexity in networks. Nature **466**(7307), 761–764 (2010)
2. Przulj, N.: Protein-protein interactions: making sense of networks via graph theoretic modeling. BioEssays **33**(2), 115–123 (2011)
3. Lee, K., Chuang, H., Beyer, A., et al.: Protein networks markedly improve prediction of subcellular localization in multiple eukaryotic species. Nucl. Acids Res. **36**, e136 (2008)
4. Ulitsky, I., Shamir, R.: Identifying functional modules using expression profiles and confidence-scored protein interactions. Bioinformatics **25**, 1158–1164 (2009)
5. Wang, C., Ding, C., Yang, Q., et al.: Consistent dissection of the protein interaction network by combining global and local metrics. Genome Biol. **8**, R271 (2007)
6. Wang, J., Du, Z., Payattakool, R., et al.: Recent advances in clustering methods for protein interaction networks. BMC Genomics **11**(Suppl 3), S10 (2010)
7. Spirin, V., Mirny, L.A.: Protein complexes and functional modules in molecular networks. Proc. Nat. Acad. Sci. U.S.A. **100**(21), 12123–12128 (2003)

8. Gillis, J., Pavlidis, P.: The impact of multifunctional genes on "Guilt by Association" analysis. PLoS One **6**(2), 16 (2011)
9. Davis, B.B., Garder, M.R.: Deep South. University of Chicago Press, USA (1941)
10. Lusseau, D., Schneider, K., Boisseau, O.J., Haase, P., Slooten, E., Dawson, S.M.: The bottlenose dolphin community of doubtful sound features a large proportion of long-lasting associations - can geographic isolation explain this unique trait. Behav. Ecol. Sociobiol. **54**(4), 396–405 (2003)
11. Zachary, W.: An information flow model for conflict and fission in small groups. J. Anthropol. Res. **33**, 452–473 (1977)
12. Scott, J., Hughes, M.: The Anatomy of Scottish Capital: Scottish Companies and Scottish Capital. Croom Helm, London (1980)
13. Zhou, T., Lu, L., Zhang, Y.C.: Predicting missing links via local information. Eur. Phys. J. B **71**, 623–630 (2009)
14. Leicht, E.A., Holme, P., Newman, M.E.J.: Vertex similarity in networks. Eur. Phys. J. **E73**, 026120 (2006)
15. Blondel, V.D., Guillaume, J.L., Lambiotte, R., Lefebvre, E.: Fast unfolding of communities in large networks. J. Stat. Mech.: Theory Exp. **2008**, P10008 (2008). Email alert: RSS feed
16. Lee, C., Reid, F., McDaid, A., Hurley, N.: Detecting highly overlapping community structure by greedy clique expansion. In: SNA-KDD 2010, pp. 33–42, February 2010
17. Palla, G., Dernyi, I., Farkas, I., Vicsek, T.: Uncovering the overlapping community structure of complex networks in nature and society. Nature **435**(9), 814–818 (2005)
18. Reguly, T., Breitkreutz, A., Boucher, L., et al.: Comprehensive curation and analysis of global interaction networks in Saccharomyces cerevisiae. J. Biol. **5**(4), 11 (2006)

Filtering Inaccurate Entity Co-references
on the Linked Open Data

John Cuzzola[1(✉)], Ebrahim Bagheri[1], and Jelena Jovanovic[2]

[1] Ryerson University, Toronto, ON, Canada
{jcuzzola, bagheri}@ryerson.ca
[2] University of Belgrade, Belgrade, Serbia
jeljov@gmail.com

Abstract. The Linked Open Data (LOD) initiative relies heavily on the inter-connections between different open RDF datasets where RDF links are used to connect resources. There has already been substantial research on identifying identity links between resources from different datasets, a process that is often referred to as co-reference resolution. These techniques often rely on probabilistic models or inference mechanisms to detect identity relations. However, recent studies have shown considerable inaccuracies in the LOD datasets that pertain to identity relations, e.g., owl:sameAs relations. In this paper, we propose a technique that evaluates existing identity links between LOD resources and identifies potentially erroneous links. Our work relies on the position and relevance of each resource with regards to the associated DBpedia categories modeled through two probabilistic category distribution and selection functions. Our experimental results show that our work is able to semantically distinguish inaccurate identity links even in cases when high syntactical similarity is observed between two resources.

Keywords: Resource co-reference · Linked open data · Semantic web · Semantic disambiguation · Identity links

1 Introduction

Linked Data relies on establishing relationships between resources, such as the equivalence (identity) relationship, that span across different ontologies and datasets [1]. Identity links such as those based on the owl:sameAs property are commonly used for establishing equivalence relationships by asserting that two different URIs refer to the same resource [10]. The importance of identity links is primarily due to the significant role they play in interconnecting different datasets of the Linked Open Data (LOD) cloud. Consequently, mistakes in these linkages may result in erroneous assertions by applications that automatically traverse these identity links [4]. In practice, identity links such as the owl:sameAs links tend to exhibit two main problems:

(P1) First, there are obvious errors where resources represented by URIs are clearly neither similar nor related. For example, based on existing information on the LOD cloud, *dbpedia:Dog* and the derogatory *dbpedia:Bitch(insult)* are considered to be the same. These kinds of mistakes atypical for humans are most likely made by an

© Springer International Publishing Switzerland 2015
Q. Chen et al. (Eds.): DEXA 2015, Part I, LNCS 9261, pp. 128–143, 2015.
DOI: 10.1007/978-3-319-22849-5_10

automated inference machine that failed to properly disambiguate the underlying semantics of the URIs to be matched. In addition, it is also not unlikely that human error in judgment could play some role in such erroneous identity links.

(P2) Second, the determination of co-reference resolution can at times be subjective and dependent on a specific application case. For example, *dbpedia:Evil*, *dbpedia: Morality*, and *dbpedia:Crime* have already been linked together through identity links. However, although these three resources are strongly related, they are not equivalent in that one can have questionable morality and not be evil; likewise, not every criminal act is morally wrong. From a strict interpretation of the owl:sameAs property, no pair of these three resources is identical.

There has already been novel and interesting work that focuses on co-reference resolution with specific attention to URIs (i.e., resources) of the LOD cloud (e.g., [3, 8, 9]). The proposed techniques attempt to identify any potential pair of URIs that can or need to be related to each other through identity links [6]. However, in this paper, we focus on a different aspect of this problem: our focus is on improving the quality of existing identity links that are already part of the LOD cloud. In other words, our objective is to ensure that an existing identity link collection contains reliable co-reference resolutions. To this end, this paper makes the following contributions:

- A novel algorithm that can identify potential mismatches between the URIs that are linked via identity links, and hence provide the means to filter out such mismatches in order to resolve (P1).
- The calculation of an upper-bound semantic score that can be used as a measure of semantic similarity of resources connected via identity links. This way, the algorithm points to cases where identity links show some degree of subjectivity and hence allows for the identification of cases represented in (P2).

2   Background

An increasing number of organizations are publishing their data on the Web as Linked (Open) Data thus continuously extending the Web of Data with new datasets. However, while the number of datasets keeps growing, linking between those datasets keeps lagging behind, thus leading to a considerable discrepancy between intra- and inter-dataset linking. In other words, while data tend to be well connected within individual datasets, linking between datasets is still not at the level required for a true Web of Linked Data. Based on a recent analysis of the crawlable subset of the LOD cloud (April 2014), Schmachtenberg et al. [13] reported that 56.11 % of the crawled datasets are connected to at least one other dataset, while the rest of the datasets are either only the targets of RDF links or are completely isolated. In addition, only a small number of datasets is highly linked, while the majority of them are only sparsely linked.

Aiming to improve the connectivity of the Web of Data, the Semantic Web research community has proposed several methods for automated or semi-automated linking of resources on the Web of Data, especially resources originating from disparate datasets [5, 14]. While these proposals promise to make a notable contribution to the realization of the Web of Linked Data, they also tend to generate erroneous links that over time can negatively impact the overall quality of the Web of Data [7]. Such quality issues are of particular concern since unlike the Web of Documents where humans can examine the meaning and usefulness of a link, on the Web of Data such an examination is left to the software agents harvesting and making use of the data.

While links of diverse semantics can be used to connect resources from disparate datasets, in this paper we focus exclusively on identity links connecting two resources that are considered either identical or closely related [15]. Such links are often referred to as sameAs links even though they do not necessarily originate from the owl:sameAs property, but also from other properties that have been used for establishing identity links on the Web of Data (e.g., skos:closeMatch, skos:exactMatch). Still, owl:sameAs links tend to be dominant among identity links. According to the aforementioned study by Schmachtenberg et al. [13], owl:sameAs is among the top three most used linking predicates in 7 out of 8 topical categories of the LOD cloud.

3 Proposed Approach

The main objective of our work is to flag and possibly remove questionable sameAs links from the LOD cloud. To achieve this objective, we define a two step process: (1) calculate a similarity score between the URIs involved in a given sameAs link, which would represent the likelihood that the URIs are in fact referring to the same resource; and (2) employ this similarity score to determine whether the sameAs link is valid and reliable or needs to be flagged for removal. Therefore questionable sameAs links can be flagged (and possibly filtered) if they do not meet some minimum threshold value; this value is chosen based on empirical studies reported in Sect. 6 (Fig. 2).

We refer to our proposed method as Semantic Co-reference Inaccuracy Detection (SCID). We begin by introducing the methods and formulas that form the foundation of our algorithm (Sect. 4), and then introduce the algorithm itself in Sect. 5.

4 The Components of SCID

Our strategy for detecting inaccurate identity links can be summarized as follows. Assume we construct a baseline vector v_x that is known to semantically represent resource x. We create v_x based on DBpedia categories that relate to x (positive categories) as well as categories that do not relate to x (negative categories). We denote this combined set of categories as S_x. Given a resource y of unknown similarity to x, we can construct v_y using the same categories S_x. We can now compare similarity of two vectors using measures of similarity such as Pearson correlation coefficient to judge if v_y is "same as" v_x. Our method relies on three key components:

1. *Frequency count statistics.* We require a listing of the most frequently occurring words within each category of DBpedia's 995,911 subject categories[1]. In Sect. 4.1 we explain the process of computing the required data.

---

[1] DBpedia version 3.9.

2. *A category distribution function.* This is the core function of our method. Given some input text obtained from resources x and y; and an arbitrary subset of DBpedia subject categories S, the function produces vectors v_x and v_y that can be compared to identify erroneous identity links. This is explained in Sect. 4.2.
3. *A category selection function.* The category distribution function requires a subset of DBpedia subject categories as input. This function, explained in Sect. 4.3, provides a mechanism for selecting this subset of categories.

Table 1. A sample of word and resource frequency counts within specific DBpedia categories (category:Color highlighted).

dcterms:subject category	stemmed word	word count	resources
Category:Eagles	eagl	428	59
Category:Eagles_(Band)	eagl	13	2
Category:Philadelphia_Eagles	eagl	70	18
Category:Fruit	orang	6	6
Category:Oranges	orang	106	20
Category:Color	orang	16	12
Category:Living_People	death	8222	6784

4.1   Frequency Count Statistics

Problem Outline: In order to train a model for inaccuracy detection, we require frequency counts over DBpedia categories for use in the category distribution function of SCID.

SCID centers around the dcterms:subject property of the DBpedia resources. This property provides 900,000+ subject-matter categories for approximately 11 million DBpedia resources. SCID is trained around five summary statistics (*features*) that rely on resource categories; these are described in detail in Sect. 4.2.

Table 1 is an example of word frequency counts for resources that belong to DBpedia categories {*Eagles, Eagles_(Band), Philadelphia_Eagles, Fruit, Orange, Color, Living_People*}. In this table, the row of the stemmed word "orang" and the category Color is highlighted; it indicates that there are 12 resources that refer to the category Color and use the word orange. Furthermore, the stem "orang" appears 16 times within this set of 12 resources; thus the average frequency of orange in this set is 16/12 or 1.3. These types of statistics are formally defined in Sect. 4.2.

4.2   The Category Distribution Function

Problem Outline: Given an input text and a set of DBpedia subject-matter categories, we require a normalized vector to represent the relevancy of each category to the input text.

The core of our method is the category distribution function $v = \rho(t,S)$ where t is an input text to be processed, and S is a specified subset from the DBpedia subject categories. The output is a vector v representing a proportionate mixture of each of the categories of S as they relate to the input text t. Table 2 illustrates the function with three DBPedia categories: *Eagles, Eagles_(band), and Philadelphia_Eagles* on three input texts. This example demonstrates that our method can associate different usages of a word with its appropriate category.

Table 2. The output of the category distribution function ρ(t,S) on S = [Eagles, Eagles_(band), Philadelphia_Eagles] for three sample inputs (t).

Category Subset (S)	Natural Language Input Text (t)		
	Sproles, who was acquired by the Eagles in the off- season, led the league with 2,696 all- purpose yards in 2011, but his rushing..	The Eagles were formed in 1971 by guitarist / singer Glenn Frey, With an eye towards his future band, he approached Henley to be her drummer.	The eagle is one the largest and most powerful birds of prey. Soaring high above the earth, spying its prey with its keen eyes.
Eagles	0.086	0.214	0.72
Eagles_Band	0.207	0.623	0.13
Philadelphia_Eagles	0.706	0.162	0.148
P(t,S)	v=[0.086,0.207,0.706]	v=[0.214,0.623,0.162]	v=[0.72,0.13,0.148]

A more interesting example is given in Table 3 with chosen categories *Fruit, Oranges*, and *Color*, which are not as disjoint as those in Table 2. Specifically, the Fruit category is defined as a category broader than Oranges (via the skos:broader property). Moreover, the mention of color in the input text adds a disambiguation challenge w.r.t. the Oranges and Color categories. Further adding to the complexity is that the input text refers to a specific color (dark pink), but that color is not orange; the term orange in this context refers to the Fruit category. The output of the distribution function is a proportional mixture of the three relevant categories.

Table 3. ρ(t,S) output when categories [Fruit, Oranges, Color] are related/ambiguous w.r.t term "orange" and the input text refers to all three categories.

Category Subset (S)	Natural Language Input Text (t)
	Cara Cara , a type of navel orange, are also available during the winter months. They are like the familiar Washington navels, but the fruit's interior is dark pink.
Fruit	0.274
Oranges	0.504
Color	0.220
P(t,S)	v = [0.274,0.504,0.220]

We now detail the sequence of calculations required for ρ(t,S). Formally, let T be a set of stemmed words from input text t; for simplicity of expression, in the following we refer to elements of the set T as words (instead of stemmed words). Let S be the set of pre-chosen categories. Let $W_{x,y}$ be the frequency count of a stemmed word x in category y.

Feature Formalization f^1: Define $f^1_{j,k}$ as the frequency of a specific word j in the category k to the count of all the words of the input text t within the category k.

$$f^1_{j,k} = \frac{W_{j,k}}{\sum_{x \in T} W_{x,k}} \text{ where } j \in T, k \in S \tag{1}$$

Conceptually, this feature is a measure of the significance of a word relative to a specific category.

Feature Formalization f^2: Let $f^2_{j,k}$ be the frequency of specific word j in the category k to the total frequency count across all the words of the input text within all the categories of S.

$$f^2_{j,k} = \frac{W_{j,k}}{\sum_{x \in T} \sum_{y \in S} W_{x,y}} \text{ where } j \in T, k \in S \tag{2}$$

Conceptually, this feature is a measure of the importance of a word (e.g., orange) relative to all selected categories (Fruit, Oranges, Color) combined.

Feature Formalization f^3: Let $D_{j,k}$ be the number of DBpedia resources that belong to the category k and contain the word j. We define $f^3_{j,k}$ as the ratio of DBpedia resources that contain word j and belong to the category k to the number of resources that contain any of the words from the input text that belongs to category k. Formally:

$$f^3_{j,k} = \frac{D_{j,k}}{\sum_{x \in T} D_{x,k}} \text{ where } j \in T, k \in S \tag{3}$$

This feature is similar to f^1 except that instead of counting the frequency of every word occurrence, f^3 counts the number of unique resources containing that word.

Feature Formalization f^4: Conceptually similar to f^2, we define $f^4_{j,k}$ as the ratio of the number of DBpedia resources containing specific word j from category k across the number of all the resources containing any word from the input text t that belongs to any category from the chosen set S. Like f^3, this feature deals with the number of distinct resources that contain the word rather than the word frequency count.

$$f^4_{j,k} = \frac{D_{j,k}}{\sum_{x \in T} \sum_{y \in S} D_{x,y}} \text{ where } j \in T, k \in S \tag{4}$$

Feature Formalization f^5: This last feature is defined as the ratio of the frequency of the word j within the category k to the total number of resources that belong to k and contain j. Formally, it is a measure of the average word frequency per resource.

$$f_{j,k}^5 = \alpha \frac{W_{j,k}}{D_{j,k}} \text{ where } j \in T, k \in S \tag{5}$$

and α is a normalizing constant such that $\sum_{y \in S} f_{j,y}^5 = 1$.

Word Importance $R_{j,k}$: We can now combine all the features to compute the importance of the word j relative to the category k. Let U_k be the total number of resources that belong to the category k globally within the DBpedia knowledge base. Let O_j be the frequency count of stemmed word j within the input text t. The importance of the word j relative to the category k becomes:

$$R_{j,k} = O_j \times \frac{D_{j,k}}{U_k} \times \frac{\sum_{i=1}^5 f_{j,k}^i}{5} \text{ where } j \in T, k \in S \tag{6}$$

The Category Distribution Function $\rho(t, S)$: We sum the importance of all the words per chosen category to construct the vector v of the category distribution function and normalize:

$$v = \rho(t, S) = \alpha \left[\sum_{x \in T} R_{x,k_1}, \ldots, \sum_{x \in T} R_{x,k_n} \right] \tag{7}$$

Where $j \in T, k \in S$ and α is a normalizing constant such that $\sum_{y=1}^n \sum_{x \in T} \alpha R_{x,k_y} = 1$.

The final artifact is Eq. 7 that produces the output seen in Tables 2 and 3. We have made available an online implementation of the category distribution function for those interested in further experimentation[2].

4.3　The Category Selection Function

Problem Outline: We require a well-defined method for selecting a subset of categories from amongst the 995,911 DBpedia categories to be used as the input set S in $\rho(t, S)$.

The category distribution function (Sect. 4.2) requires as its input a set of subject matter categories for consideration. For example, in Table 3 the input categories were explicitly given as [Fruit, Oranges, Color]. A focused selection of categories is required because the evaluation of all available DBpedia categories is not feasible for computational and practical reasons. Consequently, a strategy for selecting a suitable subset of categories is the focus of this section. Ideally, the selection should include both categories that are related to each other such as Fruit and Oranges in the case of Table 3, as well as disjoint categories such as Eagles, Eagles_(band), and Philadelphia_Eagles in the case of Table 2. We take advantage of DBpedia disambiguation resources to this end. Such resources often encompass homonyms that require examination of the context to differentiate between ambiguous resources. Suppose we wish to find the categories S that will be used to validate the <sameAs> identity links for $x = $ *dbpedia:*

---

[2] http://ls3.rnet.ryerson.ca/predicatefinder/category/.

Red. We see that x belongs to *uri = dbpedia:Red_(disambiguation)* alongside 116 other resources some of which are disambiguation resources themselves. Specifically, x and *uri* satisfy the constraint:

$$\{?uri \; dbpedia\text{-}owl : wikiPageDisambiguates \; ?x\} \tag{8}$$

Now, let N_{uri} be the set of resources linked from a DBpedia disambiguation URI. Formally, if $x \in N_{uri}$ then x satisfies Eq. 8. Next let C(x) be a function returning the set of categories for the resource x. Namely, $y \in C(x)$ when $\{?x \; dcterms:subject \; ?y\}$. Lastly, we define S_{uri} as the union of all subject categories for the resources that a disambiguation resource refers to and apply it recursively for resources within to dereference all resources to their respective categories.

$$S_{uri} = \begin{cases} \bigcup_{x \in N_{uri}} C(x), & \text{if x is a non-disambiguation URI} \\ S_x, & \text{otherwise} \end{cases} \tag{9}$$

Suppose no disambiguation resource *uri* exists for x (i.e.: no *uri* satisfies Eq. 8). In this circumstance, we create a temporary disambiguation resource uri_{temp} that references the single resource x so that Eq. 9 can still be applied.

5   The SCID Filter

In this section we apply frequency count statistics, the category distribution function, and the category selection method of Sect. 4 to construct two algorithms to filter out inaccurate identity links. Algorithm 1 produces disambiguation baseline vectors used to train our model. Once trained, Algorithm 2 details how the model is used to test the accuracy of candidate identity URI pairs.

5.1   Algorithm 1 – Constructing Disambiguation Vectors

Problem Outline: We require a method to construct a baseline vector v_{x,S_x} to disambiguate an ambiguous resource x (e.g., Eagles as: team, a band, or bird) against DBpedia subject categories (S_x) chosen by Eq. 9.

To construct these vectors we use the category distribution function (Sect. 4.2) and define $v_{x,S_x} = \rho(\gamma_x, S_x)$ as a disambiguation baseline vector for resource x where γ_x are the most frequently occurring words within the subject categories of resource x. Algorithm 1 outlines how these words γ_x are found and how v_{x,S_x} is computed.

Consider the resource dbpedia:Red with subject categories C(dbpedia: Red) = dbpedia:{Color, Optical_spectrum, Shades_of_red, Web_colors}. In Algorithm 1, we begin with this resource's stemmed words (line 1). In line 2, we collect all the resources that are also associated with any of the C(dbpedia:Red) categories and include them in our frequency counts. We discard those categories that contain more than 1000 resources because they are overly broad and are not representative of the target resource dbpedia:Red. In line 3, we keep the most frequently occurring words from this collection (75th percentile) and z-score normalize the frequency counts (line 5). We perform this normalization to balance the influence of a more frequent category (e.g., Color) against a less used category (e.g., Web_colors). Finally we compute the disambiguation baseline vector v (line 6).

Input (x): A resource URI (x) and a set of subject categories (S_x) from equation 9.

Output (v_{x,S_x}): A disambiguation baseline vector for resource x.

Algorithm:

Define: Let H(x) be a set of stemmed words appearing in resource x. Let C(x) be the set of categories associated with the resource x. Namely, $y \in C(x)$ iff (?x dcterms:subject ?y). Let $C^{-1}(y)$ be the set of resources that have category y. Namely, if $x \in C^{-1}(y)$ then $y \in C(x)$.

   1. Initialize $\beta \leftarrow H(x)$

   2. For each $y \in C(x), K \leftarrow C^{-1}(y), if\ |K| \leq 1000\ then\ \forall k \in K, \beta \leftarrow \beta + H(k)$

Define: Let $I_{75}\ (\beta)$ be the top 25% (75th percentile) of the most frequently occurring words of set β. Let $Z_{75}(\beta, h)$ be the z-score normalized frequency count of word h from set β.

   3. $\beta \leftarrow I_{75}(\beta)$

   4. Initialize $\gamma_x \leftarrow \emptyset$

   5.For each word of β, append to γ_x the word Z_{75} times.

      Namely, for each $\omega \in \beta, \gamma_x \leftarrow \gamma_x + \omega \times Z_{75}(\beta, \omega)$.

   6. Compute disambiguation baseline vector $v_{x,S_x} = \rho(\gamma_x, S_x)$.

Algorithm 1. Computation of disambiguation baseline vector v_{x,S_x}.

The categories of C(dbpedia:Red) form a subset of the category candidate list (S) for the dbpedia:Red_(disambiguation) resource (Sect. 4.3). We note C to contain the target categories with respect to the resource dbpedia:Red while S\C are the noisy categories of the disambiguation vector. The noisy categories are used to counterbalance the frequently occurring words of the target categories with the occurrence of the words in the non-related category set. This provides positive and negative category usage examples and aids in disambiguation. In the next section, we show how the disambiguation baseline vector v_{x,S_x} can be used to find identity link errors.

5.2   Algorithm 2 – Detection of Inaccurate Identity Links

Problem Outline: We require a method to utilize the disambiguation baseline vectors v for identifying likely errors in a collection of identity links.

Algorithm 2 details our method for filtering out likely errors within a collection of identity links. In line 1, using Algorithm 1 (Sect. 5.1), we compute a base vector v_{x,S_x} for a resource (x) we wish to validate. We then collect a set of all identity links for the resource, say M(x), using a database of identity links (e.g., www.sameas.org); then, for each identity link $m \in M(x)$, we calculate vector $\rho(y_m, S_x)$ where y_m is some descriptive text of m (e.g., rdfs:comment) (line 2). The text of y_m is chosen based on the origin of the candidate URI. Specifically, we use the *rdfs:comment* property when the candidate

URI is from DBpedia or OpenCyc[3]. We use *ns.common.topic.description* property when the URI is from Freebase[4], and *wn20schema:gloss* when it originates from WordNet[5]. The output of line 2 can be compared with the disambiguation baseline vector *v* of line 1 using some similarity measure to produce a *semantic relatedness score*. We create a *disambiguation ratio* by normalizing semantic relatedness scores using the largest seen score (line 3). Finally, we flag those identity links (URIs) that do not meet a threshold value set for the disambiguation ratio (line 4).

Table 4 shows the results of this algorithm applied to the *dbpedia:Port* resource. The table includes the semantic relatedness score as a measure of the semantic similarity between the candidate URI and the baseline vector (for dbpedia:Port); it is computed using Pearson correlation coefficient as the similarity measure (ß). If the normalized ß (disambiguation ratio) is less than the given threshold (i.e.: $ß < \delta$) then the resource should be flagged as a possible inaccuracy. The isSame classification (Table 4, col 1), determined by a human oracle, indicates whether the candidate URI is truly the same or semantically similar.

Input (x, δ): A URI (x) for identity validation and a minimum disambiguation threshold value(δ).

Output (Q): A set (Q) of identity links (URIs) for resource x that meet the minimum disambiguation threshold value (δ).

Algorithm:

  1. Compute disambiguation baseline vector v_{x,S_x} using Algorithm 1.

Define: Let y_x be a natural language text description for resource x. Let $Coeff(v_1, v_2)$ be the similarity measure between two vectors v_1 and v_2. Let M(x) be the set of identity links (URIs) for resource x from an identity links collection (e.g.: sameAs,org).

  2. For each $m \in M(x)$ calculate set $Q=[q_m, ...]$ where vector $q_m = \rho(y_m, S_x)$.

  3. For each $q_m \in Q$ calculate $\tau_m = Coeff(q_m, v_{x,S_x})$. Let J_{max} be the maximum τ_m encountered.

  4. Discard $q_m \in Q$ if $\frac{\tau_m}{J_{max}} < \delta$

Algorithm 2. A method for discovering inaccurate identity links.

Notice in Table 4 what appears to be an error in which the second entry of dbpedia: Port obtains a disambiguation score less than itself. This result is consistent with our algorithm in that strictly speaking our method is not comparing dbpedia:port with all other sameAs candidates. Instead, *all sameAs candidates of dbpedia:port is compared to the shared categories of dbpedia:port_(disambiguation) that dbpedia:port belongs to (S_x in Eq. 9)* thus forming a cluster of identity links within the radius of the disambiguation threshold value δ. Conceptually, v_{x,S_x} is the center of a cluster while δ is the "distance" from this center to the outermost boundary of our solution space. Those

---

[3] http://sw.opencyc.org/.

[4] https://www.freebase.com/.

[5] http://wordnet.princeton.edu/.

identity links within this boundary are members of the cluster while those outside this area are considered anomalies. Figure 1 illustrates.

Table 4. Algorithm 2 applied to the dbpedia:Port sameAs candidates

isSame [y/n]	SAMEAS URI CANDIDATES	Semantic Relatedness	Disambiguation Ratio	Natural Language Source
y	dbpedia.org:List_of_Panamax_ports	0.633	1	rdfs:comment
y	dbpedia.org:Port	0.591	0.933649289	rdfs:comment
y	www.w3.org:synset-seaport-noun-1	0.565	0.89257504	wn20schema:gloss
y	rdf.freebase.com/ns/en.port	0.539	0.85150079	ns:common.topic.description
y	rdf.freebase.com:..a8561	0.539	0.85150079	ns:common.topic.description
y	sw.opencyc.org:Seaport	0.478	0.755134281	rdfs:comment
y	dbpedia.org:River_port	0.385	0.60821485	rdfs:comment
n	dbpedia.org:Bad_conduct	0.168	0.265402844	rdfs:comment
n	rdf.freebase.com:m.02ss4j	0.168	0.265402844	ns:common.topic.description
n	rdf.freebase.com:en.military_discharge	0.168	0.265402844	ns:common.topic.description
n	dbpedia.org:IVDP	-0.048	-0.075829384	rdfs:comment
n	rdf.freebase.com:m.0143h2	-0.062	-0.097946288	ns:common.topic.description
n	rdf.freebase.com:en.port_wine	-0.062	-0.097946288	ns:common.topic.description
n	rdf.freebase.com:..120de2	-0.062	-0.097946288	ns:common.topic.description

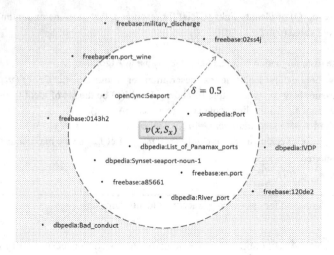

Fig. 1. Conceptual diagram for identity links clustered around vector v constructed using the shared categories of *dbpedia:Port_(disambiguation)* resource within threshold $\delta = 0.5$.

6   Experimental Evaluation

This section presents our experimental evaluation of the SCID approach on a sameAs dataset retrieved from the SameAs.org service[6]. The dataset consists of resources from five groups: Animal, City, Person, Color, and Miscellaneous (other). We decided on

---

[6] Data retrieved in October 2014 from http://www.sameas.org.

segmenting our dataset into these topical groups to be able to verify whether the performance of SCID is dependent on a specific domain or it performs the same across different non-overlapping domains. We also wanted to discover if there is a general-purpose disambiguation threshold (δ) that would be effective regardless of the topic group. Furthermore, it is important to point out that the scoping of the dataset to specific topics is a common practice within previous studies that typically focused on a narrow domain such as restaurants or people (e.g., [11, 12]). We included the miscellaneous group to further broaden our experiments.

We collected a sameAs dataset of 7,690 candidate URIs for validation (Table 5). First, we performed some necessary data cleansing on these candidates. After this pre-processing procedure, a total of 411 URIs primarily from DBpedia, Freebase, OpenCyc, and WordNet sources remained. A human oracle then identified 251 incorrect URIs (61 % errors) from these 411.

Table 5. Experimental dataset showing for each topic group: total candidate URIs (pre/post cleaning) and the number of oracle-identified errors.

Topic Group	pre-cleaning URIs	post-cleaning URIs	oracle-identified errors
Animal	759	53	34
City	2934	143	98
Person	856	41	20
Color	1021	47	25
Misc	2120	127	74
Totals	7690	411	251

The data cleansing process includes removing duplicate entries that are aliases or URI redirects for the same resource (such as those identified with the dbpedia-owl: wikiPageRedirects property). Broken-link (non-resolvable) URIs that are no longer accessible are also discarded. Furthermore, because our method relies on a natural language text description of the candidate entry, we discarded those candidates that did not have a well-defined descriptive property namely: rdfs:comment, ns.common.topic. description, or wn20schema:gloss. We ignored duplicates in cases when a candidate URI shared the same descriptive property attribute and same property value with another URI already included in the dataset. A common example is the DBpedia Lite knowledge base that often shares the same rdfs:comment predicate and value with its larger counterpart DBpedia. To illustrate the importance of this purging step, consider the concept *DBpedia:Jesus* that, according to the sameAs.org database, returns 16,889 coreferents of which 3,357 are broken/non-resolvable (20 %), 9,891 are duplicated via aliases/redirects (73 %), and only 3,641 are unique (27 %). Consequently, the large discrepancy between pre and post-cleansing of Table 5 is attributed to aliases and broken links.

Once cleansed, we computed a starting (baseline) F-score value for each of the considered resource groups (Table 5) as an initial measure of the dataset quality. For example, in the City group, the cleaned set of 143 URIs contained 98 errors giving a precision of 0.31. Since this was our starting set, we assumed a recall of 1.0, thus giving a starting F-score of 0.479. We then applied SCID to the cleaned set of candidate URIs at thresholds intervals from 0 to 0.9, and compared the new F-scores against the original baseline. The results of this experiment are summarized in Table 6. The table also provides an average of the five groups for each threshold value as well as a combined F-score tally in which all 411 candidate URIs are evaluated together (non-grouped). We can see a significant improvement from the original non-filtered F-scores for all groups, including the combined group, using any of the disambiguation thresholds with the exception of $\delta \geq 0.9$. Empirical results indicate that a threshold of 0.5 to 0.6 gives the best results with an average F-score of 0.84.

Table 6. F-scores for the five resource groups at varying threshold values (δ), including average score for each group and combined score for all candidate URIs.

	Original	$\delta \geq$ 0.9	0.8	0.7	0.6	0.5	0.4	0.3	0.2	0.1	0
Animals	0.528	0.765	0.842	**0.927**	0.905	0.884	0.884	0.884	0.844	0.792	0.633
Cities	0.479	0.276	0.5	0.674	**0.74**	0.722	0.651	0.62	0.556	0.508	0.489
People	0.677	0.551	0.645	0.727	0.8	0.857	**0.863**	0.857	0.84	0.778	0.712
Colors	0.638	0.706	0.811	**0.952**	0.889	0.857	0.83	0.8	0.786	0.733	0.677
Misc(other)	0.589	0.568	0.706	0.788	0.867	**0.881**	0.867	0.848	0.803	0.763	0.702
Average	0.5822	0.5732	0.7008	0.8136	0.8402	0.8402	0.819	0.8018	0.7658	0.7148	0.6426
Combined	0.56	0.541	0.674	0.785	0.824	**0.828**	0.793	0.768	0.719	0.669	0.617

Table 7 expands on the results of the combined group by including precision and recall metrics as well as the counts of correct and incorrect (inaccurate) candidate URIs, i.e., identity links *after* filtering. A noticeable improvement is observed from the original (baseline) F-score of 0.560 with the only drop at the $\delta \geq 0.9$ level (0.541). This drop is caused by high precision (0.898 versus 0.389) but low recall (0.388 versus 1). Nonetheless, this threshold may still be desirable as it resulted in only 7 errors remaining from the initial 251 inaccuracies. Also shown is our apparent optimal threshold of $\delta = 0.5$ preserving 200 sameAs candidates of which 149 were true positive and only 51 false positive classifications. If high recall is desired even a low threshold of 0.0 to 0.2 would result in F-score improvement over the original (non-filtered) resource set.

In Fig. 2 we provide a scatter plot of F-score versus disambiguation ratio for the complete dataset of 411 URIs, i.e., identity link candidates with oracle identified correct/incorrect entries shown in blue (true positive) and red (true negative). The plot displays the desired characteristic that inaccurate links trend towards the lower-scoring disambiguation ratio while correct links gravitate toward higher-scoring values.

Our experimentation reveals that SCID can identify a significant number of erroneous identity links independent of any specific topic (animals, cities, etc.).

A user-specified threshold of 0.5 to 0.6 appears to work well as a general purpose setting for best F-score (Fig. 2 and Tables 6 and 7).

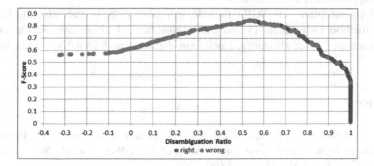

Fig. 2. Scatter plot of F-score versus disambiguation ratio for combined dataset with oracle-identified right(blue) and wrong(red) identity links (Color figure online).

Table 7. Precision, Recall, F-score statistics for the combined (non-grouped) set with correct/incorrect counts of the candidate URIs (i.e., identity links) after filtering.

	Precision	Recall	F-Score	Correct	Incorrect
Original	0.389	1.000	0.560	160	251
$\delta \geq 0.9$	0.899	0.388	0.541	62	7
$\delta \geq 0.8$	0.841	0.563	0.674	90	17
$\delta \geq 0.7$	0.832	0.744	0.785	119	24
$\delta \geq 0.6$	0.789	0.863	0.824	138	37
$\delta \geq 0.5$	0.745	0.931	0.828	149	51
$\delta \geq 0.4$	0.677	0.956	0.793	153	73
$\delta \geq 0.3$	0.626	0.994	0.768	159	95
$\delta \geq 0.2$	0.561	1.000	0.719	160	125
$\delta \geq 0.1$	0.503	1.000	0.669	160	158
$\delta \geq 0$	0.446	1.000	0.617	160	199

7 Conclusion

In this paper, we proposed a technique (SCID) to discover coreference inaccuracies in existing sameAs links. Experimental results indicate that SCID can improve the quality of an identity collection by correctly flagging questionable identity assertions. A distinguishing feature is that unlike most existing approaches (e.g., [2, 7]), SCID considers the semantics of the resources associated through identity links. The semantic relatedness and disambiguation ratio scores could provide a quantitative measure of semantic similarity for those seeking more than a binary correct/incorrect classification. This is an important advantage over the existing work that leads to future experimentation with skos:closeMatch, skos:exactMatch, and owl:equivalentClasses as the identity links. Furthermore, unlike some existing work addressing the same problem, our proposal does not depend on domain-specific rules that have to be defined by human experts (as e.g. in [12]). We position SCID not as a replacement for such rules but as a supplemental verification step.

Development plans for SCID include improvements to the relevant keywords selection method (Algorithm 1) and the exploration of alternative vector similarity methods such as Cosine similarity, Euclidean distance, and SVR/SVM (Algorithm 2).

References

1. Bizer, C., Lehmann, J., Kobilarov, G., Auer, S., Becker, C., Cyganiak, R., Hellmann, S.: DBpedia-A crystallization point for the Web of Data. Web Semant.: Sci., Serv. Agents World Wide Web **7**(3), 154–165 (2009)
2. de Melo, G.: Not quite the same: Identity constraints for the Web of Linked Data. In: des Jardins, M., Littman, M.L. (eds.) Proceedings of the Twenty-Seventh AAAI Conference on Artificial Intelligence, Menlo Park, CA, USA. AAAI Press (2013)
3. Gianluca, D., Difallah, D.E., Cudré-Mauroux, P.: ZenCrowd: leveraging probabilistic reasoning and crowdsourcing techniques for large-scale entity linking. In: Proceedings of the 21st International Conference on World Wide Web, pp. 469–478. ACM (2012)
4. Ding, L., Shinavier, J., Shangguan, Z., McGuinness, D.L.: SameAs networks and beyond: analyzing deployment status and implications of owl:sameAs in Linked Data. In: Patel-Schneider, P.F., Pan, Y., Hitzler, P., Mika, P., Zhang, L., Pan, J.Z., Horrocks, I., Glimm, B. (eds.) ISWC 2010, Part I. LNCS, vol. 6496, pp. 145–160. Springer, Heidelberg (2010)
5. Ferrara, A., Nikolov, A., Scharffe, F.: Data linking for the semantic web. Int. J. Semant. Web Inf. Syst. **7**(3), 46–76 (2011)
6. Ferrara, A., Nikolov, A., Scharffe, F.: Data linking for the semantic web. Semant. Web: Ontology Knowl. Base Enabled Tools, Serv., Appl. **2013**, 169 (2013)
7. Guéret, C., Groth, P., Stadler, C., Lehmann, J.: Assessing linked data mappings using network measures. In: Simperl, E., Cimiano, P., Polleres, A., Corcho, O., Presutti, V. (eds.) ESWC 2012. LNCS, vol. 7295, pp. 87–102. Springer, Heidelberg (2012)
8. Hogan, A., Polleres, A., Umbrich, J., Zimmermann, A.: Some entities are more equal than others: Statistical methods to consolidate linked data. In: Proceedings of the Workshop on New Forms of Reasoning for the Semantic Web: Scalable & Dynamic (NeFoRS2010) @ ESWC2010 (2010)
9. Hu, W., Chen, J., Qu, Y.: A self-training approach for resolving object coreference on the semantic web. In: Proceedings of the 20th International Conference on World Wide Web (WWW 2011), pp. 87–96. ACM, New York (2011)
10. Maali, F., Cyganiak, R., Peristeras, V.: Re-using cool URIs: entity reconciliation against LOD hubs. In: Proceedings of the Linked Data on the Web (LDOW 2011)
11. Datasets for the Identity Recognition Task. Instance Matching Track of the Ontology Alignment Evaluation Initiative 2014 Campaign. http://islab.di.unimi.it/im_oaei_2014/index.html
12. Papaleo, L., Pernelle, N., Saïs, F., Dumont, C.: Logical detection of invalid SameAs statements in RDF data. In: Janowicz, K., Schlobach, S., Lambrix, P., Hyvönen, E. (eds.) EKAW 2014. LNCS, vol. 8876, pp. 373–384. Springer, Heidelberg (2014)

13. Schmachtenberg, M., Bizer, C., Paulheim, H.: Adoption of the linked data best practices in different topical domains. In: Mika, P., et al. (eds.) ISWC 2014, Part I. LNCS, vol. 8796, pp. 245–260. Springer, Heidelberg (2014)
14. Shvaiko, P., Euzenat, J.: Ontology matching: state of the art and future challenges. IEEE Trans. Knowl. Data Eng. **25**(1), 158–176 (2013)
15. Halpin, H., Herman, I., Hayes, P.: When owl:sameAs isn't the same: an analysis of identity links on the semantic web. In: Linked Data on the Web (LDOW 2010)

Quality Metrics for Linked Open Data

Behshid Behkamal[1(✉)], Mohsen Kahani[1], and Ebrahim Bagheri[2]

[1] Computer Engineering Department, Ferdowsi University of Mashhad,
Mashhad, Iran
{behkamal,kahani}@um.ac.ir
[2] Department of Electrical and Computer Engineering, Ryerson University,
Toronto, Canada
bagheri@ryerson.ca

Abstract. The vision of the Linked Open Data (LOD) initiative is to provide a model for publishing data and meaningfully interlinking such dispersed but related data. Despite the importance of data quality for the successful growth of the LOD, only limited attention has been focused on quality of data prior to their publication on the LOD. This paper focuses on the systematic assessment of the quality of datasets prior to publication on the LOD cloud. To this end, we identify important quality deficiencies that need to be avoided and/or resolved prior to the publication of a dataset. We then propose a set of metrics to measure and identify these quality deficiencies in a dataset. This way, we enable the assessment and identification of undesirable quality characteristics of a dataset through our proposed metrics.

Keywords: Metrics · RDF datasets · Quality deficiencies · Linked open data

1 Introduction

The main goal of the Web of Data initiative is to create knowledge by interlinking dispersed but related data instead of linking related documents in the traditional Web. This massive amount of data on the LOD opens up significant challenges with regards to data quality. Since data is extracted via crowd sourcing of semi-structured sources, there are many challenges with the quality of published datasets. One of the better strategies to avoid such issues is to evaluate the quality of a dataset itself before it is published on the LOD cloud. This will help publishers to filter out low-quality data based on the quality assessment results, which in turn enables data consumers to make better and more informed decisions when using the shared datasets.

The rest of this paper is organized as follows: first, data quality research in the area of the LOD is reviewed in Sect. 2. In Sect. 3, our approach for proposing metrics starts by identifying significant quality issues in RDF datasets, followed by the development of suitable metrics to address the issues. Empirical evaluation of the developed metrics is provided in Sect. 4, and then some guidelines for quality improvement are presented in Sect. 5. Finally, the paper is concluded in Sect. 6.

© Springer International Publishing Switzerland 2015
Q. Chen et al. (Eds.): DEXA 2015, Part I, LNCS 9261, pp. 144–152, 2015.
DOI: 10.1007/978-3-319-22849-5_11

2 Related Works

In this section, we classify the related literature into two main groups: (i) quality problems of published data, and (ii) tools and applications for validation of RDF datasets.

The first group of related work investigates quality problems in published datasets. The most comprehensive work in this group discusses common errors in published datasets [1]. In another work, the quality problems of available datasets such as Geonames and DBPedia are identified using SPARQL queries [2].

The second group of works includes some tools for validating RDF datasets, each with its own error-checking functionalities. Some of them are syntax validators which accept an RDF/XML document as input and check whether the document is syntactically valid. Other kinds of online validators such as URIDebugger[1] and Vapour[2] check the dereferencability of a given URI. Also, there are some command line tools for identifying common errors in RDF documents, such as Jena Eyeball[3] and VRP[4]. Another group of tools are designed for checking common quality issues, such as Lodlaundromat[5] which accepts URLs of dirty datasets and removes syntax errors, duplicates, and blank nodes; and Luzzu[6] which is a framework allowing users to define some metrics and provides queryable quality metadata on the assessed datasets.

Generally, all of these works primarily focus on data quality problems in published datasets, and seldom provide a concrete solution for improving data quality, or attempt to identify the causes of the quality problems before the data is published. In this paper, we deliberate on the importance of filtering out poor quality data by assessing the quality of a given dataset before publishing it.

3 Our Proposed Approach for Metric Development

The objective of our work is to identify quality deficiencies of datasets and to suggest how they can be systematically evaluated before release. To this end, our approach is based on several significant quality issues identified in already published datasets on LOD. To the extent of our experience in publishing and interlinking academic data [3, 4], we found that many of the published datasets suffer from quality issues. We believe that most of these issues have roots in the deficiencies of the sources from where that data is extracted, and they can be avoided if they are identified in the initial stages of publishing. In our work when considering which quality deficiencies to consider, the main criteria for identifying and including a quality issue was based on one of the following criteria:

---

[1] http://linkeddata.informatik.hu-berlin.de/uridbg.
[2] http://validator.linkeddata.org/vapour.
[3] http://jena.sourceforge.net/Eyeball.
[4] http://139.91.183.30:9090/RDF/VRP.
[5] http://lodlaundromat.org/.
[6] http://eis-bonn.github.io/Luzzu/.

- The quality problems should have been spotted within published data and well documented in the literature;
- The quality issues should be detectable and hence avoidable in the preliminary stages of data publication, i.e., prior to their publication and release to the LOD;
- The quality deficiencies are not related to the other datasets, e.g. inconsistency with other published datasets.

Therefore, our approach for metric development starts by identifying quality deficiencies of existing datasets, specifically those that can be avoided or fixed before publishing. We will then propose a set of metrics to address the identified issues, and subsequently the proposed metrics are placed under empirical evaluation.

3.1 Identifying Quality Deficiencies

In this section, we present the quality deficiencies of data within the published datasets focusing on those which are related to the dataset itself and also detectable in the initial phase of publication. To this end, we have classified the quality issues into schema level and instance level as presented in Table 1.

Table 1. Classification of quality deficiencies

Quality deficiency	Issues	Level	Resolution method	Ref
Improper usage of vocabularies	- Not using appropriate existing vocabularies to describe the resources	Schema	Domain Expert	[1, 5]
Redefining existing classes/properties	- Redefining the classes/properties in the ontology that already exist in the vocabularies	Schema	Domain Expert	[1, 5]
Improper definition of classes/properties	- Classes with different name, but the same relations	Schema	Semi-Automated	[3]
	- Properties with different name, but the same meaning	Schema	Ontologist	[6]
	- Inadequate number of classes/properties used to describe the resources	Schema	Domain Expert	[3]
Misuse of data type	- Not using appropriate data types for the literals	Schema	Automated	[1, 2]
Errors in property values	- Missing values	Instance	Automated	[2, 5–9]
	- Out-of-range values	Instance	Automated	
	- Misspelling	Instance	Semi-Automated	
	- Inconsistent values	Instance	Automated	

(*Continued*)

Table 1. *(Continued)*

Quality deficiency	Issues	Level	Resolution method	Ref
Miss-match with the real-world	- Resources without correspondence in real-world	Instance	Domain Expert	[6, 7, 10]
Syntax errors	- Triples containing syntax errors	Instance	Validator	[1]
Misuse of data type / object property	- Improper assignment of object property to the data type property or vice versa	Instance	Validator	[1]
Improper usage of classes/properties	- Using undefined classes/properties	Instance	Semi-Automated	[1]
	- Membership of disjoint classes	Instance	Automated	
	- Misplaced classes/properties	Instance	Validator	
Redundant/similar individuals	- Individuals with similar property values, but different names	Instance	Ontologist	[3]
Invalid usage of Inverse-functional properties	- Inverse-functional properties with void values	Instance	Automated	[2]

In summary, we have identified eleven quality deficiencies characterizing eighteen quality issues at both schema and instance levels. Among all, six quality issues, which their resolution methods are domain expert or ontologist, cannot be detected by any kind of automated methods and needs the intervention of human experts. It is clear that all of these metrics are very subjective and it is hard, if not impossible, to asses them automatically. Among the remaining quality issues, only three issues can be detected and resolved by validators. To the extent of our knowledge, there is no validator to cover all of the remaining issues, particularly the issues relating to incompatibility of schema, naming, and inconsistent data. Thus, we propose a set of metrics to address the remaining quality issues. We note that the identified issues and the following proposed metrics are not meant to be comprehensive and are only limited to the issues reported in the literature.

3.2 Proposed Metrics

In this section, a set of metrics are proposed to address the quality issues extracted from Table 1 that can be resolved in an automated or semi-automated way. To achieve this, metrics, proposed in the areas of the Linked Data, relational databases, and data quality

Table 2. Proposed metrics

Name	Description	Related quality deficiencies
Miss_Vlu	The ratio of the properties defined in the schema, but not presented in dataset	Errors in property values
Out_Vlu	The ratio of the triples of dataset which contain properties with out of range values	Errors in property values
Msspl_Prp_Vlu	The ratio of the properties of dataset which contain misspelled values	Errors in property values
Und_Cls_Prp	The ratio of the triples of dataset using classes or properties without any formal definition	Improper usage of classes/properties
Dsj_Cls	The ratio of the instances of dataset being members of disjoint classes	Improper usage of classes/properties
Inc_Prp_Vlu	The ratio of the triples of dataset in which the values of properties are inconsistent	Errors in property values
FP	The ratio of the number of triples of dataset with functional properties which contain inconsistent values	Errors in property values
IFP	The ratio of the number of triples of dataset which contain invalid usage of inverse-functional properties	Invalid usage of Inverse-functional properties
Im_DT	The ratio of the number of triples of dataset which contain data type properties with inappropriate data types	Not using appropriate data types for the literals
Sml_Cls	The ratio of the classes of dataset with different names, but the same instances	Improper definition of classes

models have been considered [5, 7, 9, 11, 12]; the results of which were taken into account as guidelines for designing a useful set of metrics for our purpose. The proposed metrics are presented in Table 2.

According to the definitions presented for the metrics, it is clear that all of the metrics are defined to measure the quality problems in the scope of the RDF dataset itself, not in the context of other datasets. From the level of quality deficiency point of view, the last two metrics (Im_DT and Sml_Cls) are defined to address intrinsic quality issues at the schema level, while the others are related to the intrinsic quality problems at the instance level. According to [13], the preferred way for metric definition is to calculate the number of the undesirable outcomes divided by that of the total outcomes. Thus, all of the formulas presented for computation of quality deficiencies illustrate the undesirable outcomes using the ratio scale. The proposed metrics are theoretically validated in our recently published work [14].

4   Empirical Evaluation

The main purpose of our work is to propose a set of appropriate metrics to address the quality issues of RDF datasets before their publication. For this purpose, it is necessary to place them under empirical evaluation to observe their behavior and show their applicability in practice. Hence, we first calculated the values of the metrics for eight datasets in order to show the metric behavior over datasets of different domains and sizes. We have selected eight datasets from the EU FP6 Networked Ontology (NeOn) project.[7] We have implemented an automated tool that is able to automatically compute the metric values for any given input dataset. The code of the implemented tool and the employed datasets are publicly available[8].

Next, we manipulated the quality of these datasets by applying some heuristics and then recalculated the metric values to observe the behavior of the metrics over these changes. The aim of our dataset manipulation work was to investigate the trends of metrics over real datasets and to compare the results of applying metrics on good and poor quality data. To this end, fourteen heuristics are introduced to be used in the dataset contamination process. Some of these quality issues such as misspelling errors were made using an ontology editor, i.e. Protégé. For those quality issues such as invalid usage of inverse functional properties, errors were introduced manually. We have randomly applied the heuristics to the different datasets. The rationale for this was to measure the values for our metrics both before and after the quality issues were injected. The aim of the second experiment is to show the trends of the proposed metrics by recalculating the values of the metrics over manipulated datasets. As a result, we would ideally expect to observe meaningful changes in the values of the metrics according to the heuristics used to manipulate the datasets. In light of the results, it is observed that most of the outcomes are desirable; however, some of the values need more discussions which are presented as follows.

Some of the heuristics were not independent and as a result, the order of applying these heuristics affected the results of measuring the quality problems by the metrics. This occurs because introducing some errors into a given dataset can have a number of side effects on other metrics. Thus, the change in one quality issue can implicitly impact other quality issues and therefore, their corresponding metrics. For better investigation of metrics behavior, it is better not to concurrently apply these heuristics on the same dataset.

Another factor affecting our result was related to the ratio of heuristics done over the size of datasets. Based on this experiment, whenever the number of changes is less that 10 % of the triples, the changes of metric values cannot be properly reported.

Although in our scenario, no radical shift in the metric values was observed, but we are not going to generalize our finding about the trends of metrics, because of the limited number of datasets that we have used in this experiment. As a result, we believe more experiments need to be done to reach a valid conclusion about the reaction of metrics to the changes in the measurement subject.

---

[7] http://www.neon-project.org.

[8] https://bitbucket.org/behkamal/new-metrics-codes/src.

5 Guidelines for Quality Improvement

Based on our discussions about quality deficiencies and based on our experiments in manipulating the datasets, we propose a set of guidelines for data publishers as shown in Table 3. These solutions can be used as guidance for data publishers to improve such quality deficiencies in their datasets before publication.

As shown in Table 3, most of the quality problems measured by our metrics can be easily fixed by the publishers once they are made aware of the issues. Also, it is

Table 3. Guidelines for data publisher

No	Metrics	Related quality issues	Resolution strategy
1	Miss_Vlu	Missing property values	Checking the usage of each property defined in the schema, whenever it is used in the dataset
2	Out_Vlu	Out-of-range property values	For triples with data type properties, checking the defined range for the literal, and for object properties, checking the range of class used as predicate of triple. In both cases, ranges of values should be correct
3	Msspl_Prp_Vlu	Misspelled property values	Using a dictionary to check all of the terms used as literals in a dataset
4	Und_Cls	Using undefined classes	Where classes have been created, we suggest that the term be added to the schema or defined in a separate namespace to enable the reuse of the defined terms
5	Dsj_Cls	Membership of disjoint classes	Checking for this violation of disjoint classes can be done with a reasoner or with an appropriate query
6	Inc_Prp_Vlu	Inconsistent values of properties	Selecting correct values for properties according to the data source and removing triples contain inconsistent values for those properties
7	FP	Functional Properties with Inconsistent Values	Validating user input and checking the uniqueness and validity of functional property values
8	IFP	Invalid usage of inverse-functional properties	Validating user input and checking the uniqueness and validity of inverse-functional values
9	Im_DT	Not using appropriate data types for the literals	Syntactic fixes to the publishing framework and removing or changing the data types of the literals
10	Sml_Cls	Classes with different name, but the same instances	Checking the reported classes by an ontologist for removing similar classes if needed

understood that many of the above deficiencies are a result of not validating user input or due to not using appropriate syntactic validator for the content. Thus, a recommended approach for avoiding these quality issues is using trusted APIs to produce content as well as syntax validator to check the syntactic correctness of datasets.

6 Conclusion and Future Works

In this paper, a set of metrics has been proposed for the assessment of a dataset before its release as a part of the LOD cloud. We have shown how concrete valid metrics can be developed for RDF datasets by implementing such metrics. The proposed metrics have been validated through empirical evaluations.

We are currently focusing on the extension of our work in two main directions *(i)* we are also considering to develop statistical models for predicting the quality dimensions of a dataset using the values of the related metrics. We have undertaken similar studies for building predictive models of quality from metrics in our prior research [15]; and *(ii)* while in this paper, we have focused on quality issues of datasets which can be avoided before release, the quality issues after interlinking into the LOD remain to be further explored.

References

1. Hogan, A., Harth, A., Passant, A., Decker, S., Polleres, A.: Weaving the pedantic web. In: 3rd International Workshop on Linked Data on the Web (2010)
2. Fürber, C., Hepp, M.: Using semantic web resources for data quality management. In: Cimiano, P., Pinto, H. (eds.) EKAW 2010. LNCS, vol. 6317, pp. 211–225. Springer, Heidelberg (2010)
3. Behkamal, B., Kahani, M., Paydar, S., Dadkhah, M., Sekhavaty, E.: Publishing Persian linked data; challenges and lessons learned. In: 5th International Symposium on Telecommunications (IST), pp. 732–737. IEEE (2010)
4. Paydar, S., Kahani, M., Behkamal, B.: Publishing data of ferdowsi university of mashhad as linked data. In: Computational Intelligence and Software Engineering (2010)
5. Zaveri, A., Rula, A., Maurino, A., Pietrobon, R., Lehmann, J., Auer, S.: Quality Assessment for Linked Data: A Survey. Accepted in Semantic Web Journal (2014). http://www. semantic-web-journal.net/content/quality-assessment-linked-data-survey
6. Lei, Y., Nikolov, A., Uren, V., Motta, E.: Detecting quality problems in semantic metadata without the presence of a gold standard. In: 5th International EON Workshop at International Semantic Web Conference, pp. 51–60 (2007)
7. Bizer, C., Cyganiak, R.: Quality-driven information filtering using the WIQA policy framework. Web Semant.: Sci., Serv. Agents World Wide Web 7, 1–10 (2009)
8. Brüggemann, S., Grüning, F.: Using ontologies providing domain knowledge for data quality management. In: Pellegrini, T., Auer, S., Tochtermann, K., Schaffert, S. (eds.) Networked Knowledge - Networked Media. SCI, vol. 221, pp. 187–203. Springer, Heidelberg (2009)
9. Naumann, F., Leser, U., Freytag, J.C.: Quality-driven integration of heterogeneous information systems. In: 25th International Conference on Very Large Data Bases (VLDB 1999), Edinburgh, Scotland, UK, pp. 447–458 (1999)

10. Pipino, L.L., Lee, Y.W., Wang, R.Y.: Data quality assessment. Commun. ACM **45**, 211–218 (2002)
11. ISO: ISO/IEC 25012- Software engineering - Software product Quality Requirements and Evaluation (SQuaRE). Data quality model (2008)
12. Peralta, V.: Data freshness and data accuracy: A state of the art. Instituto de Computacion, Facultad de Ingenieria, Universidad de la Republica (2006)
13. Eppler, M.J., Wittig, D.: Conceptualizing information quality: A review of information quality frameworks from the last ten years. In: 5th International Conference on Information Quality, pp. 83–96 (2000)
14. Behkamal, B., Kahani, M., Bagheri, E., Jeremic, Z.: A Metrics-Driven approach for quality Assessment of Linked open Data. J. Theoritical Appl. Electron. Commer. Res. **9**, 64–79 (2014)
15. Bagheri, E., Gasevic, D.: Assessing the maintainability of software product line feature models using structural metrics. Softw. Qual. J. **19**, 579–612 (2011)

NoSQL, NewSQL, Data Integration

A Framework of Write Optimization
on Read-Optimized Out-of-Core Column-Store
Databases

Feng Yu[1](✉) and Wen-Chi Hou[2]

[1] Youngstown State University, Youngstown, OH 44555, USA
fyu@ysu.edu
[2] Southern Illinois University, Carbondale, IL 62901, USA
hou@cs.siu.edu

Abstract. The column-store database features a faster data reading
speed and higher data compression efficiency compared with traditional row-based databases. However, optimizing write operations in the
column-store database is one of the well-known challenges. Most existing works on write performance optimization focus on main-memory
column-store databases. In this work, we investigate optimizing write
operation (update and deletion) on out-of-core (OOC, or external memory) column-store databases. We propose a general framework to work
for both normal OOC storage or big data storage, such as Hadoop Distributed File System (HDFS). On normal OOC storage, we propose an
innovative data storage format called Timestamped Binary Association
Table (or TBAT). Based on TBAT, a new update method, called Asynchronous Out-of-Core Update (or AOC Update), is designed to replace
the traditional update. On big data storage, we further extend TBAT
onto HDFS and propose the Asynchronous Map-Only Update (or AMO
Update) to replace the traditional update. Fast selection methods are
developed in both contexts to improve data retrieving speed. A significant improvement in speed performance is shown in the extensive
experiments when performing write operations on TBAT in normal and
Map-Reduce environment.

1 Introduction

Column-store databases (also known as columnar databases or column-oriented
databases) refer to the category of databases that vertically partition data and
separately store each column. The history of column-store database can be traced
back to 1970s when transposed files were implemented in the early development
of DBMS, followed by applying vertical partitioning as a technique of table
attribute clustering. By the mid 1980s, the advantage of a fully *decomposed
storage model* (*DSM*) over the traditional row-based storage model (*NSM* or
Normalized Storage Model) was studied [7,15,21].

TAXIR (TAXonomic Information Retrieval) is the first automatic application of column-store database focusing on biological information retrieval and

© Springer International Publishing Switzerland 2015
Q. Chen et al. (Eds.): DEXA 2015, Part I, LNCS 9261, pp. 155–169, 2015.
DOI: 10.1007/978-3-319-22849-5_12

management [5,9]. KDB and Sybase IQ were the first two commercially available column-store databases developed in 93 and 95, respectively. It's not until about 2005 when many open-source and commercial implementations of column-store databases took off [1]. The well-known column-store databases include: Apache Cassandra [18], Apache HBase [12], MonetDB [4], KDB, SAP HANA [10,11], and Vertica [19].

The data storage in a column-store database is vertically partitioned and sharded by projecting each column into a separate fragment. A vertical fragment is referred as a BAT (Binary Association Table) [3,4], which is stored contiguously on a large enough disk page in order to mitigate seeking overheads across multiple ranges of data. The data in each BAT is densely patched in order to improve I/O performance, and also rapidly compressed utilizing light-weight compression schema to improve storage efficiency.

Thanks to the DSM feature, the column-store database fits well into the write-once-and-read-many environment. Given the fact that the values of each attribute are stored separately in BATs, the column-store database works especially well for OLAP and data mining queries that retrieve a large number of tuples but only considers a small collection of attributes. It can simply retrieve only the attributes included in the query prediction without the need to read the entire tuple. Thus the information retrieving speed is much faster in a column-store database. Another featured benefit of the column-store database is data compression, which can reach a higher compression rate and higher speed than traditional row-based database. One of the major reasons is that the information entropy in the data of one column is lower compared to that of row-based data.

Optimizing write operations in a column-store database has always been a challenge. The data in a column-store database is vertically decomposed into BATs and randomly distributed over the storage. Furthermore, assuming the external memory storage is employed, there is a non-trivial probability that a BAT is too large to fit into one page on the storage. Therefore, the writing on a column-store database will be significantly delayed by ad hoc access to large BATs across multiple pages. Optimizing the write operations in a column-store database is a demanding request.

Existing works majorly focus on write optimizations in a main-memory column-store database. Krueger et al. [16,17] introduced the differential update to improve the write performance in MonetDB. A special columnar data structure called *delta buffer* is introduced to temporarily store decomposed row-based input data. However, to the best of our knowledge, very few works focused on optimizing the write performance on the *out-of-core* (*OOC* or *external memory*) column-store databases. Vertica [19], a column-store database on large volume OOC storage, introduces a specially designed data storage procedure called *k-safety* to ensure ACID of update transactions on large volumes of data and improve the data importation efficiency. Nevertheless, k-safety focuses more on the transaction control rather than the write performance improvement for high velocity update query streams.

With the data volume increasing in a surprising speed, big data storage has been widely adopted as the de-facto solution. Extending the column-store databases onto the big data file systems such as Google File Systems (GFS) [13] and Hadoop Distributed File Systems (HDFS) [20] has always been a demanding request. The state-of-the-art big data column-store databases such as BigTable [6] and HBase [2] store multidimensional data labeled by a special attribute, namely timestamp, and require global indexing for fast data retrieval. However, global indexing requires extra resources of both computing and storage and can lower the speed of the write operations when rebuilding the index.

In this research, we focus on optimizing the write operations (update and deletion) on an OOC column-store database. An operation called *Asynchronous Out-of-Core Update* (or *AOC Update*) is originally designed based on a new data structure called *Timestamped Binary Association Table* (or *TBAT*). In the Map-Reduce environment, an operation called *Asynchronous Map-Only Update* (or *AMO Update*) is novelly introduced utilizing the Map-Reduce [8] environment. Furthermore, a Map-Reduce selection algorithm is developed to enable fast data retrieval. A major contribution of this work is that TBAT and AMO update allow the users to flexibly define column data type without the need of any extra data structure such as the global index.

The rest of the paper is structured as follows. Section 2 is the background introduction of column-store databases. Section 3 states the proposed method of update and deletion optimization on the OOC column-store database. The extension of update optimization in the Map-Reduce environment is developed in Sect. 4. Based on TBAT, a framework of write optimization in the OOC columns-store database is proposed in Sect. 5. Experimental results are illustrated in Sect. 6. And Sect. 7 is the conclusion and future works.

2   Background of the Column-Store Database

The data structure of a column-store database exclusively uses *BATs* (*Binary Association Tables*). A BAT is a fragment of an attribute in the original row-based storage. It usually consists of an `oid` (Object Identifier) or `ROWID`, along with a column of attribute values, which in a pair is called a *BUN* (*Binary UNits*). It is a physical model in a column-store database and the sole bulk data structure it implements. The BAT is categorized in a special group of storage models called 'DSM' (Decomposed Storage Model) [7,15].

The row-based storage data is the original user input data, called the *front-end* data or *logical* data. To input the data into a column-store database, a mapping rule should be defined from the logical data structure to the physical data structure, namely BAT.

Example 1. (From Row-Based Table to BAT) Suppose the table name is `customer`, with `id` as the primary key.

    customer(id, name, balance)
    Primary Key: id

The row-based data is shown in Fig. 1(a). In a columnar database, this logical table will be decomposed into 3 BATs namely `customer_id`, `customer_name`, customer_balance. Each BAT contains two columns: an oid and an attribute value column with the column name as the corresponding column data type.

In the example, the logical table is fully decomposed into 3 BATs, Fig. 1(b)–(d), with each BAT contains one of the attributes. This is also referred to as *full vertical fragmentation* [1]. Full vertical fragmentation has many advantages. First of all, data accessing is efficient for queries accessing many rows but with fewer columns involved in the query. Another advantage is the reduction of the workload on the CPU and memory generated by OLAP and data mining queries, which typically consider only a few columns in a logical table.

Compared to fully vertical fragmentation, the other pattern is *partial vertical fragmentation* [14]. It assumes the prior knowledge of which columns are frequently accessed together. Also it employs the attribute usage matrix to determine optimal clustering of columns into vertical fragments. However, OLAP and data mining are application areas that indicate ad-hoc queries, as a good OLAP or data mining system must be able to quickly answer queries involving attributes of arbitrary combinations. Nevertheless, the partial vertical fragmentation is useful to detect the data block location in a distributed database system.

id	name	balance
1	Alissa	100.00
2	Bob	200.00
3	Charles	300.00

(a) Row-Based Table **customer**

oid	int
101	1
102	2
103	3

(b) BAT **customer_id**

oid	varchar
101	Alissa
102	Bob
103	Charles

(c) BAT **customer_name**

oid	float
101	100.00
102	200.00
103	300.00

(d) BAT **customer_balance**

Fig. 1. **customer** Data in Row-Based and Column-Store (BAT) format

3   OOC Update Optimization

Update on a column-store database can be categorized into two types according to the residence of the target data to be changed.

In-memory Update: In-memory update is to manipulate data in main-memory write optimized storage. Seeking the oid for the target tuple in the memory is prompt and straight forward.

Out-of-Core Update (or *OOC Update*): An OOC update is to change the data stored in the external storage device. In this work, we focus on the external storage composed of disk arrays. There are two major bottlenecks of updates on out-of-core data. First is to seek the oid(s) of the target tuple(s). This could be time consuming when the tuples are in a large volume and serialized on multiple pages in the database OOC storage. Secondly, once the oid of the target tuple is retrieved, there typically will be multiple values needed to be changed. Therefore, the database system needs to access several BATs across different pages on the OOC storage, which will generate more ad hoc random access costs.

Therefore, optimizing the OOC updates is a demanding request. We propose the *Asynchronous OOC Update* (or *AOC Update*) and a specialized data structure called *Timestamped BAT* to improve the writing performance caused by OOC update in a large volume OOC column-store database.

3.1   Timestamped BAT

Traditional BATs are composed of two columns in each BUN data pair, namely oid and `attribute value`. In this work, we propose as follows a specially designed decomposed storage model with simplicity in implementation, called **Timestamped BAT** (or **TBAT**) (Fig. 2).

```
struct TBUN{
    TIMESTAMP optime,
    ROWID oid,
    ATTRIBUTE_VALUE_TYPE attrv
}
struct TBAT{
    TBUN[size_of_TBAT] tbuns
}
```

Fig. 2. TBAT data structures

As the name suggests, each tuple, TBUN, in a TBAT is only different from a BUN in a traditional BAT in its timestamp value, optime, which is used to record the time when an insert, update, or deletion operation is performed on this BAT tuple. The data type of optime is a 4 byte TIMESTAMP, for example as in MySQL[1], that occupies a relatively small space. oid is the object identifier of ROWID type, and attrv is the attribute value corresponding to the oid.

Figure 3(a) and (b) depict the example of TBATs of customer_balance and customer_id, respectively. Specifically, customer_balance and customer_id are the two TBATs decomposed from the original base table customer. The reason for all optime to be the same is because the initial data are assumed to be inserted in one batch of insertion. In addition, the oids are assumed to start from 101 in this example.

3.2   Asynchronous Out-of-Core Update

In this work, we use on the updates on a single tuple as demonstration, which can be easily extended to a general update targeting a collection of tuples.

---

[1] http://dev.mysql.com/doc/refman/5.0/en/datetime.html.

optime	oid	float
time1	101	100.00
time1	102	200.00
time1	103	300.00

(a) TBAT customer_balance

optime	oid	id
time1	101	1
time1	102	2
time1	103	3

(b) TBAT customer_id

Fig. 3. TBAT examples

In traditional BAT, an update on a tuple involves 2 phases. Phase one is to seek for the oid according to the given target ID or value. The second phase is update the target value according to the oid found in phase one. The second phase will generate random I/O on OOC storage. The TBAT is specifically designed for updates on large volume OOC storage, where both seeking oid and changing values are time consuming, due to the extra costs on data block seeking and writing.

Based on TBAT, we propose the *Asynchronous OOC Update* (or *AOC Update*). The principle of AOC update is to avoid OOC seeking and writing in every effort and to use the timestamp field of TBAT to label the newly updated data that is directly appended to the end of a TBAT. In such a manner, we don't have to frequently perform ad hoc data seeking and writing but finish this work after the system peak time. Later on, during the non-peak time, an off-line *data cleaning* will be performed on the TBAT to merge duplicated tuples with same oid but different timestamps (or optime's) and attribute values.

An Example of an AOC Update. Without loss of generality we use an example of OOC update targeting on a single tuple. It can be easily generalized to any update targeting on a collection of tuples. A SQL query of an example of AOC update is shown as follows.

```
UPDATE customer SET balance = 201.00
WHERE id = 2
```

The record with oid equals 102 in the TBAT customer_balance is the target tuple. The target value is to change the attribute value from the original value to 201.00. Instead of seeking the position to the record with oid=102, AOC update directly appends at the end of the TBAT a new tuple as (time2, 102, 201.00). The timestamp when AOC update is performed is assumed to be time2, and 201.00 is the newly updated value. The TBAT customer_balance after the AOC update is illustrated in Table 1.

Cost Analysis of the AOC Update. Assuming the database keeps the final record position in TBAT customer_balance, then the only cost of the AOC update is to look for the oid and data writing on the last position of TBAT customer_balance. Seeking oid in TBAT customer_id is inevitable and costs the same as on BAT. However, this can be overcome by utilizing indexes or

Table 1. TBAT `customer_balance` after AOC update

optime	oid	float
time1	101	100.00
time1	102	200.00
time1	103	300.00
time2	*102*	*201.00*

calculating the functional relationship between `id` and `oid`, assuming the `ids` are essentially contiguous.

Updating in TBAT `customer_balance` is much faster than on a traditional BAT since only one data appending is involved. Without the help of any index or function relationship, the complexity of one update on BAT is $O(\log n)$, if the binary search is used on a sorted BAT. However, the complexity of one AOC update is $O(1)$ when the last record position in TBAT is kept by the database. In a data intensive and high velocity data input environment, AOC updates will save significant amounts of CPU, memory, and disk overhead by just utilizing TBAT.

Selection After the AOC Update. AOC update will not hurt the data consistency on the TBAT. A straightforward proof can be given by observing the selection on a TBAT after an AOC update. We continue to use the previous example and select the `balance` of customer with `id=2` after the previous update query is executed. The selection query is as follows.

```
SELECT balance FROM customer WHERE id=2
```

Since `customer_id` is intact, seeking the `oid` of `id=2` is fast. After `oid=102` is retrieved, in TBAT `customer_balance`, two tuples will be returned

```
t1=(time1, 102, 200.00)
t2=(time2, 102, 201.00)
```

As we compare the timestamps, `time2` is later than `time1`. Then `201.00` is returned which is consistent with the last update value.

3.3   Deletion Optimization

We propose an asynchronous fashion deletion based on TBAT called *Asynchronous OOC Deletion* (or *AOC Deletion*). The AOC deletion can be derived from the AOC update by setting the timestamp to be **null** value. For instance, suppose a deletion query is issued as follows.

```
DELETE FROM customer WHERE id=2
```

The target is to delete all the data in `customer` table with `oid=102`. In traditional BAT, the deletion query will demand seeking and deleting on multiple BATs. However, in TBAT, the only operation needed is as follows.

```
UPDATE customer_id
SET timestamp=NULL where id=2
```

Namely, a tuple equals to (`NULL`, `102`, `2`) will be appended to the last position of TBAT `customer_id`, without accessing other TBATs.

After the AOC deletion, the selection query retrieving the tuples with `id=2` will obtain two tuples from TBAT `customer_id` i.e. `t1=(time1, 102, 2)` and `t2=(NULL, 102, 2)`. We can define a rule that, if there is a timestamp equals to `NULL` in any returned tuples, then it is treated as a deleted tuple. And this rule can be easily implemented.

Because any operations on `customer` table requires accessing the TBAT `customer_id` to retrieve the `oid` first, any following queries will be aware of the deletion of tuple with `oid=102` based on the `id` in the query. Thus data consistency is guaranteed after AOC deletion.

4   Update on Column-Stores in Map-Reduce

In this section, we focus on the external storage in Map-Reduce, for instance HDFS. Each file in the HDFS is sharded into multiple chunks and distributed over the cluster. Optimizing the OOC updates is a demanding request. We proposed the *Asynchronous Map-Only Update* (or *AMO Update*) to improve the writing performance caused by OOC update in an OOC column-store database on HDFS.

4.1   Update on BAT in Map-Reduce

Algorithm 1 describes a typical procedure of updates on BAT in Map-Reduce. Searching for target records in a BAT in nested loops is always not efficient. This phase can be converted into an outer join operation between the BAT and the update list (of BUNs). Algorithm 1 adjusts to the size of arbitrary update list. When the update list size is small enough to fit into the memory, a map-side outer join is performed. Otherwise, a common reduce-side outer join can be performed when the update size is too large. In both cases, the join result is retained on HDFS as the intermediate result. A filtering phase is performed that the `value` and `oid` of the target BUN is always retained if there is an update on the updating location.

4.2   Timestamped BAT in Map-Reduce

TBAT can be easily extended into Map-Reduce environment. The only difference with normal storage is that, TBAT in Map-Reduce should be considered to be

---

Algorithm 1. BAT_UPDATE_MR

Input: tbat: the TBAT file to update; update_list: the list of BUN's to update; threshold: max size of update_list to fit into memory

1: **if** sizeof(update_list) $\leq$ threshold **then**
2:     temp = bat $\bowtie_{\text{map-side}}$ update_list ON OID   ▷ map-side left outer join
3: **else**
4:     temp = bat $\bowtie_{\text{reduce-side}}$ update_list ON OID   ▷ reduce-side left outer join
5: **for all** line $\in$ temp **do**
6:     **if** line.update_list.OID != NULL **then**   ▷ if matched in the outer join
7:         output(BUN(line.update_list.OID, line.update_list.VALUE))
8:     **else**
9:         output(BUN(line.bat.OID,line.bat.VALUE))
10: **return** SUCCESS

---

sharded into multiple slips with each slip no bigger than the maximum file block size in HDFS.

The major improvement of TBAT in HDFS is that, first of all, TBAT do not require any global pre-sorting or indexing. Based on TBAT, we propose an update algorithm and a selection algorithm that can achieve efficient data updating and retrieving without any help of extra global data structure. Compared with the previous lessons learned on the distributed index, the proposed design is resilient to the similar distributed failure problems. Secondly, a user-defined attribute type `attrv` to be included in a TBUN. Therefore, the user can flexibly define arbitrary kinds of schema translated from relational database model.

4.3   Asynchronous Map-Only Update on Column-Stores

Based on TBAT, we propose the *Asynchronous Map-Only Update* (or *AMO Update*). The principle of AMO update is to avoid seeking and writing in every effort and to use the timestamp field of TBAT to label the newly updated data that is directly appended to the end of a TBAT. In such a manner, we don't have to frequently perform ad hoc data seeking and writing by simply accepting multiple versions of TBUN data with the same `oid` but different attribute values and timestamps.

The AMO update is simple and straightforward in Map-Reduce. We simply describe the procedure of the AMO update as follows. The update list of target BUNs is also assumed to be collected over distributed environment. Once the update list is submitted for execution, the mapper of the AMO update simply append the list of updating TBUNs at the end of the TBAT file. In HDFS, there is only one mapper operation involved that simply shards the update list into slips and flushes them to the distributed storage. The file append operation has been supported in Apache Hadoop since the 0.20.0 release.

Compared with the update on BAT, the cost of AMO update is significantly lower than the BAT update, since the AMO update is a map-only procedure to append update list while the BAT update involves in a more expensive outer join operation in Map-Reduce.

4.4   Map-Reduce Selection on TBAT

AMO update will not hurt the data consistency on TBAT. There could be multiple versions of the same oid data with different attriv and optime stored distributively on the HDFS. Algorithm 2 describes the selection on TBAT in Map-Reduce, given a selection range and a target oid. It can be easily extended to the selection with an input list of oids. There are two filtering phases involved in the algorithm. The MAPPER performs the first filtering on the map side, where, for each TBAT slip, the MAPPER selects only the TBUNs with their oids fall into the given selection range. The emitted the key-value pair of the MAPPER is comprised of the oid of the selected TBUN and a nested pair, which consists of the value of the TBUN and the optime or timestamp. The emitted key-value pairs will be shuffled and sorted according to the key, i.e. the oid of the TBUN selected by MAPPER, and then sent to the reducer. For each oid, the REDUCER algorithm on the reduce side will select the pair, as the target pair, with the most recent timestamp from the input list of pairs. Finally the value of the target pair is returned as one result satisfying the selection range.

Algorithm 2. TBAT_SELECTION_MR

1: **procedure** MAPPER(tbat_block: a block of TBAT file in HDFS; oid_range: range of selection for OID's)
2:     **for all** line ∈ bat_block **do**
3:         **if** line.OID ∈ oid_range **then**
4:             emit(line.OID, new pair(line.VALUE, line.TIMESTAMP))
5:                 ▷ The emitted key is the OID and the emitted value is a nested pair.
6: **procedure** REDUCER(oid: OID of a TBUN in a reduced TBAT chunk; pairs: the list of pairs associated with the same OID)
7:     target_pair=max$_{\text{TIMESTAMP}}$(pair_list)   ▷ only select the most recent pair
8:     output(oid, target_pair.VALUE)

---

5   A Write-Optimized Framework

Based on the proposed data structure and write operations, we develop a comprehensive write-optimized framework in a read-optimized column-store database. Different from traditional works, our assumption doesn't restrict the size of the data stored in the column-store database to be less than the available system memory. We assume there is a data stream constantly inputting data into the database without prior knowledge of the total size of data input.

Figure 4 illustrates our proposed framework for write optimization. The two major modules of the database in the proposed framework are the Write-Optimized Module (WOM) and Read-Optimized Module (ROM). The WOM is mainly used to host the data from the data input stream, which is assumed to be mainly from network I/O and thus is slower than the system memory I/O.

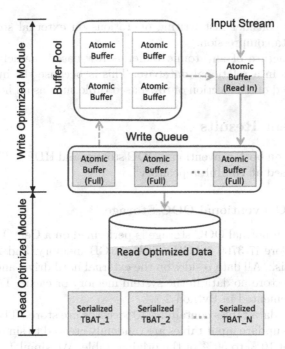

Fig. 4. A write-optimized framework

The first concern in a out-of-core column-store database is to make sure receiving the input data is non-stopped. Therefore, specialized buffers, called *atomic buffers*, are created to temporarily host a part of the inputed data stream. Since the size of the input stream data is unknown, a pool of atomic buffers is created in the system memory to constantly provide data reception. The size of each *atomic buffer* is limited. Once an atomic buffer is full, the WOM will perform a *buffer substitution*. Namely, an empty atomic buffer is retrieved from the buffer pool to seamlessly provide data reception from the input stream. At the same time, the atomic buffer that has been just filled up will be moved to the *write buffer queue* waiting to be written to the external storage. We assume the external storage is a disk array large enough to host the input data stream within a limited amount of time.

One of the key issue is to determine the size of an atomic buffer. First of all, the size of an atomic buffer cannot be too large, so that enough buffers can be allocated in the atomic buffer pool to constantly perform buffer substitutions and provide seamless data reception. Second, for each atomic buffer in the write queue, it will be decomposed into several TBAT segments and written into the external storage. The best size of each TBAT segment is the same as the data block size on the storage; as a result, there will be no extra cost for switching blocks during the writing phase on external storage. Thus, in general, if the input tuple from the data stream has m attributes, an atomic buffer will be mapped into mTBATs, the size of an atomic buffer is $m \times$ data_block_size. For simplicity interest, we only consider fast writing operations on external storage without extra cost for data compression.

The data structure of an atomic buffer is row-based, namely, a collection of tuples same as in a relational database. This is because the input stream is typically composed of a collection of row-based tuple such as ad hoc user inputs.

6   Experiment Results

We perform tests on both conventional OOC storage and HDFS. The experiment test code is released at GitHub repository[2].

6.1   Tests on Conventional OOC Storage

The experiment on normal OOC storage is performed on a CentOS 6.5 workstation with Intel Core i7-3700 3.4 GHz CPU, 16 GB memory, and 250 GB SATA 7200RPM hard disk. All data resides on the external hard drive, and the test program is forced to store no data in the system memory or cache. The experiment test code is implemented in Python 2.7.

Two synthetic data tables containing 10^4 records are stored TBAT and BAT, respectively. Five update input tables are randomly created to simulate the input stream consists of 10 % to 50 % of the original table. We simulate both updates on BATs and TBATs.

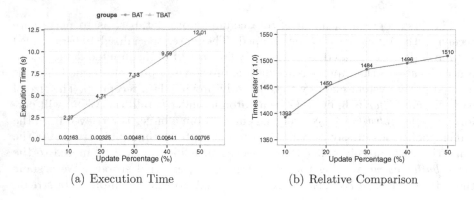

(a) Execution Time   (b) Relative Comparison

Fig. 5. Conventional OOC storage update tests

The absolute running time of AOC updates on TBAT and traditional updates on BAT are shown in Fig. 5(a). It's obvious that the AOC updates are magnitudes faster than traditional updates. From the detailed results, the average running time of AOC updates is 4.81E-03 s, while the average running time of traditional updates is 7.14 s. We also compare their running time relatively in Fig. 5(b), where the AOC update is on average 1466.436 times faster than the traditional update.

[2] https://github.com/YSU-Data-Lab/TBAT-DEXA15.

The difference between AOC updates and traditional updates grows greater with the size of the data increases due to the growth of the tuple search overhead of traditional updates.

6.2   Tests on HDFS

The experiment on HDFS is designed in order to compare the speed performance between AMO updates on TBATs and traditional updates on BATs in Map-Reduce. The experiment is performed on a Cloudera Distributed Hadoop (CDH) version 5.3 cluster with one master node and three slave nodes. The embedded Hadoop version is 2.5.0, and the total HDFS storage capacity is 310 GB with the block size of 64 MB. We use the Gigabit Ethernet as the cluster interconnection. The experiment test code is implemented in Java SE 1.7 and Apache Pig Latin version 0.12.0.

On HDFS, synthetic BAT datasets of sizes 1 GB and 10 GB are randomly generated, consisting of an `oid` column and an `attriv` (attribute value) column to simulate a large BAT. Then a TBAT is derived from the BAT with an additional `optime` (timestamp) attribute. For each dataset, five update input tables

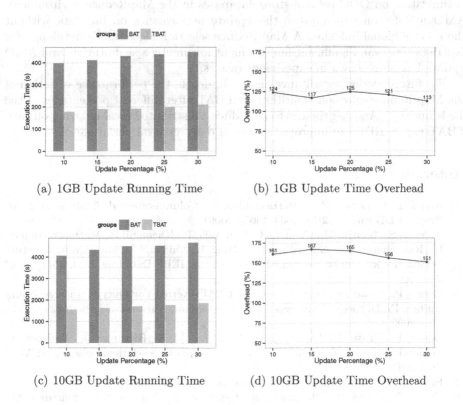

(a) 1GB Update Running Time   (b) 1GB Update Time Overhead

(c) 10GB Update Running Time   (d) 10GB Update Time Overhead

Fig. 6. HDFS update tests

are uniformly generated consists of 10 % to 30 % of the original table. We use these tables to simulate the list of update targets.

The absolute running time of AMO updates on TBAT and traditional updates on BAT are shown in Fig. 6(a) and (c). The detailed average running time of AMO updates on 1 GB and 10 GB data are 194 and 1698 s, while the average running time of traditional updates are 425 and 4413 s. The relative overhead of running time is defined as $\left| \frac{\text{time(BAT)} - \text{time(TBAT)}}{\text{time(BAT)}} \right| \times 100\%$ where time(TBAT) is the AMO update running time on TBAT. As shown in Fig. 6(b) and (d), the AMO update is averagely 120 and 160 times faster than the traditional update in the 1 GB and 10 GB tests.

7 Conclusion and Future Works

In this research, we introduce a new method called AOC update for write optimization on OOC column-store databases. AOC update employs an innovative data structure called TBAT to improve the update performance with data atomicity guaranteed. In addition, we extended AOC update to AMO update for write optimization on OOC column-store databases in the MapReduce environment. AMO update aims to improve the update performance on big data without the need of global indexing. A Map-Reduce selection algorithm is developed for fast data retrieval. Significant improvement in running speed of AOC and AMO update has been shown in experiment results.

For future works, we will investigate, in depth, the performance variation of the Map-Reduce selection algorithm on TBAT after different percentages of the file is updated. Another topic is to introduce a distributed local indexing on each TBAT slip in HDFS to improve the global data retrieval performance.

References

1. Abadi, D.J., Boncz, P.A., Harizopoulos, S.: Column-oriented database systems. Proc. VLDB Endow. **2**(2), 1664–1665 (2009)
2. Aiyer, A.S., Bautin, M., Chen, G.J., Damania, P., Khemani, P., Muthukkaruppan, K., Ranganathan, K., Spiegelberg, N., Tang, L., Vaidya, M.: Storage infrastructure behind facebook messages: using HBase at scale. IEEE Data Eng. Bull. **35**(2), 4–13 (2012)
3. Boncz, P.: Monet: A Next-Generation DBMS Kernel For Query-Intensive Applications. Ph.D. thesis, Universiteit van Amsterdam, Amsterdam, The Netherlands, May 2002
4. Boncz, P., Grust, T., Van Keulen, M., Manegold, S., Rittinger, J., Teubner, J.: Monetdb/xquery: a fast xquery processor powered by a relational engine. In: ACM SIGMOD, pp. 479–490 (2006)
5. Brill, R.: The Taxir Primer. ERIC, Washington, D.C (1971)
6. Chang, F., Dean, J., Ghemawat, S., Hsieh, W.C., Wallach, D.A., Burrows, M., Chandra, T., Fikes, A., Gruber, R.E.: Bigtable: a distributed storage system for structured data. ACM Trans. Comput. Syst. **26**(2), 4:1–4:26 (2008)

7. Copeland, G.P., Khoshafian, S.N.: A decomposition storage model. In: Proceedings of ACM SIGMOD Record, vol. 14, pp. 268–279. ACM (1985)
8. Dean, J., Ghemawat, S.: MapReduce: a flexible data processing tool. Commun. ACM **53**(1), 72–77 (2010)
9. Estabrook, G.F., Brill, R.C.: The theory of the taxir accessioner. Math. Biosci. **5**(3), 327–340 (1969)
10. Färber, F., Cha, S.K., Primsch, J., Bornhövd, C., Sigg, S., Lehner, W.: SAP HANA database: data management for modern business applications. SIGMOD Rec. **40**(4), 45–51 (2012)
11. Färber, F., May, N., Lehner, W., Große, P., Müller, I., Rauhe, H., Dees, J.: The SAP HANA database - an architecture overview. IEEE Data Eng. Bull. **35**(1), 28–33 (2012)
12. George, L.: HBase: The Definitive Guide. O'Reilly Media Inc., CA (2011)
13. Ghemawat, S., Gobioff, H., Leung, S.-T.: The google file system. SIGOPS Oper. Syst. Rev. **37**(5), 29–43 (2003)
14. Gluche, D., Grust, T., Mainberger, C., Scholl, M.: Incremental updates for materialized OQL views. In: Bry, François (ed.) DOOD 1997. LNCS, vol. 1341, pp. 52–66. Springer, Heidelberg (1997)
15. Khoshafian, S., Copeland, G.P., Jagodis, T., Boral, H., Valduriez, P.: A query processing strategy for the decomposed storage model. In: Proceedings, pp. 636. Order from IEEE Computer Society (1987)
16. Krueger, J., Grund, M., Tinnefeld, C., Plattner, H., Zeier, A., Faerber, F.: Optimizing write performance for read optimized databases. In: Kitagawa, H., Ishikawa, Y., Li, Q., Watanabe, C. (eds.) DASFAA 2010. LNCS, vol. 5982, pp. 291–305. Springer, Heidelberg (2010)
17. Krueger, J., Kim, C., Grund, M., Satish, N., Schwalb, D., Chhugani, J., Plattner, H., Dubey, P., Zeier, A.: Fast updates on read-optimized databases using multi-core cpus. Proc. VLDB Endow. **5**(1), 61–72 (2011)
18. Ladwig, G., Harth, A.: Cumulusrdf: linked data management on nested key-value stores. In: The 7th International Workshop on Scalable Semantic Web Knowledge Base Systems (SSWS 2011), p. 30 (2011)
19. Lamb, A., Fuller, M., Varadarajan, R., Tran, N., Vandiver, B., Doshi, L., Bear, C.: The vertica analytic database: C-store 7 years later. Proc. VLDB Endow. **5**(12), 1790–1801 (2012)
20. White, T.: Hadoop: The Definitive Guide, 2nd edn. O'Reilly, CA (2010)
21. Zukowski, M., Nes, N., Boncz, P.: Dsm vs. nsm: Cpu performance tradeoffs in block-oriented query processing. In: DaMoN 2008, pp. 47–54. ACM, New York (2008)

Integrating Big Data and Relational Data with a Functional SQL-like Query Language

Carlyna Bondiombouy[1], Boyan Kolev[1(✉)], Oleksandra Levchenko[1,2], and Patrick Valduriez[1]

[1] Inria and LIRMM, University of Montpellier, Montpellier, France
{carlyna.bondiombouy,boyan.kolev,
oleksandra.levchenko,patrick.valduriez}@inria.fr
[2] Odessa National Polytechnic University, Odessa, Ukraine
aleksandra.levchenko@gmail.com

Abstract. Multistore systems have been recently proposed to provide integrated access to multiple, heterogeneous data stores through a single query engine. In particular, much attention is being paid on the integration of unstructured big data typically stored in HDFS with relational data. One main solution is to use a relational query engine that allows SQL-like queries to retrieve data from HDFS, which requires the system to provide a relational view of the unstructured data and hence is not always feasible. In this paper, we introduce a functional SQL-like query language that can integrate data retrieved from different data stores and take full advantage of the functionality of the underlying data processing frameworks by allowing the ad hoc usage of user defined map/filter/reduce operators in combination with traditional SQL statements. Furthermore, the query language allows for optimization by enabling subquery rewriting so that filter conditions can be pushed inside and executed at the data store as early as possible. Our approach is validated with two data stores and a representative query that demonstrates the usability of the query language and evaluates the benefits from query optimization.

1 Introduction

A major trend in cloud computing and big data is the understanding that there is "no one size fits all" solution. Thus, there has been a blooming of different cloud data management solutions, such as NoSQL, distributed file systems (e.g. Hadoop HDFS), and big data processing frameworks (e.g. Hadoop MapReduce or Apache Spark), specialized for different kinds of data and able to perform orders of magnitude better than traditional RDBMS. However, this has led to a wide diversification of data store interfaces and the loss of a common programming paradigm. This makes it very hard for a user to integrate and analyze her data sitting in different data stores, e.g. RDBMS, NoSQL and HDFS. To address this problem, multistore systems [1, 8–10] have been

---

Work partially funded by the European Commission under the Integrated Project CoherentPaaS [5] and BACKIS program (for Oleksandra Levchenko).

Q. Chen et al. (Eds.): DEXA 2015, Part I, LNCS 9261, pp. 170–185, 2015.
DOI: 10.1007/978-3-319-22849-5_13

recently proposed to provide integrated access to multiple, heterogeneous data stores through a single query engine.

Compared to multidatabase systems [11], multistore systems trade source autonomy for efficiency, using a tightly-coupled approach. In particular, much attention is being paid on the integration of unstructured big data (e.g. produced by web applications) typically stored in HDFS with relational data, e.g. in a data warehouse. One main solution is to use a relational query engine (e.g. Apache Hive) on top of a data processing framework (e.g. Hadoop MapReduce), which allows SQL-like queries to retrieve data from HDFS. However, this requires the system to provide a relational view of the unstructured data, which is not always feasible. In case the data store is managed independently from the relational query processing system, complex data transformations need to take place (e.g. by applying specific map-reduce jobs) before the data can be processed by means of relational operators. Let us illustrate the problem, which will be the focus of this paper, with the following scenario.

Example Scenario. An editorial office needs to find appropriate reporters for a list of publications based on given keywords. It can analyze a scientific forum stored in a Hadoop cluster in the cloud to find an expert in a certain research field regarding particular keywords and the users who have mentioned them most frequently; and these results must be joined to the relational data in an RDBMS containing author and publication information. However, the forum application keeps log data about its posts in a non-tabular structure, namely in text files where a single record corresponds to one post and contains a fixed number of fields about the post itself (timestamp and username in the example) followed by a variable number of fields storing the keywords mentioned in the post.

2014-12-13, alice, storage, cloud		KW	expert	freq
2014-12-13, alice, storage, cloud		cloud	alice	2
2014-12-22, bob, cloud, virtual, app	$\rightarrow$	storage	alice	1
2014-12-24, alice, cloud		virtual	bob	1
		app	bob	1

The unstructured log data needs to be transformed into a tabular dataset containing for each keyword the expert who mentioned it most frequently (as illustrated above). Such transformation requires the use of programming techniques like chaining map/reduce operations. Then the result dataset will be ready to be joined with the publication data retrieved from the RDBMS in order to suggest an appropriate reviewer for each publication. Being able to request such data processing with a single query is the scenario that motivates our work. However, the challenge in front of the query processor is optimization, i.e. it should be able of analyzing the operator execution flow of a query and performing operation reordering to take advantage of well-known optimization techniques (e.g. selection pushdowns and use of semi-joins) in order to yield efficient query execution.

Existing solutions to integrate such unstructured and structured data do not directly apply to solve our problem, as they rely on having a relational view of the unstructured data, and hence require complex transformations. SQL engines, such as Hive, on top of distributed data processing frameworks are not always capable of querying unstructured HDFS data, thereby forcing the user to query the data by defining map/reduce functions.

Our approach is different as we propose a query language that can directly express subqueries that can take full advantage of the functionality of the underlying data processing frameworks. Furthermore, the language should allow for query optimization, so that the query operator execution sequence specified by the user may be reordered by taking into account the properties of map/filter/reduce operators together with the properties of relational operators. Finally, we want to respect the autonomy of the data stores, e.g. HDFS and RDBMS, so that they can be accessible and controlled from outside our query engine with their own interface.

In this paper, we propose a functional SQL-like query language and query engine to retrieve data from two different kinds of data stores – an RDBMS and a distributed data processing framework such as Apache Spark or Hadoop MapReduce on top of HDFS – and combine them by applying data integration operators (mostly joins). We assume that each data store is fully autonomous, i.e. the query engine has no control over the structure and organization of data in the data stores. However, users need to be aware of how data is organized across both data stores, so that they write valid queries. A single query of our language can request data to be retrieved from both stores and then a join to be performed over the retrieved datasets. The query therefore contains embedded invocations to the underlying data stores, expressed as subqueries. A subquery, addressing the data processing framework, is represented by a sequence of map/filter/reduce operations, expressed in a formal notation. On the other hand, SQL is used to express subqueries that address the relational data store as well as the main statement that performs the integration of data retrieved by all subqueries. Thus, a query benefits from both high expressivity (by allowing the ad hoc usage of user defined map/filter/reduce operators in combination with traditional SQL statements) and optimizability (by enabling subquery rewriting so that filter conditions can be pushed inside and executed at the data store as early as possible).

The rest of this paper is organized as follows. Section 2 introduces the language and its notation to express map/filter/reduce subqueries. Section 3 presents the architecture of the query engine. Section 4 elaborates the properties of map/filter/reduce operators and gives rewrite rules to perform query optimization. Section 5 is an experimental validation with structured data stored in PostgreSQL and unstructured data stored in an HDFS cluster and processed using Apache Spark. Section 6 discusses related work. Section 7 concludes.

2   Query Language

The query language is part of a more general common query language, called CloudMdsQL [4], designed in the context of the CoherentPaaS project [5] to solve the problem of querying multiple heterogeneous databases (e.g. relational and NoSQL) within a single query while preserving the expressivity of their local query mechanisms. The common language itself is SQL-based with the extended capabilities for embedding subqueries expressed in terms of each data store's native query interface. In this section, we introduce a formal notation to define Map/Filter/Reduce

(MFR) subqueries in CloudMdsQL that request data processing in an underlying big data processing framework (DPF). Notice that the data processing defined in an MFR statement is not executed by the query engine, but is translated to a sequence of invocations to API functions of the DPF. In this paper, we use Apache Spark as an example of DPF, but the concept can be generalized to a wider range of frameworks that support the MapReduce programming model (like Hadoop, CouchDB, etc.).

2.1   MFR Notation

An MFR statement represents a sequence of MFR operations on datasets. A dataset is considered simply as an abstraction for a set of tuples, where a tuple is a list of values, each of which can be a scalar value or another tuple. Although tuples can generally have any number of elements, mostly datasets that consist of key-value tuples are being processed by MFR operations. In terms of Apache Spark, a dataset corresponds to an RDD (Resilient Distributed Dataset – the basic programming unit of Spark). Each of the three major MFR operations (MAP, FILTER and REDUCE) takes as input a dataset and produces another dataset by performing the corresponding transformation. Therefore, for each operation it should be specified the transformation that needs to be applied on tuples from the input dataset to produce the output tuples. Normally, a transformation is expressed with an SQL-like expression that involves special variables.

Core Operators. The MAP operator produces key-value tuples by performing a specified transformation on the input tuples. The transformation is defined as an SQL-like expression that will be evaluated for each tuple of the input data set and should return a pair of values. The special variable TUPLE refers to the input tuple and its elements are addressed using a bracket notation. Moreover, the variables KEY and VALUE may be used as aliases to TUPLE[0] and TUPLE[1] respectively. The FILTER operator selects from the input tuples only those, for which a specified condition is evaluated to *true*. The filter condition is defined as a boolean expression using the same special variables TUPLE, KEY and VALUE. The REDUCE operator performs aggregation on values associated with the same key and produces a key-value dataset where each key is unique. The reduce transformation may be specified as an aggregate function (SUM, AVG, MIN, MAX or COUNT). Similarly to MAP, FLAT_MAP operator is introduced, which may produce numerous output tuples for a single input tuple.

Let us consider the following simple example inspired by the popular MapReduce tutorial application "word count". We assume that the input dataset for the MFR statement is a list of words. To count the words that contain the string 'cloud', we write the following composition of MFR operations:

```
MAP(KEY, 1).FILTER(KEY LIKE '%cloud%').REDUCE(SUM)
```

The first operation transforms each tuple (which has a single word as its only element) of the input dataset into a key-value pair where the word is mapped to a value of 1. The second operation selects only those key-value pairs for which the key contains the string 'cloud'. And the third one groups all tuples by key and performs a sum aggregate on the values for each key.

To process this statement, the query engine first looks for opportunities to optimize the execution by operator reordering. By applying MFR rewrite rules (explained in detail in Sect. 4.2), it finds out that the FILTER and MAP operations may be swapped so that the filtering is applied at an earlier stage. Further, it translates the sequence of operations into invocations of the underlying DPF's API. Notice that whenever a REDUCE transformation function has the associative property (like the SUM function), a combiner function call will be generated that precedes the actual reducer, so that as much data as possible will be reduced locally. In the case of Apache Spark as the DPF, the query engine generates the following Python fragment to be included in a script that will be executed in Spark's Python environment:

```
dataset.filter( lambda k: 'cloud' in k ) \
      .map( lambda k: (k, 1) ) \
      .combineByKey( lambda a, b: a + b ) \
      .reduceByKey( lambda a, b: a + b )
```

In this example, all the MFR operations are translated to their corresponding Spark functions and all transformation expressions are translated to Python anonymous functions. In fact, to increase its expressivity, the MFR notation allows direct usage of anonymous functions to specify transformation expressions. This allows user-defined mapping functions, filter predicates, or aggregates to be used in an MFR statement. The user, however, needs to be aware of how the query engine is configured to interface the DPF, in order to know which language to use for the definition of inline anonymous functions (e.g. Spark may be used with Python or Scala, CouchDB – with JavaScript, etc.).

Input/output operators are normally used for transformation of data before and after the core map/filter/reduce execution chain. The SCAN operator loads data from its storage and transforms it to a dataset ready to be consumed by a core MFR operator. The PROJECT operator converts a key-value dataset to a tabular dataset ready to be involved in relational operations.

2.2   Combining SQL and MFR

Queries that integrate data from both a relational data store and a DPF usually consist of two subqueries (one expressed in SQL that addresses the relational database and another expressed in MFR that addresses the DPF) and an integration SELECT statement. A subquery is defined as a named table expression, i.e. an expression that returns a table and has a name and signature. The signature defines the names and types of the columns of the returned relation. Thus, each query, although agnostic to the underlying data stores' schemas, is executed in the context of an ad hoc schema, formed by all named table expressions within the query.

For example, the following simple query contains two subqueries, defined by the named table expressions T1 and T2, and addressed respectively against the data stores aliased with identifiers rdb (for the SQL database) and hdfs (for the DPF):

```
T1(title string, kw string)@rdb = ( SELECT title, kw FROM tbl )
T2(word string, count int)@hdfs = {*
   SCAN(TEXT,'words.txt')
      .MAP(KEY,1).REDUCE(SUM).PROJECT(KEY,VALUE)
*}
SELECT title, kw, count FROM T1 JOIN T2 ON T1.kw = T2.word
WHERE T1.kw LIKE '%cloud%'
```

The purpose of this query is to perform relational algebra operations (expressed in the main SELECT statement) on two datasets retrieved from a relational database and a DPF. The two subqueries are sent independently for execution against their data stores in order the retrieved relations to be joined by the query engine. The SQL table expression T1 is defined by an SQL subquery. T2 is an MFR expression that requests data retrieval from a text source and data processing by the specified map/reduce operations. Both subqueries are subject to rewriting by pushing into it the filter condition kw LIKE '%cloud%', specified in the main SELECT statement, thus reducing the amount of the retrieved data by increasing the subquery selectivity and the overall efficiency. The so retrieved datasets are then converted to relations following their corresponding signatures, so that the main SELECT statement can be processed with semantic correctness. The PROJECT operator in the MFR statement provides a mapping between the dataset fields and the named table expression columns.

3 Query Engine

The dominant state-of-the-art architectural model that addresses the problem of data integration and query processing across a diverse set of data stores is the mediator/wrapper architecture. A mediator is a software module that exploits encoded knowledge about certain sets or subsets of data to create information for a higher layer of applications [11]. In addition a wrapper or an adapter is a software component that encapsulates and hides the underlying complexity of sets or subsets of data by means of well-defined interfaces (it establishes communication and a data flow between mediators and data stores). In this section, we briefly describe the architecture of our system with an overview of the required steps to process a query.

Our query engine follows the traditional mediator/wrapper architectural approach. By expressing a query between the RDBMS and parallel programming framework, the system integrates structured (Relational DB) and unstructured (distributed storage, based on HDFS) data. Figure 1 depicts the system architecture, containing a common query processor (the mediator), two wrappers, a distributed data processing framework (DPF) and an RDBMS (Relational Database Management System) data stores. The DPF performs parallel data processing over a distributed data store. In this architecture, each data source has an associated wrapper that is responsible for executing subqueries against the data store and to convert the retrieved datasets to tables matching the requested number and types of columns, so that they are ready to be consumed by relational operators at the query engine. The query processor centralizes

the information provided by the wrappers and integrates the subqueries' results, while
the wrappers transform subqueries expressed in a common language into queries for the
data stores.

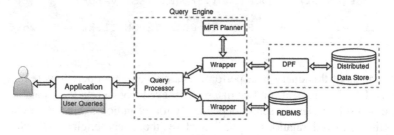

Fig. 1. Basic architecture of the query engine

Each of the wrappers needs to complete the execution of subqueries and to retrieve
the results. The wrapper of the relational database runs SQL statements against its data
store; it retrieves the datasets and delivers them to the query processor in the corresponding format. The wrapper of the distributed data processing framework needs to do
more. First and foremost it interprets a subquery written in MFR notation, then uses the
MFR planner to find optimization opportunities, and finally invokes the corresponding
sequence of methods of the DPF's API. The MFR planner decides where to insert
pushed down operations; e.g. it performs operation reordering to find the optimal place
of a filter operation in order to apply it as early as possible and thus to reduce the query
execution cost. To search for alternative operation orderings, the planner takes into
account MFR rewrite rules, introduced in next section.

4   Query Rewriting

Before its actual execution, a query may be rewritten by the query engine in order to
request early execution of some operators and thus to increase its overall efficiency.
Generally, this is done in two stages: first, the query processor determines which
operations can be pushed down for remote execution at the data stores; and second, the
MFR planner may further determine the optimal place for inclusion of pushed down
operations within the MFR operator chain by applying MFR rewrite rules.

4.1   Operation Pushdowns

Although several operations are subject to pushdowns across subqueries, in this paper
we concentrate on the inclusion of only filter operations inside an MFR subquery,
which takes place mostly in two occasions: (a) propagation of selection predicates and
(b) performance of bind joins [7], which is a method for implementing semi-joins
across heterogeneous data stores that uses subquery rewriting to push join conditions.
   Pushing a selection operation inside a subquery, either in SQL query or MFR
operation chain, is always considered beneficial, because it delegates the selection

directly to the data store, which allows to early reduce the size of data processed and retrieved from the data stores.

Using bind joins between a relational table and big data, either when requested explicitly in the query or as a result of optimization decision, allows to reduce the communication cost between the DPF and the query engine. This approach implicates that the list of distinct values of the join attribute(s) from the relation, preliminarily retrieved from the relational data store, is passed as a filter to the MFR subquery. For example, let us consider the following fragment:

SELECT H.x, R.y FROM R JOIN H ON R.id = H.id WHERE R.z = 'abc'

To process this query using the bind join method, first, the table R is retrieved from the relational data store; then, assuming that the distinct values of R.id are $r_1 \dots r_n$, the condition id IN $(r_1, \dots, r_n)$ is passed as a FILTER to the MFR subquery that retrieves the dataset H from HDFS data store. Thus, only the tuples from H that match the join criteria are retrieved.

The optimizer decides whether to use bind join in every individual case. If the number of distinct values is big, using a bind join may slower the performance as it requires data to be pushed into the subquery. In the example above, the query engine first asks the RDBMS (e.g. by running an EXPLAIN statement) for an estimation of the cardinality of data retrieved from R, after rewriting the SQL subquery by including the selection condition R.z = 'abc'. If the estimated cardinality does not exceed a certain threshold, the optimizer plans for performing a bind join that can significantly increase the MFR subquery selectivity and affect the volume of transferred data.

4.2 MFR Rewrite Rules

We enumerate the rules for reordering and rewriting of MFR operators, based on their algebraic properties.

Rule #1 (name substitution): upon pushdown, the filter is included just before the PROJECT operator and the filter predicate expression is rewritten by substituting column names with references to dataset fields as per the mapping defined by the PROJECT expressions. After this initial inclusion, other rules apply to determine whether it can be moved even farther. Example:

```
T1(a int, b int)@db1 = {* ....PROJECT(KEY, VALUE[0]) *}
SELECT a, b FROM T1 WHERE a > b
```
is rewritten to:
```
T1(a int, b int)@db1 = {* ....FILTER(KEY > VALUE[0]).PROJECT(KEY,
VALUE[0])*}
SELECT a, b FROM T1
```

Rule #2: REDUCE(< transformation >).FILTER(< predicate >) is equivalent to FILTER(< predicate >).REDUCE(< transformation >), if predicate condition is a function only of the KEY, because thus, applying the FILTER before the REDUCE will preserve the values associated to those keys that satisfy the filter condition as they would be if the FILTER was applied after the REDUCE.

Rule #3: `MAP(< expr_list >).FILTER(< predicate1 >)` is equivalent to `FILTER(< predicate2 >).MAP(< expr_list >)`, where `predicate1` is rewritten to `predicate2` by substituting `KEY` and `VALUE` as per the mapping defined in `expr_list`. Example:

```
MAP(VALUE[0], KEY).FILTER(KEY > VALUE) →
FILTER(VALUE[0] > KEY).MAP(VALUE[0], KEY)
```

Since planning a filter as early as possible always increases the efficiency, the planner always takes advantage of moving a filter by applying rules #2 and #3 whenever they are applicable.

5   Validation

In this section, we give a validation of multistore query processing and query rewriting techniques using our query engine. We reveal the steps the query engine takes to process a query using selection pushdown and bind join as optimization techniques. We also focus on the way the query engine dynamically rewrites the MFR subquery to perform a bind join.

Experimental setup. We developed the query processor in Java, based on Apache Derby's query engine. The wrappers for the data stores are Java classes implementing a common interface used by the query processor to interact with them. We use two distinct data stores: PostgreSQL as the relational database (referred to as `rdb`) and an HDFS cluster (referred to as `hdfs`) processed using the Apache Spark framework. The wrapper for `rdb` executes SQL statements against the relational database using its JDBC driver. To process subqueries against HDFS, we developed a wrapper for Apache Spark, which accepts subqueries defined using MFR notations that need to be translated to sequences of Spark operators and then executed against HDFS. The wrapper uses Spark's Python API, and thus translates each transformation to Python lambda functions. Besides, it also accepts raw Python lambda functions as transformation definitions. To evaluate the impact of optimization on query execution, we use a cluster based on the GRID5000 platform (www.grid5000.fr), with one node for the query engine, one node for `rdb` and 2 or more nodes for `hdfs`.

Datasets. We generated sample datasets in the context of the multistore query example described in Sect. 1. The `rdb` database stores structured data about publications and their keywords in the following tables:

Pubs
(id int, author varchar, title varchar)

id	author	Title
1	Ricardo	Snapshot Isolation in Cloud DBs
3	Patrick	Principles of Distributed Cloud DBs
7	Larri	Graph DBs

Pubs_KW
(pub_id int, keyword varchar)

pub_id	keyword
1	transactions
1	cloud
3	cloud
3	storage
7	graph
7	NoSQL

HDFS stores unstructured log data from a scientific forum in text files where a single record corresponds to one post and contains a timestamp and username followed by a variable number of fields storing the keywords mentioned in the post:

```
Posts (date, author, kw₁, kw₂, …, kwₙ)
 2014-11-10, alice, storage, cloud
 2014-11-10, bob, cloud, virtual, app
 2014-11-10, alice, cloud
```

Query 1. This query aims to find appropriate reviewers for publications of a certain author. It considers each publication's keywords and the experts who have mentioned them most frequently on the scientific forum. The query combines data from two data stores and can be expressed as follows.

```
T1(autor, title, kw)@rdb = ( SELECT author, title, keyword
                             FROM pubs JOIN pubs_kw ON id = pub_id )
T2(kw, expert)@hdfs = {*
  SCAN(TEXT, 'posts.txt', ',')                                    (op1)
    .FLAT_MAP( lambda data: product(data[2:], [data[1]]) )        (op2)
    .MAP( TUPLE, 1 )                                              (op3)
    .REDUCE( SUM )                                                (op4)
    .MAP( KEY[0], (KEY[1], VALUE) )                               (op5)
    .REDUCE( lambda a, b: b if b[1] > a[1] else a )               (op6)
    .PROJECT(KEY, VALUE[0])                                       (op7)
  *}
SELECT T1.title, T1.kw, T2.expert
FROM T1, T2
WHERE T1.kw = T2.kw AND T1.author = 'Patrick'
```

Query 1 contains two subqueries – against the relational database and the DPF. The first subquery is a typical SQL statement to get data about scientific publications stored in the relational database. The second subquery is an MFR operation chain that transforms the unstructured log data from the forum posts and represents the result of text analytics as a relation that maps each keyword to the person who has most frequently mentioned it. To achieve the result dataset, the MFR operations request transformations over the stored data, each of which is expressed either in a declarative way or with anonymous (lambda) Python functions.

The SCAN operation **op1** reads data from the specified text source and splits each line to an array of values. Let us recall that the produced array contains the author of the post in its second element and the mentioned keywords in the subarray starting from the third element. The following FLAT_MAP operation **op2** consumes each emitted array as a tuple and transforms each tuple using the defined Python lambda function, which performs a Cartesian product between the keywords subarray and the author, thus emitting a number of keyword-author pairs. Each of these pairs is passed to the MAP operation **op3**, which produces a new dataset, where each keyword-author pair is mapped to a value of 1. Then the REDUCE operation **op4** aggregates the number of occurrences for each keyword-author pair. The next MAP operation **op5** transforms the dataset by mapping each keyword to a pair of author-occurrences. The REDUCE **op6** finds for each keyword the author with the maximum number of occurrences, thus finding the expert who has mostly used the keyword. Finally, the PROJECT defines the mapping between the dataset fields and the columns of the returned relation.

Query Processing. First, the query processor compiles Query 1 into the preliminary execution plan, depicted in Fig. 2(a). Then, the query optimizer finds the opportunity for pushing down the condition author = 'Patrick' into the relational data store. Thus, the selection condition is included in the WHERE clause of the subquery for rdb. This pushdown implies increasing the selectivity of the subquery, which is identified by the optimizer as an opportunity to plan for performing a bind join. To further verify this opportunity, the query engine asks rdb to estimate the cardinality for the rewritten SQL subquery. Furthermore, the MFR planner seeks for opportunities to move the bind join filter condition < bj_cond > earlier in the MFR operation chain by applying the MFR rewrite rules, explained below. At this stage, although < bj_cond > is not known (as it depends on the data retrieved from rdb), the planner has all the information needed to apply the rules. After these transformations, the optimized query plan (Fig. 2(b)) is executed by the query processor.

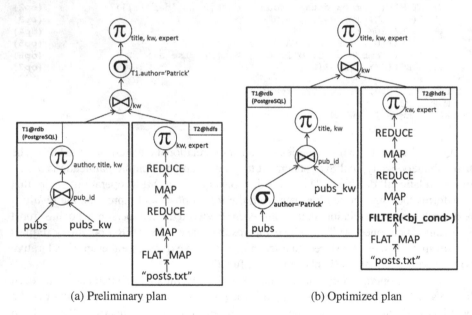

(a) Preliminary plan   (b) Optimized plan

Fig. 2. Preliminary query plan for Query 1 and its corresponding plan after transformations

To transform and execute the query plan, the query engine takes the following steps:

1. The query processor delivers to the wrapper of rdb the following SQL statement, rewritten by taking into account the pushed selection condition, for execution against the PostgreSQL data store, and waits for the corresponding result set to be

retrieved in order to compose the bind join condition that will be pushed inside the MFR subquery.

```
SELECT title, keyword
FROM pubs JOIN pubs_kw ON id = pub_id
WHERE author = 'Patrick'
```

Title	keyword
Principles...	cloud
Principles...	storage

2. As soon as the relation is retrieved, the query processor identifies a list of the distinct values of the join attribute `keyword` and derives from it the bind join condition `kw IN ('cloud', 'storage')` to push inside the subquery against `hdfs`.

3. The MFR planner for the wrapper of `hdfs` decides at which stage of the MFR sequence to insert the filter, by applying a number of rewrite rules. According to rule #1, the planner initially inserts the filter just before the PROJECT **op7** by rewriting the condition expression as follows:

`.FILTER(KEY IN ('cloud', 'storage'))`

Next, by applying consecutively rules #2 and #3, the planner moves the FILTER before the MAP **op5** by rewriting its condition expression according to rule #3:

`.FILTER(KEY[0] IN ('cloud', 'storage'))`

Analogously, rules #2 and #3 are applied again, moving the FILTER before **op3**, rewriting the expression once again, and thus settling it to its final position. After all transformations the MFR subquery is converted to the final MFR expression below.

```
# Final MFR expression:
SCAN(TEXT, 'posts.txt', ',')

.FLAT_MAP( lambda ... )
.FILTER(TUPLE[0]
        IN ('cloud', 'storage'))
.MAP( TUPLE, 1 )
.REDUCE(SUM)


.MAP(KEY[0], (KEY[1], VALUE))

.REDUCE( lambda ... )
```

```
# The corresponding Python code:
sc.textFile('posts.txt')
   .map(lambda line: line.split(','))
   .flatMap( lambda ... )
   .filter(lambda tup: tup[0]
           in ['cloud', 'storage'])
   .map(lambda tup: (tup, 1))
   .combineByKey( lambda a, b: a + b )
   .reduceByKey( lambda a, b: a + b )
   .map(lambda tup:
          (tup[0][0], (tup[0][1], tup[1]))  )
   .reduceByKey( lambda ... )
```

4. The wrapper interprets the reordered MFR sequence, translates it to the Python script above as per the Python API methods of Spark, and executes it within the Spark framework. The result of MFR query reordering and interpreting on Spark is the intermediate relation:

kw	expert
cloud	alice
storage	alice

5. The intermediate relations from steps 1 and 4 are joined to produce the final result
   that lists the suggested experts for each publication regarding the given keywords:

Title	kw	expert
Principles...	cloud	alice
Principles...	storage	alice

Query Optimization. To evaluate the impact of query rewriting and optimization on
execution time, we developed a simple analytical model, based on the execution of
representative queries on PostgreSQL and Spark/HDFS. But for lack of space, we do
not give the details here, only the simplifying assumptions. The main assumption is that
the dominant cost is incurred by network communication and I/Os. And since we use
big data sizes, we can ignore the impact of caching, assumed to be the same for
different query plans, and the cost of initializing processing, assumed to be constant.
Finally, we assume uniform data distribution to compute the selectivity of intermediate
results. The important cost parameters are fixed as follows: network (Infiniband)
bandwidth = 10 Gigabits/sec; HDD I/O bandwidth = 3 Gigabits/sec; RDB size = 0.1
Terabyte; HDFS size = 1 Terabyte in total for all nodes in the cluster. To evaluate the
size of the results retrieved from both data stores, we fixed the selectivity factor of the
WHERE clause predicate to 5 %. We assume the presence of an index on the filtered
column in PostgreSQL, so that a pushed down selection reduces also the I/O cost.

Figure 3 shows the execution times (in seconds) obtained for 3 different query
execution plans of Query 1, illustrating the benefits from applying the different optimization techniques: (1) the initial query plan without any optimization, (2) after
pushing down the selection inside the relational subquery, and (3) after pushing the
bind join condition into the MFR subquery and further moving it to an early FILTER
operation by MFR reordering. We evaluate the costs with 3 different configurations of
the HDFS cluster, with 2, 4 and 8 nodes, to assess scalability.

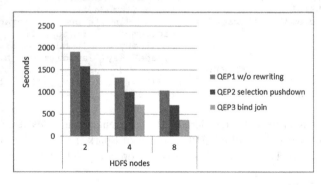

Fig. 3. Execution costs for 3 alternative plans and 3 different HDFS configurations.

The first performance benefit we observe (the difference between the costs of QEP2
and QEP1) is due to the selection pushdown into the SQL subquery as it filters the
retrieved data, reducing according to the selectivity factor both the I/O cost to read data
from the PostgreSQL database and the network cost to transfer the data to the mediator.

The difference between the costs of QEP3 and QEP2 reflects the performance gains from applying the bind join method. It reduces according to the same selectivity factor (as we assume uniform data distribution) the amount of data processed within the HDFS cluster by excluding the tuples that do not match the join criteria. This affects mostly the communication cost for shuffling data across the HDFS nodes at each reduce operation. That is why the relative benefit of the bind join optimization is higher in configurations with higher number of nodes.

The performance evaluation illustrates the query engine's ability to perform optimization and choose the most efficient execution plan.

6   Related Work

The problem of accessing heterogeneous data sources has long been studied in the context of multidatabase and data integration systems [11]. More recently, with the advent of cloud databases and big data processing frameworks, the solution has evolved towards multistore systems that provide integrated access to a number of RDBMS, NoSQL and HDFS data stores through a common query engine. We can divide multistore systems between loosely-coupled, tightly-coupled and hybrid.

Loosely-coupled multistore systems provide support for autonomous data stores. DISCO [13] is a data integration system for accessing Web data sources, using an operator-based approach. It combines a generic cost model with specific cost information provided by the data source wrappers, thus allowing flexible cost estimation. BigIntegrator [10] integrates data from NoSQL big data, such as Google's Bigtable, and relational databases. The system relies on mapping a limited set of relational operators to native queries expressed in GQL (Google Bigtable query language). With GQL, the task is achievable because it represents a subset of SQL. However, it only works for Bigtable-like systems and cannot integrate data from HDFS. Estocada [2] is a self-tuning multistore platform for providing access to datasets in native format while automatically placing fragments of the datasets across heterogeneous stores.

Tightly-coupled multistore systems trade autonomy for performance, typically in a shared-nothing cluster. Odyssey [8] enables storing and querying data within HDFS and RDBMS, using opportunistic materialized views. MISO [9] is a method for tuning the physical design of a multistore system (Hive/HDFS and RDBMS), i.e. deciding in which data store the data should reside, in order to improve the performance of big data query processing. The intermediate results of query execution are treated as opportunistic materialized views, which can then be placed in the underlying stores to optimize the evaluation of subsequent queries. JEN [14] allows joining data from two data stores, HDFS and RDBMS, with parallel join algorithms, in particular, an efficient zigzag join algorithm, and techniques to minimize data movement. As the data size grows, executing the join on the HDFS side appears to be more efficient. Polybase [6] is a feature of Microsoft SQL Server Parallel Data Warehouse to access HDFS data using SQL. It allows HDFS data to be referenced through external PDW tables and joined with native PDW tables using SQL queries.

Hybrid systems are either tightly- or loosely-coupled systems, and enable big data (e.g. in HDFS) to be accessed through a data processing framework like MapReduce or

Spark. HadoopDB [1] provides Hadoop MapReduce/HDFS access to multiple single-node RDBMS servers (e.g. PostgreSQL or MySQL) deployed across a cluster, as in a shared-nothing parallel DBMS. It interfaces MapReduce with RDBMS through database connectors that execute SQL queries to return key-value pairs. QoX [12] uses a dataflow approach for optimizing queries over relational (RDBMS and ETL) data and unstructured (HDFS) data, with a black box approach for cost modeling. SCOPE [3] is a declarative scripting language from Microsoft to specify the processing of large sequential files stored in Cosmos, a distributed computing platform. SCOPE provides selection, join and aggregation operators and allows the users to implement their own operators and user-defined functions. SCOPE expressions and predicates are translated into C#. The SELECT command and join operations are written as in SQL, but subqueries are not allowed in the FROM clause. In addition it allows implementing custom extractors, processors and reducers and combining operators for manipulating rowsets. SCOPE has been extended to combine SQL and MapReduce operators in a single language [15].

Our work fits in the hybrid system category. However, it is not limited to MapReduce, as we can support other frameworks like Spark. Furthermore, it does not give up data store's autonomy, yet allowing for optimization through the use of MFR subqueries and operator reordering. Thus, our approach is more general.

7 Conclusion

In this paper, we proposed a functional SQL-like query language and query engine to integrate data from RDBMS and big data stores (such as HDFS). Our query language can directly express subqueries that can take full advantage of the functionality of the underlying data processing frameworks. Furthermore, it allows for query optimization, so that the query operator execution sequence specified by the user may be reordered by taking into account the properties of map/filter/reduce operators together with the properties of relational operators. Finally, compared with the related work on multistore systems, our work fits in the hybrid system category. However, it is not limited to MapReduce, as most systems, and does not give up data store's autonomy, thus making our approach more general.

Our validation demonstrates that the proposed query language achieves the following requirements. First, it is highly expressive by using map/filter/reduce notation, as it was demonstrated with the hdfs subquery. Second, it is optimizable as was demonstrated in two stages: first, by pushing a selection as a WHERE clause condition inside PostgreSQL subquery; and second, by performing bind join with rewriting the MFR subquery after retrieving the dataset from the PostgreSQL database. Finally, it allows reducing the amount of processed data during the execution of the MFR sequence by rewriting/reordering MFR operators according to the determined rules. Our performance evaluation illustrates the query engine's ability to optimize a query and choose the most efficient execution strategy.

References

1. Abouzeid, A., Badja-Pawlikowski, K., Abadi, D., Silberschatz, A., Rasin, A.: HadoopDB: an architectural hybrid of MapReduce and DBMS technologies for analytical workloads. PVLDB **12**, 922–933 (2009)
2. Bugiotti, F., Bursztyn, D., Deutsch, A., Ileana, I., Manolescu, I.: Invisible glue: scalable self-tuning multi-stores. In: CIDR Conference (2015)
3. Chaiken, R., Jenkins, B., Larson, P., Ramsey, B., Shakib, D., Weaver, S., Zhou, J.: SCOPE: easy and efficient parallel processing of massive data sets. PVLDB **1**, 1265–1276 (2008)
4. CloudMdsQL. http://cloudmdsql.gforge.inria.fr
5. CoherentPaaS project. http://coherentpaas.eu
6. DeWitt, D., Halverson, A., Nehme, R., Shankar, S., Aguilar-Saborit, J., Avanes, A., Flasza, M., Gramling, J.: Split query processing in polybase. In: ACM SIGMOD Conference, pp. 1255–1266 (2013)
7. Haas, L., Kossmann, D., Wimmers, E., Yang, J.: Optimizing queries across diverse data sources. In: International Conference on Very Large Databases (VLDB), pp. 276–285 (1997)
8. Hacigümüs, H., Sankaranarayanan, J., Tatemura, J., LeFevre, J., Polyzotis, N.: Odyssey: a multi-store system for evolutionary analytics. PVLDB **6**, 1180–1181 (2013)
9. LeFevre, J., Sankaranarayanan, J., Hacigümüs, H., Tatemura, J., Polyzotis, N., Carey, M.: MISO: souping up big data query processing with a multistore system. In: ACM SIGMOD Conference, pp. 1591–1602 (2014)
10. Minpeng, Z., Tore, R.: Querying combined cloud-based and relational databases. In: International Conference on Cloud and Service Computing (CSC), pp. 330–335 (2011)
11. Özsu, T., Valduriez, P.: Principles of distributed database systems. Springer, New York (2011)
12. Simitsis, A., Wilkinson, K., Castellanos, M., Dayal, U.: Optimizing analytic data flows for multiple execution engines. In: ACM SIGMOD Conference, pp. 829–840 (2012)
13. Tomasic, A., Raschid, L., Valduriez, P.: Scaling access to heterogeneous data sources with DISCO. IEEE Trans. Knowl. Data Eng. **10**, 808–823 (1998)
14. Yuanyuan, T., Zou, T., Özcan, F., Gonscalves, R., Pirahesh, H.: Joins for hybrid warehouses: exploiting massive parallelism and enterprise data warehouses. In: EDBT/ICDT Conference, 12 p. (2015)
15. Zhou, J., Bruno, N., Wu, M., Larson, P., Chaiken, R., Shakib, D.: SCOPE: parallel databases meet MapReduce. PVLDB **21**, 611–636 (2012)

Comparative Performance Evaluation of Relational and NoSQL Databases for Spatial and Mobile Applications

Pedro O. Santos, Mirella M. Moro$^{(\boxtimes)}$, and Clodoveu A. Davis Jr.

Universidade Federal de Minas Gerais, Belo Horizonte, Brazil
{pedroo,mirella,clodoveu}@dcc.ufmg.br

Abstract. Selecting the appropriate data management infrastructure is still a hard task for the designers of mobile applications with large volumes of data. Considering NoSQL needs for such applications, this paper demonstrates how the physical implementation of the database may impact query performance. Specifically, we consider the needs of mobile users that involve constant spatial data traffic, such as querying for points of interest, map visualization, zooming and panning, routing and location tracking. We define a workload and process such queries over three types of databases: relational, document-based and graph-based. Our evaluation shows that a fair comparison requires specific workloads for each mobile feature, but that is not possible using the industry's standard benchmark tools. Overall, the paper shows that physical design must evolve to take advantage of the performance of NoSQL databases while keeping data consistency and integrity.

Keywords: Performance · Spatial databases · NoSQL · Big data

1 Introduction

Geographic information systems (GIS) that provide public services often manage very large databases with lots of concurrent user requests. GIS applications require continuous availability, low response time and high throughput. Also, the underlying geographic database manager deals with spatial objects with varying complexity, for which operations are much more expensive than in relational database management systems (RDBMS) for conventional data [3,12].

Specifically, the demand for GIS and spatially-enhanced applications, especially for mobile devices, has increased alongside the user base, making RDBMS even less suitable for the upcoming challenges. In this context, geographic location resources such as integrated Global Positioning System (GPS) modules made spatially-aware applications more attractive and useful. At the same time, such applications are more frequently used, reaching a point where most features accessed by the user contain geographic data and metadata [4].

Examples of spatially enhanced applications and GIS (in which geographic data is uploaded along with conventional data and metadata) include

© Springer International Publishing Switzerland 2015
Q. Chen et al. (Eds.): DEXA 2015, Part I, LNCS 9261, pp. 186–200, 2015.
DOI: 10.1007/978-3-319-22849-5_14

Foursquare[1] and Waze[2]. Such applications' spatial abilities make them so useful for so many people that a billion-dollar industry exists around them. Hence, the application's performance, reliability and scalability play very important roles in the user's decision to acquire and use such tools in a daily routine.

From the perspective of design and development of the databases underlying such applications, new requirements are becoming important. The trend towards replacing relational DBMS reflects such requirements [11] and propose taking out of the DBMS the responsibility for ensuring ACID properties, especially when large datasets are involved. This indicates a drive towards achieving performance gains in applications for which querying is intense and updating is relatively rare, which is the case for many spatial applications. However, maintaining the spatial database that supports such applications requires consistency assurance, which is not trivial to achieve on spatial data. Spatial integrity constraints may need to be enforced during data entry and updating, requiring the implementation of sophisticated functions in the DBMS or in the data updating application. Hence, it is still not clear how to select an adequate spatial data management tool, considering the requirements of current applications, especially mobile ones, because studies show wide differences on evaluating queries over general and spatial big data (e.g., [12] among many others).

Here, the goal is to identify strong and weak points in the performance of data management tools with spatial capabilities. We propose a methodology to compare spatial DBMSs over a range of queries common to spatial mobile applications. Then, we evaluate the performance of those DBMSs over diverse workloads, in order to demonstrate the wide range of behavior of such tools according to usual query types. The DBMSs considered are[3] PostgreSQL[4] with PostGIS[5] and pgrouting[6], MongoDB[7], and Neo4j[8] with Neo4j-Spatial[9]. Our evaluation results show that performance varies greatly in terms of algorithms and overall computational complexity, due to each systems' design. Hence, selecting a spatial DBMS heavily depends on how well it handles the *targeted* spatial data and on the most usual functions the system is expected to perform.

Next, Sect. 2 goes over related work and presents the DBMSs considered in our study. Section 3 introduces our methodology by explaining each processing stage, the dataset and queries performed. Section 4 presents the experimental evaluation with setup, implementation details, parameters and results. Section 5 concludes this paper and proposes future work.

---

[1] Foursquare: http://www.foursquare.com.

[2] Waze: http://www.waze.com.

[3] Using other DBMS requires adapting the database model and queries employed.

[4] PostgreSQL: http://www.postgresql.org/.

[5] PostGIS: http://postgis.net/.

[6] pgRouting: http://pgrouting.org/.

[7] MongoDB: http://www.mongodb.org/.

[8] Neo4j: http://www.neo4j.org/.

[9] Neo4j-Spatial: http://www.neo4j.org/develop/spatial.

2   Related Work

This section summarizes studies on query processing, spatial data extensions and mobile spatial applications in the perspective of our main contributions.

Regarding **performance evaluation of spatial databases**, recent studies on spatial joins discuss how applications rely on fast spatial processing methods (e.g., [12]). They also show there is still much room for improvement regarding spatial queries: noticeable performance gains can be achieved by changing the indexes' data structure and the access algorithms, as well as using multi-threaded implementations. In the mobile applications scenario, there are many particularities regarding data traffic, speed variations under different areas of coverage and mobile devices. Those characteristics also influence the GIS design, as immediate, limitless data requests may not be very efficient carrier-wise [7].

Even though our work does *not* focus on benchmarking, performance assessment for spatial databases is still an open issue and is the subject of several research initiatives. The Spatial Star Schema Benchmark [8] includes a data generator that is specific for spatial data warehouses, intending to create synthetic workloads. For big data, the benchmarking tool proposed by Ghazal et al. [5] also includes a synthetic data generator and a data model. Likewise, Chen et al. [2] propose a functional workload model for big data by extending TPC-C and other benchmark combined features. Even though synthetic datasets are usually favored for evaluating queries over string and numeric attributes, the equivalent synthetic generated geometries are not similar to real-world data, because the randomness of the generator results in overly squared polygons and unnatural land shapes. Furthermore, the mix of man-made and natural features usually found in GIS varies according to the region and is strongly related to the application, making it harder to accurately represent real workloads. On the relational world, Jackpine is a more generic benchmark for spatial relational databases [10], which is not ideal for evaluating mobile applications and their huge volume of data: (*i*) it covers only relational databases, while spatial Big Data and mobile applications usually employ NoSQL as well; (*ii*) it is not appropriate for measuring the effects of data volume and complexity (e.g., spatially joining regions delimited by polygons with thousands of vertices to sets of millions of points); and (*iii*) mobile applications often need network-based attributes and functions, which may be cumbersome (if not impossible) to implement over a RDBMS.

The wide variety of experimental datasets, benchmarking methodologies, and workload characterization strategies indicates how hard it is to achieve agreement as to a representative method to characterize spatial database performance. This is hardly surprising, considering how diverse the operations in such a database are: geometric operators, spatial analysis functions, topological predicate processing, and many others are components that are used in varying proportions in GIS applications. Furthermore, the inherent complexity of the underlying dataset influences performance, as parameters such as the average number of vertices per geometric shape vary depending on data source, scale, digitizing method, and other factors. In this work, we use real datasets rather than synthetically generated data, as described later.

Spatial data extensions already appear in several DBMS. However, their internal architectures, data representation choices and query processing strategies can weigh on their efficiency. Also, when focusing on the mobile systems scenario, data properties and volume vary greatly from other GIS applications. Here, we consider three DBMSs that include spatial extensions and provide a license-free installation. They also correspond to three data representation and management paradigms, with a direct effect on query processing and optimization strategies.

The first DBMS is **PostgreSQL with PostGIS.** PostgreSQL is an open source relational DBMS that supports all major operational systems, and is ACID and ANSI-SQL:2008 compliant. PostGIS is an open source geographic extension for PostgreSQL that allows spatial queries to run on SQL, over geometry types that have been standardized by the Open Geospatial Consortium (OGC[10]) and ISO. PostGIS supports an R-tree-over-GiST (Generalized Search Tree) spatial index and implements most spatial predicates and operators defined by the OGC. We also use the pgrouting library to execute routing functions in a network structure.

The second is the NoSQL platform **MongoDB.** MongoDB is an open source document-oriented DBMS with a data model based on key-value associations. It uses the JavaScript Object Notation (JSON) with spatial features based on GeoJSON (an extension that specifies geometry objects such as points, linestrings, polygons, multipoints, multilinestrings and multipolygons). GeoJSON queries on MongoDB require a spatial index, and the spatial queries can use only a small set of operators, such as inclusion, intersection and proximity.

The third DBMS is the NoSQL native graph database **Neo4j with Neo4j-Spatial.** Neo4j is a graph-oriented open source DBMS whose core features include data model flexibility based on graph elements (e.g. nodes, edges and attributes), graph traversing performance and scalability, supporting billions of nodes and relationships and implementing full ACID transactions. Neo4j Spatial is an extension that provides spatial querying over geometry attributes stored in nodes. Moreover, its spatial indexes are built over the graph structure provided by Neo4j. It also implements most OGC spatial predicates and operators.

Each of these DBMSs has positive and negative aspects for managing **mobile spatial applications**. Typical mobile applications, services and systems involve providing localized parts of a large and detailed dataset to multiple users, while allowing for data entry and spatial querying on the vicinity of the user. Such needs require efficient implementations of a relatively small number of key spatial functions, which are representative of the most important types of queries, backed by efficient indexing and retrieval. Scalability is also important, since the number of devices (smartphones, in-vehicle devices, etc.) rapidly increases. Mobile applications have become one of the biggest spatial data providers [9]. Even systems and applications that do not directly deal with spatial data can employ spatial metadata. For example, when taking a picture on a smartphone, geographical metadata is stored within the picture's file, allowing the user and

---

[10] OGC: http://www.opengeospatial.org/.

friends to trace the photo back to the location in which it was taken. Other examples include instant messaging applications, which trace the user's location and enables the other party to view it on a map. There are also websites that use the user's location to provide specific content, such as advertising, user profiling and location-based recommendations.

Finally, our evaluation differs from the related work presented here since our objective is not to propose a new benchmark, but to highlight aspects in which different spatial DBMS architectures behave differently. In that aspect, we intend to demonstrate the inherent difficulties in comparing the performance of spatial DBMSs. We designed a testing interface where SQL and NoSQL data can be equally tested (even though with different designs, features and goals). We propose a new metric (vertices per second) that is more tailored for evaluating mobile queries over spatial big data of varying complexity. This way, we are able to express the DBMS's efficiency considering the complexity of the geometric features, instead of only measuring how fast it processes a query. We also provide means to compare graph-based features obtained from spatial data. Nowadays, such features are very common in large systems, but they are usually managed using two distinct models (networks and spatial objects).

3   Comparison Methodology

Our methodology for comparing spatially-enabled DBMSs has four main objectives. First: to use real-world spatial data, for spatial queries and indexes are usually complex and expensive, making it essential to consider data as close to reality as possible. Second: to define a set of queries, insertions and updates that focus on mobile systems, as the mobile applications scenario potentially uses and produces large quantities of spatial information. Third: to create an evaluation tool to which different spatial DBMSs may be easily added, thus keeping future compatibility with new software versions and DBMS, and making it easier to perform extended comparative analyses in the future. Last: to demonstrate how current performance evaluation for GIS must be data and feature-oriented, in opposition to the current generic benchmark tools that have wide coverage but little specificity. Next, our methodology's processing stages is presented in Sect. 3.1, the dataset and workload in Sect. 3.2, the data loading procedure in Sect. 3.3, and the queries in Sect. 3.4.

3.1   Processing Stages

The methodology follows five stages: data loading, workload generation, warm-up, evaluation and shutdown. The data loading, workload generator and evaluation stages should be run separately from the others. Also, running any step should trigger the shutdown stage automatically at the end of the run.

Data Loading. This stage loads shapefiles[11] (a popular geospatial data file format) into the databases that are going to be analyzed. MongoDB imports a

---

[11] Shapefiles: http://www.esri.com/library/whitepapers/pdfs/shapefile.pdf.

different file format, GeoJSON[12], and the conversion from shapefiles is performed by the Geospatial Data Abstraction Library[13]. This step should run separately, since spatial data loading may take several hours or even days, depending on data volumes and on the DBMS's import tools.

Workload Generation. The second stage generates the workload to be processed. Here, the necessary parameters for the evaluation stage are extracted from the same spatial dataset built from the shapefiles. This procedure defines queries with the least amount of null results or sequential scans, ensuring that the DBMS uses the indexing and cache mechanisms available. The result from queries that select random entries in the shapefiles or the DBMS with loaded data is then written in a file to be read by the evaluation stage.

Warm-Up. This stage performs several queries to load indexes and other memory-related attributes that affect the DBMS's performance. When it ends, the amount of memory consumed and its memory mappers (if that is the case) are recorded. The warm-up phase is necessary as DBMSs in general try to avoid disk requests by storing index references in RAM, which is also how queries will be processed in production. The average warm up time taken by each DBMS was: PostgreSQL 0.083 h, MongoDB 0.1 h, and Neo4j 2.33 h.

Evaluation. Now, pre-defined queries are executed over each DBMS using the parameters obtained in the second stage. These queries were created according to the typical mobile applications characteristics, as presented in Sect. 3.4.

Shutdown. Finally, each DBMS's connections are shutdown, finalizing any pending transactions and releasing locks, ensuring the database is in a consistent state. It also verifies if any system resources consumed by the DBMS during that session were not freed or returned to the operating system, which may point to memory leaks and other memory management issues. The methodology also has a shutdown hook, i.e., if the evaluation process is manually canceled or fails, the shutdown procedure will still be performed by a separate thread, thus ensuring data integrity and that no open resources remain connected to the DBMS.

3.2   Dataset

In this study, we opted for using a large real-world dataset, with realistic spatial distribution (i.e., a varying density of objects through space, from crowded downtown areas to rural spaces), covering a large territory and including a variety of representation types. Also, there should be a mix of natural and man-made features in the dataset, so that the number of vertices per object varies widely.

A dataset which covers all aforementioned features is the 2013 TIGER/Line – one of the most complete and well documented datasets on the USA territory that provided as shapefiles by the U.S. Census Bureau. The USA is one of the

---

[12] GeoJSON: http://www.geojson.org/.
[13] GDAL: http://www.gdal.org/.

countries with more mobile devices connected to the Internet[14] and consequently one with the largest number of people using spatial-aware systems. Running specific queries on the 2013 TIGER dataset generates a workload that serves as a realistic evaluator when comparing different DBMSs and analyzing how they respond to queries over large spatial datasets.

3.3   Data Loading

During data loading, the dataset is read, processed and stored in disk. As mentioned earlier, we use the TIGER/Line vector dataset, which contains more than 400 million objects, thus requiring efficient data loading.

Table 1. Dataset disk space

Storage	Data	Indexes	Total
ShapeFile	136 GB	-	136 GB
PostgreSQL	132 GB	65 GB	197 GB
GeoJSON	224 GB	-	224 GB
Neo4j	271 GB	18 GB	289 GB
MongoDB	412 GB	32 GB	444 GB

Table 2. Evaluated query groups

Query group	Spatial data type
Nearby Points of Interest	Vector
Map View	Vector
Urban Routing	Network
Position Tracking	Spatio-Temporal

All three DBMSs provide their own import tool (*shp2pgsql* in PostgreSQL, *mongoimport* in MongoDB and Neo4j's *ShapeFileImporter*). *Shp2pgsql* performed very well right out of the box, translating shapefiles into PostgreSQL data relatively fast and without issues. However, the import tools provided for MongoDB and Neo4j were much slower, and new tools had to be implemented to achieve a more reasonable import time.

After completing the dataset import, spatial indexes are created for all geometry fields. Indexes are also created for all attributes used in queries, such as road type (rural and local roads, highways and interstates) for urban routing, and point types for nearby points of interest. For MongoDB and Neo4j, our import tools generated the exact same database, with the same number of objects, with varying disk usage. The disk space used by the dataset in each platform, as well as the source GeoJSON and shapefiles, are presented in Table 1.

3.4   Spatial Queries

The query sets were designed based on features that are commonly used in mobile GIS and applications, such as Google Maps, Waze, Foursquare, and many others. There are four main query groups: nearby points of interest, map view, urban routing and position tracking, as described next and summarized in Table 2.

---

[14] Cisco Visual Networking Index: http://www.cisco.com/c/en/us/solutions/collateral/ service-provider/visual-networking-index-vni/white_paper_c11-520862.html.

Nearby Points of Interest. This group contains two of the most common types of queries performed by users when looking for nearby points of interest (POI). The first is the radius-based query: a search for POIs that are closer than a given distance from a reference position (usually the user's GPS position). For example: "Find gas stations closer than 5 km from my workplace". The second is to query for the **k-nearest neighbors** (k-NN): a search for one or a group of the closest objects from a given location. Examples are: "Where is the closest bus stop to my dentist appointment?" and "What are the five nearest restaurants currently open in my neighborhood?".

Map View. This group simulates interactive user browsing, zooming and panning. The zoom usually ranges from a city block (150 m wide) to hundreds or thousands of kilometers. This group tests how well the DBMS memory manager handles the locality of reference on its indexes. An optimized indexing and caching implementation may perform better when loading spatial objects that are close to the previously loaded ones. It is also important to evaluate how the DBMS handles queries in multiple collections/tables/labels at the same time, requiring access to multiple indexes and database files.

Urban Routing. This group plays a very important role nowadays, as applications like Google Maps, Bing Maps and Waze provide a routing function. Such queries require efficient response to graph-oriented operations and determine how well their shortest path and similar algorithms behave. The urban routing algorithms are usually not performed over long distances, as there are many solutions to filter graph edges that would not likely be traversed by the algorithm, or to avoid extensive calculations.

Position Tracking. This is a feature (available in many mobile devices with or without the user's knowledge) that constantly records the user's position in order to provide location-based services (such as traffic management, fleet location, etc.) or simply to deliver location-specific advertisement. Therefore, this feature requires frequent storage of point locations. As both time and spatial attributes are indexed, index restructuring and file management play important roles. This group of queries evaluate how fast the DBMS can insert new data into an existing collection with previously built indexes.

4   Experimental Evaluation

In this section, we describe the experimental evaluation, comprising setup, database configuration, index creation, software and hardware characteristics, parameters, metrics and the performance evaluation.

4.1   Evaluation Setup

We run the experimental evaluation on an Intel Core i7 3770 K with 32 GB of non-ECC 1600 MHz RAM, 120 GB SSD for the operating system and DBMS binary files, and a 2TB 7200 RPM hard drive to the DBMS storage folders.

The operating system is Windows Server 2012 64 bits. As DBMS tuning, the Java heap size was increased for Neo4j to match the behavior observed in the other DBMSs, in which all available RAM is occupied. Spatial indexes were created in all spatial columns (PostgreSQL), fields (MongoDB) or attributes (Neo4j).

We have implemented all evaluation functions in a tool that avoids interfering with the DBMS's performance. Also, all query caching, file buffering and other strategies were handled exclusively by each DBMS. As for specific setups, Neo4j's embedded server runs on Java Virtual Machine (JVM) and was placed in a separate thread. This way, it uses memory-mapped buffers and avoids any memory sharing with the evaluation tool, thus putting the operating system in charge of its buffers. The drivers utilized are: PostgreSQL 9.3-1102 JDBC driver, MongoDB 2.12.3 Java driver and Neo4j embedded server 2.1.4.

Finally, for each group of queries, we pre-evaluated each DBMS to ensure a fair comparison. Specifically, for the **Urban Routing** group, MongoDB is not considered because the evaluated version (2.12.3) does not provide any network-related data handling. Also, given that Neo4j is a graph oriented database, its shortest path algorithms are part of Neo4j's core; i.e., it does not require the spatial extension to perform the calculation, using road segments as edges and intersections as nodes. On the other hand, for the **k-NN** queries, Neo4j Spatial results differed greatly by returning empty collections or incomplete result sets. Since Neo4j Spatial is open source, a code review indicated that its query method does *not* meet the standard k-NN implementations: in each query, it uses a density estimation to calculate the size of an enclosing polygon. Then a *within* search is performed to find elements inside the generated polygon. Therefore, this approach does not work as a true k-nearest neighbor search, because there is no guarantee that the density estimation will cover the number of objects requested. Therefore, we do not consider Neo4j in the k-NN evaluation.

4.2 Parameters and Metrics

To get a meaningful workload, each type of operation has a group of queries. Each group includes 20 parameterized queries in which parameter values vary within predefined ranges, then reinforcing the variety and size of the resulting set. It also allows better evaluating how each DBMS handles its spatial indexes in both memory and disk. Since relative positioning is not altered by cartographic projection, distance parameters are expressed in degrees to avoid unnecessary projection calculations. The parameter value ranges are defined as follows.

– For the **Nearby POI Radius-based** group, the maximum radius is 1.0 degree (approximately 110 km at the Equator) and the minimum is 0.01 (approximately 1 km). Then, each **k-NN** query varies between 1 and 10 nearby points.
– For the **Map View** group, the query window is a 16:9 rectangle (simulates a mobile screen) with width and height varying between a minimum of 0.0013 degree (approximately 150 m, a city block) and maximum of 1.0 degree.

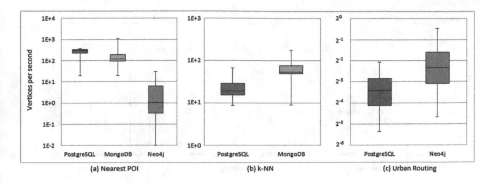

Fig. 1. Vertices per second: nearby POI and urban routing query sets

- The **Urban Routing** queries have limited maximum distances between origin and destination, where the minimum distance is 10 m (the shortest road edge available) and the maximum is 0.5 degree. According to the NHTS[15], this value is the 90th percentile of the American Commute Distance (One Way), and corresponds to 56 km. The distances may vary slightly, considering that the conversion of distances expressed in degrees to kilometers varies according to the latitude.
- During the **Position Tracing** evaluation, the spatio-temporal object is composed by the object id, its date of creation and a randomly generated point geometry within the urban routing network. The number of insertions performed simulates a daily journey of an object traveling between 1 and 60 km/h, which implies a number of points ranging from 100 to 100,000 – depending on the distance between consecutive points.

As for metrics, we consider the query execution time and number of vertices returned by the query. Specifically, the number of vertices is the sum of the vertices of the geometries of all objects, documents or tuples returned by the queries. We also propose the *vertices per second* measurement as a performance indicator. Query processing over a more complex geometry or spatial object takes considerably more processing power than operations over simple geometries or singular points, since the complexity of geometric algorithms is usually defined as a function of the number of vertices. The metrics from the main attributes are the average query time and vertices per second, which is defined as follows.

Vertices Per Second: Let t be the time in seconds spent to evaluate a query, n the number of spatial objects retrieved, and v_i the number of vertices of each object. The *vertices per second* is the sum of all object vertices returned by the query divided by its execution time: Vertices Per Second $= \frac{\sum_1^n v_i}{t}$.

---

[15] NHTS: National Household Travel Survey. Available at http://www.rita.dot.gov/bts/sites/rita.dot.gov.bts/files/subject_areas/national_household_travel_survey/index.html.

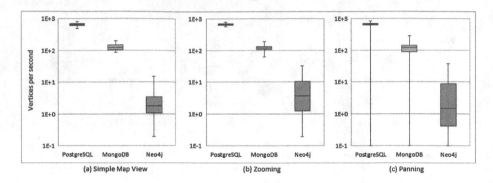

Fig. 2. Map view query set, vertices per second distribution

4.3   Performance Evaluation

In this section, we evaluate the DBMS for spatial data. As mentioned earlier, the evaluation metrics are the number of vertices per second (the higher the better) and execution time (the lower the better).

Nearby Points of Interest Radius and KNN. We start the performance evaluation with the nearby POI radius and k-NN queries. Figure 1(a) shows the results for Nearby POI Within Radius with the distribution of number of vertices per second for each DBMS. Note that PostgreSQL has the best median throughput and its scores are consistent. Then, MongoDB has a relative slower average than PostgreSQL, with maximum scores higher than PostgreSQL and minimum nearly equal. Neo4j falls into another performance level with maximum vertices per second score very close to the other two DBMS's minimum, which is recurrent in other experiment categories as explained later on.

Figure 1(b) shows the vertices per second for the **Nearby Points of Interest k Nearest Neighbors** query set (there are no results for Neo4j because its k-NN implementation does not work properly). In this group, MongoDB's performance surpasses PostgreSQL by three times. It is important to notice that the k-NN approaches used by each DBMS are different. PostgreSQL searches directly over an R-tree and uses the '< − >' operator, which returns the distance between two points (it uses the geometry centroid for polygons to speed up the calculation). Then, MongoDB searches on multiple iterations over concentrically growing circular polygons until the last neighbor is found. Overall, even though the PostgreSQL approach is more elegant and less expensive memory wise, MongoDB's index for points associated with growing concentric circles proves to be faster as there are no distance calculations to be made between each point (except on the last concentric circle iteration).

Urban Routing. Figure 1(c) shows the results for Urban Routing queries, where the average number of vertices per second processed by Neo4j is roughly twice those of PostGIS' pgRouting. Such results reveal Neo4j's greatest comparative strength: calculating the shortest path between two points in a network, as

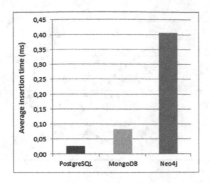

Fig. 3. Average position insert time (ms)

expected for a graph-oriented database. Also, such results may indicate that the performance bottleneck for other types of queries might be located in Neo4j's Spatial Extension, as the shortest path query is a native feature of Neo4j and does not require an R-tree-over-graph index structure.

Map View. The difference in performance is even clearer in this set of experiments. Figure 2(a) shows the results: each DBMS is separated from the other by a factor of *ten*. As an RDBMS, PostgreSQL outperforms on combining the multi-table query results while using the spatial index. MongoDB struggles to deal with multiple collections at the same time, as its memory management strategies focus on keeping the entire indexes on memory as much as possible. Such approach might not be ideal when querying a collection forces another collection's index out of the RAM. Neo4j again performs poorly, mostly because its R-tree index implementation over the graph structure forces the index to be completely loaded into memory as a graph (spatial layer), which is slower.

Zooming and panning are also part of the Map View group. In a map view, they evaluate how the accessed data and indexes are used in the next queries. It is expected that any indexes previously loaded into memory by the Map View query will remain available for the zooming and panning. Also, if any DBMS provides object caching, those objects will be expected to be retrieved faster. Hence, Figs. 2(b) and (c) show a throughput increase by zooming and panning a previously accessed portion of data. The results show the performance of PostgreSQL as being constant at a rate close to 800 vertices per second, while MongoDB reached little over 100, and Neo4j barely achieves 10 vertices per second. It is possible to notice that the minimum score for all three DBMSs falls greatly during panning. Such result indicates that the query selected a portion of the dataset that was not accessed by either viewing or zooming the map, resulting in (expensive) disk operations rather than memory accesses.

Position Tracking. Now, we analyze how the systems manage insertions into both spatial and date indexes. Figure 3 shows the results. Whereas PostgreSQL still maintains a relatively constant insertion time, both MongoDB and Neo4j oscillate between fast insertions close to instant and slow ones that are almost

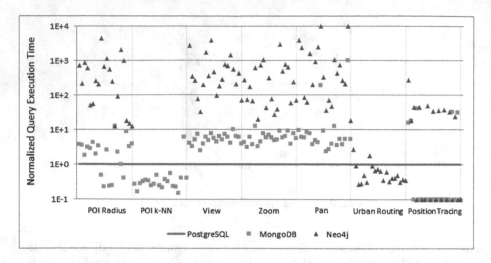

Fig. 4. Relative performance summary

100 times slower than PostgreSQL's. These results indicate that MongoDB and Neo4j implement a delayed insertion mechanism. Specifically, MongoDB considers replica sets and journal registers for writing operations. The first one guarantees the write operation will propagate insertions to all replica sets (when there is any). The second one ensures durability by confirming that any insertion operation has been logged into the on-disk journal. On the other hand, Neo4j uses a commit interval parameter: if the number of insertions is greater than or equal to the commit interval, all modifications are pushed to the disk and the commit interval is set to zero again. Such delayed insertion behavior can be seen clearly in Fig. 4, as explained next.

4.4   Relative Performance Summary

The relative performance is calculated by dividing the DBMS query execution time by PostgreSQL's, the oldest (and probably more mature) of the three. PostgreSQL is taken as reference with a normalized time of 1. Other DBMS values indicate how much faster or slower they were in comparison. Figure 4 shows the normalized execution time of 20 queries per group.

According to our evaluation results, the DBMS with best overall performance is PostgreSQL. As an established product (more mature than the other two), PostgreSQL with PostGIS is still the reference when handling spatial data, even if its relational characteristics were not particularly designed to handle the mobile applications workload we proposed. Nonetheless, despite PostgreSQL performance being consistent and superior overall, its horizontal scalability is supposedly not as easy to achieve as the two NoSQL DBMS tested. Specifically, PostgreSQL offers many scaling and table partition tools. However, such tools require implementing and configuring many variables, as well as managing and tuning the existing partitions and analyzing the gains obtained. This whole task is very time consuming, and may not compensate the extra effort.

MongoDB manages to stay relatively close to PostgreSQL in most groups, and outperforms it during the POI radius and k-NN. Such results show that its growing concentric circles implementation can be as fast as PostgreSQL's. Neo4j overall performance was weaker, but being the youngest system considered and graph-oriented, it is somehow expected not to perform well under circumstances that favor tree-indexing and very efficient disk and memory management. Moreover, despite previous results and being a graph DBMS, Neo4J has shown good performance for Urban Routing queries by outperforming PostGIS's pgRouting when using its core shortest path implementation.

We now further discuss the weaker Neo4j performance. By utilizing Resource Monitor[16] and the Java Visual[17], we found evidence that the DBMS has an underperforming file management interface. While the disk used in this experiment is capable of an average reading speed of 152 MB/s, the total disk I/O during Neo4j's benchmark peaked at 6 MB/s. Such results indicate a large amount of seek and random accesses, even though its disk usage was consistently near the 100 % mark. There are also methods that belong to the file storage management system and to the cache system that consume most of the CPU time.

In summary, our results show that PostgreSQL with PostGIS is a DBMS that works well under multiple scenarios and provides the most spatial features. MongoDB is almost as fast as PostgreSQL in some cases, performs well in the k-NN scenario and is easy to scale horizontally, but it lacks many spatial features, including networking. Neo4j has a slow performance when compared with the other DBMSs but it has a very good shortest path network implementation and can be the best option when performing routing queries.

5   Concluding Remarks

Evaluating and benchmarking spatial DBMS performances is not as simple as evaluating relational DBMS, because spatial attributes are much more complex to handle than strings, numbers and other relational data types. While several system's architectures provide different levels of performance and features, it is imperative to limit the data profile in which an analysis is made. Even in a very specific scenario (i.e., mobile applications), our evaluation achieved very heterogeneous results at both DBMS and query group levels.

Our evaluation methodology was composed by five stages (from data loading to shutdown) and performed 100 queries (20 per each of five groups) over a real geographic dataset. The overall results demonstrated how a relational DBMS performs under large amounts of spatial data, a resulting scenario that may change under heavy clustering and data distribution. Also, there is a need for data heterogeneity within the same DBMS, as each type of query runs faster

---

[16] Windows Server Resource Monitor: http://msdn.microsoft.com/en-us/library/ms191246.aspx.

[17] Java VisualVM: http://visualvm.java.net/.

in a different data structure. Indeed, how to interoperate spatial data in SQL and NoSQL databases is currently an open problem, being the subject of recent studies [1,6]. Comparing to existing studies, ours summarized the need for more specific spatial big data evaluation and more data and service-oriented benchmarks, as well as the proposition of a metric that translates the DBMS capabilities into an efficiency comparison, and not just only response time. Finally, such scenario opens new challenges for future work, such as evaluating spatial big data performance under different clustering.

Acknowledgments. This work was funded by CAPES, CNPq and FAPEMIG.

References

1. Baptista, C.S., et al.: Using OGC services to interoperate spatial data stored in SQL and NoSQL databases. In: Proceedings of GeoInfo, pp. 61–72. Campos do Jordão, Brazil (2011)
2. Chen, Y., Raab, F., Katz, R.: From TPC-C to big data benchmarks: a functional workload model. In: Rabl, T., Poess, M., Baru, C., Jacobsen, H.-A. (eds.) WBDB 2012. LNCS, vol. 8163, pp. 28–43. Springer, Heidelberg (2014)
3. Davis Jr., C.A., Fonseca, F.: Considerations from the development of a local spatial data infrastructure. Inf. Technol. Dev. **12**(4), 273–290 (2006)
4. Do, T.T., Liu, F., Hua, K.A.: When mobile objects' energy is not so tight: a new perspective on scalability issues of continuous spatial query systems. In: Wagner, R., Revell, N., Pernul, G. (eds.) DEXA 2007. LNCS, vol. 4653, pp. 445–458. Springer, Heidelberg (2007)
5. Ghazal, A., et al.: BigBench: towards an industry standard benchmark for big data analytics. In: Proceedings of SIGMOD, pp. 1197–1208 (2013)
6. Hora, A.C., Davis Jr., C.A., Moro, M.M.: Mapping network relationships from spatial database schemas to GML documents. J. Inf. Data Manage. **2**(1), 67–74 (2011)
7. Liu, C., Fruin, B.C., Samet, H.: SAC: semantic adaptive caching for spatial mobile applications. In: Proceedings of ACM SIGSPATIAL, pp. 174–183 (2013)
8. Nascimento, S.M., et al.: The spatial star schema benchmark. In: Proceedings of GeoInfo, pp. 73–84. Campos do Jordão, Brazil (2011)
9. Niedermayer, J., et al.: Probabilistic nearest neighbor queries on uncertain moving object trajectories. PVLDB **7**(3), 205–216 (2014)
10. Ray, S., Simion, B., Brown, A.D.: Jackpine: a benchmark to evaluate spatial database performance. In: Proceedings of ICDE, pp. 1139–1150 (2011)
11. Shekhar, S., Evans, M.R., Gunturi, V., Yang, K.S., Cugler, D.C.: Benchmarking spatial big data. In: Rabl, T., Poess, M., Baru, C., Jacobsen, H.-A. (eds.) WBDB 2012. LNCS, vol. 8163, pp. 81–93. Springer, Heidelberg (2014)
12. Sidlauskas, D., Jensen, C.S.: Spatial joins in main memory: implementation matters!. PVLDB **8**(1), 97–100 (2014)

Uncertain Data and Inconsistency Tolerance

Query Answering Explanation in Inconsistent Datalog+/− Knowledge Bases

Abdallah Arioua$^{(\boxtimes)}$, Nouredine Tamani, and Madalina Croitoru

University Montpellier, Montpellier, France
arioua@lirmm.fr

Abstract. The paper addresses the problem of explaining Boolean Conjunctive Query (BCQ) entailment in the presence of inconsistency within the Ontology-Based Data Access (OBDA) setting, where inconsistency is handled by the intersection of closed repairs semantics (ICR) and the ontology is represented by Datalog+/− rules. We address this problem in the case of both BCQ acceptance and failure by adopting a logical instantiation of abstract argumentation model; that is, in order to explain why the query is accepted or failed, we look for proponent or opponent sets of arguments in favor or against the query acceptance. We have also studied the computational complexity of the problem of finding an arbitrary explanation as well as all explanations.

1 Introduction

In the popular OBDA setting the domain knowledge is represented by an ontology facilitating query answering over existing data [22]. In practical systems involving large amounts of data and multiple data sources, data inconsistency might occur. Many inconsistency-tolerant semantics [6,7,17,18] have been proposed that rely on the notion of data repairs i.e. subsets of maximally consistent data with respect to the ontology. Different query answering strategies over these repairs (called semantics) are investigated in the literature (computing answers that hold in every repair corresponds to AR-semantics, computing answers that hold under the intersection of repairs corresponds to IAR-semantics or to the intersection of closed repairs corresponds to ICR-semantics). Here we only focus on the ICR-semantics. Query answering under these semantics may not be intuitively straightforward and can lead to loss of user's trust, satisfaction and may affect the system's usability [20]. Moreover, as argued by Calvanese et al. [10] *explanation facilities* should not just account for user's "Why Q?" question (why a query holds under a given inconsistency-tolerant semantics) but also for question like "Why not Q?" (why a query does not hold under a given inconsistency-tolerant semantics). Given an inconsistent OBDA setting equipped with an inconsistency-tolerant semantics (such as ICR-semantics) and given a boolean conjunctive query Q we consider two query answering problems, namely: $(\mathcal{RQ}_1)$ "why does Q hold under such semantics?" and $(\mathcal{RQ}_2)$ "why Q does not hold under such semantics?". For example, let us a consider a Unique

© Springer International Publishing Switzerland 2015
Q. Chen et al. (Eds.): DEXA 2015, Part I, LNCS 9261, pp. 203–219, 2015.
DOI: 10.1007/978-3-319-22849-5_15

Name Assumption knowledge base where employees work in departments and use offices, some employees direct departments and supervise other employees. Also a supervised employee cannot be a manager. A director of a given department cannot be supervised by an employee of the same department and two employees cannot be direct the same department. We consider factual information stating that John and Tom direct the finance department, where they both work in. Furthermore Tom is a supervisor of John and it also directs the web department. Let us now consider the query is John an employee. The knowledge base is inconsistent (e.g. Tom and John both direct finance). We can repair it, for instance considering Tom working in finance and directing the web department with John working and directing finance etc. (there are three such repairs). Every repair will entail that John is an employee. However, this is not true of the query asking whether John works in finance because we can consider a repair where this does not hold. These intuitions are formalised in the paper.

In a survey conducted in [25], knowledge base explanation that takes the form of 'Justification' has proven to be more effective than other types of explanation (Line of Reasoning and Strategy for instance). This type of explanation aims at showing the reason why certain conclusions have been drawn in particular circumstances. It is noteworthy that "Justification" is represented as *arguments* in [25]. Following this result, we build upon the logical instantiation of Dung's abstract argumentation framework for OBDA in [11] and exploit their equivalence between ICR-semantics and sceptical acceptance in argumentation framework in order to provide novel explanation facilities for ICR-semantics. In addition, we show how the ICR-semantics can be interpreted in the light of argumentation. We show that ICR-semantics behaves distinctively with respect to the question to be answered. In "Why Q?" questions, ICR-semantics ensures that a query is accepted if it is supported by a set of arguments that are well-defended. Whereas, in "Why not Q?" questions, a query fails if its supporting arguments are attacked by a set of arguments that are well-defended. We show the link between our approach and the hitting set based approach of diagnosis [23]. Our work models the hitting set as a set of arguments, as opposed to a set of faulty components. We focus on argumentation because it opens interesting and promising research avenues. The contribution of the paper lies in the following points. First, we propose an explanation for the question "why does Q hold under such semantics?" based on the set of arguments that support the entailment of the query, called *proponent set* of the query. We define the notion of a *strong proponent set* that will form the basis of the query acceptance explanation. Second, we propose an explanation for the question "why Q does not hold under such semantics?" based on the set of arguments that does not support the entailment of the query, *opponent set* of the query. Further, we show that these opponent sets play an important role in the failure of the query and form the basis of the query failure explanation. Third, we propose algorithms for finding an arbitrary explanation as well as all explanations. Furthermore, we show the relation between the former problems and the problem of finding minimal hitting sets.

2   Formal Settings and Problem Statement

In this section, we introduce: (1) OBDA setting and representation language, (2) the inconsistency-tolerant semantics and (3) the argumentation framework.

2.1   Language Specification

There are two major approaches to represent an ontology for the OBDA problem: Description Logics (DL) (such as $\mathcal{EL}$ [4] and DL-Lite [9] families) and rule-based languages (such as Datalog+/− [8] language, a generalization of Datalog that allows for existentially quantified variables in rules heads). Despite its undecidability when answering conjunctive queries, different decidable fragments of Datalog+/− are studied in the literature [5]. These fragments generalize the above mentioned DL families and overcome their limitations by allowing any predicate arity as well as cyclic structures. Here we restrict ourselves to Datalog+/− classes where the skolemised chase is finite (Finite Expansion Sets). We consider *the positive existential* conjunctive fragment of first-order logic, denoted by FOL($\wedge,\exists$), which is composed of formulas built with the connectors ($\wedge, \rightarrow$) and the quantifiers ($\exists, \forall$). We consider first-order vocabularies with constants but no other function symbol. A term t is a constant or a variable, different constants represent different values (unique name assumption), an atomic formula (or atom) is of the form $p(t_1, ..., t_n)$ where p is an n-ary predicate, and $t_1, ..., t_n$ are terms. A *ground* atom is an atom with no variables. A variable in a formula is free if it is not in the scope of a quantifier. A formula is closed if it has not free variable. We denote by $\mathbf{X}$ (with a bold font) sequences of variables $X_1, .., X_k$ with $k \geq 1$. A *conjunct* $C[\mathbf{X}]$ is a finite conjunction of atoms, where $\mathbf{X}$ is the sequence of variables occurring in C. Given an atom or a set of atoms A, $vars(A)$, $consts(A)$ and $terms(A)$ denote its set of variables, constants and terms, respectively.

An *existential rule* (rule) is a first-order formula of the form $r = \forall\mathbf{X}\forall\mathbf{Y}(H[\mathbf{X}, \mathbf{Y}]) \rightarrow \exists\mathbf{Z}C[\mathbf{Z}, \mathbf{Y}]$, with $vars(H) = \mathbf{X} \cup \mathbf{Y}$, and $vars(C) = \mathbf{Z} \cup \mathbf{Y}$ where H and C are *conjuncts* called the hypothesis and conclusion of R, respectively. We denote by $R = (H, C)$ a contracted form of a rule R. An existential rule with an empty hypothesis is called a *fact*. A fact is an existentially closed (with no free variable) conjunct.

We recall that a *homomorphism* π from a set of atoms A_1 to set of atoms A_2 is a substitution of $vars(A_1)$ by $terms(A_2)$ such that $\pi(A_1) \subseteq A_2$. Given two facts f and f' we have $f \models f'$ iff there is a homomorphism from f' to f, where $\models$ is the first-order semantic entailment. A *rule* $r = (H, C)$ is *applicable* to a set of facts F iff there exists $F' \subseteq F$ such that there is a homomorphism π from H to the conjunction of elements of F'. If a rule r is applicable to a set F, its application according to π produces a set $F \cup \{\pi(C)\}$. The new set $F \cup \{\pi(C)\}$, denoted also by $r(F)$, is called *immediate derivation* of F by r. A *negative constraint* is a first-order formula $n = \forall\mathbf{X}\ H[\mathbf{X}] \rightarrow \perp$ where $H[\mathbf{X}]$ is a conjunct called hypothesis of n and $\mathbf{X}$ sequence of variables appearing in the hypothesis. **Knowledge base.** A knowledge base $\mathcal{K} = (\mathcal{F}, \mathcal{R}, \mathcal{N})$ is composed of finite set of facts $\mathcal{F}$ and finite set of existential rules $\mathcal{R}$ and a finite set of negative constraints $\mathcal{N}$.

Example 1. The following example is inspired by [8]. In an enterprise, employees work in departments and use offices, some employees direct departments, and supervise other employees. In addition, a supervised employee cannot be a manager. A director of a given department cannot be supervised by an employee of the same department, and two employees cannot direct the same department. The following sets of (existential) rules $\mathcal{R}$ and negative constraints $\mathcal{N}$ model the corresponding ontology:

$$\mathcal{R} = \begin{cases} \forall x \forall y \, (works\_in(x, y) \rightarrow emp(x)) & (r_1) \\ \forall x \forall y \, (directs(x, y) \rightarrow emp(x)) & (r_2) \\ \forall x \forall y \, (directs(x, y) \wedge works\_in(x, y) \rightarrow manager(x)) & (r_3) \\ \forall x \, (emp(x) \rightarrow \exists y (office(y) \wedge uses(x, y))) & (r_4) \end{cases}$$

$$\mathcal{N} = \begin{cases} \forall x \forall y \, (supervises(x, y) \wedge manager(y)) \rightarrow \bot & (n_1) \\ \forall x \forall y \forall z \, (supervises(x, y) \wedge works\_in(x, z) \wedge directs(y, z) \wedge x \neq y) \rightarrow \bot & (n_2) \\ \forall x \forall y \forall z (directs(x, z) \wedge directs(y, z) \wedge x \neq y) \rightarrow \bot & (n_3) \end{cases}$$

Let us suppose the following set of facts $\mathcal{F}$ that represent explicit knowledge:

$$\mathcal{F} = \begin{cases} directs(John, d_1) \;\; (f_1) & directs(Tom, d_1) \quad (f_2) \\ directs(Tom, d_2) \;\; (f_3) & supervises(Tom, John) \;\; (f_4) \\ works\_in(John, d_1) \;\; (f_5) & works\_in(Tom, d_1) \quad (f_6) \end{cases}$$

$\mathcal{R}$-derivation. Let $F \subseteq \mathcal{F}$ be a set of facts and $\mathcal{R}$ be a set of rules. An $\mathcal{R}$-derivation of F in $\mathcal{K}$ is a finite sequence $\langle F_0, ..., F_n \rangle$ of sets of facts s.t $F_0 = F$, and for all $i \in \{0, ..., n\}$ there is a rule $r_i = (H_i, C_i) \in \mathcal{R}$ and a homomorphism π_i from H_i to F_i s.t $F_{i+1} = F_i \cup \{\pi(C_i)\}$. For a set of facts $F \subseteq \mathcal{F}$ and a query Q and a set of rules $\mathcal{R}$, we say $F, \mathcal{R} \models Q$ iff there exists an $\mathcal{R}$-derivation $\langle F_0, ..., F_n \rangle$ such that $F_n \models Q$.

Closure. Given a set of facts $F \subseteq \mathcal{F}$ and a set of rules $\mathcal{R}$, the closure of F with respect to $\mathcal{R}$, denoted by $\mathtt{Cl}_\mathcal{R}(F)$, is defined as the smallest set (with respect to $\subseteq$) which contains F and is closed under $\mathcal{R}$-derivation. By considering the skolemised chase and the finite fragments of Datalog$+/-$ this set is unique (i.e. universal model).

Finally, we say that a set of facts $F \subseteq \mathcal{F}$ and a set of rules $\mathcal{R}$ *entail* a fact f (and we write $F, \mathcal{R} \models f$) iff the closure of F by all the rules entails f (i.e. $\mathtt{Cl}_\mathcal{R}(F) \models f$).

A conjunctive query (CQ) has the form $Q(\mathbf{X}) = \exists \mathbf{Y} \Phi[\mathbf{X}, \mathbf{Y}]$ where $\Phi[\mathbf{X}, \mathbf{Y}]$ is a conjunct such that $\mathbf{X}$ and $\mathbf{Y}$ are variables appearing in Φ. A Boolean CQ (BCQ) is a CQ of the form $Q()$ with an answer yes or no. We refer to a BCQ with the answer *no* as **failed query**, whereas a query with the answer *yes* as **accepted query**.

Inconsistency-Tolerant Semantics. Given a knowledge base $\mathcal{K} = (\mathcal{F}, \mathcal{R}, \mathcal{N})$, a set $F \subseteq \mathcal{F}$ is said to be *inconsistent* iff there exists a constraint $n \in \mathcal{N}$ such that $F \models H_n$, where H_n is the hypothesis of the constraint n. A set of facts is consistent iff it is not inconsistent. A set $F \subseteq \mathcal{F}$ is $\mathcal{R}$-*inconsistent* iff there exists a constraint $n \in \mathcal{N}$ such that $\mathtt{Cl}_\mathcal{R}(F) \models H_n$. A set of facts is said to be

$\mathcal{R}$-*inconsistent* iff it is not $\mathcal{R}$-*consistent*. A knowledge base $(\mathcal{F}, \mathcal{R}, \mathcal{N})$ is said to be *inconsistent* iff $\mathcal{F}$ is $\mathcal{R}$-inconsistent.

Notice that (like in classical logic), if a knowledge base $\mathcal{K} = (\mathcal{F}, \mathcal{R}, \mathcal{N})$ is inconsistent, then everything is entailed from it. A common solution [6,17] is to construct maximal (with respect to set inclusion) consistent subsets of $\mathcal{K}$. In this finite chase case there is a finite number of such sets. They are called *repairs* and denoted by $\mathcal{R}epair(\mathcal{K})$. Once the repairs are computed, different semantics can be used for query answering over the knowledge base. In this paper we focus on (Intersection of Closed Repairs semantics) [6].

Definition 1 (ICR-Semantics). *Let* $\mathcal{K} = (\mathcal{F}, \mathcal{R}, \mathcal{N})$ *be a knowledge base and let* Q *be a query.* Q *is ICR-entailed from* $\mathcal{K}$, *written* $\mathcal{K} \models_{ICR} Q$ *iff* $\bigcap_{\mathcal{A} \in \mathcal{R}epair(\mathcal{K})} \mathrm{Cl}_{\mathcal{R}}(\mathcal{A}) \models Q$.

Example 2. The knowledge base in Example 1 is inconsistent because the set of facts $\{f_1, f_4, f_6\} \subseteq \mathcal{F}$ is inconsistent since it violates the negative constraint n_2. In this example we obtain 3 repairs. The following is one of them:

$\mathcal{A}_1 = \{directs(John, d_1), supervises(Tom, John), directs(Tom, d_2)\}$

The closure of $\mathcal{A}_1$ is:

$\mathrm{Cl}_{\mathcal{R}}(\mathcal{A}_1) = \{directs(John, d_1), supervises(Tom, John),$
$directs(Tom, d_2), manager(Tom), emp(John), emp(Tom),$
$\exists y_1 (office(y_1) \wedge uses(Tom, y_1))),$
$\exists y_2 (office(y_2) \wedge uses(John, y_2)))\}$

2.2 Problem Statement

In what follows we introduce the motivation of providing explanation facilities for Inconsistent OBDA.

Example 3. Consider the query $Q = emp(John)$ in the knowledge base of Example 1. Q is accepted under ICR-semantics since it follows from the intersection of all closed repairs. The user may be interested in knowing why this is the case. The query $Q' = works\_in(John, d_1)$ is a failed query since it does not follow from the intersection of all closed repairs. Such query failure may need an explanation when the user is expecting the acceptance of the query.

The above example shows the need for explanation facilities with the aim of helping the user to understand why a query holds or fails. Thus, we define our problem as: given an inconsistent knowledge base $\mathcal{K}$ and a boolean conjunctive query Q:

$\mathcal{R}Q_1$: why Q does hold under the ICR-semantics in the knowledge base $\mathcal{K}$?
$\mathcal{R}Q_2$: why Q' does not hold under the ICR-semantics in the knowledge base $\mathcal{K}$?

The Proposed Solution. To depict the intuition underlying our proposal, in what follows we show how an explanation can be provided by means of argumentation within the context of an inconsistent OBDA.

Example 4 (Explanation). Let us consider the query $Q = emp(John)$. The knowledge base of Example 1 is inconsistent thus we have the following consistent possibilities:

1. Tom works in d_1 and directs d_2, while John works in d_1 and directs d_1.
2. Tom directs d_2 and supervises John who directs d_1.
3. Tom works in d_1, directs d_1 and d_2, and supervises John who in turn works in d_1.

Please note that the depiction of logical rules in textual form is for illustration purposes only. Thus, in this paper we do not address the problem of generating textual explanation based on natural language processing.

From above we can always infer $Q = emp(John)$ since from (1) and (3) we can construct (including and not limited to) the argument "John is an employee because he works in d_1", and from (2) we get "John is an employee because he directs d_1". In other words, the query Q is supported from all possible "points of view". This support is in a form of arguments, these arguments are grouped in what we call a "proponent set".

As for query $Q' = works\_in(John, d_1)$, we can see that in (1) and (3) we have an argument that says "John works in d_1" which supports Q'. However, the query is not supported from all points of view, for instance in (2) we have the argument "John directs d_1 and he is supervised by Tom, so John cannot work in d_1 since he would be a manager and a manager cannot be supervised" that attacks the argument "John works in d_1'. It seems that the former argument is holding the query Q' from being accepted from all possible points of view. Note that, there may be more than one argument that jointly hold the query from being accepted. These arguments are grouped in what we call "opponent set".

In the two cases (Q and Q') we consider proponent and opponent sets as the core of our explanation. Moreover, we deepen the explanation by considering possible attacks scenario between arguments aiming at showing the user the dialectical procedure that allows the acceptance or the failure of the query. To do so, we make use of Dung's argumentation frameworks [12] adapted for the aforementioned rule-based language in [11]. First let us specify the structure of an argument under such language.

2.3   Rule-Based Dung Argumentation Framework Instantiation

As defined in [11], given a knowledge base $\mathcal{K} = (\mathcal{F}, \mathcal{R}, \mathcal{N})$, the *corresponding argumentation framework* $\mathcal{AF}_\mathcal{K}$ is a pair $(\mathtt{Arg}, \mathtt{Att})$ where $\mathtt{Arg}$ is the set of arguments that can be constructed from $\mathcal{F}$ and $\mathtt{Att}$ is an asymmetric binary relation called *attack* defined over $\mathtt{Arg} \times \mathtt{Arg}$. An argument is defined as follows.

Definition 2 (Argument *[11]*). *Given a knowledge base $\mathcal{K} = (\mathcal{F}, \mathcal{R}, \mathcal{N})$, an argument a is a tuple $a = (F_0, F_1, \ldots, F_n, C)$ where:*

- $(F_0, \ldots, F_n)$ *is an $\mathcal{R}$-derivation of F_0 in $\mathcal{K}$,*
- C *is an atom, a conjunction of atoms, the existential closure of an atom or the existential closure of a conjunction of atoms such that $F_n \models C$.*

F_0 is the support of the argument a (denoted $\mathtt{Supp}(a)$) and C is its conclusion (denoted $\mathtt{Conc}(a)$).

Example 5 (Argument). The following argument indicates that John is an employee because he directs department d_1:

$$a = (\{directs(John, d_1)\}, \{directs(John, d_1), emp(John)\}, emp(John)).$$

Definition 3 (Attack [11]). *The attack between two arguments expresses the conflict between their conclusion and support. We say that an argument a attacks an argument b iff there exists a fact $f \in \mathtt{Supp}(a)$ such that the set $\{\mathtt{Conc}(b), f\}$ is $\mathcal{R}$-inconsistent.*

Example 6 (Attack). Consider the argument a of Example 5, the following argument $b = (\{supervises(Tom, John), works\_in(Tom, d_1)\}, supervises (Tom, John) \wedge works\_in(Tom, d_1))$ attacks a, because $\{supervises(Tom, John) \wedge works\_in(Tom, d_1), directs(John, d_1)\}$ is $\mathcal{R}$-inconsistent since it violates the constraint n_2.

Admissibility, Semantics and Extensions in $\mathcal{AF}$. Let $\mathcal{K} = (\mathcal{F}, \mathcal{R}, \mathcal{N})$ be a knowledge base and $\mathcal{AF}_\mathcal{K}$ its *corresponding argumentation framework*. Let $\mathcal{E} \subseteq \mathtt{Arg}$ be a set of arguments. We say that $\mathcal{E}$ is *conflict free* iff there exist no arguments $a, b \in \mathcal{E}$ such that $(a, b) \in \mathtt{Att}$. $\mathcal{E}$ *defends* an argument a iff for every argument $b \in \mathtt{Arg}$, if we have $(b, a) \in \mathtt{Att}$ then there exists $c \in \mathcal{E}$ such that $(c, b) \in \mathtt{Att}$. $\mathcal{E}$ is *admissible* iff it is conflict free and defends all its arguments. $\mathcal{E}$ is a *preferred extension* iff it is maximal (with respect to set inclusion) admissible set (please see [12] for other types of semantics).

We denote by $\mathtt{Ext}(\mathcal{AF}_\mathcal{K})$ the set of all extensions of $\mathcal{AF}_\mathcal{K}$. An argument is sceptically accepted if it is in all extensions, credulously accepted if it is in at least one extension and not accepted if it is not in any extension.

Equivalence Between ICR and Preferred Semantics. Let $\mathcal{K} = (\mathcal{F}, \mathcal{R}, \mathcal{N})$ be a knowledge base and $\mathcal{AF}_\mathcal{K}$ the corresponding argumentation framework. A query Q is sceptically accepted under preferred semantics iff $(\bigcap_{\mathcal{E} \in \mathtt{Ext}(\mathcal{AF}_\mathcal{K})} \mathtt{Concs}(\mathcal{E})) \vDash Q$, such that $\mathtt{Concs}(\mathcal{E}) = \bigcup_{a \in \mathcal{E}} \mathtt{Conc}(a)$. The results of [11] show the equivalence between sceptically acceptance under preferred semantics and ICR-entailment:

Theorem 1 (Semantics Equivalence [11]). *Let $\mathcal{K} = (\mathcal{F}, \mathcal{R}, \mathcal{N})$ be a knowledge base, let $\mathcal{AF}_\mathcal{K}$ the corresponding argumentation framework, Q a query. $\mathcal{K} \vDash_{ICR} Q$ iff Q sceptically accepted under preferred semantics.*

3   Argumentative Explanation

In this section we introduce the argumentation-based explanation for query acceptance (Subsect. 3.1) and failure (Subsect. 3.2). Note that due to the finiteness property of our language the argumentative framework is finite.

3.1 Explaining Query Acceptance

We introduce the notion of supporting argument of a query, notion needed when selecting the arguments for or against a query:

Definition 4 (Supporting Argument). *Let $\mathcal{K}$ be a knowledge base, $\mathcal{AF}_\mathcal{K} = (\text{Arg}, \text{Att})$ the corresponding argumentation framework, Q a query. An argument $a \in \text{Arg}$ supports Q iff $\text{Conc}(a) \models Q$. The set of all supporting arguments of Q is denoted by $\mathcal{S}up(Q)$.*

A query may have more than one supporting argument with different statuses (accepted or not accepted). The arguments with an accepted status (belong to at least one extension) form a *proponent set* of the query Q. We exclude not accepted arguments since they cannot form a base for the support of the acceptance of the query.

Definition 5 (Proponent set). *Let $\mathcal{K}$ be a knowledge base, $\mathcal{AF}_\mathcal{K}$ the corresponding argumentation framework, $\text{Ext}(\mathcal{AF}_\mathcal{K})$ set of all preferred extensions of $\mathcal{AF}_\mathcal{K}$, Q a query and $\mathcal{S}up(Q)$ the set of all supporting arguments of Q. Then, the set $\mathcal{P} \subseteq \mathcal{S}up(Q)$ is a proponent set of Q iff: (1) $\forall a \in \mathcal{P}$ there exists an extension $\mathcal{E} \in \text{Ext}(\mathcal{AF}_\mathcal{K})$ such that $a \in \mathcal{E}$; (2) there exists no $\mathcal{P}'$ such that $\mathcal{P} \subset \mathcal{P}'$ and $\mathcal{P}'$ satisfies (1).*

Note that there is only one proponent set of any query Q.

Example 7. Consider a knowledge base $\mathcal{K}$ that corresponds to an argumentation framework s.t. the set of arguments is $\text{Arg} = \{a, b, c, d, e, f, g, h\}$ and the attack relation is: $\text{Att} = \{(f, a), (c, f), (e, c), (b, e), (b, f), (d, b), (c, d), (c, g), (h, g)\}$. For space reasons the knowledge base underpinning this argumentation framework is not shown. Let Q be a query. Suppose that we have $\text{Conc}(a) \models Q$, $\text{Conc}(d) \models Q$ and $\text{Conc}(g) \models Q$. This means that the set of supporting arguments of Q is $\{a, d, g\}$. The set of preferred extensions is $\text{Ext}(\mathcal{AF}_\mathcal{K}) = \{\mathcal{E}_1 = \{a, c, b, h\}, \mathcal{E}_2 = \{e, f, d, h\}\}$. The proponent set $\mathcal{P}$ of Q is $\{a, d\}$. Notice that $g \notin \mathcal{P}$ since it does not belong to any extension.

A stronger notion of support for a query is expressed in what we call strong proponent set. The arguments in this set are forced to be distributed over all extensions.

Definition 6 (Strong Proponent Set). *Let $\mathcal{K}$ be a knowledge base, $\mathcal{AF}_\mathcal{K}$ the corresponding argumentation framework, $\text{Ext}(\mathcal{AF}_\mathcal{K})$ set of all preferred extensions of $\mathcal{AF}_\mathcal{K}$, Q a query and $\mathcal{P}$ the proponent set of Q. Then, a set $\mathcal{S} \subseteq \mathcal{P}$ is a strong proponent set of Q iff: (1) $\forall \mathcal{E} \in \text{Ext}(\mathcal{AF}_\mathcal{K})$ there exists an argument $a \in \mathcal{S}$ such that $a \in \mathcal{E}$. (2) there exists no proper subset $\mathcal{S}' \subset \mathcal{S}$ such that $\mathcal{S}'$ satisfies (1). We say the proponent set $\mathcal{P}$ is strong if it has at least one strong proponent set. The set of all strong proponent sets of a query Q is denoted by $\mathcal{S}trong(Q)$. Computing $\mathcal{S}trong(Q)$ is called the strong proponent set problem.*

Example 8. Consider the same argumentation framework of Example 7 and make the slight modification such that the set of supporting arguments of Q is $\{a, g\}$. In this case, the proponent set of Q is $\mathcal{P} = \{a\}$. Note that Q has no strong proponent set since there is an extension $\mathcal{E}_2 = \{e, f, d, h\}$ for which we can never construct a set $\mathcal{S} \subseteq \mathcal{P}$ in which there is an argument that belongs to $\mathcal{E}_2$.

If a query has a strong proponent set then this indicates that the query is supported from all possible extensions. This property has a strong relation with ICR-semantics as expressed by the next proposition.

Proposition 1. *Let $\mathcal{K}$ be a knowledge base, $\mathcal{AF}_\mathcal{K}$ the corresponding argumentation framework, Q a query and $\mathcal{P}$ its proponent set. Then $\mathcal{K} \vDash_{ICR} Q$ iff $\mathcal{P}$ is strong.*

Proof. In the technical report [1].

Proposition 1 indicates that the strong proponent set represents the reason why a query is accepted under ICR-semantics. The ICR-semantics, in fact, operates such that if the query is not supported from one extension then it should not be accepted. So, strong proponent set will form the basis of an explanation for query acceptance under ICR-semantics.

Example 9 (Count. Example 7). In order to provide an explanation for the acceptance of the query Q in Example 7, we should mention the fact that the query is supported by the well-defended arguments a and d. In general, we need to show that whenever an argument is attacked, there is an argument or a set of arguments that defend it. For instance, the argument a is attacked by f but it is defended by c and b. Moreover, the argument d is attacked by c but it is defended by e.

The intuition captured in the previous example is formalized as a tree structure called a *defense tree*.

Definition 7 (Defense Tree). *Let $a \in$ Arg. A Defense Tree $\mathcal{DT}(a)$ of a is a tree of arguments such that: (1) $H(\mathcal{DT}(a)) = 3$; (2) $\mathcal{DT}(a)$'s root node is the argument a, and $\forall b, c \in$ Arg : b is a child of c in $\mathcal{DT}(a)$ iff $(b, c) \in$ Att; (3) $\{a \cup leafs(\mathcal{DT}(a))\}$ is conflict-free; (4) every internal node has a child. Notice that $H(\mathcal{DT}(a))$ indicates the height of the tree and $leafs(\mathcal{DT}(a))$ indicates the leafs of the tree.*

Example 10. Figure 1 depicts a defense tree for the arguments a and c of Example 7.

Note that, if a query Q is accepted, then all arguments in $Strong(Q)$ have a defense tree since they are accepted with respect to an extension. Given a set of arguments S, the set of all defense trees for all arguments in S is denoted $\mathcal{DT}(S)$ and defined as $\mathcal{DT}(S) = \{\mathcal{DT}(a) | a \in S\}$.

Based on this notion, an explanation in our framework is not just the strong proponent set, but also a description of how these arguments are defended.

Fig. 1. Defense trees of the arguments a (left) and d (right).

Definition 8 (Explanation for Query Acceptance). *Let $\mathcal{K}$ be an inconsistent knowledge base, Q a query such that $\mathcal{K} \models_{ICR} Q$. An explanation $E_{\mathcal{K}}^{+}(Q)$ for the acceptance of Q in $\mathcal{K}$ is the tuple $\langle S, \mathcal{DT}(S) \rangle$ such that S is a strong proponent set of Q and $\mathcal{DT}(S)$ is the set of all defense trees of arguments in S. The set of all explanations for the acceptance of Q is denoted by $\exp^{+}(Q)$.*

3.2 Explaining Query Failure

In this subsection we consider the problem of explaining why a query Q is not ICR-entailed. In this case the failure of a query is more related to "the opposition against the query" than "the support for the query". Let us start by defining an opponent set.

Definition 9 (Opponent Set). *Let $\mathcal{K}$ be an inconsistent knowledge base and $\mathcal{AF}_{\mathcal{K}} = (\text{Arg}, \text{Att})$ its corresponding argumentation framework, Q be a query and let $\mathcal{P}$ be a proponent set of Q. A finite set of arguments $\mathcal{O}$ is an opponent set of Q iff:*

1. *$\forall a \in \mathcal{O}$ there exists an extension $\mathcal{E} \in \text{Ext}(\mathcal{AF}_{\mathcal{K}})$ such that $a \in \mathcal{E}$.*
2. *For every $b \in \mathcal{P}$, there exists $a \in \mathcal{O}$ such that $(a, b) \in \text{Att}$.*
3. *$\forall a \in \mathcal{O}$ and $\forall b \in \mathcal{P}$, it never holds that a and b are in the same extension.*
4. *for all $a \in \mathcal{O}$ there exists an argument $b \in \mathcal{P}$ such that $(a, b) \in \text{Att}$.*

An opponent set is defined with respect to a proponent set since an opponent of Q is a set of arguments that are against the arguments that support Q.

As for property (1) and (2) represent minimal requirements, i.e. acceptance and attack. Property (3) insures coherence, that means an argument that belongs to an extension should not be against a query Q while there are some arguments in its extension which are supporting Q. The last property reinforces the property (2) and eliminates redundancy. This means that an opponent set has to incorporate only arguments that participate in the attack on the proponent set. Notice that, according to the definition above, there may be more than one opponent set of a query. Thus, the set of all opponent sets of a query Q is denoted by $\mathcal{OPP}(Q)$.

Example 11 (Opponent of a Query). Consider an argumentation framework $\mathcal{AF}_{\mathcal{K}}$ s.t. $\text{Ext}(\mathcal{AF}_{\mathcal{K}}) = \{\mathcal{E}_1 = \{a, b, c\}, \mathcal{E}_2 = \{d, e, f\}, \mathcal{E}_3 = \{g, h, k, m\}\}$ and its attack relation contains the following attacks among others $\{(f,a),(h,c)\}$. Consider a query Q with a proponent set $\mathcal{P} = \{a,c\}$. Then, the set of all opponent sets is $\mathcal{OPP}(Q) = \{\{f,h\}\}$. As stated in Definition 9 all the arguments of $\mathcal{P}$ are attacked (the argument f (resp. h) attacks a (resp. c)). In addition, the argument f is accepted w.r.t the extension $\mathcal{E}_2$ and the argument h is accepted w.r.t the extension $\mathcal{E}_3$. The presence of the argument f (resp. h) in $\mathcal{E}_2$ (resp. $\mathcal{E}_3$) prevents the query to be inferred from all extensions.

The relation between opponent set and the acceptance and failure of a query is presented i what follows.

Proposition 2. *Let $\mathcal{K}$ be an inconsistent knowledge base, Q be a query and let $\mathcal{P}$ be a proponent set of Q. Then, $\mathcal{P}$ is strong iff $\mathcal{OPP}(Q) = \emptyset$.*

Corollary 1. *Let $\mathcal{K}$ be an inconsistent knowledge base, Q be a query and let $\mathcal{P}$ be a non-empty proponent set of Q. Then, $\mathcal{K} \models_{ICR} Q$ iff $\mathcal{OPP}(Q) = \emptyset$.*

Proof. In [1].

Proposition 2 and Corollary 1 indicate that if the query has a support represented by the proponent set $\mathcal{P}$, and an opposition represented by opponent sets, then the query is not accepted. In this case, an opponent set of a query Q plays an important role in the failure of the query and can form the basis of the explanation.

Like in explaining query acceptance, we do not limit the explanation to the arguments that force the query to fail, we also present how these arguments are defended.

Definition 10 (Explanation for Query Failure). *Let $\mathcal{K}$ be an inconsistent knowledge base, Q a query such that $\mathcal{K} \nvDash_{ICR} Q$. An explanation $E_{\mathcal{K}}^-(Q)$ for the failure of Q in $\mathcal{K}$ is the tuple $\langle \mathcal{O}, \mathcal{DT}(\mathcal{O}) \rangle$ such that $\mathcal{O}$ is an opponent set for Q and $\mathcal{DT}(\mathcal{S})$ is the set of all defense trees of arguments in S. The set $\mathsf{exp}^-(Q)$ is the set of all $E_{\mathcal{K}}^-(Q)$.*

The selection of the set $\mathcal{O}$ (resp. $\mathcal{S}$ in Definition 8) is arbitrary and it can be defined according to some criteria, which their specification is beyond the scope of the paper.

4   Algorithms

As mentioned in Sect. 2, boolean conjunctive query answering in Datalog+/− under general existential rules is undecidable. Thus, the results and the algorithms presented in this section are considered only on families of Datalog+/− for which BCQ answering is tractable w.r.t data complexity[1] (for a detailed review see [8,14]). In addition, in these fragments the *closure* $(\mathsf{Cl}_\mathcal{R}(\mathcal{F}))$ over

---

[1] This refers to the complexity of evaluating the query over the knowledge base where the size of the query is assumed to be constant.

Algorithm 1. DEFENSETREE

1: **function** DEFENSETREE$(a, \text{Ext}(\mathcal{AF}_\mathcal{K}))$
2:     $DefTree = \emptyset$; $Attackers = \emptyset$; $Defenders = \emptyset$
3:     **for all** $\mathcal{E} \in \text{Ext}(\mathcal{AF}_\mathcal{K})$ such that $a \notin \mathcal{E}$ **do**
4:         $Attackers = Attackers \cup \{b|b \in \mathcal{E} \text{ such that } b \text{ attacks } a\}$
5:         $DefTree = DefTree \cup \{(b, a)|b \in Attackers\}$
6:     **for all** $\mathcal{E} \in \text{Ext}(\mathcal{AF}_\mathcal{K})$ **do**
7:         **for all** $b \in Attackers$ **do**
8:             $Defenders = Defenders \cup \{c|c \in \mathcal{E} \text{ such that } c \text{ attacks } b\}$
9:             $DefTree = DefTree \cup \{(c, b)|c \in Defenders\}$
10:    **Return** $(DefTree)$

---

any set of facts $\mathcal{F}$ is finite and any $\mathcal{R}$-derivation from a knowledge base is finite (see [8, 14]). Consequently, the corresponding argumentation framework $\mathcal{AF}_\mathcal{K} = (\text{Arg}(\mathcal{F}), \text{Att})$ of the knowledge base $\mathcal{K}$ is finite. Precisely, the set of all arguments $\text{Arg}(\mathcal{F})$ and the set of all extensions $\text{Ext}(\mathcal{AF}_\mathcal{K})$ are finite. Furthermore, since an argument is an $\mathcal{R}$-derivation and an $\mathcal{R}$-derivation is finite so all the arguments of $\mathcal{AF}_\mathcal{K}$ are finite.

An explanation has two components: a defense tree and strong proponent set (resp. opponent set). Since the defense tree is a common component for query acceptance and failure explanation, we start in the next subsection by presenting Algorithm 1 that computes a defense tree for a given argument. Then, before presenting algorithms for computing strong proponent set and opponent set, we show the relation between the former sets (separately) and minimal hitting sets and present Algorithms 2 and 3 that establish such relation. Next, in Algorithm 4 we present the algorithm that compute minimal hitting set in order to use it in the final algorithm (Algorithm 5) to compute an explanation.

4.1   Computing Defense Tree

The function *DefenseTree()* in Algorithm 1 takes as an input an argument a and a set of extensions. First, it computes the set of attackers of a (for-loop at line 4) then for each attacker it computes its attackers (for-loop at line 7). Finally, the function returns the defense tree as a form of a binary relation called *DefTree*. The time complexity of the algorithm is mainly related to the computation of attack between two arguments in line 5 and 9 (since all the loops induce only a polynomial overhead). This issue comes down to check if the conclusion and the hypothesis of the two arguments are unsatisfiable in $\mathcal{K}$ which can be done efficiently [8].

4.2   Computing Strong Proponent and Opponent Sets

In what follows we define the Hitting Set Problem.

Definition 11 (Hitting Set Problem). *Given a collection $\mathcal{C} = \{S_1, ..., S_m\}$ of finite nonempty subsets of a set $\mathcal{B}$ (the background set). A* hitting set *of $\mathcal{C}$ is*

Algorithm 2. REDUCESTRONGTOHITTING

1: **function** REDUCESTRONGTOHITTING($\mathbf{Ext}(\mathcal{AF}_\mathcal{K})$,$Q$)
2:     $\mathcal{B} = \emptyset; \mathcal{C} = \{\{\}\}$
3:     **for all** $\mathcal{E} \in \mathbf{Ext}(\mathcal{AF}_\mathcal{K})$ **do**
4:         $S = \{a|a \in \mathcal{E} \text{ such that } \mathbf{Conc}(a) \models Q\}$
5:         $\mathcal{C} = \mathcal{C} \cup \{S\}$
6:     $\mathcal{B} = \bigcup_{S \in \mathcal{C}} S$
7:     **Return** $(\mathcal{C}, \mathcal{B})$

Algorithm 3. REDUCEOPPONENTTOHITTING

1: **function** REDUCEOPPONENTTOHITTING($\mathbf{Ext}(\mathcal{AF}_\mathcal{K})$,$Q$)
2:     $\mathcal{B} = \emptyset; \mathcal{C} = \{\{\}\}$
3:     **for all** $\mathcal{E} \in \mathbf{Ext}(\mathcal{AF}_\mathcal{K})$ **do**
4:         $S = \{a|a \in \mathcal{E} \text{ such that } \mathbf{Conc}(a) \models Q\}$
5:     **for all** $\mathcal{E} \in \mathbf{Ext}(\mathcal{AF}_\mathcal{K})$ such that $\nexists a \in \mathcal{E}$ and $\mathbf{Conc}(a) \models Q$ **do**
6:         **for all** $a \in S$ **do**
7:             $Attackers = \{b|b \in \mathcal{E} \text{ such that } b \text{ attacks } a\}$
8:             $\mathcal{C} = \mathcal{C} \cup \{Attackers\}$
9:     $\mathcal{B} = \bigcup_{S \in \mathcal{C}} S$
10:    **Return** $(\mathcal{C}, \mathcal{B})$

---

a set $\mathcal{H} \subseteq \mathcal{B}$ such that $S_j \cap \mathcal{H} \neq \emptyset$ for all $S_j \in \mathcal{C}$. A hitting set for $\mathcal{C}$ is minimal iff no proper subset of it is a hitting set for $\mathcal{C}$. The set of all minimal hitting sets is denoted by $\mathcal{M}$. Computing $\mathcal{M}$ is known as the hitting set problem.

The hitting set problem amounts to find all minimal (w.r.t $\subseteq$) sets of $\mathcal{B}$ whose intersection with each set of $\mathcal{C}$ is non-empty.

Reducibility. Recall that, the problem of computing strong proponent sets (resp. opponent sets) is defined as the problem of computing all strong proponent sets (resp. opponent sets) for a given query Q. It turns out that the problem of computing strong proponent sets (resp. opponent sets) is polynomially reducible to the hitting set problem. As its name indicates, Algorithm 2 (resp. Algorithm 3) is the algorithm responsible for transforming any instance of the strong proponent set (resp. opponent set) problem to an instance of the hitting set problem.

Theorem 2 *[Reduction]. Let $\mathcal{K}$ be a knowledge base, $\mathcal{AF}_\mathcal{K}$ the corresponding argumentation framework, $\mathbf{Ext}(\mathcal{AF}_\mathcal{K}) = \{\mathcal{E}_1, ..., \mathcal{E}_m\}$ set of all preferred extensions of $\mathcal{AF}_\mathcal{K}$ and Q be a query.*

- *Suppose that $\mathcal{C} = \{S_1, ..., S_m\}$ and $\mathcal{B}$ are the outputs of the function **ReduceStrongToHitting** ($\mathbf{Ext}(\mathcal{AF}_\mathcal{K}), Q$), then the set of arguments $\mathcal{H}$ is a strong proponent set of Q if and only if $\mathcal{H}$ is a minimal hitting set of $\mathcal{C}$.*
- *Suppose that $\mathcal{C} = \{S_1, ..., S_m\}$ and $\mathcal{B}$ are the outputs of the function **ReduceOpponentToHitting** ($\mathbf{Ext}(\mathcal{AF}_\mathcal{K}), Q$), then the set of arguments $\mathcal{H}$ is an opponent set of Q iff $\mathcal{H}$ is a minimal hitting set of $\mathcal{C}$.*

Algorithm 4. COMPUTEMINHITTINGSET

1: **function** COMPUTEMINHITTINGSET$(\mathcal{C}, \mathcal{B} = \{m_1, ..., m_n\})$
2:     $M = \mathcal{B}$
3:     **for** $i \leftarrow 1, n$ **do**
4:         $\mathcal{H} = M\text{-}\{m_j\}$
5:         **if** $\forall S \in \mathcal{C}, \mathcal{H} \cap S \neq \emptyset$ **then** $M = \mathcal{H}$
6:     **Return** (M)

Algorithm 5. COMPUTEEXPLANATION(Q,TYPE,Ext$(\mathcal{AF}_{\mathcal{K}})$)

Input: *The query to be explained, boolean variable* type *indicating the explanation type, the set of all extensions of the corresponding argumentation framework.*
Output: *An explanation.*
1: **if** $type = true$ **then** $(\mathcal{C}, \mathcal{B}) = ReduceStrongToHitting(\text{Ext}(\mathcal{AF}_{\mathcal{K}}), Q)$
2: **else**$(\mathcal{C}, \mathcal{B}) = ReduceOpponentToHitting(\text{Ext}(\mathcal{AF}_{\mathcal{K}}), Q)$
3: $\mathcal{H} = CompMinHittingSet(\mathcal{C}, \mathcal{B})$
4: $\mathcal{DT}(\mathcal{H}) = \{DefenseTree(a, \text{Ext}(\mathcal{AF}_{\mathcal{K}})) | a \in \mathcal{H}\}$
5: **Return** $(\mathcal{H}, \mathcal{DT}(\mathcal{H}))$

Proof. In [1] pages 12-13.

Complexity of Algorithms 2 and 3. It is not hard to see that Algorithms 2 and 3 run in a polynomial time $O(n^2)$ considering the worst-case scenario, in which $n = max(|\text{Ext}(\mathcal{AF}_{\mathcal{K}})|, |S|)$ where $|S|$ is the cardinality of the set of all arguments that support Q.

4.3   Computing Explanations

In order to compute an explanation we need to compute *one* strong proponent set (or opponent set) of Q. As a result of Theorem 2, computing *one* strong proponent set is at least as hard as computing *one* minimal hitting set. Based on the polynomial algorithm (Algorithm 4) for computing *one* minimal hitting set proposed in [13], we will compute a strong proponent set (or an opponent set) by means of the reduction algorithms (Algorithms 2 and 3).

Algorithm 5 computes an explanation for query failure and query acceptance. The boolean constant *true* corresponds to acceptance of Q and *false* corresponds to the failure of Q. The algorithm uses the reduction (lines 1 and 2) to make use of the hitting set algorithm (line 3). Finally, in line 4 the algorithm returns the explanation after having computed the defense trees of the set S. The complexity of *ComputeExplanation* depends on the complexity of *ReduceStrongToHitting()* (resp. *ReduceStrongToHitting()*), *CompMinHittingSet()* and *DefenseTree()* which is polynomial.

Computing all explanations is hard to perform since one has to find all strong proponent (resp. opponent) sets to be able to compute all explanations, and finding all strong proponent (resp. opponent) sets is at least as hard as find all minimal hitting sets.

Theorem 3. *Let $\mathcal{K}$ be an inconsistent knowledge base, Q a query and $\exp^+(Q)$ (resp. $\exp^-(Q)$) the set of all explanations of the acceptance (resp. failure) of Q. Then, computing $\exp^+(Q)$ (resp. $\exp^-(Q)$) is NP-hard.*

Proof. In [1] page 14.

Despite the hardness of computing all explanations, a brute-force algorithm can be constructed as follows. We use the reduction function *ReduceStrong-ToHitting()* and *ReduceOpponentToHitting()*. Next, we use Reiter's well-known hitting set algorithm [15,23] as a subroutine called *ComputeAllMinHit()*. After that, given the reduction between the problem of hitting set and strong proponent (or opponent) set, this subroutine can compute all strong proponent (or opponent) sets. Finally, for each strong proponent (or opponent) set we compute the set of defense trees of its arguments.

5 Discussion and Conclusion

In recent years explanation has drawn a tremendous attention in the field of Description Logics, OWL Ontologies debugging and Database Systems. In the field of databases there has been work on explaining query answering and query failure [21] using causality and responsibility or cooperative architectures. In the area of DLs, the question was mainly about explaining either reasoning (subsumption and non-subsumption) or unsatisfiability and incoherence. In [19] the authors addressed the problem of explaining subsumption and non-subsumption in a coherent and satisfiable DL knowledge base using *formal proofs as explanation* while other proposals [24] have used *Axiom pinpointing* and *Concept pinpointing* as explanation to highlight contradictions within an unsatisfiable and incoherent DL KB. Another proposal [16] is the so-called *justification-oriented proofs* in which the authors proposed a *proof-like* explanation without the need for a deduction system.

In previous work [2,3] we have proposed explanation facilities for inconsistent-knowledge base, we focused in these work on the representational aspects of explanation where a dialogue between the user and the system takes place to explain query failure, in this work we are more interested in the definition of the core explanation and the computational aspects, in addition we handle query answering in its broader sense, i.e. query failure and query acceptance.

Acknowledgement. Financial support from the French National Research Agency (ANR) for the project DUR-DUR (ANR-13-ALID-0002) is gratefully acknowledged.

References

1. Arioua, A., Tamani, N., Croitoru, M.: Query answering explanation in inconsistent datalog+/- knowledge bases. Technical report, INRIA GraphiK - LIRMM, INRA, UM (2014). http://www2.lirmm.fr/~arioua/Arioua2014TR.pdf, please mind the tild

2. Arioua, A., Tamani, N., Croitoru, M.: Query failure explanation in inconsistent knowledge bases an argumentation approach: Extended abstract. In: 5th International Conference on Computational Models of Argument 2014 (2014, to appear)

3. Arioua, A., Tamani, N., Croitoru, M., Buche, P.: Query failure explanation in inconsistent knowledge bases: A dialogical approach. In: Bramer, M., Petridis, M. (eds.) Research and Development in Intelligent Systems XXXI, pp. 119–133. Springer, Heidelberg (2014)

4. Baader, F., Brandt, S., Lutz, C.: Pushing the el envelope. In: Proceedings of IJCAI 2005 (2005)

5. Baget, J.-F., Mugnier, M.-L., Rudolph, S., Thomazo, M.: Walking the complexity lines for generalized guarded existential rules. In: Proceedings of IJCAI 2011, pp. 712–717 (2011)

6. Bienvenu, M.: On the complexity of consistent query answering in the presence of simple ontologies. In: Proceedings of AAAI (2012)

7. Bienvenu, M., Rosati, R.: Tractable approximations of consistent query answering for robust ontology-based data access. In: Proceedings of IJCAI 2013, pp. 775–781. AAAI Press (2013)

8. Calì, A., Gottlob, G., Lukasiewicz, T.: A general datalog-based framework for tractable query answering over ontologies. Web Semant.: Sci. Serv. Agents World Wide Web 14, 57–83 (2012)

9. Calvanese, D., De Giacomo, G., Lembo, D., Lenzerini, M., Rosati, R.: Tractable reasoning and efficient query answering in description logics: The dl-lite family. J. Autom. Reasoning 39(3), 385–429 (2007)

10. Calvanese, D., Ortiz, M., Šimkus, M., Stefanoni, G.: Reasoning about explanations for negative query answers in dl-lite. J. Artif. Intell. Res. 48, 635–669 (2013)

11. Croitoru, M., Vesic, S.: What can argumentation do for inconsistent ontology query answering? In: Liu, W., Subrahmanian, V.S., Wijsen, J. (eds.) SUM 2013. LNCS, vol. 8078, pp. 15–29. Springer, Heidelberg (2013)

12. Dung, P.M.: On the acceptability of arguments and its fundamental role in nonmonotonic reasoning, logic programming and n-person games. Artif. Intell. 77(2), 321–357 (1995)

13. Fijany, A., Vatan, F.: New approaches for efficient solution of hitting set problem. In: Proceedings of the Winter International Synposium on Information and Communication Technologies, WISICT 2004. Trinity College Dublin (2004)

14. Gottlob, G., Pieris, A., et al.: Towards more expressive ontology languages: The query answering problem. Artif. Intell. 193, 87–128 (2012)

15. Greiner, R., Smith, B.A., Wilkerson, R.W.: A correction to the algorithm in reiter's theory of diagnosis. Artif. Intell. 41(1), 79–88 (1989)

16. Horridge, M., Parsia, B., Sattler, U.: Justification oriented proofs in OWL. In: Patel-Schneider, P.F., Pan, Y., Hitzler, P., Mika, P., Zhang, L., Pan, J.Z., Horrocks, I., Glimm, B. (eds.) ISWC 2010, Part I. LNCS, vol. 6496, pp. 354–369. Springer, Heidelberg (2010)

17. Lembo, D., Lenzerini, M., Rosati, R., Ruzzi, M., Savo, D.F.: Inconsistency-tolerant semantics for description logics. In: Hitzler, P., Lukasiewicz, T. (eds.) RR 2010. LNCS, vol. 6333, pp. 103–117. Springer, Heidelberg (2010)

18. Lukasiewicz, T., Martinez, M.V., Simari, G.I.: Complexity of inconsistency-tolerant query answering in datalog+/−. In: Meersman, R., Panetto, H., Dillon, T., Eder, J., Bellahsene, Z., Ritter, N., De Leenheer, P., Dou, D. (eds.) ODBASE 2013. LNCS, vol. 8185, pp. 488–500. Springer, Heidelberg (2013)

19. McGuinness, D.L., Borgida, A.T.: Explaining subsumption in description logics. In: Proceedings of IJCAI 1995, pp. 816–821. Morgan Kaufmann Publishers Inc. (1995)

20. McGuinness, D.L., Patel-Schneider, P.F.: Usability issues in knowledge representation systems. In: Proceedings of AAAI-1998, pp. 608–614 (1998)

21. Meliou, A., Gatterbauer, W., Moore, K.F., Suciu, D.: Why so? or why no? functional causality for explaining query answers. In: Proceedings of the International Workshop on Management of Uncertain Data (2010)

22. Poggi, A., Lembo, D., Calvanese, D., De Giacomo, G., Lenzerini, M., Rosati, R.: Linking Data to Ontologies. In: Spaccapietra, S. (ed.) Journal on Data Semantics X. LNCS, vol. 4900, pp. 133–173. Springer, Heidelberg (2008)

23. Reiter, R.: A theory of diagnosis from first principles. Artif. Intell. **32**(1), 57–95 (1987)

24. Schlobach, S., Cornet, R.: Non-standard reasoning services for the debugging of description logic terminologies. In: Proceedings of IJCAI 2003, pp. 355–360. Morgan Kaufmann Publishers Inc. (2003)

25. Ye, L.R., Johnson, P.E.: The impact of explanation facilities on user acceptance of expert systems advice. Mis Q. **19**(2), 157–172 (1995)

PARTY: A Mobile System for Efficiently Assessing the Probability of Extensions in a Debate

Bettina Fazzinga[1]([⊠]), Sergio Flesca[2], Francesco Parisi[2],
and Adriana Pietramala[3]

[1] ICAR-CNR, Rende (CS), Italy
fazzinga@icar.cnr.it
[2] DIMES, University of Calabria, Rende (CS), Italy
{flesca,fparisi}@dimes.unical.it
[3] DISCo, University of Milano-Bicocca, Milano, Italy
adriana.pietramala@disco.unimib.it

Abstract. In this paper we propose *PARTY*, a *P*robabilistic *A*bstract a*R*gumen*T*ation s*Y*stem that assesses the probability that a set of arguments is an *extension* according to a semantics. *PARTY* deals with five popular semantics, i.e., *admissible, stable, complete, grounded,* and *preferred*: it implements polynomial algorithms for computing the probability of the extensions for *admissible* and *stable* semantics and it implements an efficient Monte-Carlo simulation algorithm for estimating the probability of the extensions for the other semantics, which have been shown to be intractable in [19,20]. The experimental evaluation shows that *PARTY* is more efficient than the state-of-the art approaches and that it can be profitable executed on devices having reduced computational resources.

1 Introduction

Argumentation allows disputes to be modeled, which arise between two or more parties, each of them providing arguments to assert her reasons. Although argumentation is strongly related to philosophy and law, it has gained remarkable interest in AI as a reasoning model for representing dialogues, making decisions, and handling inconsistency and uncertainty [10,11,26]. A simple but powerful argumentation framework is that proposed in the seminal paper [13], where the *abstract argumentation framework* (AAF) was introduced. An AAF is a pair $\langle A, D \rangle$ consisting of a set A of *arguments*, and of a binary relation D over A, called *defeat* (or, equivalently, *attack*) relation. Basically, an argument is an abstract entity that may attack and/or be attacked by other arguments.

Example 1. Consider the following scenario, where the defence attorney of John wants to model the situation regarding the robbery case involving his client. The arguments of the robbery case are the following, where Harry and Joanne are potential witnesses:

© Springer International Publishing Switzerland 2015
Q. Chen et al. (Eds.): DEXA 2015, Part I, LNCS 9261, pp. 220–235, 2015.
DOI: 10.1007/978-3-319-22849-5_16

a = John says he was at home when the robbery took place, and therefore denies being involved in the robbery.

b = Harry says he saw John running out of the bank just after the robbery.

c = Joanne says she saw John entering his home probably when the robbery took place.

Both arguments a and c claim that John is not involved in the robbery, while argument b means that a potential witness instils doubts about the innocence of John.

This scenario can be modeled by the AAF $\mathcal{A}$, whose set of arguments is $\{a, b, c\}$, and whose defeat relation consists of $\delta_1 = (b, a)$, $\delta_2 = (b, c)$, and $\delta_3 = (c, b)$, meaning that argument a is attacked by b, and that b and c are mutually attacking themselves. □

Several semantics for AAFs, such as *admissible, stable, complete, grounded, preferred*, and others, have been proposed [8,13] to identify "reasonable" sets of arguments, called *extensions*. Basically, each of these semantics corresponds to some properties which "certify" whether a set of arguments can be profitably used to support a point of view in a discussion. For instance, a set S of arguments is an extension according to the admissible semantics if it has two properties: it is conflict-free (that is, there is no defeat between arguments in S), and every argument (outside S) attacking an argument in S is counterattacked by an argument in S. Intuitively enough, the fact that a set is an extension according to the admissible semantics means that, using the arguments in S, you do not contradict yourself, and you can rebut to anyone who uses any of the arguments outside S to contradict yours. The other semantics correspond to other ways of determining whether a set of arguments would be a "good point" in a dispute, and will be described in the core of the paper.

As a matter of fact, in the real world, arguments and defeats are often uncertain, thus, several proposals have been made to model uncertainty in AAFs, by considering weights, preferences, or probabilities associated with arguments and/or defeats. In this regard, [14,23,27,30] have recently extended the original Dung framework in order to achieve probabilistic abstract argumentation frameworks (PrAFs), where uncertainty of arguments and defeats is modeled by exploiting probability theory. In particular, [23] proposed a PrAF where both arguments and defeats are associated with probabilities.

Considering the PrAF proposed in [23], a fundamental problem (denoted as $\text{PROB}_{\mathcal{F}}^{sem}(S)$) is then that of *computing the probability $Pr_{\mathcal{F}}^{sem}(S)$ that a set S of arguments is an extension according to a given semantics sem*. In this setting, the notion of *possible world* is used to define the meaning of a PrAF and then the probability $Pr_{\mathcal{F}}^{sem}(S)$, as explained in what follows. Basically, given a PrAF $\mathcal{F}$, a possible world represents a (deterministic) scenario consisting of some subset of the arguments and defeats in $\mathcal{F}$. That is, a possible world can be viewed as an AAF containing exactly the arguments and the defeats occurring in the represented scenario. In [23] it was shown that a PrAF admits a unique probability distribution over the set of possible worlds, which assigns a probability value to each possible world coherently with the probabilities of arguments and defeats, and allows users to derive probabilistic conclusions from the PrAF. Once established that a PrAF admits a unique probability distribution over the set of possible worlds, the probability $Pr_{\mathcal{F}}^{sem}(S)$ is naturally defined as the sum of the probabilities of the possible worlds where the set S of arguments is an extension according to the semantics *sem*.

Example 2 (Example 1 *Continued*). The defence attorney of John is willing to reason together with her client about the possible outcomes of the trial. To this end, the attorney makes aware her client of her beliefs about the fact that the jury will hear from John, Harry and Joanne, and that the jurors will be, in turn, persuaded that Harry's argument contradicts John and Joanne ones and that Joanne's argument contradicts Harry's one. The attorney is sure that the jury will hear from John and Joanne, since she is sure that John will be allowed to claim his innocence and that Joanne will testify. Moreover, the attorney believes with 80 % probability that the jury will hear from Harry, since Harry is a bit reluctant to testify. In addition, the attorney believes that, in the case that the jury will hear from Harry, the jurors will be persuaded that Harry's argument contradicts the arguments of John and Joanne, since Harry has a high reputation in the community. Finally, according to the attorney, the probability that the jurors will be persuaded that Joanne's argument contradicts Harry's one is only 40 %, since Joanne is not sure about the exact time when she saw John.

The above described scenario can be represented by the attorney as a PrAF $\mathcal{F}_{\mathcal{A}}$ obtained from the AAF $\mathcal{A}$ introduced in Example 1 by considering the arguments a, b, and c, and the defeats δ_1, δ_2 and δ_3 as probabilistic events, having probabilities $Pr(a) = 1$, $Pr(b) = 0.8$, $Pr(c) = 1$, $Pr(\delta_1) = 1$, $Pr(\delta_2) = 1$, and $Pr(\delta_3) = 0.4$. Each possible world of $\mathcal{F}_{\mathcal{A}}$ is characterized by a probability which is implied by the probability of its arguments and defeats. For instance, the possible world $\langle \{a, c\}, \emptyset \rangle$ is the AAF representing the scenario where only John and Joanne testify, and it will happen with probability 20 % (we will explain how to compute the probability of the above-mentioned AAFs in Sect. 2.2). By reasoning on the above mentioned PrAF the attorney can deduce that the probability that John will not be considered guilty by the jury (its guiltiness is not fully proved yet) is 100 %, which corresponds to the probability that the set of arguments $\{a, c\}$ is admissible, i.e. the sum of the probabilities of the possible worlds corresponding to AAFs where $\{a, c\}$ is admissible. More details about the computation of the probability that $\{a, c\}$ is admissible are given in Sect. 2.2.   □

In scenarios such as that described in Examples 1 and 2, the attorney may want to reason about the possible outcomes of the trial using low computational resources such as those available on her mobile device (i.e., tablet/smartphone). Indeed, mobile devices are nowadays often used in many workplaces and, more importantly, the fact of carrying personally owned mobile devices, most of them characterized by the availability of low computational resources, is becoming more and more common in several companies. As a matter of fact, *Marketsand-Markets* forecasts that the global Bring Your Own Device & Enterprise Mobility

Market will grow from \$71.93 billion in 2013 to \$266.17 billion in 2019 [1], which in turn suggests that devising systems that are compliant with mobile device's specifications is of definite interest. In our setting, this entails that computing the probability $Pr_{\mathcal{F}}^{sem}(S)$ that a set of arguments is an extension according to a given semantics must be done in an efficient way. Unfortunately, as pointed out in [23], computing $Pr_{\mathcal{F}}^{sem}(S)$ by directly exploiting its definition would result in an exponential time algorithm. However, in [19, 20] we showed that $\mathrm{PROB}_{\mathcal{F}}^{sem}(S)$ is actually tractable for the admissible and stable semantics, and it is intractable for other semantics, including preferred, grounded and complete. Furthermore, by exploiting the tractability results of [19, 20], in [18], we proposed an efficient new Monte-Carlo simulation technique for estimating the probability $Pr_{\mathcal{F}}^{sem}(S)$ for the cases when $\mathrm{PROB}_{\mathcal{F}}^{sem}(S)$ is intractable.

In this paper, we present *PARTY*, a Probabilistic Abstract aRgumenTation sYstem that efficiently computes or estimates $Pr_{\mathcal{F}}^{sem}(S)$ by exploiting the results of [18–20]. Specifically, *PARTY* implements polynomial time algorithms based on the results of [19, 20] for computing $Pr_{\mathcal{F}}^{sem}(S)$ with *sem* $\in$ {*admissible, stable*} and implements the Monte-Carlo simulation algorithms presented in [18] for estimating $Pr_{\mathcal{F}}^{sem}(S)$ with *sem* $\in$ {*complete, grounded, preferred*}. We experimentally validated the efficiency of PARTY over two abstract argumentation benchmarks, showing that it outperforms the state-of-the-art approach proposed in [23] and it is particularly suitable to be executed on devices that have reduced computational resources. *PARTY* is available at http://si.deis.unical.it/~flesca/argumentation/ and, to the best of our knowledge, it is the first system addressing the efficient computation of $Pr_{\mathcal{F}}^{sem}(S)$.

2   Preliminaries

In this section, we briefly overview Dung's abstract argumentation framework and its probabilistic extension introduced in [23].

2.1   Abstract Argumentation

An *abstract argumentation framework* [13] (*AAF*) is a pair $\langle A, D \rangle$, where A is a finite set, whose elements are referred to as *arguments*, and $D \subseteq A \times A$ is a binary relation over A, whose elements are referred to as *defeats* (or *attacks*). An argument is an abstract entity whose role is entirely determined by its relationships with other arguments. Given an AAF $\mathcal{A}$, we also refer to the set of its arguments and the set of its defeats as $Arg(\mathcal{A})$ and $Def(\mathcal{A})$, respectively.

Given arguments $a, b \in A$, we say that a *defeats* b iff there is $(a, b) \in D$. Similarly, a set $S \subseteq A$ *defeats* an argument $b \in A$ iff there is $a \in S$ such that a *defeats* b.

A set $S \subseteq A$ of arguments is said to be *conflict-free* if there are no $a, b \in S$ such that a *defeats* b. An argument a is said to be *acceptable* w.r.t. $S \subseteq A$ iff $\forall b \in A$ such that b *defeats* a, there is $c \in S$ such that c *defeats* b.

Several semantics for AAFs have been proposed to identify "reasonable" sets of arguments, called *extensions*. We consider the following semantics [13]: *admissible* (ad), *stable* (st), *complete* (co), *grounded* (gr), *preferred* (pr). A set $S \subseteq A$ is

- an *admissible extension* iff S is conflict-free and all its arguments are acceptable w.r.t. S;
- a *stable extension* iff S is conflict-free and S defeats each argument in $A \setminus S$;
- a *complete extension* iff S is admissible and S contains all the arguments that are acceptable w.r.t. S;
- a *grounded extension* iff S is a minimal (w.r.t. $\subseteq$) complete set of arguments;
- a *preferred extension* iff S is a maximal (w.r.t. $\subseteq$) admissible set of arguments.

Example 3. Consider the AAF $\langle A, D \rangle$ introduced in Example 1, where the set A of arguments is $\{a, b, c\}$, and the set D of defeats is $\{\delta_1 = (a, b), \delta_2 = (b, a), \delta_3 = (c, b)\}$. The set $S = \{a, c\}$ is an admissible extension, since it is conflict-free and both a and c are acceptable w.r.t. S. It is easy to see that both $\emptyset$ and $\{c\}$ are admissible extensions as well, and that set $S' = \{b\}$ is not an admissible extension since S' does not counterattack the attack from c to b. Moreover, since $S = \{a, c\}$ is conflict-free and defeats b (which the only argument outside S), it is a stable extension. As S is a maximally admissible set of arguments, it is a preferred extension. Finally, it is easy to check that S is a complete extension, as it contains all the arguments in A that are acceptable w.r.t. S, while it is not a grounded extension since it is not a minimally complete set of arguments. $\square$

Given an AAF $\mathcal{A}$, a set $S \subseteq Arg(\mathcal{A})$ of arguments, and a semantics *sem* $\in \{\text{ad}, \text{st}, \text{co}, \text{gr}, \text{pr}\}$, we define the function $ext(\mathcal{A}, sem, S)$ which returns *true* if S is an extension according to *sem*, *false* otherwise.

2.2   Probabilistic Abstract Argumentation

We now review the *probabilistic* abstract argumentation framework proposed in [23].

Definition 1 *(PrAF).* *A probabilistic argumentation framework (PrAF) is a tuple $\langle A, P_A, D, P_D \rangle$ where $\langle A, D \rangle$ is an AAF, and P_A and P_D are, respectively, functions assigning a non-zero probability value to each argument in A and defeat in D, that is, $P_A : A \rightarrow (0, 1]$ and $P_D : D \rightarrow (0, 1]$.*

Basically, the value assigned by P_A to an argument a represents the probability that a actually occurs, whereas the value assigned by P_D to a defeat (a, b) represents the conditional probability that a defeats b given that both a and b occur.

The meaning of a PrAF is given in terms of possible worlds, each of them representing a scenario that may occur in the reality. Given a PrAF $\mathcal{F}$, a possible world is modeled by an AAF which is derived from $\mathcal{F}$ by considering only a subset of its arguments and defeats. More formally, given a PrAF $\mathcal{F} = \langle A, P_A, D, P_D \rangle$, a possible world w of $\mathcal{F}$ is an AAF $\langle A', D' \rangle$ such that $A' \subseteq A$ and $D' \subseteq D \cap (A' \times A')$. The set of the possible worlds of $\mathcal{F}$ will be denoted as $pw(\mathcal{F})$.

Example 4. Consider the PrAF $\mathcal{F} = \langle A, P_A, D, P_D \rangle$ introduced in Example 2, where $A = \{a, b, c\}$, $D = \{\delta_1 = (a, b), \delta_2 = (b, a), \delta_3 = (c, b)\}$, $P_A(a) = P_A(c) = 1$, $P_A(b) = 0.8$, $P_D(\delta_1) = P_D(\delta_2) = 1$, and $P_D(\delta_3) = 0.4$. The set $pw(\mathcal{F})$ consists the following possible worlds: $w_1 = \langle \emptyset, \emptyset \rangle$, $w_2 = \langle \{a\}, \emptyset \rangle$, $w_3 = \langle \{b\}, \emptyset \rangle$, $w_4 = \langle \{c\}, \emptyset \rangle$, $w_5 = \langle \{a, b\}, \emptyset \rangle$, $w_6 = \langle \{a, c\}, \emptyset \rangle$, $w_7 = \langle \{b, c\}, \emptyset \rangle$, $w_8 = \langle A, \emptyset \rangle$, $w_9 = \langle \{a, b\}, \{\delta_1\} \rangle$, $w_{10} = \langle \{a, b\}, \{\delta_2\} \rangle$, $w_{11} = \langle \{a, b\}, \{\delta_1, \delta_2\} \rangle$, $w_{12} = \langle \{b, c\}, \{\delta_3\} \rangle$, $w_{13} = \langle A, \{\delta_1\} \rangle$, $w_{14} = \langle A, \{\delta_2\} \rangle$, $w_{15} = \langle A, \{\delta_3\} \rangle$, $w_{16} = \langle A, \{\delta_1, \delta_2\} \rangle$, $w_{17} = \langle A, \{\delta_1, \delta_3\} \rangle$, $w_{18} = \langle A, \{\delta_2, \delta_3\} \rangle$, $w_{19} = \langle A, D \rangle$. □

An interpretation for a PrAF $\mathcal{F} = \langle A, P_A, D, P_D \rangle$ is a probability distribution function I over the set $pw(\mathcal{F})$ of the possible worlds. Assuming that arguments represent pairwise independent events, and that each defeat represents an event conditioned by the occurrence of its argument events but independent from any other event, the interpretation for the PrAF $\mathcal{F} = \langle A, P_A, D, P_D \rangle$ is as follows. For each possible world $w \in pw(\mathcal{F})$, w is assigned by I the probability:

$$I(w) = \prod_{a \in Arg(w)} P_A(a) \cdot \prod_{a \in A \setminus Arg(w)} (1 - P_A(a)) \cdot \prod_{\delta \in Def(w)} P_D(\delta) \cdot \prod_{\delta \in \overline{D}(w) \setminus Def(w)} (1 - P_D(\delta))$$

where $\overline{D}(w)$ is the set of defeats that may appear in the possible world w, that is $\overline{D}(w) = D \cap (Arg(w) \times Arg(w))$.

Example 5. Continuing our running example, the interpretation I for $\mathcal{F}$ is as follows. For each possible world $w \in pw(\mathcal{F})$ such that a or c does not belong to $Arg(w)$ it holds that $I(w) = 0$, and for each possible world w such that b belongs to $Arg(w)$ and either $\delta_1 \notin Def(w)$ or $\delta_2 \notin Def(w)$ it holds that $I(w) = 0$ as well. The probabilities of the possible worlds which are assigned a non-zero probability by I are as follows.

$I(w_6) = P_A(a) \times (1 - P_A(b)) \times P_A(c) = 0.2$;
$I(w_{16}) = P_A(a) \times P_A(b) \times P_A(c) \times P_D(\delta_1) \times P_D(\delta_2) \times (1 - P_D(\delta_3)) = 0.48$;
$I(w_{19}) = P_A(a) \times P_A(b) \times P_A(c) \times P_D(\delta_1) \times P_D(\delta_2) \times P_D(\delta_3) = 0.32$. □

The probability that a set S of arguments is an extension according to a given semantics *sem* is defined as the sum of the probabilities of the possible worlds w for which S is an extension according to *sem*, that is, the sum of the probabilities of the possible worlds w for which it holds that $ext(w, sem, S) = $ true. We point out that if $S \not\subseteq Arg(w)$ then $ext(w, sem, S)$=false for every semantics *sem*.

Definition 2 ($Pr_{\mathcal{F}}^{sem}(S)$). *Given a PrAF $\mathcal{F}$, a set S, and a semantics sem, the probability $Pr_{\mathcal{F}}^{sem}(S)$ that S is an extension according to sem is $Pr_{\mathcal{F}}^{sem}(S) = $*
$$\sum_{w \in pw(\mathcal{F}) \wedge ext(w, sem, S)} I(w).$$

Example 6. In our running example, the probability that set $S = \{a, c\}$ is an admissible extension is $Pr_{\mathcal{F}}^{ad}(S) = I(w_6) + I(w_{16}) + I(w_{19}) = 1$, which is also the probability that S is a stable extension (i.e., S is a stable and an admissible extension in all the non-zero probability possible worlds). □

In the following we will also refer to the probability that a set S of arguments is conflict-free, that is the sum of probabilities of the possible worlds w wherein S is conflict-free. Though cf is not a semantics, with a little abuse of notation, we include it in the set of the considered semantic and assume that function $ext(w, \text{cf}, S)$ returns *true* iff S is a conflict-free in w. Thus, we denote as $Pr_{\mathcal{F}}^{cf}(S)$ the probability that S is conflict-free.

Obviously, computing $Pr_{\mathcal{F}}^{sem}(S)$ by directly applying Definition 2 would require exponential time, since it relies on summing the probabilities of an exponential number of possible worlds. However, as shown in [18–20], and reported in the next section, efficient algorithms can be devised for computing $Pr_{\mathcal{F}}^{sem}(S)$ for $sem \in \{\text{cf}, \text{ad}, \text{st}\}$ and estimating $Pr_{\mathcal{F}}^{sem}(S)$ for $sem \in \{\text{co}, \text{gr}, \text{pr}\}$.

3 Computing Extensions' Probabilities in Abstract Argumentation

On the basis of the input semantics *sem*, *PARTY* either *computes* $Pr_{\mathcal{F}}^{sem}(S)$ or *estimates* $Pr_{\mathcal{F}}^{sem}(S)$. Specifically, for $sem \in \{\text{cf}, \text{ad}, \text{st}\}$, *PARTY* computes $Pr_{\mathcal{F}}^{sem}(S)$ by exploiting the tractability results of [19,20], where it has been shown that $Pr_{\mathcal{F}}^{sem}(S)$ can be computed in polynomial time without applying Definition 2 (i.e., without summing the probabilities of all the possible worlds). For $sem \in \{\text{co}, \text{gr}, \text{pr}\}$, instead, since the problem of computing $Pr_{\mathcal{F}}^{sem}(S)$ is intractable [19,20], *PARTY* efficiently *estimates* $Pr_{\mathcal{F}}^{sem}(S)$ by implementing Monte-Carlo simulation algorithms proposed in [18].

3.1 Computing $Pr_{\mathcal{F}}^{cf}(S)$, $Pr_{\mathcal{F}}^{ad}(S)$ and $Pr_{\mathcal{F}}^{st}(S)$

We now describe how *PARTY* computes the probability $Pr_{\mathcal{F}}^{sem}(S)$ for the conflict-free, the admissible and the stable semantics. As shown in [19,20], the tractability of the problem of computing $Pr_{\mathcal{F}}^{cf}(S)$, $Pr_{\mathcal{F}}^{ad}(S)$, and $Pr_{\mathcal{F}}^{st}(S)$ derives from the fact that the event that S is conflict-free, the event that S is an extension according to ad, and the event that S is an extension according to st, can be expressed by means of a formula over basic probabilistic events where the operands of every subformula containing two or more operands are either independent or mutually exclusive. Specifically, *PARTY* exploits the following results to compute $Pr_{\mathcal{F}}^{cf}(S)$, $Pr_{\mathcal{F}}^{ad}(S)$, and $Pr_{\mathcal{F}}^{st}(S)$.

Fact 1 ($\mathbf{Pr}_{\mathcal{F}}^{cf}(S)$). *Given a PrAF* $\mathcal{F} = \langle A, P_A, D, P_D \rangle$ *and a set* $S \subseteq A$ *of arguments,* $Pr_{\mathcal{F}}^{cf}(S) = \prod_{a \in S} P_A(a) \cdot \prod_{\substack{\langle a, b \rangle \in D \\ \wedge a \in S \wedge b \in S}} (1 - P_D(\langle a, b \rangle))$.

Fact 2. *Given a PrAF* $\mathcal{F} = \langle A, P_A, D, P_D \rangle$ *and a set* $S \subseteq A$ *of arguments,* $Pr_{\mathcal{F}}^{ad}(S) = Pr_{\mathcal{F}}^{cf}(S) \cdot \prod_{d \in A \backslash S} (P_1(S, d) + P_2(S, d) + P_3(S, d))$, *where:* $P_1(S, d) = 1 - P_A(d)$, $P_2(S, d) = P_A(d) \times \prod_{\substack{(d, b) \in D \\ \wedge b \in S}} (1 - P_D((d, b)))$, *and* $P_3(S, d) =$

$$P_A(d) \times \left(1 - \prod_{\substack{(d, b) \in D \\ \wedge b \in S}} (1 - P_D((d, b)))\right) \times \left(1 - \prod_{\substack{(a, d) \in D \\ \wedge a \in S}} (1 - P_D((a, d)))\right).$$

Fact 3. *Given a PrAF $\mathcal{F} = \langle A, P_A, D, P_D \rangle$, and a set $S \subseteq A$ of arguments, $Pr^{st}_{\mathcal{F}}(S) = Pr^{cf}_{\mathcal{F}}(S) \cdot \prod_{d \in A \setminus S} (P_1(S, d) + P_2(S, d))$, where: $P_1(S, d) = 1 - P_A(d)$, and $P_2(S, d) = P_A(d) \times \left(1 - \prod_{\substack{\langle a, d \rangle \in D \\ \wedge a \in S}} (1 - P_D(\langle a, d \rangle))\right)$.*

These facts entail that $Pr^{cf}_{\mathcal{F}}(S)$ can be computed in $O(|S|^2)$, and that both $Pr^{ad}_{\mathcal{F}}(S)$ and $Pr^{st}_{\mathcal{F}}(S)$ can be computed in time $O(|S| \cdot |A|)$. In fact, the polynomial-time algorithms implemented by PARTY to compute $Pr^{cf}_{\mathcal{F}}(S)$, $Pr^{ad}_{\mathcal{F}}(S)$, and $Pr^{st}_{\mathcal{F}}(S)$ evaluate the expressions of Facts 1, 2 and 3 by iterating on the arguments and defeats of $\mathcal{F}$.

3.2   Estimating $Pr^{co}_{\mathcal{F}}(S)$, $Pr^{gr}_{\mathcal{F}}(S)$ and $Pr^{pr}_{\mathcal{F}}(S)$

We now introduce the Monte-Carlo approach for estimating $Pr^{sem}_{\mathcal{F}}(S)$, with $sem \in \{\mathtt{co}, \mathtt{gr}, \mathtt{pr}\}$, adopted in PARTY. Specifically, PARTY implements three methods for estimating $Pr^{sem}_{\mathcal{F}}(S)$: the first one, called standard, was defined in [23] and it is used as term of comparison for the other two methods, called MC-CF and MC-AD, respectively, that we proposed in [18].

The standard method is a Monte-Carlo estimation algorithm consisting of (i) generating a number n of AAFs, (ii) checking if S is extension according to sem in the generated AAFs, (iii) returning as output the number x/n, where x is the number of AAFs wherein S is an extension according to sem. More formally, given a PrAF $\mathcal{F}$, a set S, a semantic sem, an error level ϵ, and a quantile $z_{1-\alpha/2}$, the standard method returns an estimate $\widehat{Pr}^{sem}_{\mathcal{F}}(S)$ of $Pr^{sem}_{\mathcal{F}}(S)$ such that $Pr^{sem}_{\mathcal{F}}(S)$ lies in the interval $\widehat{Pr}^{sem}_{\mathcal{F}}(S) \pm \epsilon$ with a confidence level implied by the quantile $z_{1-\alpha/2}$. The number n of AAFs to be sampled to achieve the required error level ϵ with confidence level implied by $z_{1-\alpha/2}$ is determined by exploiting the Agresti-Coull interval [2]. In particular, according to [2], the estimated value p of $Pr^{sem}_{\mathcal{F}}(S)$ after x successes in n samples is $p = \frac{x + (z^2_{1-\alpha/2})/2}{n + z^2_{1-\alpha/2}}$, and the number of samples ensuring that the error level is ϵ with confidence level implied by $z_{1-\alpha/2}$ is $n = \frac{z^2_{1-\alpha/2} \cdot p \cdot (1-p)}{\epsilon^2} - z^2_{1-\alpha/2}$.

Our methods MC-CF and MC-AD are both Monte-Carlo based estimation methods extending the standard one as explained in what follows. MC-CF samples only the AAFs of $pw(\mathcal{F})$ wherein S is conflict-free, while MC-AD samples only the AAFs of $pw(\mathcal{F})$ wherein S is an admissible extension. More in detail, since for the considered semantics (i.e., complete, grounded, preferred), S is an extension according to sem only if S is conflict-free and it is an admissible extension, it holds that:

1. $Pr^{sem}_{\mathcal{F}}(S) = Pr^{sem|E_{CF}(S)}_{\mathcal{F}}(S) \cdot Pr^{cf}_{\mathcal{F}}(S)$, and
2. $Pr^{sem}_{\mathcal{F}}(S) = Pr^{sem|E_{AD}(S)}_{\mathcal{F}}(S) \cdot Pr^{ad}_{\mathcal{F}}(S)$,

where $E_{CF}(S)$ (resp., $E_{AD}(S)$) denotes the event that S is conflict-free (resp., admissible). Hence, MC-CF (resp. MC-AD) estimates $Pr^{sem}_{\mathcal{F}}(S)$ by (i) first computing the probability $Pr^{cf}_{\mathcal{F}}(S)$ (resp. $Pr^{ad}_{\mathcal{F}}(S)$) using Fact 1 (resp., Fact 2), (ii) then estimating the conditional probability $Pr_{\mathcal{F}}^{sem|E_{CF}(S)}(S)$ (resp. $Pr_{\mathcal{F}}^{sem|E_{AD}(S)}(S)$) that S is an extension according to sem given that S is conflict-free (resp. an admissible extension), and finally (iii) returning as an estimate of $Pr_{\mathcal{F}}^{sem}(S)$ the value of $\widehat{Pr}_{\mathcal{F}}^{sem|E_{CF}(S)}(S) \cdot Pr_{\mathcal{F}}^{cf}(S)$ (resp. $\widehat{Pr}_{\mathcal{F}}^{sem|E_{AD}(S)}(S) \cdot Pr_{\mathcal{F}}^{ad}(S)$), where $\widehat{Pr}_{\mathcal{F}}^{sem|E_{CF}(S)}(S)$ (resp., $\widehat{Pr}_{\mathcal{F}}^{sem|E_{AD}(S)}(S)$) denotes the estimate of $Pr_{\mathcal{F}}^{sem|E_{CF}(S)}(S)$ (resp. $Pr_{\mathcal{F}}^{sem|E_{AD}(S)}(S)$). In fact, both MC-CF and MC-AD work over a reduced sample space: instead of considering the whole set of possible worlds of $\mathcal{F}$ as sample space (as done in [23]), they consider the subset of the possible worlds wherein S is conflict-free or wherein S is an admissible extension. Since drawing samples from a reduced sample space leads to better performances, both MC-CF and MC-AD turn out to be more efficient than standard, as shown in Sect. 5.

Algorithm 1. *Estimating $Pr_{\mathcal{F}}^{sem}(S)$*
Input: $\mathcal{F} = \langle A, P_A, D, P_D \rangle$; $S \subseteq A$; sem; *An error level ϵ; A quantile $z_{1-\alpha/2}$*
algType $\in \{standard, MC\text{-}CF, MC\text{-}AD\}$
Output: $\widehat{Pr}_{\mathcal{F}}^{sem}(S)$ s.t. $Pr_{\mathcal{F}}^{sem}(S) \in [\widehat{Pr}_{\mathcal{F}}^{sem}(S) - \epsilon, \widehat{Pr}_{\mathcal{F}}^{sem}(S) + \epsilon]$
1: $prob$ =computeProbability($\mathcal{F}, S$,algType)
2: $x = n = 0$;
3: **do**
4:    $\langle Arg, Def \rangle$ =generateSample($\mathcal{F}, S$, algType);
5:    **if** $ext(\langle Arg, Def \rangle, sem, S)$ **then** $x = x+1$;
6:    $n = n+1$; $p = \frac{x + z_{1-\alpha/2}^2/2}{n + z_{1-\alpha/2}^2}$; $n' = \frac{z_{1-\alpha/2}^2 \cdot p \cdot (1-p)}{\epsilon^2} \cdot (prob)^2 - z_{1-\alpha/2}^2$
7: **while** $n \leq n'$
8: **return** $x/n \cdot prob$

The three methods (standard, MC-CF, and MC-AD) for estimating $Pr_{\mathcal{F}}^{sem}(S)$ are all embedded in Algorithm 1, which returns an estimate $\widehat{Pr}_{\mathcal{F}}^{sem}(S)$ of $Pr_{\mathcal{F}}^{sem}(S)$ such that $Pr_{\mathcal{F}}^{sem}(S)$ lies in the interval $\widehat{Pr}_{\mathcal{F}}^{sem}(S) \pm \epsilon$ with a confidence level implied by the quantile $z_{1-\alpha/2}$, and such that $Pr_{\mathcal{F}}^{sem}(S)$ is computed by applying the method specified by the input parameter *algType*.

First, Algorithm 1 invokes function *computeProbability* that (i) computes $Pr_{\mathcal{F}}^{cf}(S)$ if *algType* = MC-CF, or (ii) computes $Pr_{\mathcal{F}}^{ad}(S)$ if *algType* = MC-AD or (iii) returns 1 if *algType* = *standard*. Next, at each iteration, Algorithm 1 invokes function *generateSample*, that generates an AAF $\langle Arg, Def \rangle$ of $pw(\mathcal{F})$ by adding the arguments $a \in A$ to Arg and the defeats $\delta \in D \cap Arg \times Arg$ to Def according to a certain probability which depends on the chosen estimation method. More in detail, if *algType* = *standard*, the arguments a are added to Arg according to their probability $P_A(a)$ and the defeats δ are added to Def according to their probability $P_D(\delta)$. Instead, if *algType* = MC-CF (resp., MC-AD), since MC-CF (resp., MC-AD) only samples possible worlds wherein S is conflict-free (resp., S is an admissible extension)), arguments a are added to Arg according to their probabilities $P_A(a|E_{CF}(S))$ (resp., $P_A(a|E_{AD}(S))$), that are their probabilities conditioned to the fact that the event $E_{CF}(S)$ (resp., $E_{AD}(S)$) occurs. Analogously, defeats $\delta = \langle a, b \rangle$ (such that both a and b belong to Arg) are added to Def according to their probabilities $P_D(\delta | E_{\mathrm{CF}}(S))$ (resp., $P_D(\delta | E_{\mathrm{AD}}(S))$), that are their probabilities conditioned to the fact that the event $E_{\mathrm{CF}}(S)$ (resp., $E_{\mathrm{AD}}(S)$) occurs. For more detail on this aspect, we refer the reader to [18].

After generating a possible world $\langle Arg, Def \rangle$, Algorithm 1 checks if S is an extension according to sem in $\langle Arg, Def \rangle$, and, if this is the case, it increments x's value. After that, it computes the number n' of samples to be generated according to the Agresti-Coull interval taking into account the fact that, if $alg\,Type = \texttt{MC-CF}$ (resp., $alg\,Type = \texttt{MC-AD}$), the core of Algorithm 1 does not compute $\widehat{Pr}_{\mathcal{F}}^{sem}(S)$, but it computes $\widehat{Pr}_{\mathcal{F}}^{sem|E_{\mathrm{CF}}(S)}(S)$ (resp., $\widehat{Pr}_{\mathcal{F}}^{sem|E_{\mathrm{AD}}(S)}(S)$). Then, as shown in [18], since in the case that $alg\,Type = \texttt{MC-CF}$ (resp., $alg\,Type = \texttt{MC-AD}$), the number n' of samples to be generated turns out to be $n' = \frac{z_{1-\alpha/2}^2 \cdot p \cdot (1-p) \cdot (Pr_{\mathcal{F}}^{cf}(S))^2}{\epsilon^2} - z_{1-\alpha/2}^2$ (resp., $n' = \frac{z_{1-\alpha/2}^2 \cdot p \cdot (1-p) \cdot (Pr_{\mathcal{F}}^{ad}(S))^2}{\epsilon^2} - z_{1-\alpha/2}^2$), variable n' is assigned $\frac{z_{1-\alpha/2}^2 \cdot p \cdot (1-p)}{\epsilon^2} \cdot (prob)^2 - z_{1-\alpha/2}^2$, where $prob$ is computed at Line 1. Finally, Algorithm 1 returns $x/n \cdot prob$ as output. As states below, Algorithm 1 is sound.

Fact 4. *Let* $\mathcal{F} = \langle A, P_A, D, P_D \rangle$, *and* $S \subseteq A$. *Let* ϵ *be an error level, and* $z_{1-\alpha/2}$ *a quantile. The estimate* $\widehat{Pr}_{\mathcal{F}}^{sem}(S)$ *returned by Algorithm 1 is such that* $Pr_{\mathcal{F}}^{sem}(S) \subset [\widehat{Pr}_{\mathcal{F}}^{sem}(S) - \epsilon, \widehat{Pr}_{\mathcal{F}}^{sem}(S) + \epsilon]$ *with confidence level implied by* $z_{1-\alpha/2}$.

4   The PARTY System

In this section we describe *PARTY*, our *Probabilistic Abstract aRgumenTation sYstem*. PARTY has been designed to work with devices that have reduced computational resources than standard computers and, as we will see in the experimental evaluation section, it is particularly efficient on these devices. The system is compatible with Android mobile Operating System (OS), versions ranging from 2.2 (Froyo) to 5.0 (Lollipop), and it was tested on Android 4.0.3. The system may be freely downloaded from the PARTY website http://si.deis. unical.it/~flesca/argumentation/ and it does not require internet connection.

System Architecture. PARTY's architecture is shown in Fig. 1. It consists of two main macro levels: the front-end and the back-end. The front-end, consisting of the user interface (UI), contains three modules: *PrAF Builder/Loader*, *Set S/Semantics Selector* and *Result Viewer*. The *PrAF Builder/Loader* module allows the user both to define its own *PrAF* and to load an existing *PrAF* previously stored in a *.txt* file whose structure will be described in the following. During the building phase, the user can specify sets A, S and D as well as the probability values for each argument in A and defeat in D. The *Set S/Semantics Selector* module allows the user to select a subset of arguments to be included in the set S and to select a semantics sem. Finally, the *Result Viewer* module displays the results obtained from computing/estimating the probability of extensions for the PrAF and set S of arguments given as input.

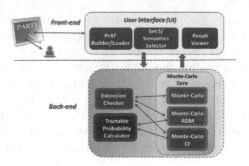

Fig. 1. PARTY architecture

The back-end of the system consists of several modules that can be grouped into three macro-main blocks: *Monte-Carlo Core*, *Extension Checker*, and *Tractable Probability Calculator*. The first macro block consists of the three sub-modules *Monte-Carlo*, *Monte-Carlo ADM*, and *Monte-Carlo CF*. The first sub-module implements the **standard** Monte-Carlo method, while *Monte-Carlo ADM* and *Monte-Carlo CF* implement MC-AD and MC-CF, respectively (see Sect. 3.2). Module *Extension Checker* computes function $ext(\langle Arg, Def \rangle, sem, S)$, which checks if S is an extension according to sem (see Algorithm 1, Line 5). Finally, module *Tractable Probability Calculator* computes $Pr_{\mathcal{F}}^{\text{cf}}(S)$, $Pr_{\mathcal{F}}^{\text{ad}}(S)$, and $Pr_{\mathcal{F}}^{\text{st}}(S)$ using Facts 1, 2, and 3, respectively.

User Interface (UI). PARTY's UI looks intuitive and easy to use, even for inexperienced users. UI consists of four main tabs: *Overview*, *PrAF Builder/Loader*, *Set S/Semantics Selector*, and *Result Viewer*. Tab *Overview* displays some information about the system, while the other three tabs logically correspond to the three UI blocks introduced above (see Fig. 1, front-end level). The tab *PrAF Builder/Loader* (Fig. 2(a)) allows the user to visually build its own *PrAF*, by representing arguments as nodes and attacks (defeats) as directed edges. Nodes are created by clicking in the white area enclosed by the red border, while the edges are created by clicking on a proper connector and dragging the source node towards the destination node. The user can, also, specify the probability value of each node and edge. Figure 2(a) shows an example of a *PrAF* built by means of PARTY UI.

The user can also choose to load a *PrAF* stored in a *.txt* file. In this case, the *Load PrAF* button enables the user to navigate the file system and select the *.txt* file that encodes a *PrAF*. The *.txt* file contains a list of pairs of the form $\langle n, pr \rangle$, representing the node name (argument) and its probability, and a list of triples of the form $\langle n_1, n_2, pr \rangle$ representing a directed edge (attack) from node n_1 to node n_2 and the probability value of the edge. Moreover, it contains the list of nodes belonging to set S. The tab *Set S/Semantics Selector* (Fig. 2(b)) gives the possibility to choose the nodes of S (drop down menu) and, for each semantic, allows the choice of the appropriate method of evaluation (exact computation

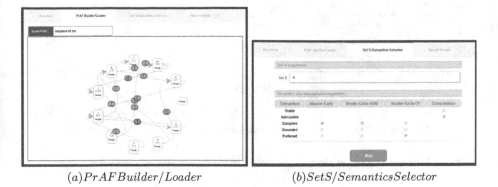

(a) $PrAFBuilder/Loader$ (b) $SetS/SemanticsSelector$

Fig. 2. PARTY -tab *PrAF Builder/Loader* and tab *Set S/Semantics Selector*

or estimate). To run a simulation, the user just clicks on the *Run* button. When the results are available, the system displays them in the *Result Viewer* tab.

5 Experimental Evaluation

The aim of this empirical study is evaluating the efficiency of PARTY in the case that it runs on devices with limited computational resources. We first evaluate the efficiency of the polynomial-time algorithms of Sect. 3.1, that compute the probability of extensions for the admissible and stable semantics, and then we assess the extent to which our Monte-Carlo simulation approaches are more efficient than the state-of-the-art technique. The choice to focus on devices that have reduced computational resources depends both on the fact they are an excellent testbed to test systems of high complexity and on the fact that such devices are nowadays widely diffused and suitable to be carried along in every situation. In fact, portable devices and efficient software applications may be very helpful in several argumentation scenarios, such as that discussed in the introduction: an attorney could assess everywhere and very easily the possible outcomes of a dispute and, based on these, finding out the most advantageous strategy for winning the dispute.

Experimental Setting. In our experimental evaluation, we run our proposed Monte-Carlo simulators (`MC-CF` and `MC-AD`), the `standard` Monte-Carlo approach (see Sect. 3.2) as well as the polynomial-time algorithms of Sect. 3.1 over two benckmarks that will referred to as eJRM and tw. The first one was made during the *eJRM* project (see http://www.ejrm.it/) and it consists of different *PrAFs*, whose sizes range from 10 and 13 defeats to 20 arguments and 28 defeats. We considered several sets S, whose number of arguments ranges from the 20 % to the 50 % of the number of arguments of the *PrAFs*. The second benchmark is available on the web page http://www.dbai.tuwien.ac.at/ proj/argumentation/dynpartix/ and it consists of different *AAFs*, whose number of arguments ranges from 25 to 100, from which we have built corresponding

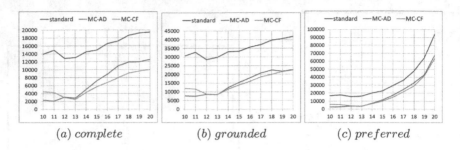

(a) complete   (b) grounded   (c) preferred

Fig. 3. Average execution time (ms) over eJRM, by varying the set of arguments.

PrAFs by randomly generating argument and defeat probabilities. We considered several sets S, whose number of arguments ranges from the 10 % to the 30 % of the number of arguments of the *PrAFs*. The experiments were performed on a device running Google's Android Operating System Ice Cream Sandwich 4.0.3, with CPU Nvidia Tegra2, 1 GHz dual core and 1 GB RAM.

Assessing PARTY Efficiency. Table 1 shows the average execution time (milliseconds) of PARTY, over both the benchmarks eJRM and tw, for the two tractable semantics stable (st) and admissible (ad), by varying the size of the set of arguments. Table 1 highlights that, on both benchmarks, the execution time is really negligible (few ms). In Fig. 3(a), (b), and (c), we report the average execution times of the three Monte-Carlo simulators embedded in PARTY for the complete, grounded, and preferred semantics, respectively, over the eJRM benchmark, by varying the size of the set of arguments. As shown, *Monte-Carlo ADM* (red line, MC-AD of Sect. 3.2) is slightly faster than *Monte-Carlo CF* (green line, MC-CF of Sect. 3.2), and both are faster than the standard approach (blue line, standard of Sect. 3.2).

In Fig. 4(a) and (b) we report the average execution times of the three Monte-Carlo simulators for the complete and grounded semantics, respectively, over the tw benchmark, by varying the size of the set of arguments. We do not report the execution times of three simulators for the preferred semantics over tw, since, for a number of arguments larger than 25, the time required by the three simulators is prohibitive. Indeed, both MC-AD and MC-CF required about 15 minutes to terminate over a PrAF of 50 arguments, and we halted the execution of standard after 25 min. These amounts of time required by the simulators in the case of the preferred semantics was rather expected, as, differently from the complete and grounded semantics, the problem of verifying if a set S is a preferred extension is coNP-complete [16]. Figure 4(a) and (b) show that both *Monte-Carlo CF*

Table 1. Tractable semantics - execution time (millis.)

		eJRM											tw			
$|A|$		10	11	12	13	14	15	16	17	18	19	20	25	50	75	100
sem	ad	0.27	0.63	0.60	0.34	0.32	0.36	0.41	0.67	0.28	0.54	0.34	1.80	3.16	6.21	5.41
	st	0.96	0.13	0.10	1.02	0.87	0.95	1.01	0.87	0.96	1.01	0.23	1.714	2.11	5.21	5.11

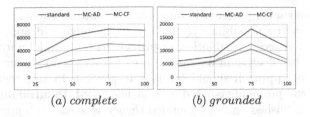

(a) *complete*   (b) *grounded*

Fig. 4. Average execution time (ms) over tw, by varying the set of arguments (Colur figure online).

(green line) and *Monte-Carlo ADM* (red line) are faster than the standard approach (blue line), and, in this case, *Monte-Carlo CF* runs slightly slower than *Monte-Carlo ADM*. Figure 4(b) shows a decrease of the required time over PrAFs consisting of more than 75 arguments. This is due to the fact that, with a large number of arguments and defeats, $Pr_{\mathcal{F}}^{gr}(S)$ is likely to be very close to 0, causing the Monte-Carlo simulators to converge very fast. Overall, the results show that the PARTY is particularly suitable to be executed on devices that have reduced computational resources. Indeed, the probabilities for the tractable semantics can be computed in a few milliseconds and the estimates of the probabilities for the intractable semantics can be obtained in a few seconds by running *Monte-Carlo ADM* or *Monte-Carlo CF*.

6   Related Work

Recently approaches for handling uncertainty in AAFs by relying on probability theory have been proposed in [14,23,27,30]. With the aim of modeling jury-based dispute resolutions, [14] proposed a PrAF where uncertainty is taken into account by specifying probability distribution functions (PDFs) over possible worlds and shown how an instance of the proposed PrAF can be obtained by specifying a probabilistic assumption-based argumentation framework (introduced by themselves). In the same spirit, [27] defined a PrAF as a PDF over the set of possible worlds, and introduced a probabilistic version of a fragment of ASPIC framework [25] that can be used to instantiate the proposed PrAF. Differently from the approaches in [14,23,27] proposed a PrAF where probabilities are directly associated with arguments and defeats, instead of being associated with possible worlds. [23] proposes a Monte-Carlo simulation approach to approximate $Pr(S)$, which has been later improved in [18] by exploiting the results of [19,20]. In [23], as well as in [14,18,19,27], $Pr(S)$ is defined as the sum of the probabilities of the possible worlds where S is an extension according to a given semantics. [30], instead, did not define a probabilist version of a classical semantics, but introduced a new probabilistic semantics, based on specifying a class of PDFs (*p-justifiable* PDFs) over sets of possible AAFs, and shown that this probabilistic semantics generalizes the complete semantics.

Besides the approaches that model uncertainty in AAFs by relying on probability theory, many proposals have been made where uncertainty is represented by exploiting weights or preferences on arguments and/or defeats [7,9,12,15,24], or by relying on the possibility theory [4–6]. Although the approaches based on weights, preferences, possibilities, or probabilities to model uncertainty have been proved to be effective in different contexts, there is no common agreement on what kind of approach should be used in general. In this regard, [21,22] observed that the probability-based approaches may take advantage from relying on a well-established and well-founded theory, whereas the approaches based on weights or preferences do not conform to well-established theories yet.

Although several systems, such as [3,17,28,29], are available for reasoning in non-probabilistic argumentation frameworks, to the best of our knowledge, PARTY is the first system running on mobile devices and efficiently addressing the computation of probability of extensions in argumentation frameworks.

7   Conclusions

In this paper we presented PARTY, which is the first system addressing the problem of efficiently computing the probability of extensions in argumentation frameworks. PARTY efficiency was experimentally validated over two abstract argumentation benchmarks, showing that PARTY outperforms the state-of-the-art approach.

References

1. http://www.marketsandmarkets.com/Market-Reports/enterprise-mobility-334.html
2. Agresti, A., Coull, B.A.: Approximate is better than "exact" for interval estimation of binomial proportions. Am. Stat. **52**(2), 119–126 (1998)
3. Alsinet, T., Béjar, R., Godo, L., Guitart, F.: Using answer set programming for an scalable implementation of defeasible argumentation. In: ICTAI, pp. 1016–1021 (2012)
4. Alsinet, T., Chesñevar, C., Godo, L., Sandri, S., Simari, G.R.: Formalizing argumentative reasoning in a possibilistic logic programming setting with fuzzy unification. Int. J. Approx. Reason. **48**(3), 711–729 (2008)
5. Alsinet, T., Chesñevar, C.I., Godo, L., Simari, G.R.: A logic programming framework for possibilistic argumentation: formalization and logical properties. Fuzzy Sets Syst. **159**(10), 1208–1228 (2008)
6. Amgoud, L., Prade, H.: Reaching agreement through argumentation: a possibilistic approach. In: KR, pp. 175–182 (2004)
7. Amgoud, L., Vesic, S.: A new approach for preference-based argumentation frameworks. Ann. Math. Artif. Intell. **63**(2), 149–183 (2011)
8. Baroni, P., Giacomin, M.: Semantics of abstract argument systems. In: Simari, G., Rahwan, I. (eds.) Argumentation inArtificial Intelligence, pp. 25–44. Springer, USA (2009)
9. Bench-Capon, T.J.M.: Persuasion in practical argument using value-based argumentation frameworks. J. Log. Comput. **13**(3), 429–448 (2003)
10. Bench-Capon, T.J.M., Dunne, P.E.: Argumentation in artificial intelligence. Artif. Intell. **171**(10–15), 619–641 (2007)

11. Besnard, P., Hunter, A. (eds.): Elements of Argumentation. The MIT Press, Cambridge (2008)
12. Coste-Marquis, S., Konieczny, S., Marquis, P., Ouali, M.A.: Weighted attacks in argumentation frameworks. In: KR (2012)
13. Dung, P.M.: On the acceptability of arguments and its fundamental role in nonmonotonic reasoning, logic programming and n-person games. Artif. Intell. **77**(2), 321–358 (1995)
14. Dung, P.M., Thang, P.M.: Towards (probabilistic) argumentation for jury-based dispute resolution. In: COMMA, pp. 171–182 (2010)
15. Dunne, P.E., Hunter, A., McBurney, P., Parsons, S., Wooldridge, M.: Weighted argument systems: basic definitions, algorithms, and complexity results. Artif. Intell. **175**(2), 457–486 (2011)
16. Dunne, P.E., Wooldridge, M.: Complexity of abstract argumentation. In: Simari, G., Rahwan, I. (eds.) Argumentation in Artificial Intelligence, pp. 85–104. Springer, USA (2009)
17. Egly, U., Gaggl, S.A., Woltran, S.: ASPARTIX: implementing argumentation frameworks using answer-set programming. In: Garcia de la Banda, M., Pontelli, E. (eds.) ICLP 2008. LNCS, vol. 5366, pp. 734–738. Springer, Heidelberg (2008)
18. Fazzinga, B., Flesca, S., Parisi, F.: Efficiently estimating the probability of extensions in abstract argumentation. In: Liu, W., Subrahmanian, V.S., Wijsen, J. (eds.) SUM 2013. LNCS, vol. 8078, pp. 106–119. Springer, Heidelberg (2013)
19. Fazzinga, B., Flesca, S., Parisi, F.: On the complexity of probabilistic abstract argumentation. In: IJCAI, pp. 898–904 (2013)
20. Fazzinga, B., Flesca, S., Parisi, F.: On the complexity of probabilistic abstract argumentation frameworks. ACM Trans. Comput. Logic **16**(3) (2015)
21. Hunter, A.: Some foundations for probabilistic abstract argumentation. In: COMMA, pp. 117–128 (2012)
22. Hunter, A.: A probabilistic approach to modelling uncertain logical arguments. Int. J. Approx. Reason. **54**(1), 47–81 (2013)
23. Li, H., Oren, N., Norman, T.J.: Probabilistic argumentation frameworks. In: Modgil, S., Oren, N., Toni, F. (eds.) TAFA 2011. LNCS, vol. 7132, pp. 1–16. Springer, Heidelberg (2012)
24. Modgil, S.: Reasoning about preferences in argumentation frameworks. Artif. Intell. **173**(9–10), 901–934 (2009)
25. Prakken, H.: An abstract framework for argumentation with structured arguments. Argument Comput. **1**(2), 93–124 (2010)
26. Rahwan, I., Simari, G.R. (eds.): Argumentation in Artificial Intelligence. Springer, USA (2009)
27. Rienstra, T.: Towards a probabilistic dung-style argumentation system. In: AT, pp. 138–152 (2012)
28. Snaith, M., Reed, C.: TOAST: online ASPIC$^+$ implementation. In: COMMA, pp. 509–510 (2012)
29. South, M., Vreeswijk, G., Fox, J.: Dungine: a java dung reasoner. In: COMMA, pp. 360–368 (2008)
30. Thimm, M.: A probabilistic semantics for abstract argumentation. In: ECAI, pp. 750–755 (2012)

Uncertain Groupings: Probabilistic Combination of Grouping Data

Brend Wanders[(✉)], Maurice van Keulen, and Paul van der Vet

Faculty EEMCS, University of Twente, Enschede, The Netherlands
{b.wanders,m.vankeulen}@utwente.nl, paul@vandervet-ca.nl

Abstract. Probabilistic approaches for data integration have much potential [7]. We view data integration as an iterative process where data understanding gradually increases as the data scientist continuously refines his view on how to deal with learned intricacies like data conflicts. This paper presents a probabilistic approach for integrating data on groupings. We focus on a bio-informatics use case concerning homology. A bio-informatician has a large number of homology data sources to choose from. To enable querying combined knowledge contained in these sources, they need to be integrated. We validate our approach by integrating three real-world biological databases on homology in three iterations.

1 Introduction

The field of bio-informatics is for an important part about combining available data sources in novel ways in a pursuit to answer new far-reaching research questions. A bio-informatician typically has a large number of data sources to choose from, created and cultivated by different research institutes. Some are curated or partially curated, while others are automatically generated based on certain biological methods.

Though bio-informaticians are knowledgeable in the field and aware of the different data sources at their disposal and methods used, they do not know the exact intricacies of each data source. Therefore, a bio-informatician typically obtains a desired integrated data set not in one attempt, but after several iterations of refinement.

Most data sources are created for a specific purpose. A bio-informatician's use typically goes beyond this foreseen use. The act of repurposing of the data, i.e., using the data for a purpose other than its intended purpose, is another source of integration complexity. For example, the quality of data in a certain attribute may be lower than required.

In short, data understanding is a continuous process, with the bio-informatician's understanding of the intricacies of data sources growing over time. It is therefore required that this evolving knowledge can be expressed and refined. We call this specification an *"integration view"*. Querying and analyzing the result of a refined integration view produces more understanding which is in turn used to further refine the integration view.

© Springer International Publishing Switzerland 2015
Q. Chen et al. (Eds.): DEXA 2015, Part I, LNCS 9261, pp. 236–250, 2015.
DOI: 10.1007/978-3-319-22849-5_17

In this paper, we focus on a particular bio-informatics scenario: homology. Several databases exist that contain homology data. In essence, homology data represents groups of proteins that are expected to have the same function in different species. Obtained by using different methods, the sources only partially agree on the homological relationships. Combining them allows for querying and analyzing the combined knowledge on homology.

Contributions. In this paper we present a technique for combining grouping data from multiple sources. The main contributions of this paper are:

- A generic probabilistic approach to combining grouping data in which an evolving view on integration can be iteratively refined.
- An experimental evaluation on a real-world bio-informatics use case.

The use case is further explained in Sect. 1.1. We then generalize the use case to the problem of integrating grouping data and elaborate on how our probabilistic integration approach addresses this problem in Sect. 1.2.

1.1 Use Case

Our real-world use case comes from bio-informatics and concerns groups of *orthologous* proteins. Proteins in the same group are expected to have the same function(s).

The main goal of orthology is to conjecture the function of a gene or protein. Suppose we have identified a protein in disease-causing bacteria that, if silenced by a medicine, will kill the bacteria. A bio-informatician will want to make sure that the medicine will not have serious side-effects in humans. A normal procedure is to try to find orthologous proteins. If such proteins exist, they may also be targeted by the medicine, thus potentially causing side-effects.

We explain orthology, and orthologous groups, with an example featuring a fictitious paperbird taxa (see Fig. 1). This example will be used throughout the paper.

The evolution of the paperbird taxa started with the Ancient Paperbird, the extinct ancestor species of the paperbird genus. Through evolution the Ancient Paperbird species split into multiple species, the three prominent ones being the Long-beaked Paperbird, the Hopping Paperbird and the Running Paperbird. The Ancient Paperbird is conjectured to have genes $K\ L\ M$. After

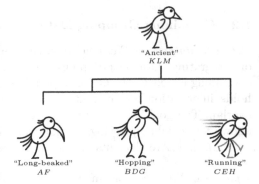

Fig. 1. Paperbirds, hypothetical phylogenetic tree annotated with species names and genes.

sequencing of their genetic code, it turns out that the Long-beaked Paperbird species has genes A F, the Hopping Paperbird species has genes B D G, and the Running Paperbird species has C E H. For the sake of the example, the functions of the different genes are known to the reader. With real taxa, the functions of genes can be ambiguous. For the paperbird species, genes A, B and C are known to influence the beak's curvature. D and E influencing the beak's length. Finally, genes F, G and H are known to influence the flexibility of the legs.

D and E are known to govern the length of the beak. Based on this, on the similarity between the two sequences, and on the conjectured function of the beak curvature function ancestor gene L, we call D and E orthologous, with L as common ancestor. Orthology relations are ternary relations between three genes: two genes in descendant species and the common ancestor gene from which they are evolved. The common ancestor is hypothetical. An orthologous group is defined as a group of genes with orthologous relations to every other member in the group. In this case, the group DE is an orthologous group. Proteins can by analogous arguments also be called orthologs. An extended review of orthology can be found in [3].

There are various computational methods for determining orthology between genes from different species [1,5]. These methods result in databases that contain groups of proteins or genes that are likely to be orthologous. Such databases are often made accessible to the scientific community. In our research, we aim to combine the insight into orthologous groupings contained in Homologene [8], PIRSF [13], and eggNOG [9]. An automatic combination of these sources may provide a continuously evolving representation of the current combined scientific insight into orthologous groupings of higher quality than any single heuristic could provide for other bio-informaticians to utilize.

One of the main problems in homology is to distinguish between orthologs and paralogs. The distinction is beyond the scope of this paper as it does not matter for our technique.

1.2   Combining Grouping Data

Problem Statement. We generalize the use case by viewing it as the problem of integrating data on groupings. We define a *data source* S_i as a database containing elements D_E^i and groups D_G^i where $\forall g \in D_G^i : g \subseteq D_E^i$. Each source holds information on different sets of proteins, i.e., the various D_E^i partially overlap. The goal is to construct a new data set with groups over $\bigcup_i D_E^i$ that allows for scalable querying for questions like 'Which elements are in a group with e?' and 'Are elements e_1 and e_2 in the same group?'.

Approach. We focus on an iterative probabilistic integration of the grouping data. It is based on the generic probabilistic data integration approach of [10] which constructs a *probabilistic database*. We call this representation an *uncertain grouping*. Being probabilistic, the above queries return possible answers with their likelihoods. Hence, an uncertain grouping is a grouping of elements for which the true grouping is unknown, but which faithfully represents the user's critical and fine-grained view on how much the data elements and query results can be trusted. Although probabilistic data integration is an active research problem [7], there is to our knowledge no work on probabilistic integration of data on groups.

Furthermore, we view integration as an iterative process. Starting from a simple integration view such as 'one-database-source-is-entirely-correct-but-it-is-unclear-which-one', one naturally discovers the limitations of this view while using the resulting data. Subsequently, more fine-grained integration rules are specified which combine the data in a better way, deals with conflicting data in a better way, and specifies better likelihoods for certain portions of the data to be correct (trust assignment). The integration view allows for an automatic re-construction of the integration result. As long as the integration result is not good enough, the process is repeated leading to handling inconsistencies and ambiguities at ever finer levels of granularity.

Outlook. The rest of this paper is laid out as follows: the next section discusses the real-world use case, followed by an overview of related work. Section 2 presents a formalization of our technique and on how an integration view evolves. Section 3 describes the experimental evaluation and discusses the results. Section 4 discusses, among other things, the complexity of the use case and the scalability of our technique. We conclude the paper with Sect. 5.

1.3   Related Work

Uncertainty forms an important aspect of data integration. Both the uncertainty created during the integration, as well as the integration of sources that contain uncertain data. [7] offers a comprehensive survey of the relevance of uncertainty management in data integration. Of special note is [6], which applies uncertain data integration in the context of biological databases by integrating heterogeneous data sources necessary for functional annotation of proteins.

Biological data sources are usually available in the form of a database. We want to have the product of the data combination available as a database as well. Probabilistic databases such as MayBMS [2] and Trio [12] allow the use of normal database techniques to apply to probabilistic data. As such, they provide a platform on which uncertain data integration can be implemented.

[4] Presents the tool ProGMAP for the comparison of orthologous protein groups from different databases. Instead of integrating protein groups, ProGMAP assists the user in comparing protein groups by providing statistical insight. Groups are compared pairwise and various visual display methods assist the user in assessing the strengths and weaknesses of each database. Our approach differs from ProGMAP in that we want to provide the user with a technique to query the combined data sources, instead of assisting the user in comparing them.

Current work in uncertain data integration is focused on entity resolution and schema integration. To the best of the authors' knowledge, no previous work using a uncertain data integration approach for the integration of classifications or groupings has been presented.

2   Probabilistic Integration of Grouping Data

In this section, we explain our iterative probabilistic integration approach in more detail starting with a running example.

2.1   Running Example

Figure 2 presents three exa-mple data sources, each containing two or three orthologous groups. We use the notation XYZ_i for a group of three elements, X, Y, and Z originating from source S_i. Observe that not every source is complete, for example, S_2 does not mention E. It depends on the source what this absence means:

S_1	ABC_1	DE_1	FG_1
S_2	AB_2	CD_2	FH_2
S_3		ABE_3	FGH_3

(a) Data sources

| S_i | Source i |
| XYZ_i | Group of 3 elements (from S_i) |

(b) Legend

Fig. 2. Running example.

- E is implicitly a group on its own,
- E is does not belong to any group, or
- it is unknown to which group E belongs.

2.2   Integration Views

From Sect. 1.1, we know that in our fictitious reality $A\,B\,C$, $D\,E$, and $F\,G\,H$ is the correct grouping. Observe that none of the sources in Fig. 2 is complete and fully correct. A bio-informatician integrating these sources, however, does not know what is the correct grouping, not even how well (s)he can trust the data. The goal is to determine based on current scientific knowledge contained in the sources, what the correct grouping is, or rather, the confidence in possible groupings.

We model an uncertain grouping as a probabilistic database adhering to the possible worlds model. In this model, an uncertain grouping is a compact repre-sentation of many possible groupings: the possible worlds. Probabilistic database technology is known to allow for scalable querying of an exponentially growing number of possible worlds [2]. Querying in a possible worlds model means that the query result is equivalent with evaluating the query on each possible world individually and combining those answers into one probabilistic answer.

Although we abstract from what an *integration view* exactly looks like, one can regard it as a set of data integration rules specifying not only how the raw data should be merged, but also which relevant alternatives exist in case of conflicts as well as what confidence to assign to certain portions of the data and such alternatives.

Our method of working with integration views is iterative, i.e., one starts with a simple view on how the data should be integrated and trusted based on

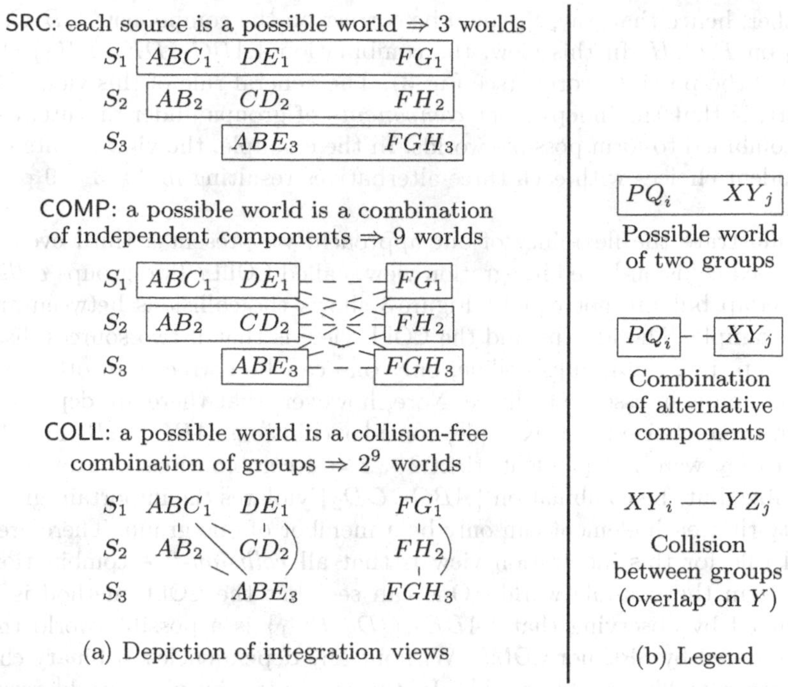

SRC: each source is a possible world ⇒ 3 worlds

S_1 | ABC_1 DE_1 | FG_1

S_2 | AB_2 CD_2 | FH_2

S_3 | ABE_3 FGH_3

COMP: a possible world is a combination
of independent components ⇒ 9 worlds

S_1 | ABC_1 DE_1 --- FG_1

S_2 | AB_2 CD_2 --- FH_2

S_3 | ABE_3 --- FGH_3

COLL: a possible world is a collision-free
combination of groups ⇒ 2^9 worlds

S_1 ABC_1 DE_1 FG_1

S_2 AB_2 CD_2 FH_2

S_3 ABE_3 FGH_3

(a) Depiction of integration views

PQ_i XY_j

Possible world
of two groups

PQ_i — XY_j

Combination
of alternative
components

XY_i —— YZ_j

Collision
between groups
(overlap on Y)

(b) Legend

Fig. 3. Example of uncertain grouping.

initial assumptions that may or may not be correct. By evaluating and using the integrated result, a bio-informatician gains more understanding in the data, which (s)he uses to adapt and refine the integration view. The reason behind this way of working is, that we believe, as we stated before, that data under-standing is a continuous process, with the bio-informaticians understanding of the intricacies of each data source growing over time. With the integration view method, the bio-informatician is able to express and refine his evolving opinion on the reliability of the data in the sources and how the data should be com-bined. He can then query and analyze the result of his actions to see how they reflect on the results. In the sequel, we illustrate the method by going through three iterations, each centered around a different integration view (SRC, COMP, and COLL, respectively) and evaluate the evolving integrated data.

Suppose we would start with taking the simplistic view of 'one-data-source-is-entirely-correct', SRC for short: the belief that one source is entirely correct, but it is unknown which one. In this view, each data source is a possible world (see Fig. 3). There is basically one *choice*: which alternative data source is the correct one: S_1, S_2, or S_3.

Other more fine-grained views on combining the data in the sources lead to more choices. For example, one could argue that the disputes among the sources around elements A, B, C, D, E and around F, G, H are independent of each other, hence that, say, S_1 could be correct on the component A, B, C, D, E and S_2 on F, G, H. In this view, the combination $\{ABC_1, DE_1, FH_2\}$ should be among the possible worlds (see Fig. 3). The general rule of this view, COMP for short, is that the independent *components* of groups under dispute, can be freely combined to form possible worlds. In the example, the view results in two independent choices with each three alternatives resulting in $3 \times 3 = 9$ possible worlds.

To illustrate the flexibility of our approach, we present a third even more fine-grained collision-based integration view, called COLL. Two groups *collide* iff they overlap but are not equal.[1] Figure 3 shows the collisions between groups in our example. The idea behind the COLL-view is that if two sources disagree on a group, i.e., the groups collide, only one can be correct.[2] In other words, each collision is in essence a choice. Note, however, that there are dependencies between these choices. For example, consider collisions ABC_1–AB_2 and DE_1–CD_2. If they were independent, then $2 \times 2 = 4$ combinations of groups would be possible, but the combination $\{ABC_1, CD_2\}$ violates the important grouping property that each element can only be a member of one group. Therefore, the general rule for this integration view is that all *collision-free* combinations of groups form the possible worlds. One can see that the COLL method is more fine-grained by observing that $\{ABE_3, CD_2, FG_1\}$ is a possible world that is not considered by SRC nor COMP. Without any dependencies, n binary choices would generate 2^n possible worlds. In the example, the view would result in $2^9 = 512$ worlds if there would be no dependencies. With dependencies, the number of possible worlds in the example is reduced to 40 (including the empty world).

Typically one would have many more considerations, sometimes rather fine-grained, that one would like to 'add' to one's integration view. For example, a bio-informatician may believe that groups CD_2 and FH_2 are extra untrustworthy, because he holds the opinion that the research group who determined those results is rather sloppy in the execution of their experiments. Or, he may have more trust in curated data, or even different levels of trust for data curated by different people or committees. Our approach can incorporate such considerations as well.

2.3   Formalization

In this section, we provide a formalization of a probabilistic database consisting of an uncertain grouping. The formalization is based on [10] which provides a generic formalization of a probabilistic database. We summarize the main concepts of [10] (DEFINITIONS) and show how it can be specialized to support uncertain groupings (SPECIALIZATIONS). In Sect. 2.4 we subsequently show how an uncertain grouping can be constructed for a certain integration view.

---

[1] This second condition 'not equal' is theoretically not necessary (See Sect. 2.4).

[2] Actually, this is a simplification as both can be incorrect (see Sect. 4).

Definition 1 (Database; Data Item). *We model a 'normal' database* $D \in \mathbb{P}\mathcal{D}$ *in an abstract way as a set of* data items.[3] *Typically, a data item* $d \in \mathcal{D}$ *would be a tuple for a relational database or a triple for an RDF store, but in essence it can be anything.*

Specialization 1 (Element; Group). We define two special kinds of data items as disjoint subsets of $\mathcal{D}$:

- *Elements* $e \in \mathcal{D}_E$, and
- *Groups* $g \in \mathcal{D}_G$, where $\mathcal{D}_G = \{g \mid g \subseteq \mathcal{D}_E\}$.

Specialization 2 (Data Source). Without loss of generality, we define a *data source* as a database D containing only elements and groups: $D = D_G \cup D_E$ with $D_G \subseteq \mathcal{D}_G$ and $D_E \subseteq \mathcal{D}_E$.

Definition 2 (Probabilistic Database). *A* probabilistic database $\bar{D}$ *is a database capable of handling huge volumes of data items and possible alternatives for these data items while still being able to efficiently query and update.* Possible world theory *views a probabilistic database as a set of possible databases* D_i, *also called* possible worlds, *each with a probability* $P(D_i)$.

Obviously, an implementation would not store the possible worlds individually, but as a compact representation capable of representing vast numbers of possible worlds in limited space. Possible world theory prescribes that a query Q on a compact representation should result in a compact answer representing all possible answers (equivalent with evaluating Q in each world individually).

Our compact representation is based on modeling uncertainty, the 'choices' of Sect. 2.2 in particular, with random events. Method SRC of the running example results in one choice: which of the three data sources is the correct one. We introduce a random variable $r \in \mathcal{R}$ with three possible assignments $(r \mapsto 1)$ representing S_1 is correct, $(r \mapsto 2)$ representing S_2 is correct, and $(r \mapsto 3)$ representing S_3 is correct.

Definition 3 (rv, rva, World Set). *We call the collection of all possible random variable assignments (rvas for short) with their probabilities a* world set $W \in \mathcal{R} \rightsquigarrow \mathcal{V} \rightsquigarrow [0..1]$. *We denote with* $P(r \mapsto v) = W(r)(v)$ *the probability of a rva; the probabilities of all alternatives for one random variable* $r \in \mathcal{R}$ *(rv for short) should add up to one.*

In the example, $W = \{r \mapsto \{1 \mapsto p_1, 2 \mapsto p_2, 3 \mapsto p_3\}\}$. Because all alternatives for one rv should add up to one, $p_1 + p_2 + p_3 = 1$.

Definition 4 (wsd). *Alternative data items are linked to the world set by means of world set descriptors (wsd)* φ. *A wsd is a conjunction of rvas* $(r_i \mapsto v_i)$. *The wsd determines for which rvas, hence for which possible worlds, the data item exists.*

---

[3] $\mathbb{P}$ denotes a power set.

Ḋ

group	φ
d_1 ABC_1	$(r_1 \mapsto 1)$
d_2 DE_1	$(r_1 \mapsto 1)$
d_3 FG_1	$(r_2 \mapsto 1)$
d_4 AB_2	$(r_1 \mapsto 2)$
d_5 CD_2	$(r_1 \mapsto 2)$
d_6 FH_2	$(r_2 \mapsto 2)$
d_7 ABE_3	$(r_1 \mapsto 3)$
d_8 FGH_3	$(r_2 \mapsto 3)$

W

rva	P	
$(r_1 \mapsto 1)$	p_1	'S_1 is correct' for component A, B, C, D, E
$(r_1 \mapsto 2)$	p_2	'S_2 is correct' for component A, B, C, D, E
$(r_1 \mapsto 3)$	p_3	'S_3 is correct' for component A, B, C, D, E
$(r_2 \mapsto 1)$	p_4	'S_1 is correct' for component F, G, H
$(r_2 \mapsto 2)$	p_5	'S_2 is correct' for component F, G, H
$(r_2 \mapsto 3)$	p_6	'S_3 is correct' for component F, G, H

Fig. 4. Probabilistic database representation $\bar{D} = (\dot{D}, W)$ for the uncertain grouping constructed under integration view COMP (see Fig. 3).

Definition 5 (Compact Representation). *The compact representation can now be defined as* $\bar{D} = (\dot{D}, W)$, *i.e., a set of data items each with a wsd* $\dot{D}$ *and a world set* W.

In the example, there are eight groups which can be linked to the appropriate rva. See Fig. 4 for an illustration. Note that in a concrete database, the data is normalized into three tables: **group** containing at least an identifier for each group, **element** containing all elements, and **group_element** describing which element belongs to which group. Only **group** is uncertain in this case, i.e., its tuples need to have the shown wsds φ.

Definition 6 (Valuation). *'Considering a case' means that we choose a value for one or more random variables and reason about the consequences of this choice. We call such a choice a valuation* θ. *If the choice involves all the variables of the world set, the valuation is total.*

Definition 7 (Possible World). *A total valuation induces a single possible world:* $\theta(\bar{D}) = \{d \mid (d, \varphi) \in \dot{D} \wedge \varphi(\theta)\}$, *where* $\varphi(\theta) = true$ *iff forall* $(r_i \mapsto v) \in \theta$, *there is no* $(r_i \mapsto v')$ *in* φ *such that* $v \neq v'$. *We denote with* $\mathrm{PWS}(\bar{D})$ *the set of all possible worlds, and with* $\mathrm{P}(D)$ *the probability of a world* D.

For example, the valuation $\theta = \{r_1 \mapsto 1, r_2 \mapsto 2\}$ induces the combination $\{ABC_1, DE_1, FH_2\}$. In this way, the concept of valuation bridges the gap between the compact representation and possible world theory.

Queries can be evaluated directly on the compact representation to obtain a compact representation of all possible answers. For example, the query "which elements are in the same group as A?" can be evaluated by selecting groups containing A, which results in 3 tuples d_1, d_4, and d_7. Observe that these tuples are mutually exclusive, because their wsds contain different values for r_1.

From the compact representation, one can derive different kinds of answers to the query, such as, the most likely answer, or the second most likely answer. For numerical queries, one can derive the minimum, maximum, expected value, standard deviation, etc. In this example, we may derive that C and E are only

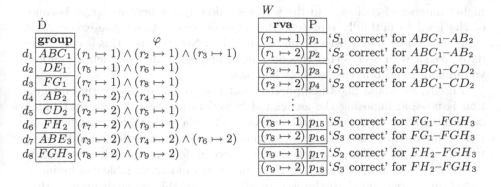

Fig. 5. Probabilistic database representation $\bar{D} = (\dot{D}, W)$ for the uncertain grouping constructed under integration view COLL (see Fig. 3).

in the same group as A if the respective group exists, i.e., under valuations $\{(r_1 \mapsto 1)\}$ and $\{(r_1 \mapsto 3)\}$, respectively. Therefore, C is homologous with A with a probability of p_1 and E is homologous with A with a probability of p_3. Observe that B is in the same group as A in all three tuples, hence it is homologous with A with a probability of $p_1 + p_2 + p_3 = 1$.

We like to emphasize that the above is a summary of the main concepts of [10] which provides a generic formalization of a probabilistic database. In addition, we have also shown how the formalization can be specialized to support uncertain groupings. For a more detailed presentation of the generic formalization, we refer to [10].

2.4   Integration Views Revisited

We argue that integration problems such as conflicts, ambiguity, trust, etc. can all be modelled in terms of choices that can be formalized with random events, which in turn can be represented in a probabilistic database with random variables and annotating tuples with world set descriptors composed of random variable assignments. In this section, we like to emphasize the flexibility of the approach.

Consider for example the probabilistic database constructed according to integration view COLL (see Fig. 5). Observe how the 9 collisions result in 9 random variables in a straightforward way. Furthermore, the concept of collision-freeness is represented in the world set descriptors. For example, tuple ABC_1 can only exist if all collisions in which it is involved fall in its favour. The possible answers to a query come with a probability for the trustworthiness of the answer, essentially the combined probability of all worlds that agree on that answer. Note that our modelling of COLL induces empty databases for valuations that would lead to one or more collisions. One could normalize the probabilities of query answers with $1 - P(\emptyset)$, the combined probability of all collision-free combinations.

Observe also how such an intricate integration view as COLL, does not produce more tuples in the **group** table, only the world set grows because of the higher number of choices, and the world set descriptors become larger because of the need to faithfully represent the dependencies between the existence of tuples caused by the collision-freeness condition. Nevertheless, this is only more data. We show in Sect. 3 that this does not cause scalability problems even in a voluminous real-world case such as homology.

Finally, we would like to emphasize that the process of discovering integration issues and imposing the associated consideration on the data by refining one's integration view, is an iterative process. We claim that such considerations can be imposed on the data by introducing more random variables and adding rvas to the wsds of the appropriate tuples. Recall, for example, the issue of the sloppy research group of Sect. 2.2. Here, one new random variable can be introduced and a rva added to the wsd of all tuples of this research group. After such a refinement, the bio-informatician obtains a database that can be directly queried so that he can examine its consequences. He thus iteratively refines his integration view until the data faithfully expresses his opinions as well as the result of any query or analysis run on this data.

3   Evaluation

Two main questions guide the evaluation: can our framework be applied in an existing probabilistic database, and if so, how well does it scale to realistic amounts of data, in particular to determine if current probabilistic database technology can cope with the amounts of uncertainty introduced by our framework. We use the probabilistic database MayBMS [2].

3.1   Experimental Setup

For the evaluation, we constructed a test set of homology data from the biological databases Homologene (release 67, [8]), PIRSF (release 2012_03, [13]), and eggNOG (release 3.0, [9]). The groupings from each were loaded into a single database for the construction of the integration views and querying. Where necessary database-specific accession numbers were converted to UniProt accession numbers. This ensures that identical proteins in different groups are correctly referenced.

Two query classes can be distinguished among commonly executed queries:

1. **Single**: "Which proteins are homologous with X?" with X a known protein.
2. **Pair**: "Are X and Y homologues?" with X and Y known proteins.

Based on these two classes we generate query suites based on sampling proteins from the combined database:

1. 1000 single and 1000 pair queries. All pairs are guaranteed to have a homologous relation. This suite is used to determine average query execution times for all integration views.
2. 100 single queries and 200 pair queries. For the latter, 100 queries have a homologous relation and the other 100 do not.

Mean query times per 'single' query   Mean query times per 'pair' query

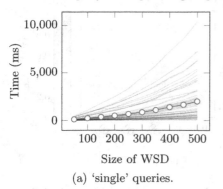
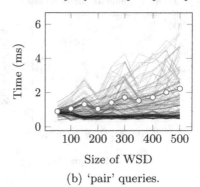

(a) 'single' queries.   (b) 'pair' queries.

Fig. 6. Mean query time (in white-red) and distinct query times (in gray) (Color figure online)

Random variable assignments for the integration views SRC, COMP and COLL were generated according to our integration approach. Probabilities were assigned uniformly over the rvas.

Because of experimenting with an existing system (MayBMS), we accept some technical limitations inherent in these systems. Overcoming these limitations is not the focus of our work and a note on them can be found in [11]. One of the limitations is that the wsd of a tuple can at most contain 500 rvas. Larger wsd were truncated to 500 rvas. Additional integration views based on COLL were generated with wsds of sizes 450, 400, ..., 100, 50. These integration views are referred to as COLLN, with N being the size of the wsd. No size indication means COLL500.

The experiments were conducted on an Intel i7 x86-64bit with 7.7 GB ram running Linux 3.2.0. Compilation was done with gcc 4.6.3.

3.2 Experiments

Experiment 1: Mean Query Times. Based on query suite 1, each query is executed 10 times. Mean query time per integration view is calculated form the latter 9 measurements; the first is discarded to prevent adverse effects of caching.

SRC mean: 18.627 ms, std.dev.: 26.864
COMP mean: 19.061 ms, std.dev.: 27.569
COLL mean: 23488.197 ms, std.dev.: 93184.375.

Preliminary results show that the amount of uncertainty of each integration view has a large impact on the mean execution time. Large standard deviations indicate large variations of query times within each integration view. The following experiments investigate the cause of this variation.

Experiment 2: World Set Descriptor Size. The goal here is to determine the impact of wsd size on query execution time. Query suite 2 is used on integration views COLL50, COLL100, ..., COLL500.

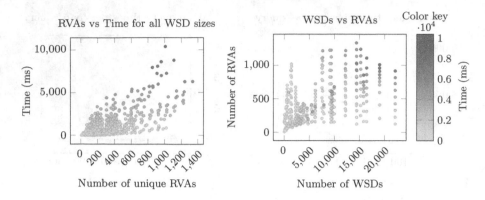

Fig. 7. Impact of number of rvas of wsds involved in query answering ('single').

Figure 6 presents the trend in mean query time with growing wsd for both query classes separately. The 'pair' queries are orders of magnitude faster than the 'single' queries due to smaller amounts of uncertainty per query result. The two drops in Fig. 6(b) at COLL200 and COLL350 are most likely due to favourable alignment of data in memory.

Experiment 3: Numbers of wsds and rvas. The goal here is to investigate the impact of the number of wsds and rvas involved in answering a query on the query time. Query suite 2 is used on integration views COLL50, ..., COLL500.

As can be seen in Fig. 7, the framework and MayBMS handle the real-world uncertainty well. For a large part, queries are executed within 2 s. The slower queries are slow due to a combination of a large number of unique rvas and wsds. Based on further analysis of what execution time is spent on for the integration view with a large amount of uncertainty (COLL), we conclude that most time is consumed by confidence computation.

Further Remarks. The wsd size is used as an artificial bound on the amount of uncertainty. Both SRC and COMP feature only a single rva hence are effectively equivalent wrt execution time. Due to technical limitations, COLL has a maximum of 500 rvas per wsd. This did not hinder the experiments, because we simulated data sets with less uncertainty by truncating the wsds to ever smaller sizes anyway.

We changed the representation of wsds in MayBMS to allow for 500 rvas per wsd, instead of being limited to normal 30. (see Appendix A of [11] for more details). Conversion of representation can be done during integration view construction or querying. Overhead of conversion during querying was shown to impact queries involving large wsds the most but still negligible.

We encountered three measurements that qualify as outliers. Two occurred for 'pair' queries with small execution times. As the experiments were conducted on a normal workstation, we strongly suspect that another program interfered with query execution. One outlier occurred during the measurements of 'single' queries, specifically for protein F6ZHU6 (a UniProt identifier). This protein is related to muscle activity and is a member of an abnormally large number of orthologous groups, the cause of which is further discussed in [11].

While conducting the experiments, a small number of queries did not finish. We suspect the method we use to interface with MayBMS to be the cause. Because our implementation is intended as a research prototype we have not spent significant effort on finding the cause, as it is not scientifically relevant.

4   Discussion

Complexity From Practice. An unsuspecting bioinformatician him/herself would perhaps, just like us, initially also assume that groups within one source are non-overlapping. For homology databases, one discovers that this is not true. According to bioinformatician A. Kuzniar this overlap is due to a subset-superset relation between the two groups.

Open World versus Closed World. Consider, for example, source S_1 and the fact that it doesn't mention H. Should this be interpreted (closed world assumption) as a statement that H is not orthologous to any protein, in particular, F and G? Or (open world assumption) that S_1 doesn't make a statement at all about H, i.e., it might be orthologous to any protein?

Considering only sources S_1 and S_2 — note that S_2 doesn't mention G — one could hold the view that it is possible for G and H to be orthologous as both are possibly orthologous to F according to the respective sources. There is, however, no possible world in the uncertain grouping of S_1 and S_2 where G and H are in the same group using any of the integration view methods presented. Hence, the integration views of Sect. 2.2 all follow a closed world assumption.

The technical report [11] contains a detailed discussion on both these topics, and continues with the topic of confidence precision and alternative representations of the group data.

5   Conclusions

Motivated by the real-world use case of homology, we propose a generic technique for combining groupings. Proteins in a homologous group are expected to have the same function in different species. Homology data is relevant when, e.g., a medicine is being developed and the potential for side-effects has to be determined. We combine 3 different biological databases containing homology data.

In e-science as well as business analytics, data understanding is a continuous process with the analist's understanding of the intricacies and quality of data sources growing over time. We propose a generic probabilistic approach to combining grouping data in which an evolving view on integration can be iteratively queried and refined. Such an 'integration view' models complications such as conflicts, ambiguity, and trust as probabilistic data.

Experiments show that our approach scales with existing probabilistic database technology. The evaluation is based on realistic amounts of data obtained from the combination of 3 biological databases, yielding 776 thousand groups with a total of 14 million members and 2.8 million random variables.

Acknowledgements. We would like to thank the late Tjeerd Boerman for his work on the use case and his initial concept of groupings. We would also like to thank Arnold Kuzniar for his insights and feedback on our use of biological databases and Ivor Wanders for his reviewing and editing assistance.

References

1. Altenhoff, A., Dessimoz, C.: Phylogenetic and functional assessment of orthologs inference projects and methods. PLoS Comput. Biol. **5**, e1000262 (2009)
2. Antova, L., Koch, C., Olteanu, D.: $10^{(10^6)}$ worlds and beyond: efficient representation and processing of incomplete information. VLDB J. **18**(5), 1021–1040 (2009)
3. Koonin, E.: Orthologs, paralogs, and evolutionary genomics. Annu. Rev. Genet. **39**, 309–338 (2005)
4. Kuzniar, A., Lin, K., He, Y., Nijveen, H., Pongor, S., Leunissen, J.A.M.: Progmap: an integrated annotation resource for protein orthology. Nucleic Acids Res. **37**(suppl. 2), W428–W434 (2009)
5. Kuzniar, A., van Ham, R., Pongor, S., Leunissen, J.: The quest for orthologs: finding the corresponding gene across genomes. Trends Genet. **24**, 539–551 (2008)
6. Louie, B., Detwiler, L., Dalvi, N., Shaker, R., Tarczy-Hornoch, P., Suciu, D.: Incorporating uncertainty metrics into a general-purpose data integration system. In: 19th International Conference on Scientific and Statistical Database Management. SSBDM 2007, p. 19 (2007)
7. Magnani, M., Montesi, D.: A survey on uncertainty management in data integration. J. Data Inf. Qual. **2**(1), 5:1–5:33 (2010)
8. NCBI Resource Coordinators. Database resources of the national center for biotechnology information. **41**(D1), D8–D20 (2013)
9. Powell, S., Szklarczyk, D., Trachana, K., Roth, A., Kuhn, M., Muller, J., Arnold, R., Rattei, T., Letunic, I., Doerks, T., et al.: eggNOG v3. 0: orthologous groups covering 1133 organisms at 41 different taxonomic ranges. Nucleic Acids Res. **40**, D284–D289 (2011)
10. van Keulen, M.: Managing uncertainty: the road towards better data interoperability. IT - Inf. Technol. **54**(3), 138–146 (2012)
11. Wanders, B., van Keulen, M., van der Vet, P.E.: Uncertain groupings: probabilistic combination of grouping data. Technical report TR-CTIT-14-12, Centre for Telematics and Information Technology, University of Twente, Enschede (2014)
12. Widom, J.: Trio: A system for integrated management of data, accuracy, and lineage. Technical report 2004–40, Stanford InfoLab (2004)
13. Wu, C.H., Nikolskaya, A., Huang, H., Yeh, L.-S.L., Natale, D.A., Vinayaka, C.R., Hu, Z.-Z., Mazumder, R., Kumar, S., Kourtesis, P., Ledley, R.S., Suzek, B.E., Arminski, L., Chen, Y., Zhang, J., Cardenas, J.L., Chung, S., Castro-Alvear, J., Dinkov, G., Barker, W.C.: Pirsf: family classification system at the protein information resource. Nucleic Acids Res. **32**(suppl. 1), D112–D114 (2004)

Database System Architecture

Cost-Model Oblivious Database Tuning with Reinforcement Learning

Debabrota Basu[1]([✉]), Qian Lin[1], Weidong Chen[1], Hoang Tam Vo[3],
Zihong Yuan[1], Pierre Senellart[1,2], and Stéphane Bressan[1]

[1] School of Computing, National University of Singapore, Singapore, Singapore
debabrota.basu@u.nus.edu
[2] Institut Mines–Télécom, Télécom ParisTech, CNRS LTCI, Paris, France
[3] SAP Research and Innovation, Singapore, Singapore

Abstract. In this paper, we propose a learning approach to adaptive performance tuning of database applications. The objective is to validate the opportunity to devise a tuning strategy that does not need prior knowledge of a cost model. Instead, the cost model is learned through reinforcement learning. We instantiate our approach to the use case of index tuning. We model the execution of queries and updates as a Markov decision process whose states are database configurations, actions are configuration changes, and rewards are functions of the cost of configuration change and query and update evaluation. During the reinforcement learning process, we face two important challenges: not only the unavailability of a cost model, but also the size of the state space. To address the latter, we devise strategies to prune the state space, both in the general case and for the use case of index tuning. We empirically and comparatively evaluate our approach on a standard OLTP dataset. We show that our approach is competitive with state-of-the-art adaptive index tuning, which is dependent on a cost model.

1 Introduction

In a recent SIGMOD blog entry [10], Guy Lohman asked "Is query optimization a 'solved' problem?". He argued that current query optimizers and their cost models can be critically wrong. Instead of relying on wrong cost models, the author and his colleagues have proposed in [19] a *learning* optimizer.

In this paper, we propose a learning approach to performance tuning of database applications. By performance tuning, we mean selection of an optimal physical database configuration in view of the workload. In general, configurations differ in the indexes, materialized views, partitions, replicas, and other parameters. While most existing tuning systems and literature [6,17,18] rely on a predefined cost model, the objective of this work is to validate the opportunity for a tuning strategy to do without.

To achieve this, we propose a formulation of database tuning as a *reinforcement learning* problem (see Sect. 3). The execution of queries and updates is modeled as a Markov decision process whose states are database configurations,

Q. Chen et al. (Eds.): DEXA 2015, Part I, LNCS 9261, pp. 253–268, 2015.
DOI: 10.1007/978-3-319-22849-5_18

actions are configuration changes and rewards are functions of the cost of configuration change and query/update evaluation. This formulation does not rely on a pre-existing cost model, rather it learns it.

We present a solution to the reinforcement learning formulation that tackles the curse of dimensionality (Sect. 4). To do this, we reduce the search space by exploiting the quasi-metric properties of the configuration change cost, and we approximate the cumulative cost with a linear model.

We instantiate our approach to the use case of index tuning (Sect. 5). We use this case to demonstrate the validity of a cost-model oblivious database tuning with reinforcement learning, through experimental evaluation on a TPC-C workload [14] (see Sect. 6). We compare the performance with the *Work Function Index Tuning* (WFIT) Algorithm [18]. Results show that our approach is competitive yet does not need to know a cost model.

Related work is discussed in Sect. 2.

2   Related Work

Our work is intertwined with mainly two lines of research. Our methodology is designed to deal with the problem of automated database configuration. Using our approach described in Sect. 4, we have proposed COREIL, an algorithm to solve the problem of index tuning. Traditionally, most of the works proposed in the field of automated database configuration are conducted in an offline manner. In offline methodologies, database administrators identify and update representative workloads from the database queries based on these representative workloads, new database configurations are realized to create new beneficial indexes [1], smart vertical partition for reducing I/O costs [16], or possibly for engendering a combination of index selection, partitioning and replication for both stand-alone databases [11] and parallel databases [2].

But with increasing complexity and agility of database applications and the introduction of modern database environments such as database as a service, the aforementioned tasks of database administrators are becoming more tedious and problematic. Therefore it is desirable to design automated solutions of the database design problem that are able to continuously monitor the incoming queries or the changes in workload and can react readily by adapting the database configuration. An online approach for physical design tuning is proposed in [6], that progressively chooses an optimal solution at each step through case-by-case analysis of potential benefits. Similarly, [17] proposes a self-regulating framework for continuous online physical tuning where effective indexes are created and deleted in response to the shifting workload. In one of the most recent proposals for semi-automated index tuning, WFIT [18], the authors have proposed a method based on the Work function algorithm and the feedbacks from manual changes of configurations. To evaluate the cost of executing a query workload with the new indexes as well as the cost of configuration transition, i.e. for profiling indexes' benefit, most of the aforementioned online algorithms like WFIT exploit the what-if optimizer [7] which returns such estimated costs.

As COREIL is able to learn the estimated cost of queries gradually through subsequent iterations, it is applicable to a wider range of database management systems which may not implement what-if-optimizer or expose its interface to the users.

For designing and tuning online automated databases, our proposed approach uses the more general structure of *reinforcement learning* [20] that offers a rich pool of techniques available in literature. Markov decision processes (MDPs) [13] are one such model where each action leads to a new state and a given reward according to a probability distribution that must be learned. On the basis of such a cumulative reward, these processes decide the next action to perform for the optimal performance. Though use of Markov decision process for modelling data cleaning tasks has been proposed in [4], its application in data management is limited because of typically huge state space and complex structures of data in each state (in our case, indexes). Some recent research works like [3] have also approached dynamic index selection based on data mining and optimization algorithms. But in our proposed method, COREIL, we tackle the issues of using reinforcement learning in database applications. Other complications like delayed rewards obtained after a long sequence of state transitions and partial observability [21] of current state due to uncertainty, are also less prevalent in the proposed structure of COREIL.

3 Problem Definition

Let R be a logical database schema. We can consider R to be the set of its possible database instances. Let S be the set of physical database configurations of instances of R. For a given database instance, two configurations s and s' may differ in the indexes, materialized views, partitions, replicas, and other parameters. For a given instance, two configurations will be logically equivalent if they yield the same results for all queries and updates.

The cost of changing configuration from $s \in S$ to $s' \in S$ is denoted by the function $\delta(s, s')$. The function $\delta(s, s')$ is not necessarily symmetric as the cost of changing configuration from s to s' and the reverse may not be the same. On the other hand, it is a non-negative function and also verifies the identity of indiscernibles (there is no free configuration change) and the triangle inequality (it is always cheaper to do a direct configuration change). Therefore, it is a quasi-metric on S.

Let Q be a workload set, defined as a schedule of queries and updates (for brevity, we refer to both as queries). Without significant loss of generality, we consider the schedule to be sequential and the issue of concurrency control orthogonal to the current presentation. Query q_t is the tth query in the schedule, which is executed at time t.

The cost of executing query $q \in Q$ on configuration $s \in S$ is denoted by the function $cost(s, q)$. We model a query q_t as a random variable, whose generating distribution may not be known *a priori*: q_t is only observable at time t.

Let s_0 be the initial configuration of the database. At any time t the configuration is changed from s_{t-1} to s_t with the following events in order:

1. Arrival of query q_t. We call $\hat{q}_t$ the observation of q_t at time t.
2. Choice of the configuration $s_t \in S$ based on $\hat{q}_1, \hat{q}_2, \ldots, \hat{q}_t$ and s_{t-1}.
3. Change of configuration from s_{t-1} to s_t. If no configuration change occurs at time t, then $s_t = s_{t-1}$.
4. Execution of query $\hat{q}_t$ under the configuration s_t.

The cost of configuration change and query execution at time t, referred as the *per-stage cost*, is

$$C(s_{t-1}, s_t, \hat{q}_t) := \delta(s_{t-1}, s_t) + cost(s_t, \hat{q}_t)$$

We can phrase in other words the stochastic decision process of choosing the configuration changes as a *Markov decision process* (MDP) [13] where states are database configurations, actions are configuration changes, and penalties (negative rewards) are the per-stage cost of the action. Note that transitions from one state to another on an action are deterministic (in contrast to the general framework of MDPs, there is no uncertainty associated with the new configuration when a configuration change is decided). On the other hand, penalties are both *stochastic* (they depend on the query, a random variable) and *uncertain* (the cost of a query in a configuration is not known in advance, in the absence of a reliable cost model).

Ideally, the problem would be to find the sequence of configurations that minimizes the sum of future per-stage costs; of course, assuming an infinite horizon [20], this sum is infinite. One practical way to circumvent this problem is to introduce a *discount factor* γ that gives more importance to immediate costs than to costs distant in the future, and to try and minimize a *cumulative cost* defined with γ. Under Markov assumption, a sequence of configuration changes is determined by a *policy* $\pi \colon S \times Q \to S$, which, given the current configuration s_{t-1} and a query $\hat{q}_t$, returns a configuration $s_t := \pi(s_{t-1}, \hat{q}_t)$.

We define the *cost-to-go* function V^π for a policy π as:

$$V^\pi(s) := \mathbb{E}\left[\sum_{t=1}^{\infty} \gamma^{t-1} C(s_{t-1}, s_t, \hat{q}_t)\right] \quad \text{satisfying} \quad \begin{cases} s_0 = s \\ s_t = \pi(s_{t-1}, \hat{q}_t), t \geq 1 \end{cases} \quad (1)$$

where $0 < \gamma < 1$ is the discount factor. The value of $V^\pi(s)$ represents the expected cumulative cost for the following policy π from the current configuration s.

Let $\mathcal{U}$ be the set of all policies for a given database schema. Our problem can now be formally phrased as to minimize the expected cumulative cost, i.e., to find an optimal policy π^* such that $\pi^* := \arg\min_{\pi \in \mathcal{U}} V^\pi(s_0)$.

4   Adaptive Performance Tuning

4.1   Algorithm Framework

In order to find the optimal policy π^*, we start from an arbitrary policy π, compute an estimation of its cost-to-go function, and incrementally attempt to improve it using the current estimate of the cost-to-go function $\overline{V}$ for each $s \in S$. This strategy is known as *policy iteration* [20] in reinforcement learning literature.

Assuming the probability distribution of q_t is known in advance, we improve the cost-to-go function $\overline{V}^{\pi_t}$ of the policy π_t at iteration t using

$$\overline{V}^{\pi_t}(s) = \min_{s' \in S} \left(\delta(s, s') + \mathbb{E}\left[cost(s', q)\right] + \gamma \overline{V}^{\pi_{t-1}}(s') \right) \tag{2}$$

We obtain the updated policy as $\arg\min_{\pi_t \in \mathcal{U}} \overline{V}^{\pi_t}(s)$. The algorithm terminates when there is no change in the policy. The proof of optimality and convergence of policy iteration can be found in [12].

Unfortunately, policy iteration suffers from several problems. First, there may not be any proper model available beforehand for the cost function $cost(s, q)$. Second, the curse of dimensionality [12] makes the direct computation of $\overline{V}$ hard. Third, the probability distribution of queries is not assumed to be known *a priori*, making it impossible to compute the expected cost of query execution $\mathbb{E}\left[cost(s', q)\right]$.

Algorithm 1. Algorithm Framework

1: Initialization: an arbitrary policy π_0 and a cost model C_0
2: Repeat till convergence
3:     $\overline{V}^{\pi_{t-1}} \leftarrow$ approximate using a linear projection over $\phi(s)$
4:     $C^{t-1} \leftarrow$ approximate using a linear projection over $\eta(s, \hat{q}_t)$
5:     $\pi_t \leftarrow \arg\min_{s \in S'} \left(C^{t-1} + \gamma \overline{V}^{\pi_{t-1}}(s) \right)$
6: End

The basic framework of our algorithm is shown in Algorithm 1. Initial policy π_0 and cost model C_0 can be intialized arbitrarily or using some intelligent heuristics. In line 5 of Algorithm 1, we have tried to overcome the issues at the root of the curse of dimensionality by juxtaposing the original problem with approximated per-stage cost and cost-to-go function. Firstly, we map a configuration to a vector of associated feature $\phi(s)$. Then, we approximate the cost-to-go function by a linear model $\theta^T \phi(s)$ with parameter θ. It is extracted from a reduced subspace S' of configuration space S that makes the search for optimal policy computationally cheaper. Finally, we learn the per-stage cost $C(s, s', \hat{q})$ by a linear model $\zeta^T \eta(s, \hat{q})$ with parameter ζ. This method does not need any prior knowledge of the cost model, rather it learns the model iteratively. Thus, we have resolved shortcomings of policy iteration and the need of predefined cost model for the performance tuning problem in our algorithm. These methods are depicted and analyzed in the following sections.

4.2 Reducing the Search Space

To reduce the size of search space in line 5 of c 1, we filter the configurations that satisfy certain necessary conditions deduced from an optimal policy.

Proposition 1. *Let s be any configuration and $\hat{q}$ be any observed query. Let π^* be an optimal policy. If $\pi^*(s, \hat{q}) = s'$, then $cost(s, \hat{q}) - cost(s', \hat{q}) \geq 0$. Furthermore, if $\delta(s, s') > 0$, i.e., if the configurations certainly change after query, then $cost(s, \hat{q}) - cost(s', \hat{q}) > 0$.*

Proof. Since $\pi^*(s, \hat{q}) = s'$, we have

$$\delta(s, s') + cost(s', \hat{q}) + \gamma V(s')$$
$$\leq cost(s, \hat{q}) + \gamma V(s)$$
$$= cost(s, \hat{q}) + \gamma \mathbb{E}\left[\min_{s''} (\delta(s, s'') + cost(s'', \hat{q}) + \gamma V(s''))\right]$$
$$\leq cost(s, \hat{q}) + \gamma \delta(s, s') + \gamma V(s'),$$

where the second inequality is obtained by exploiting triangle inequality $\delta(s, s'') \leq \delta(s, s') + \delta(s', s'')$, as δ is a quasi-metric on S.

This infers that

$$cost(s, \hat{q}) - cost(s', \hat{q}) \geq (1 - \gamma)\delta(s, s') \geq 0.$$

The assertion follows.   □

By Proposition 1, if π^* is an optimal policy and $s' = \pi^*(s, \hat{q}) \neq s$, then $cost(s, \hat{q}) > cost(s', \hat{q})$. Thus, we can define a reduced subspace as

$$S_{s, \hat{q}} = \{s' \in S \mid cost(s, \hat{q}) > cost(s', \hat{q})\}.$$

Hence, at each time t, we can solve

$$\pi_t = \underset{s \in S_{s_{t-1}, \hat{q}_t}}{\arg\min} \left(\delta(s_{t-1}, s) + cost(s, \hat{q}_t) + \gamma \overline{V}^{\pi_{t-1}}(s)\right). \tag{3}$$

Next, we design an algorithm that converges to an optimal policy through searching in the reduced set $S_{s, \hat{q}}$.

4.3   Modified Policy Iteration with Cost Model Learning

We calculate the optimal policy using the *least square policy iteration* (LSPI) [9]. If for any policy π, there exists a vector $\boldsymbol{\theta}$ such that we can approximate $V^\pi(s) = \boldsymbol{\theta}^T \phi(s)$ for any configuration s, then LSPI converges to the optimal policy. This mathematical guarantee makes LSPI an useful tool to solve the MDP as defined in Sect. 3. But the LSPI algorithm needs a predefined cost model to update the policy and evaluate the cost-to-go function. It is always not obvious that any form of cost model would be available and as mentioned in Sect. 1, pre-defined cost models may be critically wrong. This motivates us to develop another form of the algorithm, where the cost model can be equivalently obtained through learning.

Algorithm 2. Recursive least squares estimation.

1: **procedure** RLSE($\hat{\epsilon}^t, \overline{B}^{t-1}, \zeta^{t-1}, \eta^t$)
2: $\quad \gamma^t \leftarrow 1 + (\eta^t)^T \overline{B}^{t-1} \eta^t$
3: $\quad \overline{B}^t \leftarrow \overline{B}^{t-1} - \frac{1}{\gamma^t}(\overline{B}^{t-1}\eta^t(\eta^t)^T\overline{B}^{t-1})$
4: $\quad \zeta^t \leftarrow \zeta^{t-1} - \frac{1}{\gamma^t}\overline{B}^{t-1}\eta^t\hat{\epsilon}^t$
5: $\quad$ **return** B^t, ζ^t.
6: **end procedure**

Algorithm 3. Least squares policy iteration with RLSE.

1: Initialize the configuration s_0.
2: Initialize $\theta^0 = \theta = 0$ and $B^0 = \epsilon I$.
3: Initialize $\zeta^0 = 0$ and $\overline{B}^0 = \epsilon I$.
4: **for** t=1,2,3,... **do**
5: $\quad$ Let $\hat{q}_t$ be the just received query.
6: $\quad\quad s_t \leftarrow \underset{s \in S_{s_{t-1}, \hat{q}_t}}{\arg\min} \ (\zeta^{t-1})^T \eta(s_{t-1}, q(s_{t-1}, s)) + (\zeta^{t-1})^T \eta(s, \hat{q}_t) + \gamma\theta^T\phi(s)$
7: $\quad$ Change the configuration to s_t.
8: $\quad$ Execute query $\hat{q}_t$.
9: $\quad \hat{C}^t \leftarrow \delta(s_{t-1}, s_t) + cost(s_t, \hat{q}_t)$.
10: $\quad \hat{\epsilon}^t \leftarrow (\zeta^{t-1})^T\eta(s_{t-1}, \hat{q}_t) - cost(s_{t-1}, \hat{q}_t)$
11: $\quad B^t \leftarrow B^{t-1} - \frac{B^{t-1}\phi(s_{t-1})(\phi(s_{t-1}) - \gamma\phi(s_t))^T B^{t-1}}{1 + (\phi(s_{t-1}) - \gamma\phi(s_t))^T B^{t-1}\phi(s_{t-1})}$.
12: $\quad \theta^t \leftarrow \theta^{t-1} + \frac{(\hat{C}^t - (\phi(s_{t-1}) - \gamma\phi(s_t))^T\theta^{t-1})B^{t-1}\phi(s_{t-1})}{1 + (\phi(s_{t-1}) - \gamma\phi(s_t))^T B^{t-1}\phi(s_{t-1})}$.
13: $\quad (\overline{B}^t, \zeta^t) \leftarrow RLSE(\hat{\epsilon}^t, \overline{B}^{t-1}, \zeta^{t-1}, \eta^t)$
14: $\quad$ **if** (θ^t) converges **then**
15: $\quad\quad \theta \leftarrow \theta^t$.
16: $\quad$ **end if**
17: **end for**

Assume that there exists a feature mapping η such that $cost(s, q) \approx \zeta^T\eta(s, q)$ for some vector ζ. Changing the configuration from s to s' can be considered as executing a special query $q(s, s')$. Therefore we approximate

$$\delta(s, s') = cost(s, q(s, s')) \approx \zeta^T\eta(s, q(s, s')).$$

The vector ζ can be updated iteratively using the well-known *recursive least squares estimation* (RLSE) [22] as shown in Algorithm 2, where $\eta^t = \eta(s_{t-1}, \hat{q}_t)$ and $\hat{\epsilon}^t = (\zeta^{t-1})^T\eta^t - cost(s_{t-1}, \hat{q}_t)$ is the prediction error. Combining RLSE with LSPI, we get our cost-model oblivious algorithm as shown in Algorithm 3.

In Algorithm 3, the vector θ determines the current policy. We can make decision by solving the equation in line 6. The values of $\delta(s_{t-1}, s)$ and $cost(s, \hat{q}_t)$ are obtained from the cost model. The vector θ^t is used to approximate the cost-to-go function following the current policy. If θ^t converges, then we update the current policy (line 14–16).

To check the efficiency and effectiveness of this algorithm, instead of using any heuristics we have initialzed policy π_0 as initial configuration s_0 and the cost-model C_0 as 0 shown in the lines 1–3 of Algorithm 3.

5   Case Study: Index Tuning

In this section, we present COREIL, an algorithm for tuning the configurations differing in their secondary indexes and handling the configuration changes corresponding to the creation and deletion of indexes, which instantiates Algorithm 3.

5.1   Reducing the Search Space

Let I be the set of indexes that can be created. Each configuration $s \in S$ is an element of the power set 2^I. For example, 7 attributes in a schema of R yield a total of 13699 indexes and a total of 2^{13699} possible configurations. Such a large search space invalidates a naive brute-force search for the optimal policy.

For any query $\hat{q}$, let $r(\hat{q})$ be a function that returns a set of recommended indexes. This function may be already provided by the database system (e.g., as with IBM DB2), or it can be implemented externally [1]. Let $d(\hat{q}) \subseteq I$ be the set of indexes being modified (update, insertion or deletion) by $\hat{q}$. We can define the reduced search space as

$$S_{s,\hat{q}} = \{s' \in S \mid (s - d(\hat{q})) \subseteq s' \subseteq (s \cup r(\hat{q}))\}. \tag{4}$$

Deleting indexes in $d(\hat{q})$ will reduce the index maintenance overhead and creating indexes in $r(q)$ will reduce the query execution cost. Note that the definition of $S_{s,\hat{q}}$ here is a subset of the one defined in Sect. 4.2 which deals with the general configurations.

Note that for tree-structured indexes (e.g., B$^+$-tree), we could further consider the *prefix closure* of indexes for optimization. For any configuration $s \in 2^I$, define the prefix closure of s as

$$\langle s \rangle = \{i \in I \mid i \text{ is a prefix of an index } j \text{ for some } j \in s\}. \tag{5}$$

Thus in Eq. (4), we use $\langle r(\hat{q}) \langle$ to replace $r(\hat{q})$ for better approximation. The intuition is that in case of $i \notin s$ but $i \subseteq \langle s \rangle$ we can leverage the prefix index to answer the query.

5.2   Defining the Feature Mapping ϕ

Let V be the cost-to-go function following a policy. As mentioned earlier, Algorithm 3 relies on a proper feature mapping ϕ that approximates the cost-to-go function as $V(s) \approx \theta^T \phi(s)$ for some vector θ. The challenge lies in how to define ϕ under the scenario of index tuning. In COREIL, we define it as

$$\phi_{s'}(s) := \begin{cases} 1, & \text{if } s' \subseteq s \\ -1, & \text{otherwise.} \end{cases}$$

for each $s, s' \in S$. Let $\phi = (\phi_{s'})_{s' \in S}$. Note that $\phi_\emptyset$ is an intercept term since $\phi_\emptyset(s) = 1$ for all $s \in S$. The following proposition shows the effectiveness of ϕ for capturing the values of the cost-to-go function V.

Proposition 2. *There exists a unique* $\boldsymbol{\theta} = (\theta_{s'})_{s' \in S}$ *which approximates the value function as*

$$V(s) = \sum_{s' \in S} \theta_{s'} \phi_{s'}(s) = \boldsymbol{\theta}^T \phi(s). \tag{6}$$

Proof. Suppose $S = \{s^1, s^2, \ldots, s^{|S|}\}$. Note that we use superscripts to denote the ordering of elements in S.

Let $\boldsymbol{V} = (V(s))_{s \in S}^T$ and M be a $|S| \times |S|$ matrix such that

$$M_{i,j} = \phi_{s^j}(s^i).$$

Let $\boldsymbol{\theta}$ be a $|S|$-dimension column vector such that $M\boldsymbol{\theta} = \boldsymbol{V}$. If M is invertible then $\boldsymbol{\theta} = M^{-1}\boldsymbol{V}$ and thus Eq. (6) holds.

We now show that M is invertible. Let $\boldsymbol{\psi}$ be a $|S| \times |S|$ matrix such that

$$\boldsymbol{\psi}_{i,j} = M_{i,j} + 1.$$

We claim that $\boldsymbol{\psi}$ is invertible and its inverse is the matrix $\boldsymbol{\tau}$ such that

$$\boldsymbol{\tau}_{i,j} = (-1)^{|s^i| - |s^j|} \boldsymbol{\psi}_{i,j}.$$

To see this, consider

$$(\boldsymbol{\tau}\boldsymbol{\psi})_{i,j} = \sum_{1 \le k \le |S|} (-1)^{|s^i| - |s^k|} \boldsymbol{\psi}_{i,k} \boldsymbol{\psi}_{k,j}$$

$$= \sum_{s_j \subseteq s_k \subseteq s_i} (-1)^{|s^i| - |s^k|}.$$

Therefore $(\boldsymbol{\tau}\boldsymbol{\psi})_{i,j} = 1$ if and only if $i = j$. By the Sherman-Morrison formula, M is also invertible. $\qquad\square$

However, for any configuration s, $\boldsymbol{\theta}(s)$ is a $|2^I|$-dimensional vector. To reduce the dimensionality, the cost-to-go function can be approximated by $V(s) \approx \sum_{s' \in S, |s'| \le N} \theta_{s'} \phi_{s'}(s)$ for some integer N. Here we assume that the collaborative benefit among indexes could be negligible if the number of indexes exceeds N. In particular when $N = 1$, we have

$$V(s) \approx \theta_0 + \sum_{i \in I} \theta_i \phi_i(s). \tag{7}$$

where we ignore all the collaborative benefits among indexes in a configuration. This is reasonable since any index in a database management system is often of individual contribution for answering queries [15]. Therefore, we derive ϕ from Eq. (7) as $\phi(s) = (1, (\phi_i(s))_{i \in I}^T)^T$. By using this feature mapping ϕ, COREIL approximates the cost-to-go function $V(s) \approx \boldsymbol{\theta}^T \phi(s)$ for some vector $\boldsymbol{\theta}$.

5.3   Defining the Feature Mapping η

A good feature mapping for approximating functions δ and *cost* must take into account both the benefit from the current configuration and the maintenance overhead of the configuration.

To capture the difference between the index set recommended by the database system and that of the current configuration, we define a function $\beta(s, \hat{q}) = (1, (\beta_i(s, \hat{q}))_{i \in I}^T)^T$, where

$$
\beta_i(s, \hat{q}) := \begin{cases} 0, & i \notin r(\hat{q}) \\ 1, & i \in r(\hat{q}) \text{ and } i \in s \\ -1, & i \in r(\hat{q}) \text{ and } i \notin s. \end{cases}
$$

If the execution of query $\hat{q}$ cannot benefit from index i then $\beta_i(s, \hat{q})$ always equals zero; otherwise, $\beta_i(s, \hat{q})$ equals 1 or –1 depending on whether s contains i or not. For tree-structured indexes, we could further consider the prefix closure of indexes as defined in Eq. (5) for optimization.

On the other hand, to capture whether a query (update, insertion or deletion) modifies any index in the current configuration, we define a function $\alpha(s, \hat{q}) = (\alpha_i(s, \hat{q}))_{i \in I}$ where

$$
\alpha_i(s, \hat{q}) = \begin{cases} 1, & \text{if } i \in s \text{ and } \hat{q} \text{ modify } i \\ 0, & \text{otherwise.} \end{cases}
$$

Note that if $\hat{q}$ is a selection query, α trivially returns 0.

By combining β and α, we get the feature mapping $\eta = (\beta^T, \alpha^T)^T$ used in COREIL. It can be used to approximate the functions δ and *cost* as described in Sect. 4.3.

6   Performance Evaluation

In this section, we present an empirical evaluation of COREIL. We implement a prototype of COREIL in Java and compare its performance with that of the state-of-the-art WFIT Algorithm [18]. WFIT is based on the Work Function Algorithm [5]. To determine the change of configuration, it considers all the queries seen so far and solves a deterministic problem towards minimizing the total processing cost.

6.1   Experimental Setup

We conduct all the experiments on a server running IBM DB2 10.5. The server is equipped with Intel i7-2600 Quad-Core @ 3.40 GHz and 4 GB RAM. We measure wall-clock times for execution of all components. Specially, for execution of workload queries or index creating/dropping, we measure the response time of processing corresponding SQL statement in DB2. Additionally, WFIT uses the what-if optimizer of DB2 to evaluate the cost. In this setup, each query is executed only once and all the queries were generated from one execution history. The scale factor (SF) used here is 2.

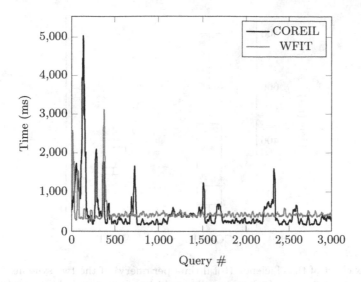

Fig. 1. Evolution of the efficiency (total time per query) of the two systems from the beginning of the workload (smoothed by averaging over a moving window of size 20)

6.2 Dataset and Workload

The dataset and workload is conforming to the TPC-C specification [14] and generated by the OLTP-Bench tool [8]. The 5 types of transactions in TPC-C are distributed as NewOrder 45 %, Payment 43 4 transactions are associated with 3 ∼ 5 SQL statements (query/update). Note that [18] additionally uses the dataset NREF in its experiments. However, this dataset and workload are not publicly available.

6.3 Efficiency

Figure 1 shows the total cost of processing TPC-C queries for online index tuning of COREIL and WFIT. Total cost consists of the overhead of corresponding tuning algorithm, cost of configuration change and that of query execution. Results show that, after convergence, COREIL has lower processing cost most of the time. But COREIL converges slower than WFIT, which is expected since it does not rely on the what-if optimizer to guide the index creations. With respect to the whole execution set, the average processing cost of COREIL (451 ms) is competitive to that of WFIT (452 ms). However, if we calculate the average processing cost of the 500th query forwards, the average performance of COREIL (357 ms) outperforms that of WFIT (423 ms). To obtain further insight from these data, we study the distribution of the processing time per query, as shown in Fig. 2. As can be seen, although COREIL exhibits larger variance in the processing cost, its median is significantly lower that that of WFIT. All these results confirms that COREIL has better efficiency than WFIT under a long term execution.

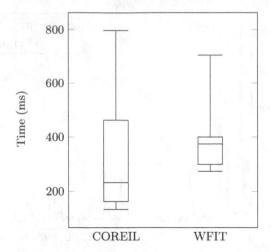

Fig. 2. Box chart of the efficiency (total time per query) of the two systems. We show in both cases the 9th and 91th percentiles (whiskers), first and third quartiles (box) and median (horizontal rule).

Figures 3 and 4 show analysis of the overhead of corresponding tuning algorithm and cost of configuration change respectively. By comparing Fig. 1 with Fig. 3, we can see that the overhead of the tuning algorithm dominates the total cost and the overhead of COREIL is significantly lower than that of WFIT. In addition, WFIT tends to make costlier configuration changes than COREIL, which is reflected in a higher time for configuration change. This would be discussed further in the micro-analysis. Note that both methods converge rather quickly and no configuration change happens beyond the 700th query.

6.4   Effectiveness

To verify the effectiveness of indexes created by the tuning algorithms, we extract the cost of query execution from the total cost. Figure 5 (note the logarithmic y-axis) indicates that the set of indexes created by COREIL shows competitive effectiveness with that created by WFIT, though WFIT is more effective in general and exhibits less variance after convergence. Again, this is to be expected since COREIL does not have access to any cost model for the queries. As previously noted, the total running time is lower for COREIL than WFIT, as overhead rather than query execution dominates running time for both systems.

We have also performed a micro-analysis to check whether the indexes created by the algorithms are reasonable. We observe that WFIT creates more indexes with longer compound attributes, whereas COREIL is more parsimonious in creating indexes. For instance, WFIT creates a 14-attribute index as shown below.

```
[S_W_ID, S_I_ID, S_DIST_10, S_DIST_09, S_DIST_08, S_DIST_07,
```

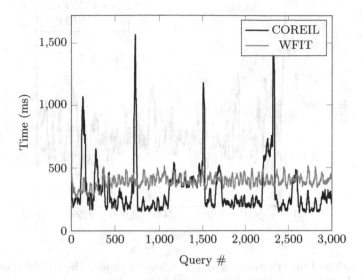

Fig. 3. Evolution of the overhead (time of the optimization itself) of the two systems from the beginning of the workload (smoothed by averaging over a moving window of size 20)

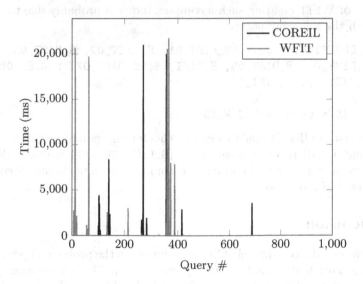

Fig. 4. Evolution of the time taken by configuration change (index creation and destruction) of the two systems from the beginning of the workload; no configuration change happens past query #1000

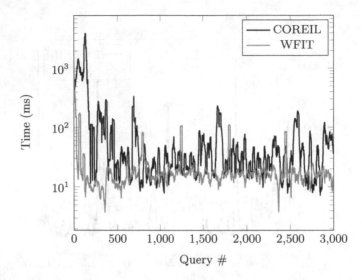

Fig. 5. Evolution of the effectiveness (query execution time in the DBMS alone) of the two systems from the beginning of the workload (smoothed by averaging over a moving window of size 20); logarithmic y-axis

```
S_DIST_06, S_DIST_05, S_DIST_04, S_DIST_03, S_DIST_02,
S_DIST_01, S_DATA, S_QUANTITY]
```

The reason of WFIT creating such a complex index is probably due to multiple queries with the following pattern.

```
SELECT S_QUANTITY, S_DATA, S_DIST_01, S_DIST_02, S_DIST_03,
       S_DIST_04, S_DIST_05, S_DIST_06, S_DIST_07, S_DIST_08,
       S_DIST_09, S_DIST_10
FROM STOCK
WHERE S_I_ID = 69082 AND S_W_ID = 1;
```

In contrast, COREIL tends to create shorter compound-attribute indexes. For example, COREIL created an index [S_I_ID, S_W_ID] which is definitely beneficial to answer the query above and is competitive in performance compared with the one created by WFIT.

7   Conclusion

We have presented a cost-model oblivious solution to the problem of performance tuning. We have first formalized this problem as a Markov decision process. We have devised and presented a solution, which addresses the curse of dimensionality. We have instantiated the problem to the case of index tuning and implemented the COREIL algorithm to solve it. Experiments show competitive performance with respect to the state-of-the-art WFIT algorithm, despite COREIL being cost-model oblivious.

Now that we have validated the possibility for cost-model oblivious database tuning, we intend in future work to study the trade-off for COREIL between efficiency and effectiveness in the case of index tuning. To show universality and robustness of COREIL, we are planning to run further tests on other datasets like TPC-E, TPC-H and benchmark for online index tuning. To find out its sensitivity on setup, we want to experiment with varying scale factors and and other parameters. Furthermore, we want to extend our approach to other aspects of database configuration, including partitioning and replication. This is not straightforward, as the solution will require heuristics that help curb the combinatorial explosion of the configuration space as well as may need some intelligent initialization technique.

Acknowledgement. This research is funded by the National Research Foundation Singapore under its Campus for Research Excellence and Technological Enterprise (CREATE) programme with the SP2 project of the Energy and Environmental Sustainability Solutions for Megacities - E2S2 programme.

References

1. Agrawal, S., Chaudhuri, S., Narasayya, V.R.: Automated selection of materialized views and indexes in SQL databases. In: VLDB (2000)
2. Agrawal, S., Narasayya, V., Yang, B.: Integrating vertical and horizontal partitioning into automated physical database design. In: SIGMOD (2004)
3. Azefack, S., Aouiche, K., Darmont, J.: Dynamic index selection in data warehouses. CoRR abs/0809.1965 (2008). http://arxiv.org/abs/0809.1965
4. Benedikt, M., Bohannon, P., Bruns, G.: Data cleaning for decision support. In: CleanDB (2006)
5. Borodin, A., El-Yaniv, R.: Online Computation and Competitive Analysis. Cambridge University Press, Cambridge (1998)
6. Bruno, N., Chaudhuri, S.: An online approach to physical design tuning. In: ICDE (2007)
7. Chaudhuri, S., Narasayya, V.: Autoadmin: What-if index analysis utility. In: SIGMOD (1998)
8. Difallah, D.E., Pavlo, A., Curino, C., Cudre-Mauroux, P.: OLTP-Bench: an extensible testbed for benchmarking relational databases. Proc. VLDB Endow. **7**(4), 277–288 (2013)
9. Lagoudakis, M.G., Parr, R.: Least-squares policy iteration. J. Mach. Learn. Res. **4**, 1107–1149 (2003)
10. Lohman, G.M.: Is query optimization a "solved" problem? (2014). http://wp.sigmod.org/?p=1075
11. Papadomanolakis, S., Dash, D., Ailamaki, A.: Efficient use of the query optimizer for automated physical design. In: VLDB (2007)
12. Powell, W.B.: Approximate Dynamic Programming: Solving the Curses of Dimensionality. Wiley-Interscience, New York (2007)
13. Puterman, M.L.: Markov Decision Processes: Discrete Stochastic Dynamic Programming. Wiley, New York (2009)
14. Raab, F.: TPC-C - the standard benchmark for online transaction processing (OLTP). In: Gray, J. (ed.) The Benchmark Handbook. Morgan Kaufmann, San Francisco (1993)

15. Ramakrishnan, R., Gehrke, J., Gehrke, J.: Database Management Systems, vol. 3. McGraw-Hill, New York (2003)
16. Rasin, A., Zdonik, S.: An automatic physical design tool for clustered column-stores. In: EDBT (2013)
17. Schnaitter, K., Abiteboul, S., Milo, T., Polyzotis, N.: On-line index selection for shifting workloads. In: SMDB (2007)
18. Schnaitter, K., Polyzotis, N.: Semi-automatic index tuning: keeping DBAs in the loop. Proc. VLDB Endow. 5(5), 478–489 (2012)
19. Stillger, M., Lohman, G.M., Markl, V., Kandil, M.: LEO - DB2's LEarning Optimizer. In: VLDB (2001)
20. Sutton, R.S., Barto, A.G.: Reinforcement Learning. MIT Press, Cambridge (1998)
21. White, D.J.: Markov Decision Processes. Wiley, New York (1993)
22. Young, P.: Recursive least squares estimation. In: Young, P. (ed.) Recursive Estimation and Time-Series Analysis, pp. 29–46. Springer, Berlin Heidelberg (2011)

Towards Making Database Systems PCM-Compliant

Vishesh Garg, Abhimanyu Singh, and Jayant R. Haritsa[✉]

Database Systems Laboratory, SERC/CSA,
Indian Institute of Science, Bangalore, India
{vishesh,abhimanyu,haritsa}@dsl.serc.iisc.ernet.in

Abstract. Phase Change Memory (PCM) is a new *non-volatile* memory technology that is comparable to traditional DRAM with regard to read latency, and markedly superior with regard to storage density and idle power consumption. Due to these desirable characteristics, PCM is expected to play a significant role in the next generation of computing systems. However, it also has limitations in the form of expensive writes and limited write endurance. Accordingly, recent research has investigated how database engines may be redesigned to suit DBMS deployments on the new technology.

In this paper, we address the pragmatic goal of minimally altering current implementations of database operators to make them PCM-conscious, the objective being to facilitate an easy transition to the new technology. Specifically, we target the implementations of the "workhorse" database operators: *sort, hash join* and *group-by*, and rework them to substantively improve the write performance without compromising on execution times. Concurrently, we provide simple but effective *estimators* of the writes incurred by the new techniques, and these estimators are leveraged for integration with the query optimizer.

Our new techniques are evaluated on TPC-H benchmark queries with regard to the following metrics: number of writes, response times and wear distribution. The experimental results indicate that the PCM-conscious operators collectively reduce the number of writes by a factor of 2 to 3, while concurrently improving the query response times by about 20 % to 30 %. When combined with the appropriate plan choices, the improvements are even higher. In essence, our algorithms provide both short-term and long-term benefits. These outcomes augur well for database engines that wish to leverage the impending transition to PCM-based computing.

1 Introduction

Phase Change Memory (PCM) is a recently developed non-volatile memory technology, constructed from chalcogenide glass material, that stores data by switching between amorphous (*binary 0*) and crystalline (*binary 1*) states. Broadly speaking, it is expected to provide an attractive combination of the best features of conventional disks (persistence, capacity) and of DRAM (access speed).

© Springer International Publishing Switzerland 2015
Q. Chen et al. (Eds.): DEXA 2015, Part I, LNCS 9261, pp. 269–284, 2015.
DOI: 10.1007/978-3-319-22849-5_19

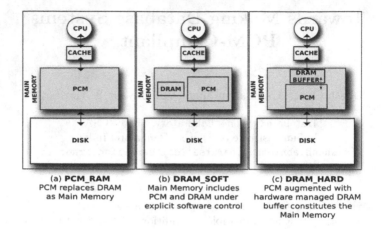

Fig. 1. PCM-based architectural options [3]

For instance, it is about 2 to 4 times denser than DRAM, while providing a DRAM-comparable read latency. On the other hand, it consumes much less energy than magnetic hard disks while providing substantively smaller write latency. Due to this suite of desirable features, PCM technology is expected to play a prominent role in the next generation of computing systems, either augmenting or replacing current components in the memory hierarchy [7,10,15].

A limitation of PCM, however, is that there is a significant difference between the read and write behaviors in terms of energy, latency and bandwidth. A PCM write, for example, consumes 6 times more energy than a read. Further, PCM has limited write endurance since a memory cell becomes unusable after the number of writes to the cell exceeds a threshold determined by the underlying glass material. Consequently, several database researchers have, in recent times, focused their attention on devising new implementations of the core database operators that are adapted to the idiosyncrasies of the PCM environment (e.g. [3,12]).

Architectural Model

The prior database work (which we have analyzed in detail in [5]) has primarily focused on computing architectures wherein either (a) PCM completely replaces the DRAM memory [3]; or (b) PCM and DRAM co-exist side-by-side and are independently controlled by the software [12]. We hereafter refer to these options as **PCM_RAM** and **DRAM_SOFT**, respectively.

However, a third option that is gaining favor in the architecture community, and also mooted in [3] from the database perspective, is where the PCM is augmented with a small hardware-managed DRAM buffer [10]. In this model, which we refer to as **DRAM_HARD**, the address space of the application maps to PCM, and the DRAM buffer can simply be visualized as yet another level of the existing cache hierarchy. For ease of comparison, these various configurations are pictorially shown in Fig. 1.

There are several practical advantages of the DRAM_HARD configuration: First, the write latency drawback of PCM_RAM can be largely concealed by the intermediate DRAM buffer [10]. Second, existing applications can be used *as is* but still manage to take advantage of both the DRAM and the PCM. This is in stark contrast to the DRAM_SOFT model which requires incorporating additional machinery, either in the program or in the OS, to distinguish between data mapped to DRAM and to PCM – for example, by having separate address space mappings for the different memories.

Our Work

In this paper, we propose minimalist reworkings, that are tuned to the DRAM_HARD model, of current implementations of database operators. In particular, we focus on the "workhorse" operators: *sort, hash join* and *group-by*. The proposed modifications are not only easy to implement but are attractive from the performance perspective also, simultaneously reducing *both* PCM writes and query response times. The new implementations are evaluated on Multi2sim [11], a state-of-the-art architectural simulator, after incorporating major extensions to support modelling of the DRAM_HARD configuration. Their performance is evaluated on *complete* TPC-H benchmark queries. This is a noteworthy point since earlier studies of PCM databases had only considered operator performance in isolation. But, it is possible that optimizing a specific operator may turn out to be detrimental to downstream operators that follow it in the query execution plan. For instance, the proposal in [3] to keep leaf nodes unsorted in B^+ indexes – while this saves on writes, it is detrimental to the running times of *subsequent* operators that leverage index ordering – for instance, *join filters*. Finally, we include the metric of *wear distribution* in our evaluation to ensure that the reduction in writes is not achieved at the cost of skew in wear-out of PCM cells.

Our simulation results indicate that the customized implementations collectively offer substantive benefits with regard to PCM writes – the number is typically brought down *by a factor of two to three*. Concurrently, the query response times are also brought down by about *20–30 percent*. As a sample case in point, for TPC-H Query 19, savings of 64 % in PCM writes are achieved with a concomitant 32 % reduction in CPU cycles.

Fully leveraging the new implementations requires integration with the query optimizer, an issue that has been largely overlooked in the prior literature. We take a first step here by providing simple but effective statistical *estimators* for the number of writes incurred by the new operators under uniform data distribution scenarios, and incorporating these estimators in the query optimizer's cost model. Sample results demonstrating that the resultant plan choices provide substantively improved performance are provided in our experimental study.

Overall, the above outcomes augur well for the impending migration of database engines to PCM-based computing platforms.

Organization

The remainder of this paper is organized as follows: We define the problem framework in Sect. 2. The design of the new PCM-conscious database operators,

Table 1. Notations used in operator analysis

Term	Description
D	DRAM size
K	DRAM Associativity
N_R, N_S	Row cardinalities of input relations R and S, respectively
L_R, L_S	Tuple lengths of input relations R and S, respectively
P	Pointer size
H	Size of each hash table entry
A	Size of aggregate field (for group-by operator)
N_j, N_g	Output tuple cardinalities of join and group-by operators, respectively
L_j, L_g	Output tuple lengths of join and group-by operators, respectively

and an analysis of their PCM writes, are presented in Sects. 3, 4 and 5. Our experimental framework and the simulation results are reported in Sects. 6 and 7, respectively. This is followed by a discussion in Sect. 8 on integration with the query optimizer. Finally, Sect. 9 summarizes our conclusions and outlines future research avenues.

2   Problem Framework

In this section, we overview the problem framework, the assumptions made in our analysis, and the notations used in the sequel.

We model the DRAM_HARD memory organization shown in Fig. 1(c). The DRAM buffer is of size D, and organized in a *K-way set-associative* manner, like the L1/L2 processor cache memories. Moreover, its operation is identical to that of an *inclusive* cache in the memory hierarchy, that is, a new DRAM line is fetched from PCM each time there is a DRAM miss. The last level cache in turn fetches its data from the DRAM buffer.

We assume that the writes to PCM are in word-sized units (4B) and are incurred only when a data block is evicted from DRAM to PCM. A *data-comparison write (DCW)* scheme [14] is used for the writing of PCM memory blocks during eviction from DRAM – in this scheme, the memory controller compares the existing PCM block to the newly evicted DRAM block, and selectively writes back only the modified words. Further, *N-Chance* [4] is used as the DRAM eviction policy due to its preference for evicting non-dirty entries, thereby saving on writes. The failure recovery mechanism for updates is orthogonal to our work and is therefore not discussed in this paper.

As described above, the simulator implements a realistic DRAM buffer. However, in our write analyses and estimators, we assume for tractability that there are no conflict misses in the DRAM. Thus, for any operation dealing with data whose size is within the DRAM capacity, our analysis assumes no evictions and consequently no writes. The experimental evaluation in Sect. 7.3 indicates the impact of this assumption to be only marginal.

With regard to the operators, we use R to denote the input relation for the *sort* and *group-by* unary operators. Whereas, for the binary *hash join* operator, R is used to denote the smaller relation, on which the hash table is constructed, while S denotes the probing relation.

In this paper, we assume that all input relations are *completely PCM-resident*. Further, for presentation simplicity, we assume that the sort, hash join and group-by expressions are on singleton attributes – the extension to multiple attributes is straightforward.

A summary of the main notation used in the analysis of the following sections is provided in Table 1.

3   The *Sort* Operator

The *quicksort* algorithm is the most commonly used sorting algorithm in database systems. In the single-pivot quicksort algorithm with n elements, the average number of swaps is of the order of $0.3nln(n)$ [13]. If the initial array is much larger than the DRAM size, it would entail evictions from the DRAM during the swapping process of partitioning. These evictions might lead to PCM writes if the evicted DRAM lines are *dirty*, which is likely since elements are being swapped. If the resulting partition sizes continue to be larger than DRAM, partitioning them in turn will again cause DRAM evictions and consequent writes. Clearly, this trend of writes will continue in the recursion tree until the partition sizes become small enough to fit within DRAM.

From the above discussion, it is clear that it would be desirable for the sorting algorithm to converge fast to partition sizes within DRAM size with fewer number of swaps. For uniformly-distributed data[1], these requirements are satisfied by *flashsort* [8]. Specifically, flashsort can potentially form DRAM-sized partitions in a *single* partitioning step with at most N_R swaps. The sorting is done in-place with a time complexity of $O(N_R log_2 N_R)$ with constant extra space.

The flashsort algorithm proceeds in three phases: *Classification*, *Permutation* and *Short-range Ordering*. The Classification phase divides the input data into p partitions, where p is an input parameter. Specifically, an element with value v is assigned to $Partition(v)$, computed as $1 + \lfloor \frac{(p-1)(v-v_{min})}{v_{max}-v_{min}} \rfloor$, where v_{min} and v_{max} are the smallest and largest values in the array, respectively. The number of elements in each such partition is counted to derive the boundary information. Next, the Permutation phase moves the elements to their respective partitions. Finally, the individual partitions are sorted in the Short-range Ordering phase to obtain the overall sorted array.

We choose the number of partitions p to be $\lceil c \times \frac{N_R L_R}{D} \rceil$, where $c \geq 1$ is a multiplier to cater to the space requirements of additional data structures constructed during sorting. In our experience, setting $c = 2$ works well in practice. The resulting partitions, each having size less than D, are finally sorted in the Short-range Ordering phase using quicksort.

---

[1] In [5], we present a modified flashsort algorithm, called *multi-pivot flashsort*, for skewed data.

PCM Write Analysis: Though the partition boundary counters are continuously updated during the Classification phase, they are expected to incur very few PCM writes. This is because the updates are all in quick succession, making it unlikely for the counters to be evicted from DRAM during the update process. Next, while in the Permutation phase, there are no more than $N_R L_R$ writes since each tuple is written at most once while placing it inside its partition boundaries. Since each partition is within the DRAM size, its Short-range Ordering phase will finish in the DRAM itself, and then there will be another $N_R L_R$ writes upon eventual eviction of sorted partitions to PCM.

Thus, the number of word-writes incurred by this algorithm is estimated by

$$W_{sort} = \frac{2 N_R L_R}{4} = \frac{N_R L_R}{2}. \tag{1}$$

4   The *Hash Join* Operator

Hash join is perhaps the most commonly used join algorithm in database systems. Here, a hash table is built on the smaller relation, and tuples from the larger relation are used to probe for matching values in the join column. Since we assume that all tables are completely PCM-resident, the join here *does not* require any initial partitioning stage. Instead, we directly proceed to the join phase. Thus, during the progress of hash join, writes will be incurred during the building of the hash table, and also during the writing of the join results.

Each entry in the hash table consists of a pointer to the corresponding build tuple, and the hash value for the join column. Due to the absence of prior knowledge about the distribution of join column values for the build relation, the hash table is expanded dynamically according to the input. Typically, for each insertion in a bucket, a new space is allocated, and connected to existing entries using a pointer. Thus, such an approach incurs an additional pointer write each time a new entry is inserted.

Our first modification is to use a well-known technique of allocating space to hash buckets in units of *pages* [6]. A page is of fixed size and contains a sequence of contiguous fixed-size hash-entries. When a page overflows, a new page is allocated and linked to the overflowing page via a pointer. Thus, unlike the conventional hash table wherein each *pair* of entries is connected using pointers, the interconnecting pointer here is only at page granularity. Note that although open-addressing is another alternative for avoiding pointers, probing for a join attribute value would have to search through the *entire table* each time, since the inner table may contain *multiple* tuples with the same join attribute value.

A control bitmap is used to indicate whether each entry in a page is vacant or occupied, information that is required during both insertion and search in the hash table. Each time a bucket runs out of space, a new page is allocated to the bucket. Though such an approach may lead to space wastage when some of the pages are not fully occupied, we save on the numerous pointer writes that are otherwise incurred when space is allocated on a per-entry basis.

Secondly, we can reduce the writes incurred due to storing of the hash values in the hash table by restricting the length of each hash value to just a single byte. In this manner, we trade-off precision for fewer writes. If the hash function distributes the values in each bucket in a perfectly uniform manner, it would be able to distinguish between $2^8 = 256$ join column values in a bucket. This would be sufficient if the number of distinct values mapping to each bucket turn out to be less than this value. Otherwise, we would have to incur the penalty (in terms of latency) of reading the actual join column values from PCM due to the possibility of false positives.

PCM Write Analysis: We ignore the writes incurred while initializing each hash table bucket since they are negligible in comparison to inserting the actual entries. Assuming there are E_{page} entries per page, there would now be one pointer for each E_{page} set of entries. Additionally, for each insertion, a bit write would be incurred due to the bitmap update. The join tuples would also incur writes to the tune of $N_j \times L_j$. Thus, the total number of word-writes for hash join would be

$$W_{hj} = \frac{N_R \times (H + \frac{P}{E_{page}} + \frac{1}{8}) + N_j \times L_j}{4}$$

Since in practice both $\frac{P}{E_{page}}$ and $\frac{1}{8}$ are small as compared to H,

$$W_{hj} \approx \frac{N_R \times H + N_j \times L_j}{4}. \tag{2}$$

5 The *Group-By* Operator

We now turn our attention to the group-by operator which typically forms the basis for aggregate function computations in SQL queries. Common methods for implementing group-by include *sorting* and *hashing* – the specific choice of method depends both on the constraints associated with the operator expression itself, as well as on the downstream operators in the plan tree. We discuss below the PCM-conscious modifications of both implementations, which share a common number of *output* tuple writes, namely $N_g \times L_g$.

5.1 Hash-Based Grouping

A hash table entry for group-by, as compared to the corresponding entry in hash join, has an additional field containing the aggregate value. For each new tuple in the input array, a bucket index is obtained after hashing the value of the column present in the group-by expression. Subsequently, a search is made in the bucket indicated by the index. If a tuple matching the group-by column value is found, the aggregate field value is updated; else, a new entry is created in the bucket. Thus, unlike hash join, where each build tuple had its individual entry, here the grouped tuples share a common entry with an aggregate field that is constantly updated over the course of the algorithm.

Since the hash table construction for group-by is identical to that of the hash join operator, the PCM-related modifications described in Sect. 4 can be applied here as well. That is, we employ a page-based hash table organization, and a reduced hash value size, to reduce the writes to PCM.

PCM Write Analysis: From the above discussion, it is easy to see that the total number of word-writes incurred for the PCM-conscious hash-based group-by is given by

$$W_{gb\_ht} = \frac{N_g \times H + N_R \times A + N_g \times L_g}{4}. \tag{3}$$

5.2   Sort-Based Grouping

Sorting may be used for group-by when a fully ordered operator such as *order by* or *merge join* appears downstream in the plan tree. Another use case is for queries with a *distinct* clause in the aggregate expression, in order to identify the duplicates that have to be discarded from the aggregate.

Sorting-based group-by differs in a key aspect from sorting itself in that the sorted tuples do not have to be written out. Instead, it is the aggregated tuples that are finally passed on to the next operator in the plan tree. Hence, we can modify the flashsort algorithm of Sect. 3 to use *pointers* in both the Permutation and Short-range Ordering phases, subsequently leveraging these pointers to perform aggregation on the sorted tuples.

PCM Write Aanalysis: The full tuple writes of $2N_R L_R$ which were incurred in the flashsort scheme, are now replaced by $2N_R \times P$ since pointers are used during both the Classification and Short-range Ordering phases. Thus, the total number of word-writes for this algorithm for uniformly distributed data would be

$$W_{gb\_sort} = \frac{2N_R \times P + N_g \times L_g}{4}. \tag{4}$$

6   Simulation Testbed

This section details our experimental settings in terms of the hardware parameters, the database and query workload, and the performance metrics on which we evaluated the PCM-conscious operator implementations.

6.1   Architectural Platform

Since PCM memory is as yet not commercially available, we have taken recourse to a simulated hardware environment to evaluate the impact of the PCM-conscious operators. For this purpose, we chose Multi2sim [11], an open-source application-only[2] simulator.

We evaluated the PCM-conscious algorithms on Multi2sim in cycle-accurate simulation mode. Since it does not have native support for PCM, we made a major extension to its existing memory module to model PCM with a hardware-controlled DRAM buffer as main memory. Furthermore, we added separate data tracking functionality for the DRAM and PCM-resident data, to implement the DCW scheme (Sect. 2) of DRAM line write-back to PCM. Likewise, we made several other enhancements to Multi2sim for PCM modelling, and these are enumerated in [5].

---

[2] Simulates only the application layer without the OS stack.

Table 2. Experimental setup

Simulator	Multi2sim-4.2 with added support for PCM
L1D cache (private)	32 KB, 64B line, 4-way set-associative, 4 cycle latency, write-back, LRU
L1I cache (private)	32 KB, 64B line, 4-way set-associative, 4 cycle latency, write-back, LRU
L2 cache (private)	256 KB, 64B line, 4-way set-associative, 11 cycle latency, write-back, LRU
DRAM buffer (private)	4 MB, 256B line, 8-way set-associative, 200 cycle latency, write-back, N-Chance (N = 4)
Main Memory	2 GB PCM, 4 KB page, 1024 cycle read latency (per 256B line), 64 cycle write latency (per 4B modified word)

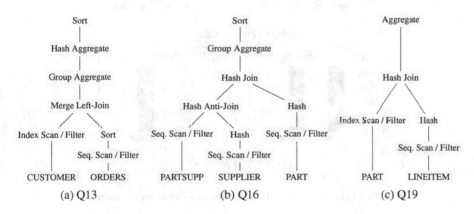

Fig. 2. Query execution plan trees

The specific configurations of the memory hierarchy *(L1 Data, L1 Instruction, L2, DRAM Buffer, PCM)* used for evaluation in our experiments are enumerated in Table 2. These values are scaled-down versions, w.r.t. number of lines, of the hardware simulation parameters used in [9] – the reason for the scaling-down is to ensure that the simulator running times are not impractically long. However, we have been careful to ensure that the *ratios* between the capacities of adjacent levels in the hierarchy are maintained as per the original configurations in [9].

6.2 Database and Queries

For the data, we used the default 1 GB database generated by the TPC-H [1] benchmark. This size is certainly very small compared to the database sizes typically encountered in modern applications – however, we again chose this reduced value to ensure viable simulation running times. Furthermore, the database is significantly larger than the simulated DRAM (4 MB), representative of most real-world scenarios.

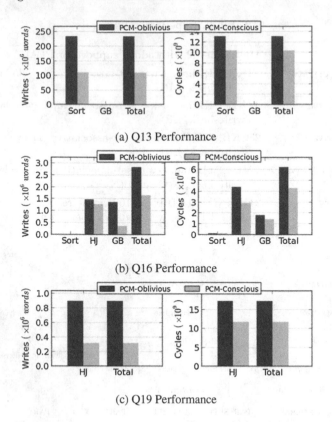

(a) Q13 Performance

(b) Q16 Performance

(c) Q19 Performance

Fig. 3. Performance of TPC-H queries (Color figure online)

For simulating our suite of database operators – *sort, hash join* and *group-by* – we created a separate library consisting of their native PostgreSQL [2] implementations. To this library, we added the PCM-conscious versions described in the previous sections.

While we experimented with several of the TPC-H queries, results for three queries: Query 13 (Q13), Query 16 (Q16) and Query 19 (Q19), that cover a diverse spectrum of the experimental space, are presented here. For each of the queries, we first identified the execution plan recommended by the PostgreSQL query optimizer with the native operators, and then forcibly used the same execution plan for their PCM-conscious replacements as well. This was done in order to maintain fairness in the comparison of the PCM-oblivious and PCM-conscious algorithms, though it is possible that a *better* plan is available for the PCM-conscious configuration – we return to this issue later in Sect. 8. The execution plans associated with the three queries are shown in Fig. 2.

6.3 Performance Metrics

We measured the following performance metrics for each of the queries:

PCM Writes: The total number of word (4B) updates that are applied to the PCM memory during the query execution.

CPU Cycles: The total number of CPU cycles required to execute the query.

Wear Distribution: The frequency distribution of writes measured on a per-256B-block basis.

7 Experimental Results

Based on the above framework, we conducted a wide variety of experiments and present a representative set of results here. We begin by profiling the PCM writes and CPU cycles behavior of the native and PCM-conscious executions for Q13, Q16 and Q19 – these results are shown in Fig. 3. In each of these figures, we provide both the total and the break-ups on a per-operator basis, with *GB* and *HJ* labels denoting group-by and hash join operators, respectively.

Focusing our attention first on Q13 in Fig. 3(a), we find that the bulk of the overall writes and cycles are consumed by the sort operator. Comparing the performance of the Native (blue bar) and PCM-conscious (green bar) implementations, we observe a very significant savings (53 %) on writes, and an appreciable decrease (20 %) on cycles.

Turning our attention to Q16 in Fig. 3(b), we find that here it is the group-by operator that primarily influences the overall writes performance, whereas the hash join determines the cycles behavior. Again, there are substantial savings in both writes (40 %) and cycles (30 %) delivered by the PCM-conscious approach.

Finally, moving on to Q19 in Fig. 3(c), where hash join is essentially the only operator, the savings are around 64 % with regard to writes and 32 % in cycles.

7.1 Operator-Wise Analysis

We now analyse the savings due to each operator independently and show their correspondence to the analyses in Sects. 3, 4 and 5.

Sort. For Q13, as already mentioned, we observed savings of 53 % in writes and 20 % in cycles. In the case of Q16, the data at the sorting stage was found to be much less than the DRAM size. Hence, both the native and PCM-conscious executions used the standard sort routine, and as a result, the cycles and writes for both implementations match exactly.

Hash Join. Each entry in the hash table consisted of a pointer to the build tuple and a hash value field. New memory allocation to each bucket was done in units of pages, with each page holding up to 64 entries. A search for the matching join column began with the first tuple in the corresponding bucket, and went on till the last tuple in that bucket, simultaneously writing out the join tuples for successful matches. For Q16, we observed a 12 % improvement in writes and 31 % in cycles due to the PCM-conscious hash join, as shown in Fig. 3(b). The high savings in cycles was the result of the caching effect due to page-wise allocation. These improvements were even higher with Q19 – specifically, 65 %

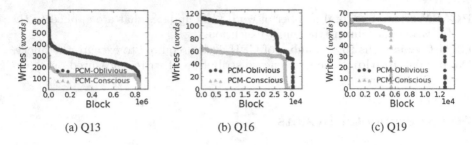

(a) Q13   (b) Q16   (c) Q19

Fig. 4. Queries wear distribution

writes and 32 % cycles, as shown in Fig. 3(c). The source of the enhancement was the 3 bytes of writes saved due to single-byte hash values[3], and additionally, the page-based aggregation of hash table entries.

Group-By. In Q16, the aggregate operator in the group-by has an associated *distinct* clause. Thus, our group-by algorithm utilized sort-based grouping to carry out the aggregation. Both the partitioning and sorting were carried out through pointers, thereby reducing the writes significantly. Consequently, we obtain savings of 74 % in writes and 20 % in cycles, as shown in Fig. 3(b). When we consider Q13, however, the grouping algorithm employed was hash-based. Here, the hash table consisted of very few entries which led to the overhead of the page metadata construction overshadowing the savings obtained in other aspects. Specifically, only marginal improvements of about 4 % and 1 % were obtained for writes and cycles, as shown in Fig. 3(a).

7.2   Lifetime Analysis

The above experiments have shown that PCM-conscious operators can certainly provide substantive improvements in both writes and cycles. However, the question still remains as to whether these improvements have been purchased at the expense of *longevity* of the memory. That is, are the writes skewed towards particular memory locations? To answer this, we show in Fig. 4, the maximum number of writes across all memory blocks for the three TPC-H queries (as mentioned earlier, we track writes at the block-level–256 bytes–in our modified simulator). The x-axis displays the block numbers in decreasing order of writes.

We observe here that the maximum number of writes is considerably more for the native systems as compared to the PCM-conscious processing. This conclusively demonstrates that the improvement is with regard to *both* average-case and worst-case behavior.

7.3   Validating Write Estimators

We now move on to validating the estimators (presented in Sects. 3, 4 and 5) for the number of writes incurred by the various database operators.

---

[3] The hash values of all entries within a bucket are placed contiguously.

Sort. The size of the *orders* table is approximately 214 MB. The flashsort algorithm incurred writes of 110.6M. Replacing the values for $N_R L_R$ with the table size in Eq. 1, we get the writes as $W_{sort} = (214 \times 10^6)/2 = 107$M. Thus the estimate is close to the number of observed word-writes.

Hash Join. For the hash join in Q19, the values of N_R, H, N_j, L_j are 0.2M, 5 bytes, 120 and 8 bytes, respectively. Substituting the parameter values in Eq. 2, the writes are given by: $W_{hj} = (0.2 \times 10^6 \times 5 + 120 \times 8)/4 \approx 0.25$M which is close to the actual word-writes of 0.32M.

Group-By. The values of the parameters N_R, L_R, P, N_g and L_g for Q16 are 119056, 48 bytes, 4 bytes, 18341 and 48 bytes, respectively. The grouping algorithm used was sort-based grouping. Using Eq. 4 results in: $W_{gb\_sort} = (2 \times 119056 \times 4 + 18341 \times 48)/4 = 0.46$M. This closely corresponds to the observed word-writes of 0.36M.

A summary of the above results is provided in Table 3. It is clear that our estimators predict the write cardinality with an acceptable degree of accuracy for the PCM-conscious implementations, making them suitable for incorporation in the query optimizer.

8 Query Optimizer Integration

In the earlier sections, given a user query, the modified operator implementations were used for the *standard* plan choice of the PostgreSQL optimizer. That is, while the execution engine was PCM-conscious, the presence of PCM was completely *opaque* to the optimizer. However, given the read-write asymmetry of PCM in terms of both latency and wear factor, it is possible that alternative plans, capable of providing better performance profiles, may exist in the plan search space. To discover such plans, the database query optimizer needs to incorporate PCM awareness in both the operator cost models and the plan enumeration algorithms.

Current query optimizers typically choose plans using a latency-based costing mechanism. We revise these models to account for the additional latency incurred during writes. Additionally, we introduce a new metric of *write cost* in the operator cost model, representing the incurred writes for a plan in the PCM

Table 3. Validation of write estimators

Operator	Estimated word-writes (in millions) (e)	Observed word-writes (in millions) (o)	Error factor $(\frac{e-o}{o})$
Sort	107	110.6	−0.03
Hash Join	0.25	0.32	−0.22
Group-By	0.46	0.36	0.27

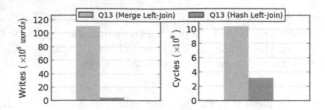

(a) Performance of Alternative Plans

Metric	Opt(PCM-O) Exec(PCM-O)	Opt(PCM-O) Exec(PCM-C)	Opt(PCM-C) Exec(PCM-C)	Opt(PCM-C) Exec(PCM-O)
Mega Word-Writes	233.6	110.6	4.66	12.8
Giga Cycles	13.1	10.4	3.2	4.5

(b) Overall performance comparison

Fig. 5. Integration with query optimization and processing engine

environment, using the estimators described in Sects. 3, 4 and 5. We henceforth refer to the latency cost and the write cost of a plan as **LC** and **WC**, respectively.

A new user-defined parameter, called the *latency slack*, is incorporated in the query optimizer. This slack, denoted by λ, represents the maximum relative slowdown, compared to the LC-optimal query plan, that is acceptable to the user in lieu of getting better write performance. Specifically, if the LC of the LC-optimal execution plan P_o is C_o and the LC of an alternate plan P_i is C_i, the user is willing to accept P_i as the final execution plan if $C_i \leq (1+\lambda)C_o$. The P_i with the least WC satisfying this equation is considered the WC-optimal plan.

With the new metric in place, we need to revise the plan enumeration process during the planning phase. This is because the native optimizer propagates only the LC-optimal (and interesting order) plans through the internal nodes of the dynamic programming lattice, which may lead to pruning of potential WC-optimal plans. On the other hand, propagating the *entire* list of sub-plans at each internal node can end up in an exponential blow-up of the search space. As an intermediate option between these two extremes, we use a heuristic propagation mechanism at each internal node, employing an algorithmic parameter, *local threshold* λ_l ($\geq \lambda$). Specifically, let p_i and p_o be a generic sub-plan and the LC-optimal sub-plan at a node, respectively, with c_i and c_o being their corresponding LC values. Now, along with the LC-optimal and interesting order sub-plans, we also propagate p_i with the *least* WC that satisfies $c_i \leq (1 + \lambda_l)c_o$. We observed that setting $\lambda_l = \lambda$ delivered reasonably good results in this respect.

In light of these modifications, let us revisit Query Q13, for which the default plan was shown in Fig. 2(a). With just the revised latency costs (i.e. $\lambda = 0$), the optimizer identified a new execution plan wherein the merge left-join between the *customer* and *orders* tables is replaced by a hash left-join. The relative performance of these two alternatives with regard to PCM writes and CPU cycles are shown in Fig. 5(a). We observe here that there is a *huge difference* in both the query response times as well as write overheads between the plans. Specifically, the alternative plan reduces the writes by well over an order of magnitude! As we gradually increased the latency slack value, initially there was no change in plans. However, when the slack was made as large as 5, the hash left-join gave way to a nested-loop left-join, clearly indicating that the nested-loop join provides write savings only by incurring a steep increase in latency cost.

To put matters into perspective, Fig. 5(b) summarizes the relative performance benefits obtained as the database layers are gradually made PCM-conscious (in the figure, the labels Opt and Exec refer to Optimizer and Executor, respectively, while PCM-O and PCM-C refer to PCM-Oblivious and PCM-Conscious, respectively). For the sake of completeness, we have also added results for the case when the Optimizer is PCM-C but the Executor is PCM-O (last column). The results clearly indicate that future query optimizers for PCM-based architectures need to incorporate PCM-Consciousness at *both* the Optimizer and the Executor levels in order to obtain the best query performance.

9 Conclusion

Designing database query execution algorithms for PCM platforms requires a change in perspective from the traditional assumptions of symmetric read and write overheads. We presented here a variety of minimally modified algorithms for the workhorse database operators: *sort, hash join* and *group-by*, which were constructed with a view towards simultaneously reducing both the number of writes and the response time. Through detailed experimentation on complete TPC-H benchmark queries, we showed that substantial improvements on these metrics can be obtained as compared to their contemporary PCM-oblivious counterparts. Collaterally, the PCM cell lifetimes are also greatly extended by the new approaches.

Using our write estimators for uniformly distributed data, we presented a redesigned database optimizer, thereby incorporating PCM-consciousness in all layers of the database engine. We also presented initial results showing how this can influence plan choices, and improve the write performance by a substantial margin. While our experiments were conducted on a PCM simulator, the cycle-accurate nature of the simulator makes it likely that similar performance will be exhibited in the real world as well. In our future work, we would like to design write estimators that leverage the metadata statistics to accurately predict writes for skewed data. Additionally, we wish to design multi-objective optimization algorithms for query plan selection with provable performance guarantees.

Overall, the results of this paper augur well for an easy migration of current database engines to leverage the benefits of tomorrow's PCM-based computing platforms.

References

1. http://www.tpc.org/tpch
2. http://www.postgresql.org

3. Chen, S., Gibbons, P.B., Nath, S.: Rethinking database algorithms for phase change memory. In: Proceedings of 5th Biennial Conference on Innovative Data Systems Research (CIDR) (2011)

4. Ferreira, A., Zhou, M., Bock, S., Childers, B., Melhem, R., Mosse, D.: Increasing PCM main memory lifetime. In: Proceedings of 13th Conference on Design, Automation and Test in Europe (DATE) (2010)

5. Garg, V., Singh, A., Haritsa, J.R.: On improving write performance in PCM databases. Technical report TR-2015-01, DSL/SERC, IISc, dsl.serc.iisc.ernet.in/publications/report/TR/TR-2015-01.pdf (2015)

6. Larson, P.-A.: Grouping and duplicate elimination: Benefits of early aggregation. Microsoft Technical report (1997)

7. Lee, B.C., Ipek, E., Mutlu, O., Burger, D.: Architecting phase change memory as a scalable dram alternative. In: Proceeding of 36th International Symposium on Computer Architecture (ISCA) (2009)

8. Neubert, K.-D.: The flashsort1 algorithm (1998). http://www.drdobbs.com/database/the-flashsort1-algorithm/184410496

9. Qureshi, M.K., Karidis, J., Franceschini, M., Srinivasan, V., Lastras, L., Abali, B.: Enhancing lifetime and security of PCM-based main memory with start-gap wear leveling. In: Proceedings of 42nd International Symposium on Microarchitecture (MICRO) (2009)

10. Qureshi, M.K., Srinivasan, V., Rivers, J.A.: Scalable high performance main memory system using phase-change memory technology. In: Proceedings of 36th International Symposium on Computer Architecture (ISCA) (2009)

11. Ubal, R., Jang, B., Mistry, P., Schaa, D., Kaeli, D.: Multi2sim: a simulation framework for CPU-GPU computing. In: Proceedings of 21st International Conference on Parallel Architectures and Compilation Techniques (PACT) (2012)

12. Viglas, S.D.: Write-limited sorts and joins for persistent memory. In: Proceedings of 40th International Conference on Very Large Data Bases (VLDB) (2014)

13. Wild, S., Nebel, M.E.: Average case analysis of Java 7's dual pivot quicksort. In: Epstein, L., Ferragina, P. (eds.) ESA 2012. LNCS, vol. 7501, pp. 825–836. Springer, Heidelberg (2012)

14. Yang, B.-D., Lee, J.-E., Kim, J.-S., Cho, J., Lee, S.-Y., Yu, B.-G.: A low power phase-change random access memory using a data-comparison write scheme. In: Proceedings of 2007 IEEE International Symposium on Circuits and Systems (ISCAS) (2007)

15. Zhou, P., Zhao, B., Yang, J., Zhang, Y.: A durable and energy efficient main memory using phase change memory technology. In: Proceedings of 36th International Symposium on Computer Architecture (ISCA) (2009)

Workload-Aware Self-Tuning Histograms of String Data

Nickolas Zoulis[1], Effrosyni Mavroudi[2], Anna Lykoura[3],
Angelos Charalambidis[4], and Stasinos Konstantopoulos[4(✉)]

[1] Computer Science, Athens University of Economics and Business, Athens, Greece
[2] Electrical and Computer Engineering,
National Technical University of Athens, Kesariani, Greece
[3] Applied Mathematical and Physical Science,
National Technical University of Athens, Kesariani, Greece
[4] Institute of Informatics and Telecommunications,
NCSR 'Demokritos', Athens, Greece
{acharal,konstant}@iit.demokritos.gr

Abstract. In this paper we extend STHoles, a very successful algorithm that uses query results to build and maintain multi-dimensional
histograms of numerical data. Our contribution is the formal definition
of extensions of all relevant concepts; such that they are independent of
the domain of the data, but subsume STHoles concepts as their numerical specialization. At the same time, we also derive specializations for
the string domain and implement these into a prototype that we use
to empirically validate our approach. Our current implementation uses
string prefixes as the machinery for describing string ranges. Although
weaker than regular expressions, prefixes can be very efficiently applied
and can capture interesting ranges in hierarchically structured string
domains, such as those of filesystem pathnames and URIs. In fact, we
base the empirical validation of the approach on existing, publicly available Semantic Web data where we demonstrate convergence to accurate
and efficient histograms.

1 Introduction

Query optimizers in query processing systems typically rely on *histograms*, data
structures that approximate data distribution, in order to be able to apply their
cost model. Histograms can be constructed by scanning the database tables and
aggregating the values of the attributes in the table; and similarly maintained
in the face of database updates.

This histogram lifecycle, however, cannot be efficiently applied to large-scale
and frequently updated databases, such as, for example, stores of sensor data.
An alternative approach is taken by *adaptive* query processing systems that
update their histograms by observing and analysing the results of the queries that
constitute the client-requested workload, as opposed to maintenance workload
only for updating the histograms. The relevant databases literature focuses on

© Springer International Publishing Switzerland 2015
Q. Chen et al. (Eds.): DEXA 2015, Part I, LNCS 9261, pp. 285–299, 2015.
DOI: 10.1007/978-3-319-22849-5_20

numerical attributes, exploiting the concept of an *interval* as a description of a set of numerical values that is *succinct* and that has a *length* that can be used to estimate the cardinality of many different intervals that have roughly the same density.

In the work described here, we investigate how to extend adaptive query processing so that it can be applied to the domain of *strings*, typically treated as purely categorical symbols that can only be described by enumeration. This, however, disregards the fact that there are several classes of strings that have an internal structure and that can be handled in a more sophisticated manner. Specifically, we use string *prefixes* to expresses 'intervals', i.e., sub-spaces of the overall string space that are interesting from the point of view of providing query optimization statistics. Although weaker than regular expressions, prefixes can be very efficiently applied and can capture interesting ranges in hierarchically-structured string domains, such as that of URIs.

This attention on URIs is motivated by their prominent position in the increasingly popular *Semantic Web* and *Linked Data* infrastructures for publishing data. In fact, these paradigms motivate adaptive query processing for a further reason besides the scale of the data: *distributed querying* engines often concentrate loose federations of publicly-readable remote data sources over which the distributed querying engine cannot effect that histograms are maintained and published. Furthermore, the URIs of large-scale datasets are not hand-crafted names but are automatically generated following naming conventions, usually hierarchical. These observations both motivate extending adaptive query processing to Semantic Web data stores and also present an opportunity for our string prefix extension.

In the remainder, we review self-tuning histograms and optimized querying of Semantic Web data (Sect. 2) where we identify STHoles as our starting point, a very successful algorithm for multi-dimensional histograms of numerical data. In Sect. 3 we proceed to formalize the key concepts in STHoles in a way that subsumes STHoles as its specialization for numerical intervals; and to provide an extension to string prefixes. We then present experimental results using our prototype implementation (Sect. 4) and conclude (Sect. 5).

2   Background

In their simplest form, self-tuning histograms comprise a set of *buckets* where each bucket holds a range of value of a numerical attribute a and the number of results that fall within this range. Buckets are progressively refined from query execution feedback after each selection on a, using the difference between the actual result sizes and those estimated by the statistics in the buckets of a. Since actual values are not uniformly distributed along the range of values that a can have, high frequency buckets are split to smaller (i.e., with narrower range) and more accurate ones; to maintain memory usage, buckets with similar frequencies are merged to reclaim memory space. STGrid [1] extends these ideas to multi-dimensional self-tuning histograms that use query workloads to refine a grid-based histogram structure. These self-tuning histograms are a low-cost alternative to traditional histograms with comparable accuracy. However, since the splitting (or merging) of each bucket entails the splitting (or merging) of several other buckets that could be far away from and unrelated to the original one, overall accuracy is degraded in order to satisfy the grid-partitioning constraint.

To alleviate the poor bucket layout problem of STGrid, STHoles [2] allows buckets to overlap. This more flexible data structure allows STHoles to exploit feedback in a truly multi-dimensional way and is adopted by many subsequent algorithms [3,4], including the one presented here. STHoles allows for inclusion relationships between buckets, resulting in a tree-structured histogram where each node represents a bucket. Holes are sub-regions of a bucket with different tuple density and are buckets themselves. To refine an STHoles histogram, query results are used to count how many tuples fall inside each bucket of the current histogram. Each partial intersection of query results and a bucket can be used to refine the histogram by drilling new holes, whenever the query results diverge from the prediction made through the bucket's statistics.

In order to maintain a constant number of buckets, buckets with close tuple densities are merged to make space for new holes. A penalty function measures the difference in approximation accuracy between the old and the new histogram to choose which buckets to merge. Parent-child merges are useful to eliminate buckets that become too similar to their parents; sibling merges are useful to extrapolate frequency distributions to yet unseen regions in the data domain and also to consolidate buckets with similar density that cover nearby regions.

STHoles and, in general, workload-aware self-tuning histograms have been successfully used in relational databases as a low-overhead alternative to statically re-scanning database tables. The resulting histogram is focused towards the current workload, providing more accurate statistics for data regions that are being queried more frequently. Furthermore, they are able to adapt to changes in data distribution and thus are well-suited for datasets with frequently changing contents. They are, however, for the most part targeting numerical attributes, since they exploit the idea that a value range is an indication of the size of the range. Turning our attention to the Semantic Web, the *Resource Description Framework (RDF)* is the dominant standard for expressing information. RDF information is a graph where *properties* (labelled edges) link two resources to each other or one resource to a *literal* (a concrete value). The relevance of this discussion to self-tuning histograms is that RDF uses URIs as abstract symbols that denote resources. Given this prominent role of URIs in RDF data, extending self-tuning histograms to string attributes can have a significant impact in optimizing querying of RDF datasets.

There has been relatively limited amount of work around string selectivity estimation in the field of relational databases. Chaudhuri et al. [5] proposed to collect multiple candidate identifying substrings of a string using, for example, a Markov estimator and build a regression tree as a combination function of their estimated selectivities, in order to alleviate the selectivity underestimation problem of queries involving string predicates in previous methods, which used independence and Markov assumptions. In 2005, Lim et al. [6] introduced CXHist, which is a workload-aware histogram for selectivity estimation supporting a broad class of XML string-based queries. CXHist is the first histogram based on classification that uses feature distributions to summarize queries and quantize their selectivities into buckets and a naive-Bayes classifier to capture the mapping between queries and their selectivity.

Within the Semantic Web community itself, the SWOOGLE search engine collects metadata, such as classes, class instances and properties for web documents and relations between documents [7]. LODStats computes several schema-level statistical values for large-scale RDF datasets using an approach based on *statement streams* [8]. Work more closely related to ours includes RDFStats [9], which is a generator for statistics of RDF sources like SPARQL endpoints. They generate different statistical items such as instances per class and histograms. Unlike our approach, they generate different static histograms (i.e. that must be rebuilt to reflect any changes in the RDF source) per class, property and XML data type. For range estimations on strings, RDFStats mentions three possibilities: (a) one bucket for each distinct string, resulting in large histograms; (b) reducing strings to prefixes; or (c) using a hash function to reduce the number of distinct strings, although no appropriate general-purpose hash function has been identified. However, as Harth et al. [10] have also noted in relation to Q-Trees for indexing RDF triples, hashing URIs is a purely syntactic mapping from URIs to numerical coordinates and fails to take into account the semantic similarity between resources; and no universally good function has been identified.

As URIs the most prominent datatype in the Semantic Web, not having developed an extension that can naturally handle URI strings has left Semantic Web data outside the scope of many developments in self-tuning histograms.

3   Self-Tuning String Histograms

In this section we first introduce some basic definitions and then proceed to establish a new histogram structure that extends the structure of the STHoles algorithm with the ability to cover strings. We also present the algorithms that construct and refine this new structure. In our treatment, string ranges are specified by prefixes. Prefixes can naturally express ranges of semantically related resources, given the natural tendency to group together relevant items in hierarchical structures such as pathnames and URIs. Although this mostly contributes to applying selftuning histogram algorithms to Semantic Web data, we discuss how our structure is also amenable to regular expressions or other string patterns so that future extensions can address challenges from the wider data management community.

3.1   Preliminaries

Let D be a *dimension*, any subset of D be a *range* in D, and $\mathcal{P}(D)$ the set of all possible ranges in D. A range can be defined either implicitly by constraints over the values of D or explicitly by enumeration. Note that $D \in \mathcal{P}(D)$, meaning that a range does not *need* to impose a restriction but can also include the whole dimension. Let H be a histogram of n dimensions $D_1, \ldots D_n$. Let $V(H)$ be the set of all possible n-dimensional vectors $(r_1, \ldots r_n)$ where $\forall i \in [1, n] : r_i \in \mathcal{P}(D_i)$

A histogram is represented as an inclusion hierarchy of *buckets*; we shall use B_H to denote the set of buckets of a histogram H.

Definition 1. *Each bucket $b \in B_H$ is an entity of histogram H such that:*

- *b is associated with a* box$(b) \in V(H)$, *the vector that specifies the set of tuples that the bucket describes.*
- *b is associated with a* size(b) *which indicates the number of tuples that match* box(b).
- *b is associated with n values* dvc$(b, D_i), i = 1 \ldots n$ *which indicate the number of distinct values appearing in dimension D_i of the tuples that match* box(b).

We define the density of a bucket b to be the quantity

$$\text{density}(b) = \frac{\text{size}(b)}{\prod_{i:r_i \in \text{box}(b)} \text{dvc}(b, D_i)}.$$

Definition 2. *Every histogram implicitly includes a bucket $b_\top$ such that* box$(b_\top) = (D_1, \ldots D_n)$ *that is, the bucket that imposes no restrictions in any of the dimensions of H and includes all tuples. We call this the* top bucket $b_\top$.

The implication of Definition 2 is that the overall size of the dataset and the number of distinct values in each dimension should be known (or at least approximated) regardless of what query feedback has been received. In our implementation we assume the *root bucket* (the top-most bucket of the hierarchy) as an approximation of the top.

Let $\mathcal{Q}_H$ be the set of all possible queries over the tables covered by H. Regardless of how they are syntactically expressed, we perceive $\mathcal{Q}_H$ as the set of all possible restrictions over the dimensions of H; thus:

Definition 3. *Each query $q \in \mathcal{Q}_H$ is an entity of histogram H such that:*

- *q is associated with a* box$(q) \in V(H)$, *the vector that specifies the restrictions expressed by the query*
- *q is associated with a* size(q) *which indicates the number of tuples that are returned by executing q.*

In our approach, we assume *string prefixes* as the description language for implicitly defining string ranges. The motivation is discussed in the Introduction (Sect. 1). We use $\text{Pref}(r)$ to denote the set of prefixes that specify a string range. We define intersection between ranges and between boxes as follows:

Definition 4. *Given two ranges of the same dimension, $r_1, r_2 \in \mathcal{P}(D)$, their range intersection $r_1 \sqcap r_2$ is defined as follows:*

1. *If r_1, r_2 are string ranges defined by sets of prefixes, then $r_1 \sqcap r_2 = \{p | (p_1, p_2) \in$ $\mathrm{Pref}(r_1) \times \mathrm{Pref}(r_2) \wedge (p = p_1 = p_2 \vee$ one of p_1, p_2 is a prefix of the other and p is the longest (more specific) of the two)\}*
2. *If one of the ranges is a string range defined by sets of prefixes (say r_1 without loss of generality) and the other is an explicit set of strings (say r_2), then $r_1 \sqcap r_2 = \{v | v \in r_2 \wedge \exists p \in r_1 : p$ is a prefix of $v\}$*
3. *In any other case, $r_1 \sqcap r_2 = r_1 \cap r_2$.*

Definition 5. *Given two boxes $v_1, v_2 \in V(H)$ from the n-dimensional histogram H, let $v_1 = (r_{1,1}, \ldots r_{1,n})$ and $v_2 = (r_{2,1}, \ldots r_{2,n})$. We define box intersection:*

$$v_1 \sqcap v_2 = (r_{1,1} \sqcap r_{2,1}, \ldots r_{1,n} \sqcap r_{2,n}).$$

Definition 6. *Given two boxes $v_1, v_2 \in V(H)$ from the n-dimensional histogram H, let $v_1 = (r_{1,1}, \ldots r_{1,n})$ and $v_2 = (r_{2,1}, \ldots r_{2,n})$. We say that v_1 encloses v_2 iff $\forall i \in [1, n]$ at least one of the following holds:*

1. *$r_{2,i} \subseteq r_{1,i} \subset D_i$, that is, none of the ranges is the complete dimension and $r_{2,i}$ is contained within $r_{1,i}$.*
2. *$r_{2,i} = D_i$ and $r_{1,i} \subset D_i$, that is, if one of the ranges is the complete dimension then it is enclosed by the one that is not.*
3. *$r_{2,i} = r_{1,i} = D_i$, that is, both ranges are the complete dimension.*

It should be noted that we have defined an unrestricted dimension as *being enclosed by* (rather than *enclosing*) a restriction. The rationale behind this will be revisited in conjunction with *bucket merging* (Sect. 3.4).

Definition 7. *Given two boxes $v_1, v_2 \in V(H)$ from histogram H, v_1 tightly encloses v_2 iff v_1 encloses v_2 and there is no $u \in V(H)$ such that $v_1 \supsetneq u \supsetneq v_2$.*

Definition 8. *Given a query $q \in \mathbb{Q}_H$, we associate with q the best fit, the set of buckets $\mathrm{bf}(q) \subseteq \mathbb{B}_H$ such that*

$$\forall b \in \mathrm{bf}(q) : \mathrm{box}(b) \text{ tightly encloses } \mathrm{box}(q).$$

Lemma 1. *For every query there is always a non-empty best fit.*

Proof. There is always at least one bucket that *encloses* any $\mathrm{box}(q)$, the *top bucket* $b_\top$ (Definition 2). If there is no other bucket that *encloses* $\mathrm{box}(q)$, then $b_\top$ *tightly encloses* $\mathrm{box}(q)$ (Definition 7) and thus $\mathrm{bf}(q) = \{b_\top\}$, which is non-empty. If there are other buckets that *enclose* $\mathrm{box}(q)$, then there is also at least one that *tightly encloses* $\mathrm{box}(q)$, so $\mathrm{bf}(q)$ is non-empty.

3.2 Cardinality Estimation

Being able to predict the size of querying results is important input for query execution optimizers, but the specifics of how this optimization is performed is outside the scope of this paper. We will here proceed to define metrics over the values associated with the buckets of H in order to predict $size(q), q \in \mathcal{Q}_H$, the number of results returned by q. More specifically, we will define an extension of self-tuning histograms that handles string values in a more sophisticated manner than explicit ranges.

In the literature, numerical intervals are used to succinctly define ranges and efficiently decide if a query is enclosed by a bucket or not. The numerical difference between the interval's starting and ending value is sometimes used to define *range length* and, in multi-dimensional buckets, *bucket volume*: an estimator of the number of tuples in a bucket. Categorical dimensions (including strings) are, on the other hand, either not handled at all or only explicitly defined as sets of values, with bucket volume defined using the cardinality of the value range. Given this, we define range length as follows:

Definition 9. *Given a histogram dimension D and a range $r \in \mathcal{P}(D)$ we define the function* $length : \mathcal{P}(D) \to \mathbb{R}$ *as follows:*

1. *Unrestricted ranges that span the whole dimension have length 0.*
2. *If r is a numerical range defined by interval $[x, y]$, then $length(r) = y - x + 1$. The addition of the unit term guarantees that the length cannot be zero even if $x = y$, i.e., even if the numerical range is a single number.*
3. *If r is a string range defined by a set of prefixes $\mathrm{Pref}(r)$, then $length(r) = |1 + \mathrm{Pref}(r)|$, the number of prefixes defining the range.*
4. *In any other case, including explicitly defined numerical and string ranges, then $length(r) = |r|$, the number of distinct values in the range.*

What should be noted in Definition 9 is that the only situation in which the length of a range can be 1 is when it is an individual value. This is important for Definition 10 immediately below:

Definition 10. *Given a histogram H, we define the function* $est_H : V(H) \to \mathbb{R}$ *as follows:*

$$est_H(q) = \sum_{b \in bf(q)} \frac{size(b)}{\prod_{i:length(r_i)=1} dvc(b, D_i)}$$

We propose this function as an estimator of the number of tuples that lie inside q, given a histogram.

3.3 Histogram Construction and Refinement

The construction of the histogram follows the same high level steps as the STHoles algorithm. In particular, we start with an empty histogram. For each query q in the workload, we identify *candidate buckets* b_i that intersect with q.

If the root bucket b_0 does not fully enclose q, we expand its bounding box so that it covers q and we update its statistics as follows:

$$size(b_0) := size(b_0) + size(q)$$
$$dvc(b_0, D_i) := \max\{dvc(b_0, D_i), dvc(q, D_i)\}$$

Algorithm 1. Refinement of a histogram H given a set of queries W.

---

procedure REFINE(H,W)
    for all queries $q \in W$ **do**
        if q is not contained in H **then**
            expand H's root bucket so that it contains q
        for all buckets b_i such that $q \sqcap b_i \neq \emptyset$ **do**
            $(c_i, T_{c_i}, d_{c_i}) \leftarrow$ SHRINKBUCKET(b_i, q)
            if estimation is not accurate **then**
                DRILLHOLE($b_i, c_i, T_{c_i}, d_{c_i}$)
    while H has too many buckets **do**
        Let b_1, b_2 in H with the lowest penalty$_H(b_1, b_2)$
        MERGE(b_1, b_2)

Algorithm 2. Drilling a hole in bucket b, given a candidate hole c and the counted cardinality T_c and distinct values $D_c(i)$ for each dimension D_i.

---

procedure DRILLHOLE(b, c, T_c, $d_c(\cdot)$)
    if box(b) = box(c) **then**
        size(b) $\leftarrow T_c$
        dvc(b, D_i) $\leftarrow D_c(i)$ $\forall i \in attributes$
    else
        Add a new child b_n of b to the histogram
        box(b_n) $\leftarrow c$
        size(b_n) $\leftarrow T_c$
        dvc(b_n, D_i) $\leftarrow d_c(i)$ $\forall i \in attributes$
        Migrate all children of b that are enclosed by c
        so they become children of b_n

---

where size($\cdot$) is the number of tuples and dvc($\cdot$) is the number of distinct values inside buckets and query results (Definition 1).

For each candidate bucket b_i we compute $b_i \sqcap q$ and these intersections constitute *candidate holes* c_i. We then shrink each candidate hole to the largest sub-region that does not intersect with the box of any other bucket, we count the exact number of tuples from the result stream that lie inside the shrunk hole and the distinct values count. Then, we determine whether the current density of the candidate bucket is close to the actual density of the candidate hole. If not, we 'drill' the candidate hole as a new histogram bucket and we move all children of b_i that are enclosed by c_i to the new bucket (Algorithms 1 and 2).

A point of divergence from STHoles is when shrinking candidate holes. Let X be the set of all buckets that partially intersect with candidate hole c_i. STHoles selects at each step the pair $\langle x, j \rangle$ that comprises bucket $x \in X$ and dimension j such that shrinking c_i along j by excluding x has as a result the smallest reduction of c_i. Instead of checking for the optimal $\langle x, j \rangle$ our method selects the first pair where shrinking c_i along j by excluding x results in the smallest *relative* reduction of c_i's *length* in that dimension, the intuition being that often excluding x will give similar relative reduction along all dimensions. We then shrink c_i, we

---

Algorithm 3. Shrink a bucket that is enclosed by the intersection of b and q and does not partially intersect any other bucket.

---

function SHRINKBUCKET(b, q)

    $c \leftarrow \text{box}(q) \cap \text{box}(b)$

    $\mathcal{P} \leftarrow \{b_i \in children(b) \mid c \cap \text{box}(b_i) \neq \emptyset \land \text{box}(b_i) \not\subseteq c\}$

    while $\mathcal{P} \neq \emptyset$ **do**

        Get first bucket $b_i \in \mathcal{P}$ and dimension j

        such that shrinking c along j by excluding b_i results

        in the smallest reduction of c.

        Shrink c along j

        $\mathcal{P} \leftarrow \{b_i \in children(b) \mid c \cap \text{box}(b_i) \neq \emptyset \land \text{box}(b_i) \not\subseteq c\}$

    Count from the result the number of tuples in c, T_c

    for all attributes i **do**

    Count from the result the number of

    distinct values of the ith attribute in c, $d_c(i)$.

  return $(c, T_c, d_c(\cdot))$

---

update participants and repeat the procedure until there are no participants left (Algorithm 3). This may result in a suboptimal shrink, but we avoid examining all possible combinations at each step. Furthermore, in STHoles the number of tuples in this shrunk subregion is estimated assuming uniformity; instead, we measure exactly the number of tuples and distinct values per dimension.

3.4 Bucket Merging

In order to limit the number of buckets and memory usage, buckets are *merged* to make space for drilling new holes. Following STHoles, our method looks for *parent-child* or *sibling* buckets that can be merged with minimal impact on the cardinality estimations. We diverge from STHoles when computing the box, size, and dvc associated with the merged bucket as well as in the *penalty* measure that guides the merging process towards merges that have the smallest impact.

Let b_1, b_2 be two buckets in the n-dimensional histogram H and let H' be the histogram after the merge and b_m the bucket in H' that replaces b_1 and b_2. In the *parent-child* case, one of the two buckets, let that be b_1, *tightly encloses* the other. In this case, we merge b_2 into b_1, so that $\text{box}(b_m) \equiv \text{box}(b_1)$.

Our *penalty* measure should be proportionate to the absolute differences between the cardinality estimates made from the two original buckets and the cardinality estimates made from the resulting bucket. Let Q_1, Q_2 be two sets of queries that would have b_1 and b_2 resp. be involved in their cardinality estimation. That is: $\forall k \in \{1, 2\} \forall q \in Q_k : b_k \in \text{bf}(q)$

Furthermore, Q_k contain all and only the n queries so that for each dimension $D_i, i = 1, \ldots n$ of H there is exactly one query that:

- has a specific value for the dimension D_i and this value is inside the range of $\text{box}(b_k)$ for D_i. It is not important what this particular value is, since the estimation is the same for all values inside the bucket's box's range.
- is completely unspecified for all other dimensions.

We define the penalty to be the sum over all $q \in Q_k$ of the estimation error differences between H and H'. As seen in Definition 10, only the terms that involve b_k are significant for calculating the difference, and thus:

$$\text{penalty}_H(b_1, b_2) = \sum_{q \in Q_1} |\text{est}_H(q) - \text{est}_{H'}(q)| + \sum_{q \in Q_2} |\text{est}_H(q) - \text{est}_{H'}(q)|$$

$$= \sum_{i \in [1,n]} \left| \frac{\text{size}(b_1)}{\text{dvc}(b_1, i)} - \frac{\text{size}(b_m)}{\text{dvc}(b_m, i)} \right| + \sum_{i \in [1,n]} \left| \frac{\text{size}(b_2)}{\text{dvc}(b_2, i)} - \frac{\text{size}(b_m)}{\text{dvc}(b_m, i)} \right|$$

$$= \sum_{i \in [1,n]} \left| \frac{\text{size}(b_1)}{\text{dvc}(b_1, i)} - \frac{\text{size}(b_2)}{\text{dvc}(b_2, i)} \right|$$

In *sibling* merges, let b_p be the bucket that *tightly encloses* both b_1 and b_2. The merged bucket encloses siblings b_1 and b_2 and is also extended to enclose any further siblings b_c that partially overlap with either b_1 or b_2. The size of b_m is estimated by adding; the distinct values count of b_m is estimated by the maximum distinct values count among the merged buckets:

1. box(b_p) *tightly encloses* box(b_m)
2. box(b_m) *tightly encloses* both buckets b_1, b_2
3. box(b_m) *tightly encloses* the boxes of all children of b_p that either of b_1, b_2 partially interstects with. That is, box(b_m) encloses box(b_c) for all b_c such that (i) b_p *tightly encloses* b_c; and (ii) box(b_1) *partially overlaps* box(b_c) or box(b_2) *partially overlaps* box(b_c)
4. $\text{size}(b_m) = \sum_{k=1,2,c_1,\ldots} \text{size}(b_k)$
5. $\text{dvc}(b_m) = \max_{k=1,2,c_1,\ldots} \text{dvc}(b_k)$.

In Point 3 above, it should be stressed that the partially intersecting buckets b_c are *not* merged into b_m, but that the latter is expanded so that it can assume b_c as its children. This is because in some algorithms (including STHoles), box(b_m) can become larger than box(b_1)$\cup$box(b_2) in order to have a 'rectangular' description with a single interval in each dimension. As a result, it might cut across other buckets; box(b_m) should then be extended so as to subsume those as children. In order to avoid, however, dropping informative restrictions, STHoles only extends box(b_m) along dimensions where the boxes of b_c do have a restriction. In order to capture this, we have defined the *encloses* relation (Definition 6) in a way that makes unrestricted dimensions *enclosed by* (rather than enclosing) restrictions. In analogy to the parent-child case, we define the penalty to be the sum over all $q \in Q_k$ of the estimation error differences between H and H':

$$\text{penalty}_H(b_1, b_2) = \sum_{q \in Q_1} |\text{est}_H(q) - \text{est}_{H'}(q)| + \sum_{q \in Q_2} |\text{est}_H(q) - \text{est}_{H'}(q)| =$$

$$\sum_{i \in [1,n]} \left| \frac{\text{size}(b_1)}{\text{dvc}(b_1, i)} - \frac{\sum_{k=1,2,c_1,\ldots} \text{size}(b_k)}{\max_{k=1,2,c_1,\ldots} \text{dvc}(b_k, i)} \right| + \sum_{i \in [1,n]} \left| \frac{\text{size}(b_2)}{\text{dvc}(b_2, i)} - \frac{\sum_{k=1,2,c_1,\ldots} \text{size}(b_k)}{\max_{k=1,2,c_1,\ldots} \text{dvc}(b_k, i)} \right|.$$

3.5   Discussion

We have defined a multi-dimensional histogram over numerical, string, and categorical data. The core added value of this work is that we introduce the notion of *descriptions* in string dimensions, akin to intervals for numerical dimensions. This has considerable advantages for RDF stores and, more generally, in the Semantic Web and Linked Open Data domain, where URIs have a prominent role and offer the opportunity to exploit the hierarchical structure of their string representation.

We propose *prefixes* as the formalism for expressing string ranges, motivated by its applicability to URI structure. This is no loss of generality, since it is straightforward to use more expressive pattern formalisms (such as regular expressions) without altering the core method but at a considerable computational cost. The only requirement is that *intersection* and some notion of *length* can be defined. Length, in particular, can be used in the way STHoles uses it as an indication of a bucket's size relative to the size of its parent bucket, although that also depends on the specifics of the estimation functions and is not required. In the function we propose (Definition 10), the only requirement is that length equals 1 for fixed singleton values and more than 1 for intervals and prefixes.

This is related to the fact that for range queries we return the statistics of the bucket that more tightly encloses the query, instead of returning an estimation based on the ratio of the volume occupied by the query to the volume of the overall bucket. In other words, we use *length* more as a metric of the size of description, rather than a metric of the bucket size (the number of tuples that fit this description). To compensate, we exactly measure in query results (rather than estimate) bucket size when shrinking buckets, compensating for the extra computational time by avoiding examining all combinations of buckets × dimensions (cf. Sect. 3.3).

Furthermore, we can afford a larger number of buckets in the same amount of memory (and thus more accurate estimations) than if strings were treated as categorical values.

A further improvement would be to define the *length* of string prefixes in a way that multi-dimensional volume calculated from these lengths accurately estimates the number of tuples in the buckets. An even more ambitious future goal is to define the length of string ranges in a way that it can be combined with numerical range length, so that multi-dimensional and heterogeneous (strings and numbers) buckets can be assigned a meaningful volume.

A limitation of our algorithm is that when we merge two sibling buckets we assign to the resulting bucket the sum of the sizes of the merged buckets and of the children of the resulting bucket, which is an overestimation of the real size. Furthermore, we also assign as distinct value count the maximum of the distinct value counts of these buckets, which is an underestimation of the real distinct value count. These estimations will persist until subsequent workload queries effect an update of merged bucket's statistics and will be used in cardinality estimations. Besides empirically testing and tuning these estimators, we are also planning to extend the theoretical framework so that estimated values are represented as ranges or distributions, and subsequent calculations take into account the whole range or the distribution parameters rather than a single value.

In general, and despite these limitations, our framework is an accurate theoretical account of STHoles, a state-of-the-art algorithm for self-tuning multi-dimensional numerical histograms, and an extension to heterogeneous numerical/string histograms that is backwards-compatible with STHoles.

4   Experiments

To empirically validate our approach, the algorithm presented above has been implemented in Java as the *STRHist* module of the *Semagrow Stack*, an optimized distributed querying system for the Semantic Web.[1] The execution flow of the Semagrow Stack starts with client queries, analysed to build an optimal *query plan*. The optimizer relies on cardinality statistics (produced by STRHist) in order to provide an execution plan for the Semagrow *Query Execution Engine*. This engine, besides joining results and serving them to the client application, also forwards to STRHist measurements collected during query execution. STRHist analyses these query feedback logs in batches to maintain the histograms that are used by the optimizer. The histogram is persisted in RDF stores using the Sevod vocabulary [11], but STRHist effects updates on an in-memory representation. This representation is a tree of *bucket* objects, starting for a single root. For our proof-of-concept system we assume that all the buckets fit into the main memory of a single machine.

The expected behaviour of the algorithm is to improve estimates by adding buckets that punch holes and add sub-buckets in areas where there is a difference between the actual statistics and the histogram estimates. Considering how client applications access some 'areas' more heavily than others, the algorithm zooms into such critical regions to provide more accurate statistics. Naturally, the more interesting observations relate to the effect of merges as soon as the available space is exhausted.

We applied STRHist to the AGRIS bibliographic database on agricultural research and technology maintained by the *Food and Agriculture Organization of the UN*. AGRIS comprises approximately 23 million RDF triples describing 4 million distinct publications with standard bibliographic attributes.[2] AGRIS consolidates data from more than 150 institutions from 65 countries. Bibliography items are denoted by URIs that are constructed following a convention that includes the location of the contributing institution and incorporation AGRIS. As scientific output increases through the years and since there is considerable variation in the countries, there are interesting generalizations to be captured by patterns over publication URIs.

---

[1] *STRHist* is available at https://bitbucket.org/acharal/strhist

  For more details on Semagrow, please see http://www.semagrow.eu.

[2] Please see http://agris.fao.org for more details on AGRIS. The AGRIS site mentions 7 million distinct publications, but this includes recent additions that are not in end-2013 data dump used for these experiments.

Table 1. Estimation error (RMS and absolute) versus training batch and merges (Parent-Child and Sibling merges). Configured for a maximum of 100 buckets.

Training	Error		Merges			Training	Error		Merges		
Batch	RMS	Abs	P-C	Sib	Total	Batch	RMS	Abs	P-C	Sib	Total
01	0.769	5.402	0	0	0	13	1.426	9.660	393	305	698
02	0.509	3.959	0	0	0	14	1.442	9.794	470	322	792
03	0.653	4.938	12	39	51	15	1.444	9.835	552	341	893
04	0.784	5.691	27	77	104	16	1.517	10.320	620	369	989
05	0.806	5.876	39	107	146	17	1.536	10.464	685	395	1080
06	0.958	6.897	73	135	208	18	1.506	10.134	755	409	1164
07	1.076	7.505	112	160	272	19	1.506	10.124	810	430	1240
08	1.279	8.814	184	176	360	20	1.484	10.031	878	451	1329
09	1.294	8.773	211	203	414	21	1.438	09.495	949	468	1417
10	1.260	8.598	235	232	467	22	1.450	09.371	1031	492	1523
11	1.377	9.474	285	260	545	23	1.452	09.412	1121	511	1632
12	1.447	9.938	345	280	625	24	1.448	09.361	1207	534	1741

We define a 3-dimensional histogram over subject, predicate and object variables. Subject URIs are represented as strings[3] while predicate URIs are treated as categorical values, since there is always a small number of distinct predicates. Each bucket is composed of a 3-dimensional subject/predicate/object bounding box, a size indicating the number of triples contained in the bucket, and the number of distinct subjects, predicates and objects.

We automatically generated a workload of query feedback as follows: we randomly select a URI in the dataset and construct a prefix by trimming the URI string to a random length. The query feedback record is constructed by selecting the fragment of the dataset that satisfies the aforementioned prefix. The evaluation workload W is generated by randomly selecting subject URIs from the dataset DB and given a histogram H over D we estimate the size of the results of hypothetical queries that select rows that have this particular value. We then measure the *average absolute estimation error* and the *root mean square error* of histogram H on the evaluation workload W:

$$\mathrm{err}_{H,D}^{ABS}(W) = \frac{1}{|W|} \sum_{q \in W} |\mathrm{est}_H(q) - \mathrm{act}_D(q)|$$

$$\mathrm{err}_{H,D}^{RMS}(W) = \frac{1}{|W|} \sqrt{\sum_{q \in W} (\mathrm{est}_H(q) - \mathrm{act}_D(q))^2}$$

where $\mathrm{est}_H(q)$ is the cardinality estimation for query q and $\mathrm{act}_D(q)$ is the actual number of tuples in D that satisfy q.

We created 24 batches of 50 training queries, totalling 1200 training queries, followed by a set of 100 evaluation queries used to compare the estimations

---

[3] We use the *canonical string representation* of URIs as defined in Sect. 2, IETF RFC 7320 (http://tools.ietf.org/html/rfc7320).

Table 2. Estimation error (RMS and absolute) versus memory usage, after executing all 24 batches.

Max	Error		Num of	Max	Error		Num of
Buckets	RMS	Abs	Merges	Buckets	RMS	Abs	Merges
100	1.44844	9.36082	1741	600	0.50146	3.13402	561
200	1.06211	6.26804	1223	700	0.41762	2.75258	467
300	0.83218	4.72165	954	800	0.39674	2.73196	347
400	0.66517	4.01031	796	900	0.38835	2.61856	78
500	0.59339	3.63918	666	1000	0.28196	2.00000	0

against the actual size of the query results. Table 1 gives the estimation error over time and the number of merges performed; the system is configured for a maximum of 100 buckets. In the first batches the error naturally decreases, as the histogram is only getting more informed. When the merges start to happen as a result of the limited resources the estimation error increases. However, we can observe that after a while the estimation error stabilizes. Another observation is that the parent-child merges are eventually preferred more than the sibling merges. In Table 2, we give the results after all training batches but for different memory sizes. As expected, the average decreases as the available buckets increase showing that the algorithm is able to utilise available memory to improve estimations.

5   Conclusions

In this paper, we have presented an algorithm for building and maintaining multi-dimensional histograms exploiting query feedback. Our algorithm is based on STHoles algorithm, but can also handle URIs by using prefixes as a succinct description of string ranges and store additional statistics (distinct value counts) as well.

As future work, we would like to use more expressive pattern formalisms (such as regular expressions), in order to describe sub-spaces of the URI space and better exploit naming conventions of URIs. It would be also interesting to identify clusters of URIs in the result stream, and therefore describe a string range using a set of prefixes/regular expressions, instead of a single prefix/regular expression. We will also try to lift some of the limitations described above. Furthermore, our algorithm needs to be evaluated in terms of accuracy, convergence, efficiency and adaptability to dataset changes, using automatically generated queries to form a workload for experimentation. Query feedback will be obtained from different real data sources in order to test the response of the system. Among them, there will be constrained and unconstrained queries with focus on the variability of subject and object values against constant predicates. Last but not least, we plan to provide a more scalable and robust implementation of the algorithm, which will be able to serve detailed and accurate data summaries and statistics about extremely large datasets.

Acknowledgements. The work described here was partially carried out at the 2014 edition of the *International Research-Centred Summer School*, held at NCSR 'Demokritos', Athens, Greece, 3–30 July 2014. For more details please see http://irss.iit. demokritos.gr

The research leading to these results has received funding from the European Union's Seventh Framework Programme (FP7/2007–2013) under grant agreement No. 318497. More details at http://www.semagrow.eu.

References

1. Aboulnaga, A., Chaudhuri, S.: Self-tuning histograms: Building histograms without looking at data. In: Proceedings of the 1999 ACM International Conference on Management of Data (SIGMOD 1999), pp. 181–192. ACM (1999)
2. Bruno, N., Chaudhuri, S., Gravano, L.: STHoles: a multidimensional workload-aware histogram. In: Proceedings of 2001 ACM International Conference on Management of Data (SIGMOD 2001), pp. 211–222. ACM (2001)
3. Srivastava, U., Haas, P.J., Markl, V., Kutsch, M., Tran, T.M.: ISOMER: Consistent histogram construction using query feedback. In: Proceedings of the 22nd International Conference on Data Engineering (ICDE 2006). IEEE Computer Society (2006)
4. Roh, Y.J., Kim, J.H., Chung, Y.D., Son, J.H., Kim, M.H.: Hierarchically organized skew-tolerant histograms for geographic data objects. In: Proceedings of 2010 ACM International Conference on Management of Data (SIGMOD 2010), pp. 627–638. ACM (2010)
5. Chaudhuri, S., Ganti, V., Gravano, L.: Selectivity estimation for string predicates: Overcoming the underestimation problem. In: Proceedings of 20th International Conference on Data Engineering (ICDE 2004). IEEE Computer Society (2004)
6. Lim, L., Wang, M., Vitter, J.S.: CXHist: An on-line classification-based histogram for XML string selectivity estimation. In: Proceedings of the 31st International Conference on Very Large Data Bases (VLDB 2005), Trondheim, Norway, 30 August – 2 September 2005, pp. 1187–1198 (2005)
7. Ding, L., Finin, T., Joshi, A., Pan, R., Cost, R.S., Peng, Y., Reddivari, P., Doshi, V., Sachs, J.: Swoogle: A search and metadata engine for the Semantic Web. In: Proceedings of the 13th ACM International Conference on Information and Knowledge Management (CIKM 2004), pp. 652–659. ACM (2004)
8. Auer, S., Demter, J., Martin, M., Lehmann, J.: LODStats – an extensible framework for high-performance dataset analytics. In: ten Teije, A., Völker, J., Handschuh, S., Stuckenschmidt, H., d'Acquin, M., Nikolov, A., Aussenac-Gilles, N., Hernandez, N. (eds.) EKAW 2012. LNCS, vol. 7603, pp. 353–362. Springer, Heidelberg (2012)
9. Langegger, A., Wöss, W.: RDFStats - an extensible RDF statistics generator and library. In: Proceedings of DEXA 2009, pp. 79–83 (2009)
10. Harth, A., Hose, K., Karnstedt, M., Polleres, A., Sattler, K.U., Umbrich, J.: Data summaries for on-demand queries over linked data. In: Proceedings of 19th International World Wide Web Conference (WWW 2010), Raleigh, NC, USA, 26–30 April 2010 (2010)
11. Charalambidis, A., Konstantopoulos, S., Karkaletsis, V.: Dataset descriptions for optimizing federated querying. In: Poster Track, Companion Volume to the Procedings of the 24th Intl World Wide Web Conference (WWW 2015), Florence, Italy, 18–22 May 2015. ACM (2015)

Data Mining I

Data Partitioning for Fast Mining of Frequent Itemsets in Massively Distributed Environments

Saber Salah, Reza Akbarinia, and Florent Masseglia[✉]

Zenith Team, INRIA and LIRMM, University of Montpellier, Montpellier, France
{saber.salah,reza.akbarinia,florent.masseglia}@inria.fr

Abstract. Frequent itemset mining (FIM) is one of the fundamental cornerstones in data mining. While, the problem of FIM has been thoroughly studied, few of both standard and improved solutions scale. This is mainly the case when (i) the amount of data tends to be very large and/or (ii) the minimum support (*MinSup*) threshold is very low. In this paper, we propose a highly scalable, parallel frequent itemset mining (PFIM) algorithm, namely Parallel Absolute Top Down (PATD). PATD algorithm renders the mining process of very large databases (up to Terabytes of data) simple and compact. Its mining process is made up of only one parallel job, which dramatically reduces the mining runtime, the communication cost and the energy power consumption overhead, in a distributed computational platform. Based on a clever and efficient data partitioning strategy, namely Item Based Data Partitioning (IBDP), PATD algorithm mines each data partition independently, relying on an absolute minimum support (*AMinSup*) instead of a relative one. PATD has been extensively evaluated using real-world data sets. Our experimental results suggest that PATD algorithm is significantly more efficient and scalable than alternative approaches.

Keywords: Machine learning · Data mining · Frequent itemset · Big data · MapReduce

1 Introduction

Since a few decades, the amount of data in the world and our lives seems ever-increasing. Nowadays, we are completely overwhelmed with data, it comes from different sources, such as social networks, S sensors, etc. With the availability of inexpensive storage and the progress that has been made in data capture technology, several organizations have set up very large databases, known as Big Data [1]. The processing of this massive amount of data, helps leveraging and uncovering hidden relationships, and brings up new, and useful information. Itemsets are one of these tackled levers and consist in frequent correlations of features. Their discovery is known as Frequent itemset mining (FIM for short), and presents an essential and fundamental role in many domains. In business

S. Slah—This work has been partially supported by the Inria Project Lab Hemera.

© Springer International Publishing Switzerland 2015
Q. Chen et al. (Eds.): DEXA 2015, Part I, LNCS 9261, pp. 303–318, 2015.
DOI: 10.1007/978-3-319-22849-5_21

and e-commerce, for instance, FIM techniques can be applied to recommend new items, such as books and different other products. In science and engineering, FIM can be used to analyze such different scientific parameters (e.g., based on their regularities). Finally, FIM methods can help to perform other data mining tasks such as text mining [2], for instance, and, as it will be better illustrated by our experiments in Sect. 4, FIM can be used to figure out frequent co-occurrences of words in a very large-scale text database. However, the manipulation and processing of large-scale databases have opened up new challenges in data mining [3]. First, the data is no longer located in one computer, instead, it is distributed over several machines. Thus, a parallel and efficient design of FIM algorithms must be taken into account. Second, parallel frequent itemset mining (PFIM for short) algorithms should scale with very large data and therefore very low $MinSup$ threshold. Fortunately, with the availability of powerful programming models, such as MapReduce [4] or Spark [5], the parallelism of most FIM algorithms can be elegantly achieved. They have gained increasing popularity, as shown by the tremendous success of Hadoop [6], an open-source implementation. Despite the robust parallelism setting that these solutions offer, PFIM algorithms remain holding major crucial challenges. With very low $MinSup$, and very large data, as will be illustrated by our experiments, most of standard PFIM algorithms do not scale. Hence, the problem of mining large-scale databases does not only depend on the parallelism design of FIM algorithms. In fact, PFIM algorithms have brought the same regular issues and challenges of their sequential implementations. For instance, given best FIM algorithm X and its parallel version X'. Consider a very low $MinSup$ δ and a database $\mathcal{D}$. If X runs out of memory in a local mode, then, with a large database $\mathcal{D}'$, X' might also exceed available memory in a distributed mode. Thus, the parallelism, all alone, does not guarantee a successful and exhaustive mining of large-scale databases and, to improve PFIM algorithms in MapReduce, other issues should be taken into account. Our claim is that the data placement is one of these issues. We investigate an efficient combination between a mining process (i.e., a PFIM algorithm) and an efficient placement of data, and study its impact on the global mining process.

We have designed and developed a powerful data partitioning technique, namely Item Based Data Partitioning (IBDP for short). One of the drawbacks of existing PFIM algorithms is to settle for a disjoint placement. IBDP allows, for a given item i to be placed in more than one mapper if necessary. Taking the advantages from this clever data partitioning strategy, we have designed and developed a MapReduce based PFIM algorithm, namely Parallel Absolute Top Down Algorithm (PATD for short), which is capable to mine a very large-scale database in just one simple and fast MapReduce job. We have evaluated the performance of PATD through extensive experiments over two massive datasets (up to one Terabyte and half a billion Web pages). Our results show that PATD scales very well on large databases with very low minimum support, compared to other PFIM alternative algorithms.

The rest of the paper is organized as follows. Section 2 gives an overview of FIM problem, basic used notations, and the necessary background. In Sect. 3, we propose our PATD algorithm and we depict its whole core working process. Section 4 reports on our experimental validation over real-world data sets. Section 5 discusses related work, and Sect. 6 concludes.

2   Definitions and Background

In this section, we set up the basic notations and terminology, that we are going to adopt in the rest of the paper.

The problem of FIM was first introduced in [7], and then manifold algorithms have been proposed to solve it. In Definition 1, we adopt the notations used in [7].

Definition 1. *Let $\mathcal{I} = \{i_1, i_2, \ldots, i_n\}$ be a set of literals called items. An Itemset X is a set of items from $\mathcal{I}$, i.e. $X \subseteq \mathcal{I}$. The size of the itemset X is the number of items in it. A transaction T is a set of elements such that $T \subseteq \mathcal{I}$ and $T \neq \emptyset$. A transaction T supports the item $x \in \mathcal{I}$ if $x \in T$. A transaction T supports the itemset $X \subseteq \mathcal{I}$ if it supports any item $x \in X$, i.e. $X \subseteq T$. A database $\mathcal{D}$ is a set of transactions. The support of the itemset X in the database $\mathcal{D}$ is the number of transactions $T \in \mathcal{D}$ that contain X. An itemset $X \subseteq \mathcal{I}$ is frequent in $\mathcal{D}$ if its support is equal or higher than a MinSup threshold. A maximal frequent itemset is a frequent itemset that has no frequent superset.*

The FIM problem consists of extracting all frequent itemset from a database $\mathcal{D}$ with a minimum support $MinSup$ specified as a parameter.

Example 1. Let consider a database $\mathcal{D}$ with 5 transactions as shown in Table 1. The items in each presented transaction are delimited by commas. With a minimum support of 3, there will be no frequent items (and no frequent itemsets). With a minimum support of 2, there will be 8 frequent itemsets: $\{\{\{a\}, \{b\}, \{c\}, \{f\}, \{g\}, \{a, c\}, \{b, c\}, \{f, g\}\}\}$.

Table 1. Database $\mathcal{D}$

TID	Transaction
T_1	a, b, c
T_2	a, c, d
T_3	b, c
T_4	e, f, g
T_5	a, f, g

In this paper, we focus on parallel frequent itemset mining problem, where the data is distributed over several computational machines. We have adopted MapReduce as a programming model to illustrate our mining approach, however, we strongly believe that our proposal would have good performance results in other parallel frameworks too (*e.g.* Spark).

3 Parallel Absolute Top down Algorithm

As we briefly mentioned in Sect. 1, using an efficient data placement technique, could significantly improve the performance of PFIM algorithms in MapReduce. This is particularly the case, when the logic and the principle of a parallel mining process is highly sensitive to its data. For instance, let consider the case when most of the workload of a PFIM algorithm is being performed on the mappers. In this case, the way the data is exposed to the mappers, could contribute to the efficiency and the performance of the whole mining process (i.e., invoked PFIM algorithm).

In this context, we point out to the data placement, as a custom placement of database transactions in MapReduce. To this end, we use different data partitioning methods. We illustrate the impact of data placement techniques on the performance of PFIM algorithms, by considering particular PFIM algorithms which are based on two MapReduce jobs schema (2-Jobs schema for short).

In this section, first, we investigate the impact of partitioning data (i.e., impact of data placement) on 2-Jobs schema. Second, we introduce our IBDP method for data partitioning, and then we detail its working logic and principle. Finally, we introduce PATD algorithm and elucidate its design and core mining process in MapReduce.

3.1 Impact of Partitioning Data on 2-Jobs Schema

Performing a mining process in two steps was first proposed in [8] and it was designated for centralized environments. SON algorithm [8] divides a mining process as follows:

- **Step 1:** Divide the input database $\mathcal{D}$ into n data chunks (i.e., data splits), where $\mathcal{D} = \{P_1, P_2, \ldots, P_n\}$. Then, mine each data chunk (P_i) in the memory, based on a local minimum support $(LMinSup)$, and a specific FIM algorithm. Thus, the first step of SON algorithm is to determine a list of local frequent itemsets (LFI).
- **Step 2:** From previous step result, proceed by filtering the local frequent itemsets in LFI list, based on a global minimum support $GMinSup$. This may be done with a scan on $\mathcal{D}$ and checking the frequency of each itemset is LFI. The main idea is that any frequent itemset on $\mathcal{D}$ will be frequent on at least one chunk P_i and will be found in LFI. Then, return a list of global frequent itemsets (GFI) which is a subset of LFI $(GFI \subseteq LFI)$.

In a massively distributed environment, the main bottleneck of such 2-Jobs schema PFIM algorithm is its first execution phase, where a FIM algorithm has to be executed on the chunks. The choice of this algorithm is therefore crucial. Relying on SON mining principle, we have implemented a parallel version of CDAR [9] and Apriori [7] algorithms on MapReduce, namely, Parallel Two Round CDAR (P2RC) and Parallel Two Round Apriori (P2RA) respectively. Each version makes use of CDAR or Apriori on the chunks in the first phase. P2RC divides the mining process into two MapReduce jobs as follows:

■ **Job 1:** In the first phase, the principle of CDAR (see [9] for more details) is adapted to a distributed environment. A global minimum support $GMinSup$ Δ is passed to each mapper. The latter deduces a local minimum support $LMinSup$ δ from Δ and its input data split (i.e., number of transaction in the input split). Then, each mapper divides its input data split (S) into n data partitions, $S = \{S_1, S_2, \ldots, S_n\}$. Each partition S_i in S holds only transactions that have length i, where the length of a transaction is the number of items in it. Then, the mapper starts mining the data partitions $S_i...S_n$ according to transaction lengths in decreasing order. A transaction in each partition accounts for an itemset. If a transaction T is frequent ($Support(T) \geq \delta$) in partition S_{i+1}, then it will be stored in a list of frequent itemsets L. Otherwise, T will be stored in a temporary data structure $Temp$. After checking the frequency of all transactions T in S_{i+1}, the process continues by generating i subsets of all T in $Temp$ and adds the i generated subsets to partition S_i. The same mining process is carried out until visiting all partitions S_i in S. Before counting the $Support$ of a transaction T, an inclusion test of T in L is performed. If the test returns true, the computation of the $Support$ of T will not be considered as T is already in L which means frequent. Each mapper emits all its local frequent itemsets to the reducer. The reducer writes all local frequent itemsets to the distributed file system.

■ **Job 2:** Each mapper takes a data split S and a list of local frequent itemsets LFI. Each mapper determines the inclusion of LFI elements in each transaction of S. If there is an inclusion, then the mapper emits the itemset as a key and one as value (key: itemset, value: 1). A global minimum support $GMinSup$ Δ is passed to the reducer. The reducer simply iterates over the values of each received key, and sums them up in variable sum. If ($sum \geq \Delta$), then the itemset under consideration is globally frequent.

As illustrated above, the main workload of P2RC algorithm is done on the mappers independently. Intuitively, the mapper that holds more homogeneous data (i.e., homogeneous transactions) will be faster. Actually, by referring to the mining principle of CDAR, a mapper that holds homogeneous transactions (i.e., similar transactions) allows for more itemset inclusions which in turn results in less subsets generation. Thus, placing each bucket of similar transactions (non-overlapping data partitions) on the mappers would improve the performance of P2RC algorithm. This data placement technique can be achieved by means of different data partitioning methods.

In contrast, the partitioning of data based on transaction similarities (STDP for short: Similar Transaction Data Partitioning), logically would not improve the performance of Parallel Two Round Apriori (P2RA), instead it should lower it. In this case, each mapper would hold a partition of data (i.e., data split) of similar transactions which allows for a high number of frequent itemsets in each mapper. This results in a higher number of itemset candidates generation. Interestingly, using a simple Random Transaction Data Partitioning (RTDP for short) to randomly place data on the mappers, should give the best performance of P2RA. Our experiments given in Sect. 4 clearly illustrate this intuition.

P2RC performs two MapReduce jobs to determine all frequent itemsets. Thus, PFIM algorithms that depend on SON process design duplicate the mining results. Also, at their first mining step (i.e., first MapReduce job), 2-Jobs schema PFIM algorithms output itemsets that are locally frequent, and there is no guarantee to be globally frequent. Hence, these algorithms amplify the number of transferred data (i.e., itemsets) between mappers and reducers.

To cover the above-mentioned issues, our major challenge is to limit the mining process to one simple job. This would guarantee low data communications, less energy power consumption, and a fast mining process. In a distributed computational environment, we take the full advantage of the available massive storage space, CPU(s) etc.

3.2   IBDP: An Overlapping Data Partitioning Strategy

Our claim is that duplicating the data on the mappers allows for a better accuracy in the first job and therefore leads to less infrequent itemsets (meaning less communications and fast processing). Consider a data placement with a high overlap (i.e., placement of data partitions that share several transactions), with for instance 10 overlapping data partitions, each holding 50 % of the database. Obviously, there will be less globally infrequent itemsets in the first job (i.e., if an itemset is frequent on a mapper, then it is highly likely to be frequent on the whole database.). Unfortunately, this approach has some drawbacks, we still need a second job to filter the local frequent itemsets and check their global frequency. Furthermore, such a thoughtless placement is absolutely not plausible, given the massive data sets we are dealing with. However, we take advantage of this duplication opportunity and propose IBDP, an efficient strategy for partitioning the data over all mappers, with an optimal amount of duplicated data, allowing for an exhaustive mining in just one MapReduce job. The goal of IBDP is to replace part of the mining process by a clever placement strategy and optimal data duplication.

The main idea of IBDP is to consider the different groups of frequent itemsets that are usually extracted. Let us consider a minimum threshold Δ and X, a frequent itemset according to Δ on $\mathcal{D}$. Let S_X be the subset of $\mathcal{D}$ restricted to the transactions supporting X. The first expectation is to have $|S_X| \ll |\mathcal{D}|$ since we are working with very low minimum thresholds. The second expectation is that X can be extracted from S_X with Δ as a minimum threshold. The goal of IBDP is a follows: for each frequent itemset X, build S_X the subset from which the extraction of X can be done in one job. Fortunately, itemsets usually share a lot of items between each other. For instance, with Wikipedia articles, there will be a group of itemsets related to the *Olympic games*, another group of itemsets related to *Algorithms*, etc. IBDP exploits these affinities between itemsets. It divides the search space by building subsets of $\mathcal{D}$ that correspond to these groups of itemsets, optimizing the size of duplicated data.

More precisely, given a database of transactions $\mathcal{D}$, and its representation in the form of a set $\mathcal{S}$ of n of non-overlapping data partitions $\mathcal{S} = \{S_1, S_2, \ldots, S_n\}$. Each one of these non-overlapping data partitions (i.e., $\bigcap_{i=1}^{n} S_i = \emptyset$)holds a

Algorithm 1. IBDP

1 //**Job1**
Input: Non-overlapping data partitions $\mathcal{S} = \{S_1, S_2, \ldots, S_n\}$ of a database $\mathcal{D}$
Output: Centroids

2 //*Map Task 1*
3 **map**(*key: Split Name*: $\mathcal{K}_1$, *value* = Transaction (Text Line): $\mathcal{V}_1$)
4   - Tokenize $\mathcal{V}_1$, to separate all items
5   **emit** (*key: Item, value: Split Name*)

6 //*Reduce Task 1*
7 **reduce**(*key: Item, list(values)*)
8   **while** *values.hasNext()* **do**
9   **emit** (*key:(Split Name) values.next (Item)*)

10 //**Job2**
Input: Database $\mathcal{D}$
Output: Overlapping Data Partitions
11 //*Map Task 2*
12 - Read previous job1 result once in a (key, values) data structure (DS), where
key: SplitName and values: Items
13 **map**(*key: Null*: $\mathcal{K}_1$, *value* = Transaction (Text Line): $\mathcal{V}_1$)
14   **for** *SplitName in DS* **do** **if** *Items.Item* $\cap \mathcal{V}_1 \neq \emptyset$ **then**
15   **emit** (*key: SplitName, value. $\mathcal{V}_1$*)
16

17 //*Reduce Task 2*
18 **reduce**(*key: SplitName, list(values)*)
19   **while** *values.hasNext()* **do**
20   **emit** (*key: (SplitName), values.next: (Transaction)*)

set of similar transactions (the union of all elements in $\mathcal{S}$ is $\mathcal{D}$, $\bigcup_{i=1}^{n} S_i = \mathcal{D}$).
For each non-overlapping data partition S_i in $\mathcal{S}$ we extract its "centroid". The
centroid of S_i contains the different items and their number of occurrences in S_i.
Only the items having a maximum number of occurrences over the whole set of
partitions are kept for each centroid. Once the centroids are built, IBDP simply
intercepts each centroid of S_i with each transaction in $\mathcal{D}$. If a transaction in $\mathcal{D}$
shares an item with a centroid of S_i, then the intersection of this transaction
and the centroid will be placed in an overlapping data partition called S_i'. If
we have n non-overlapping data partitions (i.e., n centroids), IBDP generates n
overlapping data partitions and distributes them on the mappers.

The core working process of IBDP data partitioning and its parallel design
on MapReduce, are given in Algorithm 1, while its principle is illustrated by
Example 2.

■ **Job 1** *Centroids*: Each mapper takes a transaction (line of text) from non-
overlapping data partitions as a value, $\mathcal{S} = \{S_1, S_2, \ldots, S_n\}$, and the name of

the split being processed as a key. Then, it tokenizes each transaction (value) to determine different items and emits each item as a key coupled with its split name as value. The reducer aggregates over the keys (items) and emits each key (item) coupled with its different value (split name) in the list of values (split names).

■ **Job 2** *Overlapping Partitions*: The format of the MapReduce output is set to "MultiFileOutput" in the driver class. In this case, the keys will denote the name of each overlapping data partition output (we override the "generateFileNameForKeyValue" function in MapReduce to return a string as key). In the map function, first we store (once) the previous MapReduce job (Centroids) in a (key, value) data structure (e.g. MultiHashMap etc.). The key in the used data structure is the split name, and the value is a list of items. Then, each mapper takes a transaction (line of text) from the database $\mathcal{D}$, and for each key in the used data structure, if there is an intersection between the values (list of items) and the transaction being processed, then the mapper emits the key as the split name (in the used data structure) and value as the transaction of $\mathcal{D}$. The reducer simply aggregates over the keys (split names) and writes each transaction of $\mathcal{D}$ to an overlapping data partition file.

Example 2. Figure 1 shows a database $\mathcal{D}$ with 5 transactions. In this example, we have two non-overlapping data partitions at step (1) and thus two centroids at step (2). The centroids are filtered in order to keep only the items having the maximum number of occurrences (3). IBDP intercepts each one of these two centroids with all transactions in $\mathcal{D}$. This results in two overlapping data partitions in (4) where the intersections only are kept in (5). Finally, the maximal frequent itemsets are extracted in (6). Redundancy is used for the counting process of different itemsets. For instance, transaction efg is duplicated in both partitions in (5) where the upper version participates to the frequency counting of a and the lower version participates to the frequency counting of fg.

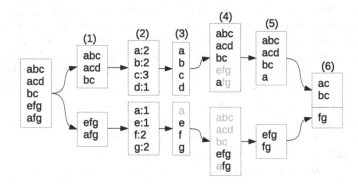

Fig. 1. Data partitioning process: (1) partitions of similar transactions are built; (2) centroids are extracted; (3) and filtered; (4) transaction are placed and filtered; (5) to keep only the intersection of original transactions and centroids; (6) local frequent itemsets are also globally frequent.

3.3   1-Job Schema: Complete Approach

We take the full advantage from IBDP data partitioning strategy and propose a powerful and robust 1-Job Schema PFIM algorithm namely PATD. PATD algorithm limits the mining process of very large database to one simple MapReduce job and exploits the natural design of MapReduce framework. Given a set of overlapping data partitions ($\mathcal{S} = \{S_1, S_2, \ldots, S_m\}$) of a database $\mathcal{D}$ and an absolute minimum support $AMinSup$ Δ, the PATD algorithm mines each overlapping data partition S_i independently. At each mapper m_i, $i = 1, \ldots, n$, PATD performs CDAR algorithm on S_i. The mining process is based on the same $AMinSup$ Δ for all mappers, i.e., each overlapping data partition S_i is mined based on Δ. The mining process is carried out in parallel on all mappers. The mining result (i.e., frequent itemsets) of each mapper m_i is sent to the reducer. The latter receives each frequent itemsets as its key and null as its value. The reducer aggregates over the keys (frequent itemsets) and writes the final result to a distributed file system.

The main activities of mappers and reducers in PATD algorithm are as follows:

■ **Mapper:** Each mapper is given a S_i, $i = 1 \ldots m$ overlapping data partition, and a global minimum support (i.e., $AMinSup$). The latter performs CDAR algorithm on S_i. Then, it emits each frequent itemset as a key and null for its value, to the reducer.
■ **Reducer:** The reducer simply aggregates over the keys (frequent itemsets received from all mappers) and writes the final result to a distributed file system.

As illustrated in the mappers and reducers logic, PATD performs the mining process in one simple and efficient MapReduce job. These properties of PATD are drawn from the use of the robust and efficient data partitioning strategy IBDP.

Example 3. Lets take the example of Fig. 1. Given an absolute minimum support $\Delta = 2$ (i.e., an itemset is considered frequent, if it appears at least in *two* transactions in $\mathcal{D}$). Following PATD mining principle, each mapper is given an overlapping data partition S_i as value. In our example, we have two overlapping data partitions (5). We consider *two* mappers m_1 and m_2, each one performs a complete CDAR with $\Delta = 2$. In Fig. 1 (5) from bottom-up : mapper m_1 mines first overlapping data partition and returns {fg} as a frequent itemset. Alike, mapper m_2 mines second overlapping data partition and returns {{ac}, {bc}}. All the results are sent to the reducer, the reducer aggregates over the keys (frequent itemsets) and outputs the final result to a distributed file system.

3.4   Proof of Correctness

To prove the correctness of PATD algorithm, it is sufficient to prove that if an itemset x is frequent, then it is frequent in at least one of the partitions produced by IBDP. Since, each partition is locally mined by one mapper, then x will be found as frequent by one of the mappers. Thus, the correctness proof is done by the following lemma.

Lemma 1. *Given a database* $\mathcal{D} = \{T_1, T_2, \ldots, T_n\}$ *and an absolute minimum support* Δ, $\forall$ *itemset* x *in* $\mathcal{D}$ *we have:* $Support_{\mathcal{D}}(x) \geq \Delta \Leftrightarrow \exists \, \mathcal{P} \setminus Support_{\mathcal{P}}(x) \geq \Delta$ *where* $\mathcal{P}$ *denotes one of the data partitions obtained by performing IBDP on* $\mathcal{D}$.

Proof.

We first prove that if $Support_{\mathcal{D}}(x) \geq \Delta$ then $\exists \, \mathcal{P} \setminus Support_{\mathcal{P}}(x) \geq \Delta$.

Let denote by $\mathcal{X}$ the set of all unique items of $\mathcal{D}$. The intersection of all transactions $\{T_1, T_2, \ldots, T_n\}$ with $\mathcal{X}$ is $\mathcal{D}$. Thus, in this particular case, $Support_{\mathcal{D}}(x) \geq \Delta \Rightarrow \exists \, \mathcal{D} \setminus Support_{\mathcal{D}}(x) \geq \Delta$. If the set of unique items $\mathcal{X}$ is partitioned into k partitions, then the intersection of each one of these k partitions with all $\{T_1, T_2, \ldots, T_n\}$ in $\mathcal{D}$, would result in a new data partition $\mathcal{P}$. Let denote by $\Pi = \{P_1, P_2, \ldots, P_k\}$, the set of all these new data partitions. For any given itemset x in $\mathcal{D}$, its total occurrence will be in one partition of Π, because, all items in $\mathcal{X}$ are shared among these partitions in Π. Therefore, $Support_{\mathcal{D}}(x) \geq \Delta \Rightarrow \exists \, I_P \setminus Support_{I_P}(x) \geq \Delta$.

Next, we prove the inverse, i.e. if $\exists \, \mathcal{P} \setminus Support_{\mathcal{P}}(x) \geq \Delta$ then $Support_{\mathcal{D}}(x) \geq \Delta$.

This is done simply by using the fact that each partition $\mathcal{P}$ is a subset of $\mathcal{D}$. Hence, if the support of x in $\mathcal{P}$ is higher than Δ, then this will be the case in $\mathcal{D}$. Thus, we have: if $\exists \, \mathcal{P} \setminus Support_{\mathcal{P}}(x) \geq \Delta \Rightarrow Support_{\mathcal{D}}(x) \geq \Delta$.

Therefore, we conclude that: $Support_{\mathcal{D}}(x) \geq \Delta \Leftrightarrow \exists \, \mathcal{P} \setminus Support_{\mathcal{P}}(x) \geq \Delta$.

4   Experiments

To assess the performance of PATD algorithm, we have carried out extensive experimental evaluations. In Sect. 4.1, we depict our experimental setup, and in Sect. 4.3 we investigate and discuss the results of our different experiments.

4.1   Experimental Setup

We implemented PATD, and all other presented algorithms on top of Hadoop-MapReduce, using Java programming language version 1.7 and Hadoop version 1.0.3. For comparing PATD performance with other PFIM alternatives, we implemented two bunches of algorithms. First, we followed SON algorithm design and implemented Parallel Two Round Apriori (P2RA) and Parallel Two Round CDAR (P2RC). These two PFIM algorithms are based on random transaction data partitioning (RTDP) and similar transaction data partitioning (STDP), respectively. Second, we designed and implemented a parallel version of standard Apriori algorithm [7], namely Parallel Apriori (PA). For comparison with PFP-Growth [10], we adopted the default implementation provided in the Mahout [11] machine learning library (Version 0.7).

We carried out all our experiments on the Grid5000 [12] platform, which is a platform for large-scale data processing. We have used a cluster of 16 and 48 machines respectively for Wikipedia and ClueWeb data set experiments. Each machine is equipped with Linux operating system, 64 Gigabytes of main memory, Intel Xeon $X3440$ 4 core CPUs, and 320 Gigabytes SATA hard disk.

4.2 Real World Datasets

To better evaluate the performance of PATD algorithm, we used two real-world data sets. The first one is the 2014 English Wikipedia articles [13] having a total size of 49 Gigabytes, and composed of 5 million articles. The second one is a sample of ClueWeb English data set [14] with size of one Terabyte and having 632 million articles. For each data set, we performed a data cleaning task. We removed all English stop words from all articles, we obtained data sets where each article represents a transaction (items are the corresponding words in the article) to each invoked PFIM algorithm in our experiments.

We vary the *MinSup* parameter value for each PFIM algorithm. We evaluate each algorithm based on its response time, its total amount of transferred data,

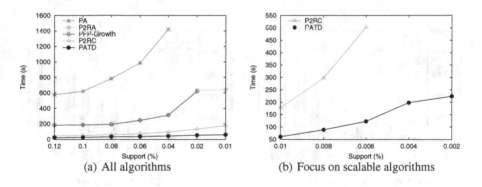

Fig. 2. Runtime and scalability on English Wikipedia data set

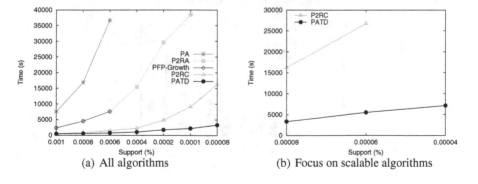

Fig. 3. Runtime and scalability on ClueWeb data set

and its energy power consumption. In particular, we consider these three different measurements, when *MinSup* is very low.

4.3   Runtime and Scalability

Figures 2 and 3 give a complete view of our experiments on both English Wikipedia and ClueWeb data sets. Figures 2(a) and (b) report our experimental results on the whole English Wikipedia data set. Figure 2(a) gives an entire view on algorithms performances for a minimum support varying from 0.12 % to 0.01 %. We see that PA algorithm runtime grows exponentially, and gets quickly very high compared to other presented PFIM algorithms. This exponential runtime reaches its highest value with 0.04 % threshold. Below this threshold, PA needs more resources (e.g. memory) than what exists in our tested machines, thus, it is impossible to extract frequent itemsets with this algorithm. Another interesting observation is that P2RA performance tends to be close to PFP-Growth until a minimum support of 0.02 %. P2RA algorithm continues scaling with 0.01 % while PFP-Growth does not. Although, P2RC scales with low minimum support values, PATD outperforms this algorithm in terms of running time. In particular, with a minimum support of 0.01 % PATD algorithm outper-

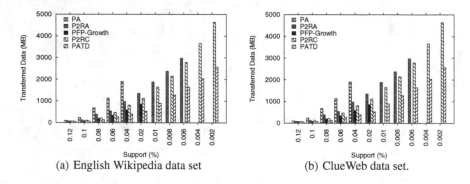

(a) English Wikipedia data set   (b) ClueWeb data set.

Fig. 4. Data communication

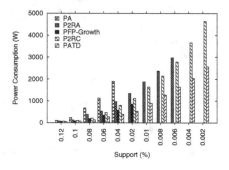

Fig. 5. Energy consumption

forms all other presented PFIM algorithms. This difference in the performance is better illustrated in Fig. 2(b).

Figure 2(b) focuses on the differences between the four algorithms that scale in Fig. 2(a). Although P2RC continues to scale with 0.002 %, it is outperformed by PATD in terms of running time. With 0.002 % threshold, we observe a big difference in the response time between PATD and P2RC. This very good performance of PATD is due to its clever and simple mining principle, and its simple MapReduce job property that allows a low mining time.

In Figs. 3(a) and (b), similar experiments have been conducted on the ClueWeb data set. We observe that the same order between all algorithms is kept compared to Figs. 2(a) and (b). There are three bunches of algorithms. One made of PA which cannot reasonably applied to this data set, whatever the minimum support. In the second bunch, we see that PFP-Growth suffers from the same limitations as could be observed on the Wikipedia data set in Fig. 2(a), and it follows a behavior that is very similar to that of P2RA, until it becomes impossible to execute. P2RA continues scaling until stops executing with a minimum support of 0.0001 %. In the third bunch of algorithms, we see P2RC and PATD scale until 0.00008 %. We decreased the minimum support parameter, and we zoom on these two algorithms. As shown in Fig. 3(b), we observe a very good performance of PATD compared to P2RC. The P2RC algorithm becomes inoperative with a minimum support below 0.00006 %, while PATD continues scaling very well. This big difference in the performance behavior between PATD and all other presented algorithms shows the high capability of PATD in terms of scaling and response time. With both, Gigabytes and Terabytes of data, PATD gives a very good and significant performance. Whatever, the data size, the number of transactions, and the minimum support, PATD scales and achieves very good results.

4.4   Data Communication and Energy Consumption

Let's now study the amount of transferred data for each PFIM algorithm. Figure 4(a) shows the transferred data (in mega bytes) of each presented algorithm on English Wikipedia data set. We observe that PA has the highest peak, this is simply due to its several round of MapReduce executions. In other hand, we see that P2RA, P2RC and PFP-Growth represent smaller peaks. Among all the presented algorithms in Fig. 4(a), we clearly distinguish PATD algorithm. We can see that whatever the used *MinSup*, PATD does not allow much data transfer compared to other algorithms. This is because PATD does not rely on chains of jobs like other presented alternatives. In addition, contrary to other PFIM algorithms, PATD limits the mappers from emitting non frequent itemsets. Therefore, PATD algorithm does not allow the transmission of useless data (itemsets).

In Fig. 4(b), we report the results of the same experiment on ClueWeb data set. We observe that PATD algorithm always has the lowest peak in terms of transferred data comparing to other algorithms.

We also measured the energy consumption of the compared algorithms during their execution. We used the Grid5000 tools that measure the power consumption of the nodes during a job execution. Figure 5 shows the total amount of the power consumption of each presented PFIM algorithm. We observe in Fig. 5, that the consumption increases when decreasing the minimum support for each algorithm. We see that PATD still gives a lower consumption comparing to other algorithms. Taking the advantage from its parallel design, PATD allows a high parallel computational execution. This impacts the mining runtime to be fast and allows for a fast convergence of the algorithm and thus, a less consumption of the energy. PATD also transfers less data over the network, and this is another reason for its lower energy consumption.

5   Related Work

In the literature, several endeavors have been made to improve the performance of FIM algorithms [15,16].

Recently, and due to the explosive data growth, an efficient parallel design of FIM algorithms has been highly required. FP-Growth algorithm [16] has shown an efficient scale-up compared to other FIM algorithms, it has been worth to come up with a parallel version of FP-Growth [10] (i.e. PFP-Growth). Although, PFP-Growth is distinguishable with its fast mining process, PFP-Growth has accounted for different flaws. In particular, with very low minimum support, PFP-Growth may run out of memory as illustrated by our experiments in Sect. 4. PARMA algorithm [17] uses an approximation in order to determine the list of frequent itemsets. It has shown better running time and scale-up than PFP-Growth. However, PARMA algorithm does not return an exhaustive list of frequent itemsets, it only approximates them.

A parallel version of Apriori algorithm [18] requires n MapReduce jobs, in order to determine frequent itemsets of size n. However, the algorithm is not efficient because it requires multiple database scans. In order to overcome conventional FIM issues and limits, a novel technique, namely CDAR has been proposed in [9]. This algorithm uses a transactions grouping based approach to determine the list of frequent itemsets. CDAR avoids the generation of candidates and renders the mining process more simple, by dividing the database into groups of transactions. Although, CDAR [9] has shown an efficient performance behavior, yet there has been no proposed parallel version of it.

Another FIM technique, called SON, has been proposed in [8], which consists of dividing the database into n partitions. The mining process starts by searching the local frequent itemsets in each partition independently. Then, the algorithm compares the whole list of local frequent itemsets against the entire database to figure out a final list of global frequent itemsets.

In this work, we have focused on the data placement as a fundamental and essential mining factor in MapReduce. We proposed the PATD algorithm that not only reduces the total response time of FIM process, but also its communication cost and energy consumption.

6 Conclusion

We proposed a reliable and efficient MapReduce based parallel frequent item-set algorithm, namely PATD, that has shown significantly efficient in terms of runtime and scalability, data communication as well as energy consumption. PATD takes the advantage of the efficient data partitioning technique IBDP. IBDP allows for an optimized data placement on MapReduce. It allows PATD algorithm to exhaustively and quickly mine very large databases. Such ability to use very low minimum supports is mandatory when dealing with Big Data and particularly hundreds of Gigabytes like what we have done in our experi-ments. Our results show that PATD algorithm outperforms other existing PFIM alternatives, and makes the difference between an inoperative and a successful extraction.

References

1. Labrinidis, A., Jagadish, H.V.: Challenges and opportunities with big data. Proc. VLDB Endow. **5**(12), 2032–2033 (2012)
2. Berry, M.: Survey of Text Mining Clustering, Classification, and Retrieval. Springer, New York (2004)
3. Fan, W., Bifet, A.: Mining big data: current status, and forecast to the future. SIGKDD Explor. Newsl. **14**(2), 1–5 (2013)
4. Dean, J., Ghemawat, S.: Mapreduce: simplified data processing on large clusters. Commun. ACM **51**(1), 107–113 (2008)
5. Zaharia, M., Chowdhury, M., Franklin, M.J., Shenker, S., Stoica, I.: Spark: cluster computing with working sets. In: Proceedings of the 2Nd USENIX Conference on Hot Topics in Cloud Computing, HotCloud 2010, CA, USA, p. 10. Berkeley (2010)
6. Hadoop
7. Agrawal, R., Srikant, R.: Fast algorithms for mining association rules in large data-bases. In: Bocca, J.B., Jarke, M., Zaniolo, C. (eds.) Proceedings of International Conference on Very Large Data Bases (VLDB), pp. 487–499. Santiago de Chile, Chile (1994)
8. Savasere, A., Omiecinski, E., Navathe, S.B. An efficient algorithm for mining asso-ciation rules in large databases. In: Proceedings of International Conference on Very Large Data Bases (VLDB), pp. 432–444 (1995)
9. Tsay, Y.-J., Chang-Chien, Y.-W.: An efficient cluster and decomposition algorithm for mining association rules. Inf. Sci. Inf. Comput. Sci. **160**(1–4), 161–171 (2004)
10. Li, H., Wang, Y., Zhang, D., Zhang, M., Chang, E.Y. Pfp: parallel fp-growth for query recommendation. In: Pu, P., Bridge, D.G., Mobasher, B., Ricci, F. (eds.) Pro-ceedings of the ACM Conference on Recommender Systems (RecSys), Lausanne, Switzerland, pp. 107–114. ACM (2008)
11. Owen, S.: Mahout in Action. Manning Publications Co., Shelter Island (2012)
12. Grid5000. https://www.grid5000.fr/mediawiki/index.php/Grid5000:Home
13. English wikipedia articles (2014). http://dumps.wikimedia.org/enwiki/latest
14. The clueweb09 dataset (2009). http://www.lemurproject.org/clueweb09.php/
15. Song, W., Yang, B., Zhangyan, X.: Index-bittablefi: an improved algorithm for mining frequent itemsets. Knowl. Based Syst. **21**(6), 507–513 (2008)

16. Han, J., Pei, J., Yin, J.: Mining frequent patterns without candidate generation. SIGMODREC ACM SIGMOD Rec. **29**, 1–12 (2000)
17. Riondato, M., DeBrabant, J.A., Fonseca, R., Upfal, E.: Parma: a parallel randomized algorithm for approximate association rules mining in mapreduce. In: 21st ACM International Conference on Information and Knowledge Management (CIKM), Maui, HI, USA, pp. 85–94. ACM (2012)
18. Anand, R.: Mining of Massive Datasets. Cambridge University Press, New York (2012)

Does Multilevel Semantic Representation Improve Text Categorization?

Cheng Wang[(⊠)], Haojin Yang, and Christoph Meinel

Hasso Plattner Institute, University of Potsdam, Prof.-Dr.-Helmert-Str. 2-3,
14482 Potsdam, Germany
{cheng.wang,haojin.yang,christoph.meinel}@hpi.de

Abstract. This paper presents a novel approach for text categorization by fusing "Bag-of-words" (BOW) word feature and multilevel semantic feature (SF). By extending Online LDA (OLDA) as multilevel topic model for learning a semantic space with different topic granularity, multilevel semantic features are extracted for representing text component. The effectiveness of our approach is evaluated on both large scale Wikipedia corpus and middle-sized 20newsgroups dataset. The former experiment shows that our approach is able to preform semantic feature extraction on large scale dataset. It also demonstrates the topics generated from different topic level have different semantic scopes, which is more appropriate to represent text content. Our classification experiments on 20newsgroups achieved 82.19 % accuracy, which illustrates the effectiveness of fusing BOW and SF features. The further investigation on word and semantic feature fusion proves that Support Vector Machine (SVM) is more sensitive to semantic feature than Naive Bayes (NB), K Nearest Neighbor(KNN), Decision Tree (DT). It is shown that appropriately fusing low-level word feature and high-level semantic feature can achieve equally well or even better result than state-of-the-art with reduced feature dimension and computational complexity.

1 Introduction

Text categorization (TC) plays an important role in text mining and information retrieval field. The fast increasing of Web data poses challenges to web mining and information retrieval communities. TC generally deals with the problem by assigning specific label c in $C = \{c_1, ..., c_j\}$ to given document d according pre-trained classifier with training examples $D = \{d_1, ...d_n\}$ [19]. This sort of problem usually requires highly descriptive and expressive representation to achieve desired classification result. In text representation, traditional approach is word frequency-driven representation such as TF-IDF weighting [22], denoted as

$$tf * idf = t(w)log\frac{D}{d(w)} \tag{1}$$

where $t(w)$ is generated by the maximum number of occurrences of words in a document. Based on that, bag of words (BOW) approach is a common way to

© Springer International Publishing Switzerland 2015
Q. Chen et al. (Eds.): DEXA 2015, Part I, LNCS 9261, pp. 319–333, 2015.
DOI: 10.1007/978-3-319-22849-5_22

represent document, but it thresholds to discover informative words in a document for semantically generating topics. Besides, BOW features usually have a very high dimensionality (eg. $10^4 \sim 10^5$). For improving efficiency, Latent Semantic Indexing (LSI) [10] and Probabilistic Latent Semantic Analysis(PLSA) were proposed [11] for dimension reduction. Recently, Latent Dirichlet Allocation (LDA) [4] provides a way to generate latent topics from given corpus. It has been extended and applied to solve several different problems, such as text based problems [3,14,16], image modality [15,18,21], video analysis [5,17] and cross-modal or multimodal media retrieval [12]. For fitting topics from massive corpus, Online LDA [9] uses online variational inference for LDA parameter estimation, which has been proved that can generate equally well or even better result than tradition approximate posterior inference algorithms for sampling topics.

Text representation with topic semantic space can definitely reduce feature dimension comparing to BOW feature. In this research we work towards representing text using topical feature (semantic feature) which extracted from a pre-constructed semantic space. We extend Online LDA to multi-level scenario and represent text with more meaningful semantic feature. The basic idea is to fit topics using different topic granularity. The motivations to propose multi-level Online LDA (ML-OLDA) as follows. (1) Current text representation approaches mainly rely on BOW based word feature representation, considering time efficiency in practical application, semantic representation can definitely reduce feature dimension [1]. (2) Most of topic modeling work exploring the methodology of topic generation. The method of representing document as topical feature in semantic level for practical applications is less discussed. To leverage multilevel topics for feature representation of document is non-trivial issue. (3) Different topic granularity actually have different semantic scope, for example, we have two topics, topic A = {"language", "greek", "english", "ancient", "latin"} and topic B = {"language", "english", "word", "term", "form"}. We see that the two topics cover different semantic scope, topic A is more global and general than topic B.

To address those problems, we propose to utilize online LDA for fitting topics with different topic granularities. The main contributions of our work can be summarized as follows

(1) This work presents a novel approach for learning multi-level semantic feature in text representation.
(2) This work investigates the availability and effectiveness of fusing word feature and semantic feature in improving text categorization performance.
(3) Our extensive experiments reveal that not all classifier is sensitive to semantic feature, appropriately fusing different features can achieve state-of-the art result with high time efficiency.

The rest of this paper is structured as follows, Sect. 2 introduces related works including standard LDA, Online LDA model and topic based categorization approaches. Section 3 mainly describes our extension version of Online-LDA for generating multi-level topics. And introduces the method to learn a semantic space with our proposed ML-OLDA approach and document semantic feature extraction with constructed semantic space. Section 4 presents experiment and evaluation on both Wikipedia and 20newsgroups, some conclusions are draw from our empirical experiments. Section 5 gives result comparison and discussion regarding text categorization. Section 6 states the conclusion of our work and future work.

2 Related Work

This section gives a brief overview of LDA model and online variational Bayes(VB) algorithm for fitting topics from massive data. And then introduces our approach based on Online LDA for fitting topics using different topic granularities.

2.1 LDA and Online LDA

Latent Dirichlet Allocation (LDA) is a generative probabilistic model for discovering latent topics from given corpus. The generative process can be decomposed into doc-topic and topic-word generative process. As shown in Fig. 1.(a), for a given document D, the topic proportion θ follows a Dirichlet distribution with prior probability α . For given θ, the specific topic z_n is draw from a multinomial distribution. Similarly, in topic-word distribution, a word w_n follows multinomial distribution with φ that is drawn from a Dirichlet distribution with prior probability β. Therefore the joint probability of words and topics can be described as

$$p(\mathbf{w}, \mathbf{z}) = \int p(\theta) \prod_{n=1}^{N} p(w_n \mid z_n) p(z_n \mid \theta) d\theta \qquad (2)$$

where w_n means the n-th word in document, z_n means the topic of n-th word. N is the number of words in corpus. The topic generation procedure can be simply described as follows:

– sample component $\theta_d \sim Dir(\alpha)$
– sample component $\varphi_k \sim Dir(\beta)$
– for each word in document d
  – sample topic $z_n \sim Multinomial(\theta_d)$
  – sample word $w_n \sim Multinomial(\varphi_{z_n})$.

Online variational inference [9] extends this approach to online fashion for LDA, Online LDA (OLDA) provides a way for fitting the parameters to variational posterior over topic distribution. Generally, simplified description of online variational Bayes for LDA can be written as

– randomly initialize topic set λ
– define $\rho_t \triangleq (\tau_0 + t)^{-\kappa}$, where $\kappa \in (0.5, 1]$ used to control the rate for forgetting old value $\tilde{\lambda}$, and $\tau_0 \geq 0$ used to slow down the early algorithm.
– for each batch t

– do E step to optimize per-document variational parameters at batch t
– do M step to update $\lambda = (1 - \rho_t)\lambda + \rho_t\tilde{\lambda}$.

The goal of OLDA is to find a good topic set λ, updating of λ based on its previous value and $\tilde{\lambda}$. The core advantage of OLDA is that it has the ability to process massive data collection.

2.2   Topic Based Text Categorization

In recent years, topic models have been widely used to text classification and social media analysis [6,13,23–25]. Traditional methods for generating topics are usually with given topic numbers. Recently, generating topics with different granularity has been used to online reviews modeling [20], which defines two levels of granularity: global and local. The proposed MG-LDA model is applied to extract ratable aspects of document. And it claims that multi-grain model is more appropriate for online reviews than term frequency analysis with minimal clustering. In [7], M.G. Chen et al. proposed to construct a topic space by running LDA over the whole dataset with different topic numbers, Multi-granularity topics form new features for representing short text and achieved surprising results on short text classification. In [24], J. Zhu et al. proposed sparse topical coding (STC) to discover latent representation of large collections of data by using sparsity-inducing regularization. Based on STC, Z. N. Zhang et al. [23] proposed a novel sparse online topic model to learn topic dictionary from large data corpora.

3   Methodology

The premise of semantic feature extraction is to construct a semantic space, where semantic feature can be extracted for newly arrived document.

3.1   Problem Formulation

The problem of semantic representation involves two stages: (1) To learn a topic dictionary as semantic space. (2) Semantic feature extraction for given document based on pre-constructed semantic space.

For given corpus $D = \{d_1, d_2......d_n\}$ which contains n documents can be divided into D_{train} and D_{test}. In the first stage, we need to learn topics T^s from L levels by using different topic granularities on D_{train}, that is, $T^s = \{T^1, ...T^L\}$, where $T^l \subseteq T^s$, $\forall l = 1, 2..L$ denotes the sub-topic set generated from l^{th} topic level and $T^l = \{t_1^l, ...t_m^l\}$ has m topics. In the second stage, assume the semantic space we constructed as $T^s \in \mathbb{R}^{n_t \times n_k}$, where n_t is topic number and n_k is the number of words within each topic. For given document $d_i \in D$, $\forall i = 1, 2...n$ is denoted as a word vector $\mathbf{w}_{d,i} \in \mathbb{R}^{n_d}$, n_d is the number of words in document d_i. In order to extract the topic feature of document d_i, we project $\mathbf{w}_{d,i}$ into semantic space T^s and represent it as a n^t dimension vector $V_i = \{v_1^{t^1}...v_m^{t^L}\}$. Then all documents can be represented as topic feature and applied to text categorization.

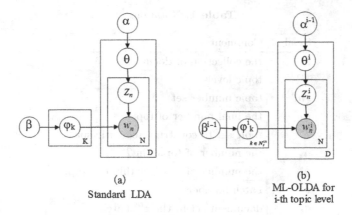

(a)
Standard LDA

(b)
ML-OLDA for
i-th topic level

Fig. 1. LDA and ML-OLDA

3.2 ML-OLDA

So far, both LDA and OLDA generate topics in single level for given topic number. To extend them in multi-level scenarios where multiple topic numbers are used for topic generation. Hierarchical Topic Model [8] is proposed to achieve this goal by learning topics in a hierarchy way. In this work, we use a straightforward approach for fitting topics with OLDA with multiple topic numbers, named ML-OLDA which considers topic learning in hierarchy. For each topic granularity, generated topics are regarded as in same topic (semantic) level. As shown in Fig. 1.(b) ML-OLDA adaptively reset number of topics for i^{th} topic level by selecting N_T from predefined topic number set $\{N_T^{(i)}\}_{i=1}^L$. And use parameters α and β in previous topic level for initializing the next topic level until reach the number of topic K_{min} that is predefined for terminating topic generation procedure. Parameter estimation method in [9] is used in this study. In fitting topics, the whole dataset is divided into several document chunks according the batch size. The main steps of ML-OLDA described as follows

- for each topic level if $K < K_{min}$
- for each document chunk c in i^{th} topic level
  - sample topic distribution $\theta_c^i \sim Dir(\alpha^{i-1})$
  - sample component $\varphi_k^i \sim Dir(\beta^{i-1})$

- for each word n in document chunk c
  - sample topic $z_{kn}^i \sim Multinomial(\theta_c^i)$
  - sample word $w_{kn}^i \sim Multinomial(\varphi_{z_k^i})$

- updating topic set
- combine learned topics.

Table 1. Notations

Symbol	Comment
D	the collection of document
$\{l\}_{l=1}^{L}$	topic levels
$\{n_t^{(l)}\}$	topic number set
n_t	the final number of topic
T^s	topic set to construct semantic space
n_s	the number of topics in T^s
n_k	the number of words within one topic
z	batch number
d_z	document set in the zth batch
n_b	batch size
V_d	semantic feature of document d
$\mathbf{w_o^{(t)}}$	overlap word between d and topic t

Learned topic sets in each topic level are combined to final topic set, which covers both general and specific semantic topics. The key contribution of ML-OLDA is using of multi granularities topics to learn semantic space. Clearly, ML-OLDA is not a principled approach for generating multi-granularity topics, but it provides new way to represent document as semantic feature and improve the text categorization performance.

3.3   Learning Semantic Space

In learning semantic space T^s we generate topics with different topic granularity so that learned topics from multiple semantic levels. Some key notations are given in Table 1 for introducing our algorithm. To build a semantic space that contains n_t topics. We firstly use a topic number set $\{n_t^{(l)}\}_{l=1}^{L}$ to control the topic granularity, the value of $n_t^{(l)}$ is vary at each topic level.

To make semantic space more expressive and discriminative, we designed two tricks that can selectively add topic to semantic space and remove redundant topics. The first trick is to remove those low-ranking topics according to topic weight. The topic weight W of topic $t_m=\{w_1^{t_m}, ...w_n^{t_m}\}$ can be computed by summing each word weight within topic t_m as,

$$W(t_m) = \sum_{i=1}^{N} v_i^m \tag{3}$$

where v_i^m denotes the weight of ith word $w_i^{t_m}$ within topic t_m.

In addition, to make the topics that generated from multi levels more discriminative and less redundant. We consider use average KL-divergence to measure the distance between existed topic sets T^s and newly generated topic set T^l at the lth topic level.

$$D(T^s, T^l) = \sum_{t_n \in T^s, t_m \in T^l} KL(p(\mathbf{w}^{t_n} \mid t_n), p(\mathbf{w}^{t_m} \mid t_m))/(K_s \times K_l) \quad (4)$$

where $t_n \in T^s$, $t_m \in T^l$, $\mathbf{w}^{t_n}$ and $\mathbf{w}^{t_m}$ are the word vector within topic t_n and t_m respectively, and

$$KL(p(\mathbf{w}^{t_n} \mid t_n), p(\mathbf{w}^{t_m} \mid t_m)) = \frac{1}{2}[KL(p(\mathbf{w}^{t_n} \mid t_n) \parallel p(\mathbf{w}^{t_m} \mid t_m))$$
$$+ KL(p(\mathbf{w}^{t_m} \mid t_m) \parallel p(\mathbf{w}^{t_n} \mid t_n))] \quad (5)$$

where $KL(p(\mathbf{w}^{t_n} \mid t_n) \parallel p(\mathbf{w}^{t_m} \mid t_m))$ represents the divergence between $p(\mathbf{w}^{t_n} \mid t_n)$ and $p(\mathbf{w}^{t_m} \mid t_m)$.

After remove low weight topics, a subset of optimized topics is selected according to the distance between newly generated topic set T^l and existed topic space T^s. At the end of each topic level, we combine those selected topic subsets T_{sub} and insert to T^s. The whole procedure terminates if the number of topic in T^s reaches predefined topic number. The algorithm can be summarized as in Algorithm 1.

Algorithm 1. Learning Semantic Space with ML-OLDA

Input: $D, \{l\}_{l=1}^{L}, \{N_T^{(l)}\}, n_t, n_k$
Output: $T^s \in \mathbb{R}^{n_t \times n_k}$

1: $T^s \triangleq \varnothing$
2: $n_s \triangleq 0$
3: **while** $n_s < n_t$ **do**
4:    **for** each $l \in L$ **do**
5:       $n_c = \frac{n}{n_b} + 1$
6:       **while** $(z \leq n_c))$ **do**
7:          $T^l \Leftarrow Fittingtopic(d_z, n_t^{(l)}, n_k)$
8:          $z = z + 1$
9:       **end while**
10:      **for** each $t \in T^l$ **do**
11:        **if** $W(t) < 0.01$ **then**
12:           $Remove(t)$
13:        **end if**
14:      **end for**
15:      $T_{sub} = argmax_{T_{sub} \in T^l} KL(T^s, T^l)$
16:    **end for**
17:    $T^s = T^s \cup T_{sub}$
18:    $n_s \Leftarrow Numberof(T^s)$
19: **end while**

3.4   Topical Feature Extraction

The semantic space that learned with Algorithm 1 can be considered as "bag of topics", based on that the topical (semantic) features can be extracted for representing given document input. For given document d and pre-constructed semantic space T^s, the feature extraction procedure can be described as in Algorithm 2. The input document d is converted into word set after removing stop words. The calculation of possibilities of d over each topic in semantic space is performed by summing the weight of those overlap words between $\mathbf{w_{d,N}}$ and words in each topic $t \in T^s$. Finally, we combine those calculated topic weights as semantic feature for representing the document. As shown in Fig. 2, two extracted feature vectors are given for representing two different documents respectively.

Algorithm 2. Topical Feature Extraction

Input: $d, T^s \in \mathbb{R}^{n_t \times n_k}$
Output: $V_d \in \mathbb{R}^{n_t}$
1: $V_d \triangleq \varnothing$
2: $\mathbf{w_{d,N}} \Leftarrow Split(d)$
3: **for** each $t \in T^s$ **do**
4:    $\mathbf{w_o^{(t)}} = \mathbf{w_{d,N}} \cap \mathbf{w}^{(t)}$
5:    $v^{(t)} \Leftarrow W(\mathbf{w_o^{(t)}})$ # according to equation (3)
6:    $V_d = V_d \cup v^{(t)}$
7: **end for**

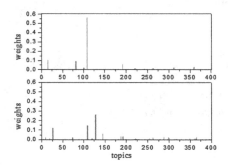

Fig. 2. Illustrative examples of extracted topic features. The topic feature of document "102632" (top) and "66322" (bottom) respectively. In this example, $T^s \in \mathbb{R}^{400}$ is a topic space that contains 400 topics.

4   Experiments

In this section, we conducted experiments on both large scale dataset, 3.3 million Wikipedia articles, and middle-sized dataset-20 newsgroups. In the first experiment, we prove that our approach can easily apply to fit topics from massive data and produce multi-level topics with different topic granularity. In the second experiment, we verify the effectiveness of ML-OLDA based semantic feature in the task of text categorization. The comparison between our approach and baselines suggests multi-level semantic representation can achieve comparable result with state-of-the-art by fusing BOW and semantic feature. We further investigated the feature fusion of BOW feature and semantic feature in improving text categorization performance. All of our experiments were conducted on a desktop with 3.2 GHz processors and 8 GB RAM. The first experiment is conducted online due to the scale of Wikipedia dataset. The second experiment is conducted on our pre-collected 20newsgroups dataset for evaluating the effectiveness of multi-level semantic feature in improving the performance of text categorization.

4.1 Qualitative Experiment on Wikipedia

We employ a similar method to access wikipedia articles as in [9] and use its implemented online Variational Bayes for parameter inference. We simply apply a two-level OLDA for fitting 64 topics and 256 topics respectively[1]. By comparing topics in 64-topic level (T_{64}) with 256-topic level (T_{256}), found that the topics generated in T_{64} are more generic and global than topics generated in T_{256}. This confirms the results in [7] that multi-granularity topic space can be beneficial compare to single-granularity topic space and more appropriate for representation task [20]. For example, considering "language" topic, the most important word within topic generated in T_{256} are "language","greek","english","ancient", "latin" while in T_{64} are "language","english", "word", "term","form". Table 2 demonstrates that word weight of the topics in T_{256} are higher than them in T_{64}. It makes those topic tend to be focus and specific and representing more specific or detailed semantic meaning. From this perspective, we argue that multi-level topics can cover various semantics which can be used to enrich text representation. Therefore the representation of document can be more informative, expressive and discriminative by combining multi-level topics.

4.2 Text Categorization on 20newsgroups Dataset

This subsection mainly focuses on applying extracted multilevel semantic feature to text categorization tasks. 20newsgroups[2] was used in our research, it consists of total 18846 documents and covers 20 categories. The training subset contains 11314 documents and test subset contains 7532 documents. The goals of this experiment are: (1) To examine the availability of fusing word feature (BOW feature) and topical feature (semantic feature) in improving text classification performance. (2) To investigate the appropriate feature fusion approach for achieving better classification result.

---

[1] The number of topics in different topic level can be arbitrary and should not be same in order to observe topics generated with different topic granularities.

[2] http://qwone.com/~jason/20newsgroups/.

Table 2. Generated topics from two topic levels. In second column topic number is 64 (T_{64}), in second column topic number is 256 (T_{256})

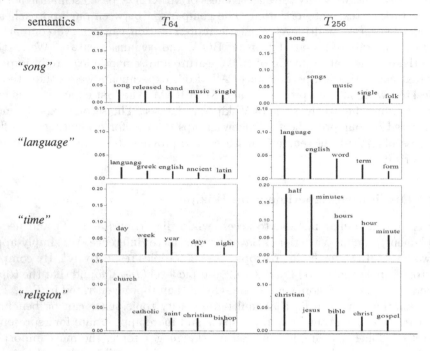

Availability of Feature Fusion. In order to examine the effectiveness of learned semantic feature (SF for short) on 20newsgroups. Four mainstream classification algorithms, K-Nearest-Neighbor (KNN), Support Vector Machine (SVM) with RBF kernel, Naive Bayesian (NB) and Decision Tree (DT) are adopted in experiments. For feature fusion strategy, there are many approaches described in [2]. Although we believe non-linear feature fusion would produce better result, for simplicity, we consider linear weighted fusion in this work. And assume BOW feature and SF feature have equal weight for representing text component. Our experiments show the average classification accuracy is range from 53.4 % (KNN) to 69 % (SVM) if we only consider SF feature. The result is not comparable with BOW based approaches. However, when combining word feature (BOW for short) and SF, some classifiers can achieve equally well or even better result than only BOW feature based approach with less time consumption. Because semantic feature is adopted, it largely reduces input dimension for training classifiers compare to high dimensionality BOW feature.

Table 3 presents the performances of different feature and classifier combinations. The first group of experiment untiled BOW feature only and the number of features is fixed at 20k. In the second group of experiment we combine BOW and SF features that extracted with pre-constructed semantic space. The topics in semantic space are obtained from 3 topic levels and totally 400 topics. Our

Table 3. Classification on 20newsgroups with different combinations, the best performed approaches are boldly marked.

Feature+Method	Precision	Recall	F1-score	Training(s)	Test(s)
BOW+KNN	0.75	0.74	0.74	10.4	667
BOW+SF+KNN	0.72	0.71	0.71	9	492
BOW+SVM	0.78	0.77	0.77	496	44
BOW+SF+SVM	**0.83**	**0.82**	**0.82**	**120**	**19**
BOW+NB	0.81	0.80	0.80	6.9	46.45
BOW+SF+NB	**0.81**	**0.81**	**0.81**	**4.1**	**23.8**
BOW+DT	0.65	0.65	0.65	16.3	0.8
BOW+SF+DT	0.62	0.62	0.62	8.33	0.46

experiments show that the fusion of BOW and SF feature is not effective for all classifiers. Multi-level semantic feature positively impacts the performance of SVM and NB classifier. For SVM, the precision rate, recall rate and f1-score are increased by 5 % respectively. But NB achieved sightly improvements. On the other hand, SF negatively affects KNN and DT classifier, because all measured results dropped compare to the methods that without considering SF. We should note that DT classifier performs poor in any case and SF makes it no difference in classification. From time efficiency aspect, SF feature involved combinations require less training as well as test time.

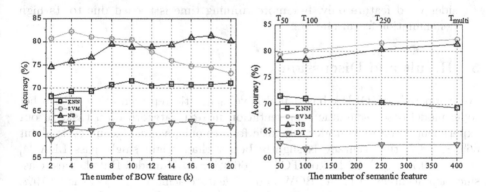

Fig. 3. Classification accuracy, *Left*: BOW feature with fix SF number. *Right*: on single and multi topic level

Exploring of Feature Fusion. For further investigate the feature fusion setting for training classifier. We fine-tuned the number of BOW and SF feature separately in order to explore the sensitiveness of different classifiers to semantic feature. To this end, we conducted two experiments to find effective feature fusion approach, meanwhile, to verify the effectiveness of multi-level semantic representation. The experiments on feature fusion approaches include:

(1) Classification on variable number of BOW feature with fixed SF feature. We used the whole semantic space, named T_{multi} which is pre-constructed using 400 multi-level topics that consists of 50-topic level (T_{50}), 100-topic level (T_{100}) and 250-topic level (T_{250}). By increasing the number BOW feature from $2k$ to $20k$, we observe the classification accuracy of different methods. Figure 3 (Left) shows that with fixed semantic feature number, SVM (RBF-kernel) is much sensitive than others. And achieved the best result with the setting of fusing $4k$ word features with 400 semantic features. The performance of KNN and NB are improving gradually along with increasing of BOW feature.

(2) Classification on single topic level with fixed BOW feature, To study the effectiveness of multi-level semantic feature, we fixed the BOW feature number at the scale that classifiers achieve their corresponding best classification accuracy in Fig. 3 (Left), Fig. 3 (Right) presents that SVM and NB achieved classification accuracy 82.19 % and 81.13 % respectively when using T_{multi} as semantic space for multilevel semantic feature extraction. However, the accuracy of KNN and DT are decreasing from topic level T_{50} to T_{multi}, which means KNN and DT are weak in processing semantic feature involved classification problem. The two experiments prove that SVM and NB have the capability to handle semantic feature involved categorization tasks, and SVM (RBF-kernel) is much more sensitive than other classifiers. DT performs poorly in most case and KNN achieves outstanding result when consider word feature only, but more running time is needed due to its high denomination feature inputs.

5   Result and Discussion

We compared ML-OLDA based multilevel semantic representation with other topic model-based semantic representation in text categorization task. In our experiment when we consider semantic feature only, average accuracy is about $66\% \pm 0.8\%$, the accuracy is slightly better than origin single pass LDA [4] $(61.7\%\pm0.7\%)$ and similar to STC [24] $(66.3\%\pm1.0\%)$ model. But the best classification accuracy of fusing BOW and SF feature achieved by SVM is 82.19%, improved by 0.59% from MedSTC [23]. Figure 4 denotes the performance comparison as well as the confusion matrix of best performed classifier SVM.

Our study shows multilevel semantic representation improves text categorization performance through combining with word feature. What this work emphasized is multilevel semantic feature representation that derived from ML-OLDA in the context of text categorization. This is not to say ML-OLDA outperforms existed topic models in topic generation, but our straightforward approach empirically proved its capability in improving text categorization performance. Although our best result is sightly better than MedSTC model, note that we

Approach	Accuracy(%)
MedSTC [23]	81.6
STC [24]	66.3 ± 1.0
LDA [4]	61.7 ± 0.7
ML-OLDA(SF)	**66.0 ± 0.8**
ML-OLDA(BOW+SF)	**82.19**

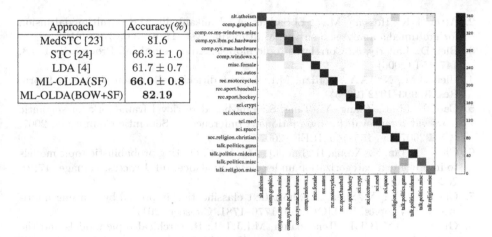

Fig. 4. The confusion matrix of SVM (rbf) classifier with fused BOW and SF feature on 20newsgroups

here only compare the classification accuracy that achieved by those approaches, because LDA, STC and MedSTC considers topical feature only. But we should note, in our classification evaluation, by fusing BOW and SF feature that generated from ML-OLDA achieved equally well performance with much less time consumption.

6   Conclusion

We have presented ML-OLDA approach for learning semantic space and extracting multilevel semantic feature in improving text categorization. The multiple topic level based method is effective in building more informative and discriminative topics. Semantic feature can effectively reduce feature dimension without sacrificing classification accuracy. Our further study on fusing word and semantic feature suggests SVM classifier is more efficient in processing high-level semantic feature in text categorization.

Our future work will focus on exploring other properties of multilevel semantic feature that generated with ML-OLDA and apply it to large-scale text representation in information retrieval. Due to semantic feature is high-level feature, we also consider generalize ML-OLDA to multimedia retrieval and cross-modal representation, such as text-image based cross representation. Meanwhile, we will try some more complicate and reasonable non-linear feature fusion approaches to obtain better feature representations in various tasks.

References

1. Aggarwal, C.C., Zhai, C.: A survey of text classification algorithms. In: Aggarwal, C.C., Zhai, C. (eds.) Mining Text Data, pp. 163–222. Springer, US (2012)

2. Atrey, P.K., Hossain, M.A., El Saddik, A., Kankanhalli, M.S.: Multimodal fusion for multimedia analysis: a survey. Multimedia Syst. **16**(6), 345–379 (2010)
3. Blei, D., Lafferty, J.: Correlated topic models. Adv. Neural Inf. Process. Syst. **18**, 147–154 (2006)
4. Blei, D.M., Ng, A.Y., Jordan, M.I.: Latent dirichlet allocation. J. Mach. Learn. Res. **3**, 993–1022 (2003)
5. Cao, J., Li, J., Zhang, Y., Tang, S.: Lda-based retrieval framework for semantic news video retrieval. In: International Conference on Semantic Computing, 2007, ICSC 2007, pp. 155–160. IEEE (2007)
6. Chen, E., Lin, Y., Xiong, H., Luo, Q., Ma, H.: Exploiting probabilistic topic models to improve text categorization under class imbalance. Inf. Process. Manage. **47**(2), 202–214 (2011)
7. Chen, M., Jin, X., Shen, D.: Short text classification improved by learning multi-granularity topics. In: IJCAI, pp. 1776–1781. Citeseer (2011)
8. Griffiths, D.M.B.T.L., Tenenbaum, M.I.J.J.B.: Hierarchical topic models and the nested chinese restaurant process. Adv. Neural Inf. Process. Syst. **16**, 17–24 (2004)
9. Hoffman, M., Bach, F.R., Blei, D.M.: Online learning for latent dirichlet allocation. In: advances in neural information processing systems, pp. 856–864 (2010)
10. Hofmann, T.: Probabilistic latent semantic indexing. In: Proceedings of the 22nd Annual International ACM SIGIR Conference on Research and Development in Information Retrieval, pp. 50–57. ACM (1999)
11. Hofmann, T.: Unsupervised learning by probabilistic latent semantic analysis. Mach. Learn. **42**(1–2), 177–196 (2001)
12. Jia, Y., Salzmann, M., Darrell, T.: Learning cross-modality similarity for multinomial data. In: 2011 IEEE International Conference on Computer Vision (ICCV), pp. 2407–2414. IEEE (2011)
13. Kwak, H., Lee, C., Park, H., Moon, S.: What is twitter, a social network or a news media? In: Proceedings of the 19th International Conference on World Wide Web, pp. 591–600. ACM (2010)
14. Li, L., Roth, B., Sporleder, C.: Topic models for word sense disambiguation and token-based idiom detection. In: Proceedings of the 48th Annual Meeting of the Association for Computational Linguistics, pp. 1138–1147. Association for Computational Linguistics (2010)
15. Lienou, M., Maître, H., Datcu, M.: Semantic annotation of satellite images using latent dirichlet allocation. IEEE Geosci. Remote Sens. Lett. **7**(1), 28–32 (2010)
16. Mcauliffe, J.D., Blei, D.M.: Supervised topic models. In: Platt, J.C., Koller, D., Singer, Y., Roweis, S. (eds.) Advances in Neural Information Processing Systems, pp. 121–128. MIT Press, Cambridge (2008)
17. Lo Presti, L., Sclaroff, S., La Cascia, M.: Object matching in distributed video surveillance systems by LDA-based appearance descriptors. In: Foggia, P., Sansone, C., Vento, M. (eds.) ICIAP 2009. LNCS, vol. 5716, pp. 547–557. Springer, Heidelberg (2009)
18. Putthividhy, D., Attias, H.T., Nagarajan, S.S.: Topic regression multi-modal latent dirichlet allocation for image annotation. In: IEEE Conference on Computer Vision and Pattern Recognition (CVPR), 2010, pp. 3408–3415. IEEE (2010)
19. Sebastiani, F.: Machine learning in automated text categorization. ACM Comput. Surv. (CSUR) **34**(1), 1–47 (2002)
20. Titov, I., McDonald, R.: Modeling online reviews with multi-grain topic models. In: Proceedings of the 17th International Conference on World Wide Web, pp. 111–120. ACM (2008)

21. Wang, C., Blei, D., Li, F.-F.: Simultaneous image classification and annotation. In: IEEE Conference on Computer Vision and Pattern Recognition, 2009, CVPR 2009, pp. 1903–1910. IEEE (2009)
22. Wu, H.C., Luk, R.W.P., Wong, K.F., Kwok, K.L.: Interpreting tf-idf term weights as making relevance decisions. ACM Trans. Inf. Syst. (TOIS) 26(3), 13 (2008)
23. Zhang, A., Zhu, J., Zhang, B.: Sparse online topic models. In: Proceedings of the 22nd International Conference on World Wide Web, pp. 1489–1500. International World Wide Web Conferences Steering Committee (2013)
24. Zhu, J., Xing, E.P.: Sparse topical coding (2012). arXiv preprint arXiv:1202.3778
25. Zhu, Y., Li, L., Luo, L.: Learning to classify short text with topic model and external knowledge. In: Wang, M. (ed.) KSEM 2013. LNCS, vol. 8041, pp. 493–503. Springer, Heidelberg (2013)

Parallel Canopy Clustering on GPUs

Yusuke Kozawa[1]([✉]), Fumitaka Hayashi[1], Toshiyuki Amagasa[2],
and Hiroyuki Kitagawa[2]

[1] Graduate School of Systems and Information Engineering, University of Tsukuba,
Tsukuba, Japan
{kyusuke,hayashi}@kde.cs.tsukuba.ac.jp
[2] Faculty of Engineering, Information and Systems, University of Tsukuba,
Tsukuba, Japan
{amagasa,kitagawa}@cs.tsukuba.ac.jp

Abstract. Canopy clustering is a preprocessing method for standard
clustering algorithms such as k-means and hierarchical agglomerative
clustering. Canopy clustering can greatly reduce the computational cost
of clustering algorithms. However, canopy clustering itself may also take
a vast amount of time for handling massive data, if we naïvely implement
it. To address this problem, we present efficient algorithms and implementations of canopy clustering on GPUs, which have evolved recently as
general-purpose many-core processors. We not only accelerate the computation of original canopy clustering, but also propose an algorithm
using grid index. This algorithm partitions the data into cells to reduce
redundant computations and, at the same time, to exploit the parallelism of GPUs. Experiments show that the proposed implementations
on the GPU is 2 times faster on average than multi-threaded, SIMD
implementations on two octa-core CPUs.

1 Introduction

Clustering is one of the prominent data mining tasks and has been paid significant attention during the last decade because of the increasing importance
of efficient and effective data analytics due to the growing amount of available
data. Since clustering algorithms usually require high computational cost, a large
number of acceleration techniques have been proposed. One of such techniques
is canopy clustering [12], which is a preprocessing method for another clustering algorithm. Canopy clustering first divides a dataset into rougher groups,
called canopies, than desired clusters, by using an inexpensive distance measure.
Canopy clustering reduces computational cost by restricting distance computations in clustering algorithms only between points within the same canopy.
For example, clustering of visual image data is accelerated by canopy clustering
recently [4].

Meanwhile, GPUs (graphics processing units) have attracted many
researchers and developers for accelerating data-mining algorithms as well as

---

F. Hayashi—Currently working at International Laboratory Corporation.

© Springer International Publishing Switzerland 2015
Q. Chen et al. (Eds.): DEXA 2015, Part I, LNCS 9261, pp. 334–348, 2015.
DOI: 10.1007/978-3-319-22849-5_23

database processing [7,14]. Although GPUs are originally developed for processing graphics, they have been recently utilized to accelerate data-parallel applications, by exploiting hundreds to thousands of GPU's simple processing units. Standard clustering algorithms on GPUs such as k-means have been already well studied [9,10,18]. However, the acceleration of canopy clustering by GPUs has not been attempted. Speeding up canopy clustering on GPUs can lead to the reduction of total processing time of clustering.

This paper presents efficient algorithms and implementations of canopy clustering for GPUs. First we implement original canopy clustering [12] on GPUs by exploiting efficient implementations of data-parallel primitives [7]. The implementation is further accelerated by a technique called kernel fusion [20], which can decrease the amount of memory accesses. Then we present an algorithm using grid index in order to exploit the spatial locality of data points. The grid index also enables the GPU to efficiently create multiple canopies in parallel. Experiments show that the grid-indexed implementation outperforms a multi-threaded, SIMD implementation on two octa-core CPUs by a factor of 2 on average.

The rest of this paper is organized as follows: Sect. 2 reviews related work on clustering, especially parallel clustering and GPU clustering, and Sect. 3 briefly describes the preliminaries on GPU computing. Section 4 explains the algorithm of canopy clustering and the GPU implementation. Section 5 describes the algorithm of canopy clustering using grid index and its GPU implementation. Then Sect. 6 presents the experimental evaluation of these algorithms. Finally Sect. 7 concludes this paper with future directions.

2   Related Work

Clustering is a task to separate data points from a dataset such that similar points belong to the same group called *cluster* [6]. In general, standard clustering algorithms require high computational cost. For example, k-means [6] has the time complexity of $O(nkt)$, where n is the number of data points, k is the number of clusters, and t is the number of iterations. As another example hierarchical agglomerative clustering [6] has the time complexity of $O(n^3)$ with n data points. Thus there are a large number of techniques to accelerate the processing of clustering algorithms. One of such techniques is canopy clustering, which is described in Sect. 4. Another option for acceleration is to parallelize the computation of clustering algorithms. The following first reviews parallel clustering on CPUs and then describes clustering on GPUs.

2.1   Parallel Clustering

Significant efforts have been made to parallelize clustering algorithms. For instance, Dash et al. [3] proposed a parallel algorithm of hierarchical agglomerative clustering for shared-memory architectures that employs partially overlapping partitioning. Patwary et al. [15] parallelized the DBSCAN algorithm [6] by using the disjoint-set data structure and a tree-based bottom-up scheme to build clusters. More recently, Li et al. [11] presented an approach of k-means for the MapReduce framework. They utilize three techniques: locality sensitive hashing, a novel center initialization algorithm, and a pruning scheme for avoiding unnecessary distance computations. To the best of our knowledge, only Soroush et al. [17] dealt with the parallelization of canopy clustering. They use canopy clustering as a running example for their new storage manager for parallel array processing, ArrayStore, and implement canopy clustering on top of ArrayStore. Their work is different from ours in that their objective is to develop a novel storage manager for parallel array processing, while our objective is to accelerate the processing of canopy clustering by using GPUs.

2.2   GPU Clustering

Clustering on GPUs has been extensively studied since the introduction of CUDA [13], which is described in Sect. 3 n particular, the k-means algorithm is well studied because of its popularity [9,10,18]. Wasif et al. [18] implemented all the steps of k-means entirely on the GPU, and evaluated the implementation not only on a single GPU but also on multi-GPU platforms. Kohlhoff et al. [9] also presented an efficient implementation of k-means for GPUs by utilizing parallel sorting as preprocessing. Another implementation of k-means for GPUs is proposed by Li et al. [10]. They take into account the dimensionality and adaptively employ two schemes for low-dimensional data and high-dimensional data, respectively.

There also exists work on the acceleration of other algorithms such as DBSCAN and hierarchical agglomerative clustering [2,16,19]. Böhm et al. [2] proposed CUDA-DClust, a parallel algorithm for density-based clustering on GPUs. They further accelerated this algorithm by using an index structure suited for GPU processing. Another variant of DBSCAN, called Mr. Scan, is introduced by Welton et al. [19]. This algorithm is implemented on the MRNet tree-based distribution network with GPU-configured nodes, and is capable of clustering 6.5 billion points in 17.3 min. Hierarchical agglomerative clustering was accelerated Shalom et al. [16]. They use partially overlapping partitions in order to efficiently parallelize the computation on GPUs.

As can be seen, there exist much existing work on the acceleration of clustering algorithms by GPUs, but there is no work on the acceleration of canopy clustering on GPUs, which is a target of this work.

3   GPU Computing

GPU computing means the use of GPUs for accelerating general-purpose computations rather than processing graphics [14]. GPUs have recently evolved as many-core processors and have been utilized to accelerate a wide range of applications such as data analytics and scientific computations. The advantages of GPUs over traditional CPUs are their high computational performance with a large number of simple processing units, relatively low price, and low power consumption. In order to develop programs for GPUs, the de-facto standard framework is CUDA (Compute Unified Device Architecture) provided by NVIDIA [13].

The CUDA GPU architecture is made up of multiple *streaming multiprocessors (SMs)*, which in turn consist of many simple processing units called *scalar processors (SPs)*. CUDA provides fine-grained parallelism by launching a massive number of lightweight threads. A large number of threads are grouped into a *thread block* (or a *block* for short). Typical numbers of threads per block are 128 and 256. Threads within a block run concurrently on an SM, sharing the resources of the SM. On the other hand, an SM can accommodate multiple blocks simultaneously, maintaining its resources among blocks and scheduling the threads of blocks. Blocks comprise a *grid*, which is generated each time when functions to be executed on GPUs are called; such functions are named *kernels*.

Meanwhile, GPUs have several kinds of memory. The largest memory on GPUs is *global memory*. For instance, NVIDIA Tesla K40 has the global memory of 12 GB. It can be accessible from all threads of a grid and has a high bandwidth although the access latency is high. SMs also contain on-chip memory, which can be accessed much faster than global memory although the size is small (64 KB). This memory can be used as *shared memory*, which is shared by threads within a block and can be used to exchange data among the threads.

There are important accessing patterns to global memory, called *coalesced accesses* [13]. In general, global memory accesses are serviced with one or more memory transactions, depending on the access pattern. If all threads of a *warp*, which is a group of 32 threads, access an aligned and contiguous region of 128 bytes, then these accesses are coalesced into one transaction. Otherwise, the accesses are coalesced into not one but multiple transactions and the performance deteriorates compared to the single-transaction case.

4   Simple Canopy Clustering

Canopy clustering was proposed by McCallum et al. [12] and is employed as a preprocessing step of major clustering algorithms such as k-means and hierarchical agglomerative clustering. Canopy clustering has been paid attention because it can greatly reduce the computational cost of large-scale clustering processing, which takes considerably long time if simply implemented. Canopy clustering firstly divides a dataset into rougher groups, called *canopies*, than desired clusters, by using a simple and inexpensive distance. Having created canopies, it makes more rigorous clusters by using clustering algorithms such as k-means. The main idea of canopy clustering is to reduce unnecessary computations by creating rough groups and restricting distance computations only between points within the same canopy.

Section 4.1 explains the algorithm of canopy clustering in more detail. Section 4.2 describes the GPU implementation of simple canopy clustering.

Algorithm 1. Simple Canopy Clustering.

Input: A set S of data points x_i, thresholds T_1 and T_2

1 $C \leftarrow \emptyset$ ▷ C is a set of canopies
2 $\Sigma \leftarrow S$ ▷ Σ is a set of center candidates
3 **while** $\Sigma \neq \emptyset$ **do**
4   │ $c \leftarrow$ get a point from Σ at random   ▷ c is a center
5   │ $C \leftarrow \emptyset$
6   │ **for** $x \in S$ **do**
7   │ │ **if** $d(x, c) \leq T_1$ **then**
8   │ │ │ $C \leftarrow C \cup \{x\}$ ▷ a canopy C includes a point x
9   │ │ **end**
10   │ │ **if** $d(x, c) \leq T_2$ **then**
11   │ │ │ $\Sigma \leftarrow \Sigma - \{x\}$ ▷ remove x from the candidates
12   │ │ **end**
13   │ **end**
14   │ $\mathcal{C} \leftarrow \mathcal{C} \cup \{C\}$
15 **end**
16 **return** $\mathcal{C}$

4.1 Algorithm

A pseudo-code of canopy clustering is shown in Algorithm 1. There are three inputs: a set of data points and two thresholds T_1 and T_2 ($< T_1$). The first threshold T_1 influences the number of points included in canopies, and T_2 affects the number of canopies created. At first, the algorithm initializes the candidates of centers as the input dataset (Line 2). Then a center point c is picked from the candidates at random (Line 4). A canopy around this center point is to be created in the subsequent steps (Lines 5–13). A canopy includes a data point x if the distance $d(x, c)$ between x and c is less than or equal to threshold T_1 (Lines 7–9). At the same time, the inequality $d(x, c) \leq T_2$ is evaluated and, if it holds, the data point x is deleted from the center candidates (Lines 10–12). Such canopy creation steps continue until there are no center candidates (Line 3), and finally a set of canopies is returned (Line 16).

Let us consider the example shown in Fig. 1, where nine 2-dimensional points are plotted. A center c is randomly selected from the nine points, and it turns out to be the point x_3 as shown in Fig. 1b. A canopy C_1 around this center is created as follows: First, the data points other than c are tested whether their distances against c are less than T_1. If a point satisfy the condition, it is included in the canopy. In the example, the canopy C_1 contains the points x_2, x_4, x_5, and x_7. Second, the data points are checked whether they should be deleted from the center candidates. If the distance $d(c, x)$ between the center c and a point x is less than or equal to a threshold T_2, this point is excluded from the candidates. Hence the points x_2, x_3, and x_5 are removed from the set of candidates in the example. The next canopy is also constructed in the same way, as shown in Fig. 1c, where the point x_6 becomes the next center. In this case,

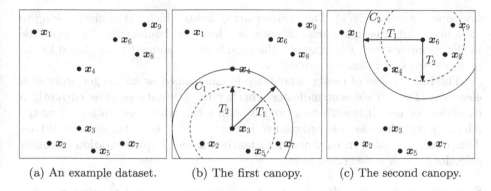

(a) An example dataset.   (b) The first canopy.   (c) The second canopy.

Fig. 1. An example of canopy clustering.

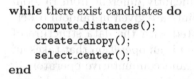

```
while there exist candidates do
    compute_distances();
    create_canopy();
    select_center();
end
```

Fig. 2. GPU implementation of the while loop in Algorithm 1.

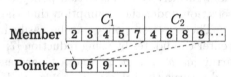

Fig. 3. Two arrays representing canopies.

the canopy C_2 includes the four points x_4, x_6, x_8, and x_9, and the points x_6, x_8, and x_9 are no longer center candidates. The same canopy-creation procedure continues while center candidates exist.

4.2 Implementation

We divide the iteration of Algorithm 1 into three kernels: select_center, compute_distances, and create_canopy. In addition, for the convenience of checking termination condition, we slightly change the while loop as shown in Fig. 2. That is, centers are selected at the last of iterations rather than the first.

The following sections explain data structures for easing efficient implementations and then describe implementations of the three kernels. We also introduce an optimization technique called kernel fusion.

Data Structures. We need to maintain the following data on global memory: input data points, resulting canopies, and information of center candidates.

The input data is a set of n d-dimensional data points, which can be represented by an $n \times d$ matrix. This data is read-only and is accessed only when distances are computed. To enable coalesced accesses in distance computations, the matrix is stored in a column-major format.

Resulting canopies are stored as two 1-dimensional arrays in a similar way to sparse-matrix formats [1]. Figure 3 shows the two arrays with regard to the previous example (Fig. 1). The member array holds which data points belong to canopies, while the index array maintains where the boundaries of canopies lie in the member array. For example, the members of canopy C_1 are stored in the member array at indexes 0 through 4.

The information of center candidates is maintained as an integer array of n elements. The ith element indicates whether the ith data point is currently a candidate or not. If the ith data point is still a candidate, we store $i - 1$ as the ith element; otherwise n is assigned to indicate that it is no longer a candidate. This format enables an easy and sufficiently efficient implementation of center selection, which is described next.

Selecting Centers. Before constructing a canopy, we need to select a center from the set of center candidates. To this end, we decide to choose the data point of minimum index among candidates as a center, for the ease of implementation. Intuitively this selection policy is not significantly different from random selection under the assumption that the data points are given in random order.

The selection policy can be easily implemented with the data structure of center candidates by using reduction [7]. *Reduction* is an operation that takes an array $[a_0, a_1, \ldots, a_{n-1}]$ and a binary associative and commutative operator $\oplus$, and returns the value $a_0 \oplus \cdots \oplus a_{n-1}$. If we use the `min` operation as the binary operator of reduction and apply the reduction to the array of center candidates, the above selection policy is realized. If the result of reduction is less than n, then there still exist candidates; otherwise there is no more candidate and thus the execution should be terminated. We implemented the reduce primitive based on the description of NVIDIA's guide [5].

Computing Distances. Having selected a center, we compute distances between the center and points. In this paper, we focus on the Euclidean distance although other distances are also applicable. Parallelization of distance computations is realized by simply assigning one thread to one distance computation. Since the data matrix on global memory is stored in column-major order, the threads can access the data points in a coalesced manner, thereby achieving high effective memory bandwidth. Having finished the distance computations, threads test the distances against the two threadsholds T_1 and T_2. The threads store the information whether the distance is less than or equal to T_1 into a T_1-flag array: The ith element of this array equals 1 if $d(c, x_i) \leq T_1$; otherwise the element is 0. Meanwhile, if the distance is less than or equal to T_2, the candidate is excluded from the set of center candidates, by assigning n to the corresponding position of center-candidate array.

Creating Canopies. This step gathers the indexes of data points included in the canopy, on the basis of the result of the last step. To this end, this step uses the procedure called *filter* [7], which consists of the following two steps: exclusive scan over the T_1-flag array and gathering the indexes of data points included in the canopy. Exclusive scan is an operation that takes an array $[a_0, a_1, \ldots, a_{n-1}]$ of n elements, a binary associative and commutative operator $\oplus$, and an identity I;

it returns the array $[I, a_0, a_0 \oplus a_1, \ldots, a_0 \oplus a_1 \oplus \cdots \oplus a_{n-2}]$. Applying exclusive scan with addition over the T_1-flag array gives us the array index of indexes that indicate where the elements should be outputted. By using this array, we store i to a result at the index of index$[i]$ if the ith element of T_1-flag array equals 1.

Kernel Fusion. We here introduce an optimization technique called *kernel fusion* [20]. Kernel fusion is, as the name indicates, to combine multiple kernels into a single kernel, thereby reducing the number of kernel calls and achieving the better utilization of resources. The above-explained kernels perform few computations and thus they are bandwidth-bound kernels. By introducing kernel fusion, we can expect performance improvements because it typically reduces the amount of accesses to global memory.

We combine the three kernels (i.e., compute_distances, create_canopy, and select_centers). In one kernel, a block deals with a chunk of input data, computes distances of the points in the chunk against the center, creates a part of canopy, and selects the next center. Having finished the distance computations, a block creates a part of canopy on shared memory, and adds the number of canopy members to a counter on global memory, which is initialized to zero in advance. For this addition, we use an atomic function, which returns the original value of target variable. The returned value can be used to the output index of canopy-member array. Thereby blocks write out canopy members to global memory in parallel. Meanwhile, center selection is done by using an atomic min function: First a block-level candidate is selected on shared memory by the atomic function, and then one thread in a block updates the candidate information on global memory, which is initialized to n. One important point to note is that the canopy members are not properly sorted with the above scheme because of the atomic function. Thus, we sort the canopy members after all the canopies are generated, by using segmented sort of Modern GPU[1], which sorts multiples segments in one array segment by segment.

5   Canopy Clustering with Grid Index

The simple canopy clustering algorithm has the problem that it performs a significant number of redundant computations, because it does not make use of spatial locality of data points. This section introduces an accelerated algorithm of canopy clustering by using a grid index structure. Section 5.1 describes the grid index, and then Sect. 5.2 explains implementation details.

5.1   Grid Index

The grid index is constructed by partitioning the data space into equi-width hypercubes, which we call *grid cells*. A grid cell $G_{i_1, i_2, \ldots, i_d}$ in d-dimensional space can be defined as a set $\{x \in S \mid i_j \cdot T_1 \leq x_j < (i_j + 1) \cdot T_1, 1 \leq j \leq d\}$, where

---

[1] http://nvlabs.github.io/moderngpu/.

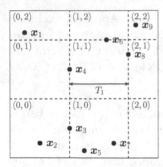

$$G_{0,0} = \{x_2\},\ \mathcal{N}(G_{0,0}) = \{G_{1,0}, G_{1,1}\}$$
$$G_{0,2} = \{x_1\},\ \mathcal{N}(G_{0,2}) = \{G_{1,1}, G_{1,2}\}$$
$$G_{1,0} = \{x_3, x_5, x_7\},\ \mathcal{N}(G_{1,0}) = \{G_{0,0}, G_{1,1}, G_{2,1}\}$$
$$G_{1,1} = \{x_4\},$$
$$\mathcal{N}(G_{1,1}) = \{G_{0,0}, G_{0,2}, G_{1,0}, G_{1,2}, G_{2,1}, G_{2,2}\}$$
$$G_{1,2} = \{x_6\},\ \mathcal{N}(G_{1,2}) = \{G_{0,2}, G_{1,1}, G_{2,1}, G_{2,2}\}$$
$$G_{2,1} = \{x_8\},\ \mathcal{N}(G_{2,1}) = \{G_{1,0}, G_{1,1}, G_{1,2}, G_{2,2}\}$$
$$G_{2,2} = \{x_9\},\ \mathcal{N}(G_{2,2}) = \{G_{1,1}, G_{1,2}, G_{2,1}\}$$

(a) Partitioned data space.

(b) Corresponding grid cells.

Fig. 4. An example of grid index.

S is a set of data points. The subscript $(i_1, i_2, ..., i_d)$ is called *cell coordinates*. A threshold T_1 is chosen as the width of grid cells, so that canopy creation is accomplished by computing the distances only between data points in a cell and its neighbor cells. A cell is called a *neighbor* of another cell if each dimension of the cell coordinates does not differ more than one, and the information of neighbors is also associated to grid cells. The other cells do not need to be taken into consideration in canopy creation because, if two cells are not neighbors, the distance between two points in each cell must be greater than T_1.

Figure 4 shows an example of partitioned data space and the corresponding grid cells. In Fig. 4b, $\mathcal{N}(G)$ means a set of neighbors of G. By using the grid index, we can create a canopy as follows: Let us consider that the point x_3 is selected as a center. Since this point belongs to the cell $G_{1,0}$, the distances are computed against the points in $G_{1,0}$ and its neighbor cells. In this case, the neighbors are $G_{0,0}$, $G_{1,1}$, and $G_{2,1}$. Distances are computed against five points (i.e., x_2, x_4, x_5, x_7, and x_8). Thus we can omit the distance computations with regard to the other three points (i.e., x_1, x_6, and x_9).

5.2   Implementation

This section presents how to efficiently implement canopy clustering with grid index on GPUs. It is especially important to efficiently construct a grid index on GPUs and to create multiple canopies in parallel. The following explains how to construct a grid index, how to select multiple centers, and how to create multiple canopies in parallel, one by one.

Constructing Grid Index on the GPU. A grid index can be represented by two sparse binary matrices: a matrix of cell members and a matrix of neighbor information. A sparse binary matrix, in turn, can be expressed by two 1-dimensional arrays as used for canopies. Thus we store a grid index on the GPU as four 1-dimensional arrays in total. Such a grid index is efficiently constructed on the GPU by exploiting fast data-parallel primitives in three steps:

computing cell coordinates from data points and sorting them; building the cell-member matrix on the basis of the coordinates; and constructing the neighbor matrix on the basis of the coordinates.

Computing and Sorting Cell Coordinates. The first step computes the cell coordinates of each data point (i.e., the coordinates of the cell to which the point should belong), and these coordinates are stored on global memory in column-major order. Then the coordinates are sorted in the dictionary order with data-point indexes as associated data. Based on an example of Thrust[2], we realize the dictionary-order sort by sorting dimension by dimension from the last: If there are d dimensions, first the dth dimension with data indexes is the target of sorting, and then the $(d-1)$th dimension is subject to next sorting. Before the sorting of $(d-1)$th dimension, the coordinates of $(d-1)$th dimension is not consistent with the order of dth dimension. Thus it is necessary to arrange the order according to the data-index array. Having arranged the coordinates, we sort the $(d-1)$th dimension. The sorting procedure continues until the sorting of first dimension completes. Finally, the dimensions other than the first need to be arranged in the final dictionary order. As for sorting, we used the sort function of Thrust library. We also used the gather function of Thrust for arranging the dimensions according to the data indices.

Building the Cell-Member Matrix. Having sorted the coordinates, we next construct the cell-member matrix. The array of cell members is already obtained at this time, because it is the array of data indexes sorted in the first step. Thus it is sufficient to compute the array of pointers in order to finish the construction of cell-member matrix. This array can be easily obtained by examining the sorted coordinates: If the coordinates of a point are different from the coordinates of the next point, the index of the next point indicates the boundary of same cell members.

This step can be easily parallelized on the GPU by assigning threads to points. A thread checks the coordinates of the assigned point and the next point, and stores the index to an array on global memory if the two points have different coordinates. For enabling concurrent writes to the resulting array, we prepare a global counter of current index for the pointer array. This counter is first set to 1, because we initialize the first element of pointer array to 0. If a thread wants to write an index to the pointer array, the thread increments the counter by 1 with an atomic function. By using the value returned from the atomic function as the output index of pointer array, the thread writes the boundary pointer value. Since we use an atomic function, the write order is not deterministic. This means that the elements in the pointer array is not properly ordered. Thus we sort the pointer array so that adjacent elements of this array stands for the boundaries of cell members.

Constructing the Cell-Neighbor Matrix. The last step is to construct a cell-neighbor matrix. To this end, firstly the unique cell coordinates are gathered from the sorted coordinates created in the first step. With the coordinates, this

---

[2] https://code.google.com/p/thrust/.

step is performed in two passes: The first pass just counts the number of neighbors of each cell, and the second pass constructs a cell-neighbor matrix according to this information.

The first pass simply checks all of the combinations of coordinates. This is parallelized by assigning one block to the coordinates of one cell. Within the block, one thread compares the assigned coordinates and coordinates of another cell. Each block outputs the counting result to a corresponding location of pre-allocated array on global memory. Then exclusive scan is performed over this array, and the output becomes the pointer array of neighbor matrix.

The second pass similarly checks all of the combinations, but this time outputs cell-neighbor indices to global memory. Since we have already obtained the pointer array, it is known where a block should output the neighbor indices. In order to enable threads to write concurrently, a counter is used again. The counter is prepared on shared memory and is initialized to 0. If a thread have found a neighbor, the counter is incremented and the neighbor index is stored to the array on global memory.

Selecting Multiple Centers. The selection of multiple centers is performed in two phases: selection and verification phases. The selection phase chooses a canopy center per grid cell, so that there are canopy centers as many as the grid cells. The verification phase checks whether or not the selected centers violate a condition derived from the canopy-creation process: The distance of two canopy centers is not less than a threshold T_2. If a pair of centers violates this condition, a point of the pair is invalidated as canopy center at current iteration.

The selection phase selects centers by reduction in a similar way as in Sect. 4.2. More specifically, one block deals with one cell and selects a center from the points in the cell. A block performs the reduction on shared memory and one thread of the block outputs the selected center to a pre-allocated array of as many elements as the number of cells. Note that the information of candidates is maintained on global memory as in Sect. 4.2.

The verification phase checks the above-described condition of centers. In order to check multiple centers in parallel, we employ a rather pessimistic scheme: A center is invalidated if one of the other centers is closer than the threshold T_2, although the other centers may be invalidated by other blocks concurrently. To realize this scheme, one block checks the validity of one center, and one thread within the block compares the center against another center. This scheme may over-invalidate centers, but experiments in Sect. 6 indicate that it can provide a sufficient level of parallelism.

Creating Multiple Canopies. In order to generate multiple canopies in parallel, while basically the same procedure to simple canopy clustering can be employed, the difficulty lies in how to assign the work to thread blocks for the creation of canopies. A naïve scheme is to make one canopy by one block, but it could suffer from severe load imbalance, because the required number of distance computations largely varies canopy by canopy. To alleviate this problem, we assign one block to either a cell that contains a center or one of the neighbors of such a cell. Blocks compute distances between a center and the points of assigned cell, and generate a part of canopy. For example, let us consider to create a canopy around the point x_3 in Fig. 4. In this case, one block handles the cell $G_{1,0}$ and other three blocks are responsible for the neighbor cells $G_{0,0}$, $G_{1,1}$, and $G_{2,1}$ with respect to the cell $G_{1,0}$.

6   Experiments

This section evaluates the performance of our proposed implementations. Section 6.1 summarizes the experimental settings and Sect. 6.2 compares CPU and GPU implementations with and without grid index.

6.1   Experimental Settings

We used CUDA 6.5 for GPU implementations and conducted experiments on an NVIDIA Tesla K40 GPU, which has global memory of 12 GB and 15 SMs, each of which consists of 192 SPs at 745 MHz. The machine has two Intel Xeon Processor E5-2687W v2 at 3.40 GHz with memory of 128 GB. The OS is CentOS release 6.5 (Final).

We compared four implementations: `cpu-simple`, a CPU implementation of simple canopy clustering; `cpu-grid`, a CPU implementation with grid index; `gpu-simple`, a GPU implementation of simple canopy clustering including kernel fusion; and `gpu-grid`, a GPU implementation with grid index. We parallelized the CPU implementations by ourselves using OpenMP and AVX, which is SIMD functionality on Intel CPUs [8]. The multi-threaded, SIMD implementations on two octa-core CPUs are up to 17 times faster than serial implementations.

We used synthetic datasets for evaluation because we had difficulty in finding suitable real-world datasets. The synthetic datasets are generated as follows: The number of clusters is given as an input, and each cluster is represented by a tuple of d normal distributions, which are independent with each other. The means of normal distributions are generated by the uniform distribution between 0 and 1, and the standard deviations are given by the normal distribution with mean 0.01 and standard deviation 0.005. With these clusters, the given number of points are generated one by one: First we determine which cluster a new point belongs to in a uniformly-random manner, and then the point's coordinates are given by the normal distributions of the cluster. We set the number of clusters to 100 in the experiments. The dimensionality d was varied from 2 to 6; this may seem small, but canopies are often constructed on the basis of a small number of attributes [12]. Thus the dimensionality is not very small for canopy clustering.

6.2   Results

Varying Dataset Sizes. Figure 5 shows the elapsed times of the four implementations with varying the size of datasets. Note that the elapsed time of GPU

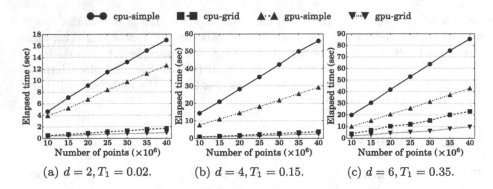

(a) $d = 2, T_1 = 0.02$.　　(b) $d = 4, T_1 = 0.15$.　　(c) $d = 6, T_1 = 0.35$.

Fig. 5. Elapsed times with varying the size of synthetic datasets.

implementations includes CPU–GPU data-transfer time, but does not include GPU–CPU data-transfer time because we assume that the result of canopy clustering is used in later clustering on GPUs. For this experiment, the threshold T_1 is chosen such that the number of canopies become around 3,000, and the other threshold T_2 is fixed to $0.7T_1$ for the experiment purpose; in practice, the thresholds should be determined according to the needs of applications.

Overall, each kind of GPU implementations is consistently faster than the corresponding CPU implementation. The `gpu-simple` implementation outperforms `cpu-simple` by a factor of 1.4 on two-dimensional datasets, 1.9 on four-dimensional datasets, and 2.0 on six-dimensional datasets. As the dimensionality increases, the GPU implementation gets faster than the CPU implementation. This is because canopy clustering of higher-dimensional datasets results in more bandwidth-bound kernels. Consequently the wide memory bandwidth of GPUs become more effective in higher-dimensional cases.

The GPU implementation using grid index (i.e., `gpu-grid`) is also faster than `cpu-grid`, leading to speedup of 1.3–1.8 in two dimensions, 1.1–1.6 in four dimensions, and 1.9–2.5 in six dimensions. From this result, we believe that our scheme of multiple center selection (Sect. 5.2) achieves to provide sufficient level of parallelism for GPUs. However, `gpu-grid` becomes less efficient in four-dimensional case. This can be considered due to load imbalance: Since we assign one block to one grid cell, the processing time of blocks varies depending on the size of grid cells, which can be largely different by dataset characteristics and thresholds. This load-imbalance problem should be remedied in our future implementations.

Varying Thresholds. We here fix the size of datasets to 10 million and change the thresholds T_1 and T_2, which affect the number of canopy members and the number of canopies, respectively. Figure 6 shows the elapsed times of the four implementations. The value of T_1 is varied so that the number of canopies lies in roughly the range of 1,000 and 10,000. The threshold T_2 is fixed to $0.7T_1$ as before.

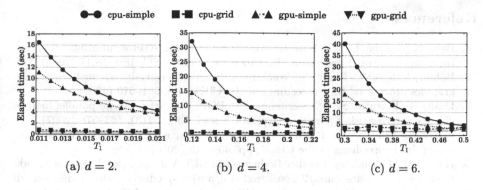

Fig. 6. Elapsed times with varying the thresholds T_1 and T_2.

The implementations using grid index get faster and faster as the thresholds decrease, compared to the implementations without grid index. This is because more canopies are generated when thresholds are smaller. The implementations with grid index carry out distance computations between a center and a limited number of points, while the implementations without grid index perform distance computations between a center and all of the other points at each iteration. In other words, the more canopies are generated, the more efficient the implementation using grid indices are, compared to the simple canopy clustering. Meanwhile, in six-dime national case (Fig. 6(c)), the speedup ratios between gpu-grid and cpu-grid range from 1.33 to 2.01. Such instability comes from the load-imbalance problem as before, because the organization of grid index differs according to the thresholds.

7   Conclusions

This paper has presented GPU implementations of canopy clustering, parallelizing simple canopy clustering and canopy clustering with grid index. The implementation of simple canopy clustering is accelerated by the kernel-fusion technique. Canopy clustering using grid index exploits the spatial locality of input data points, thereby reducing the number of distance computations and providing high parallelism to the GPU. Experiments showed that the kernel-fused implementation and the GPU implementation with grid index outperform multi-threaded, SIMD implementations on two octa-core CPUs by a factor of up to 2. One future direction is to alleviate the load-imbalance problem of implementations with grid index. A simple solution would be to assign multiple blocks to one grid cell adaptively on the basis of the number of cell members.

Acknowledgments. This research was partly supported by the Grant-in-Aid for Scientific Research (B) (#26280037) from Japan Society for the Promotion of Science.

References

1. Bell, N., Garland, M.: Implementing sparse matrix-vector multiplication on throughput-oriented processors. In: Proceedings of the SC, 18:1–18:11 (2009)
2. Böhm, C., Noll, R., Plant, C., Wackersreuther, B.: Density-based clustering using graphics processors. In: Proceedings of the CIKM, pp. 661–670 (2009)
3. Dash, M., Petrutiu, S., Scheuermann, P.: pPOP: Fast yet accurate parallel hierarchical clustering using partitioning. Data Knowl. Eng. **61**(3), 563–578 (2007)
4. Fan, Z.G., Wu, Y., Wu, B.: Maximum normalized spacing for efficient visual clustering. In: Proceedings of the CIKM, pp. 409–418 (2010)
5. Harris, M.: Optimizing Parallel Reduction in CUDA. http://developer.download. nvidia.com/compute/cuda/2_2/sdk/website/projects/reduction/doc/reduction.pdf
6. Han, J., Kamber, M., Pei, J.: Data Mining: Concepts and Techniques, 3rd edn. Morgan Kaufmann, Burlington (2011)
7. He, B., Lu, M., Yang, K., Fang, R., Govindaraju, N.K., Luo, Q., Sander, P.V.: Relational Query Coprocessing on Graphics Processors. ACM Trans. Database Syst. **34**(4), 21:1–21:39 (2009)
8. Lomont, C.: Introduction to Intel® Advanced Vector Extensions. https://software. intel.com/en-us/articles/introduction-to-intel-advanced-vector-extensions
9. Kohlhoff, K.J., Pande, V.S., Altman, R.B.: K-Means for parallel architectures using All-Prefix-sum sorting and updating steps. IEEE Trans. Parallel Distrib. Syst. **24**(8), 1602–1612 (2013)
10. Li, Y., Zhao, K., Chu, X., Liu, J.: Speeding up k-Means algorithm by GPUs. J. Comput. Syst. Sci. **79**(2), 216–229 (2013)
11. Li, Q., Wang, P., Wang, W., Hu, H., Li, Z., Li, J.: An efficient K-means clustering algorithm on MapReduce. In: Bhowmick, S.S., Dyreson, C.E., Jensen, C.S., Lee, M.L., Muliantara, A., Thalheim, B. (eds.) DASFAA 2014, Part I. LNCS, vol. 8421, pp. 357–371. Springer, Heidelberg (2014)
12. McCallum, A., Nigam, K., Ungar, L.H.: Efficient clustering of High-dimensional data sets with application to reference matching. In: Proceedings of the KDD, pp. 169–178 (2000)
13. NVIDIA: CUDA C Programming Guide. http://developer.download.nvidia.com/ compute/DevZone/docs/html/C/doc/CUDA_C_Programming_Guide.pdf
14. Owens, J.D., Houston, M., Luebke, D., Green, S., Stone, J.E., Phillips, J.C.: Proc. IEEE GPU Comput. **96**(5), 879–899 (2008)
15. Patwary, M.A., Palsetia, D., Agrawal, A., Liao, W.k., Manne, F., Choudhary, A.: A new scalable parallel DBSCAN algorithm using the disjoint-set data structure. In: SC, pp. 62:1–62:11 (2012)
16. Shalom, S.A.A., Dash, M.: Efficient partitioning based hierarchical agglomerative clustering using graphics accelerators with CUDA. Int. J. Artif. Intell. Appl. **4**(2), 13–33 (2013)
17. Soroush, E., Balazinska, M., Wang, D.: ArrayStore: a storage manager for complex parallel array processing. In: SIGMOD, pp. 253–264 (2011)
18. Wasif, M., Narayanan, P.: Scalable clustering using multiple GPUs. In: HiPC, pp. 1–10 (2011)
19. Welton, B., Samanas, E., Miller, B.P.: Mr. Scan: Extreme scale density-based clustering using a tree-based network of GPGPU nodes. In: SC, 84:1–84:11 (2013)
20. Wu, H., Diamos, G., Cadambi, S., Yalamanchili, S.: Kernel weaver: automatically fusing database primitives for efficient GPU computation. In: MICRO, pp. 107–118 (2012)

Query Processing and Optimization

Efficient Storage and Query Processing of Large String in Oracle

George Eadon[1], Eugene Inseok Chong[1(✉)], and Ananth Raghavan[2]

[1] Oracle, 1 Oracle Drive, Nashua, NH 03062, USA
{george.eadon, eugene.chong}@oracle.com
[2] Oracle, 500 Oracle Parkway, Redwood Shores, CA 94065, USA
ananth.raghavan@oracle.com

Abstract. Variable size strings are a fundamental data type in RDBMS and used in virtually all database components and applications including XMLs, blogging, customer service comments, e-commerce product descriptions, etc. Many applications could require large strings to store XML documents, JSON documents, customer support history, blog entries, or HTML documents. Many social network applications as well as web 3.0 applications also require large string type. A naïve implementation would simply increase the size of the traditional variable strings, but this will incur performance problems due to row chaining. Using Large Object type (LOB) will enable users to store large string without row chaining, but it is difficult to manipulate LOBs and many built-in operators for strings are not applicable to LOBs. Oracle 12c provides a capability of storing large strings without the row-chaining problem while eliminating LOB's deficiencies. In addition, users can control the data placement and storage format based on their application workload. However, reading the large string from storage for each reference to the string would be inefficient for queries that reference the strings frequently. This paper presents an efficient processing strategy for queries involving large strings, while supporting theoretically unlimited size of the strings. It illustrates how seemingly simple conceptual work involves careful design and extensive engineering work to have a scalable and efficient implementation. The solution has been implemented in Oracle 12c, and the performance results show its efficiency.

Keywords: Large string · Database · Operator evaluation · SQL · Query processing

1 Introduction

Variable size strings are a core data type of RDBMS, and they are used in virtually all database applications. With a plethora of unstructured/semi-structured data [7] LOB storage mechanism has gained a significant popularity. However, there have been limitations imposed on this type of data in terms of ease of data manipulation; in particular, LOBs do not generally work with built-in operators or built-in SQL features such as SELECT DISTINCT. The traditional limit of 4000 bytes on variable strings is too small to handle the unstructured/semi-structured data. The need for larger size string is apparent and many applications want to store XML documents [10], JSON

© Springer International Publishing Switzerland 2015
Q. Chen et al. (Eds.): DEXA 2015, Part I, LNCS 9261, pp. 351–366, 2015.
DOI: 10.1007/978-3-319-22849-5_24

documents [11], customer service history, medical records, or product descriptions directly in the variable string. Social networks [8, 9] and web 3.0 applications [6] as well as big data applications are also heavy users of variable size strings.

The motivation of this support comes from a large number of customers requesting the feature. They want to store documents, discussion threads, HTML pages, etc. and do operations such as compare, order by, group by, and fetch directly on them. Typically, this data exceeds the maximum size of 4000 bytes allowed. When the data is fetched, the customers want to apply all built-in operators on the data seamlessly. The customers do not want to store the data as LOB because of the inability of applying the built-in operators and poor performance compared to string columns. Some customers store data in character form up to 4000 characters and they want to convert data into a Unicode without data loss, which could extend up to 16000 bytes.

Historically Oracle has supported 32767 byte variable strings in its procedural language PL/SQL, but only supported 4000 bytes for VARCHAR (variable-length character) and NVARCHAR (variable-length character in national character set) types in SQL, and 2000 bytes for RAW (variable-length binary) in SQL. With the changes described in this paper Oracle will support up to 32767 bytes for VARCHAR, NVARCHAR, and RAW data types in SQL. This 32767 byte limitation is due to the size of the column length field, which is a signed 16-bit type, and support up to 2 GB is possible if the length field is increased to a 32-bit type. However, our design is based on the vision that the size of the string is limited only by LOB storage; hence it provides practically unlimited size of the string. Oracle currently supports the maximum LOB size of 8 terabytes to 128 terabytes, depending upon the chunk size [5].

There are several ways to implement this large variable string. The simplest is to increase the size of the traditional VARCHAR. In this case a number of data blocks required to accommodate the large string will be chained together. Each chained block requires a separate I/O to access. Therefore it incurs more I/O to access the data. Secondly, accessing other data placed after the large string will require more I/Os because we need to scan data blocks for the large string first to get to the data we seek. This approach is unacceptable due to its poor I/O performance. As can be seen in the experiments section, the select performance of columns other than the large size strings can be an order of magnitude slower.

Some users have used a 2-column approach without database kernel change, i.e. they define 2 columns when they create a table: one column for a traditional variable string (VARCHAR) and the other for a Large Object (BLOB/CLOB). Small strings are stored in VARCHAR column with null LOBs, and the large strings are stored in LOBs with null VARCHAR. However, this approach is undesirable because it unnecessarily increases the number of columns in the table, and makes it difficult for user to use the large strings because LOBs are hard to manipulate compared to traditional VARCHAR – e.g. built-in SQL operations like SELECT DISTINCT may not work for LOBs.

Our approach stores a small string in-line within a row up to 4000 bytes by default, and then larger strings out-of-line in a BLOB. All LOBs are handled internally by our own function so that from a user's perspective the column is still of a large variable string type and all built-in operators will work for the string. This way we do not have an overhead of chained data blocks and users do not need to manipulate LOBs. Again, this approach can be implemented in three ways: 2-columns, 1-column, or hybrid.

The 2-column scheme uses hidden BLOB storage. That is, when a table is created a traditional scalar column and an implicit hidden BLOB column are created. This hidden BLOB column cannot be manipulated by users. Though it is stored as two-column, it will act as one logical column for all SQL processing.

The advantage of the 2-column scheme is that it does not change the storage format of the data, i.e. VARCHAR type is stored as VARCHAR format and BLOB column is stored as BLOB format. The disadvantage of the 2-column scheme is that we increase the number of internal columns of a table and we increase implementation complexity of the SQL compiler because a single logical column manifests as two physical columns. The 1-column scheme stores the string in a BLOB. A value smaller than 4000 bytes is stored in an inline BLOB, whereas larger values are stored in an out-of-line BLOB. The hybrid scheme allows the storage format to vary from row to row, using flags in the row format to indicate which storage format was used for each value; values smaller than 4000 bytes are stored in traditional format while larger values are stored in BLOB format. The disadvantage of the hybrid scheme is that it requires changes to the on-disk storage format.

We have chosen the 1-column scheme because it does not increase the number of internal columns of a table and it has a consistent data type. But it will bring more complexity because we introduce a new storage format for large strings, i.e. a column of type VARCHAR is stored as BLOB. In addition, we have created a parameter that sets the threshold at which VARCHAR, NVARCHAR, and RAW columns switch from traditional format to BLOB storage, giving the users flexibility of choosing the threshold other than the default 4000 bytes.

Once the string is stored in a BLOB, we need to read the BLOB as efficiently as possible to get its string value. Unlike a regular BLOB column where the in-memory locator is built to manipulate the LOB columns later on, we bypass the in-memory locator build and privilege checking because the locator is internally generated and internally consumed without ever being returned to the client. We built an internal function to read the string value from BLOB.

The large string stored as BLOB has two representations: LOB locator and the scalar value stored in the LOB. The secondary indexes and queries use the scalar value whereas DMLs may use locator for efficient processing in addition to the scalar value.

Reading the large string from storage may require a significant time and it could affect the query performance negatively. Therefore, we need to read the value when it is absolutely necessary and only once. To achieve this, we prepend all large strings in the WHERE clause and append all large strings not in the WHERE clause to the filter. Any large string appended to the filter is read only when a row satisfies the filter condition, avoiding unnecessary evaluation of the large string.

An index built on a large string may not contain the whole value due to index key size limitations, or a function such as hash should be applied to contain the whole key. Therefore, some extra processing is required to utilize the index when a user query filters on the large string column, and the processing should be done seamlessly without user's awareness.

There are a number of optimization issues: (1) determining the threshold where we switch between traditional inline storage format and BLOB format; (2) operator evaluation for the BLOB to string conversion so that it is evaluated only once, instead of evaluating every time the operator is encountered; (3) efficient management of memory because the memory requirement is increased with the large string; and (4) efficient handling of indexes due to index key size limitation and the space requirement of indexes.

We will examine each issue in the following sections and propose efficient solutions.

1.1   Related Work

Many commercial vendors support large strings. Among major vendors, IBM DB2 [1] supports up to 32704 bytes, and Microsoft SQL Server [2] supports up to 8000 bytes while VARCHAR(MAX) will support up to about 2 GB by converting it to Binary Large Object (BLOB). Open source database MySQL [12] allows up to 65,635 bytes, but the maximum length is limited by the maximum row size allowed, which is 65,535 bytes. PostgreSQL [13] supports up to 1 GB if the length is not specified.

Unlike SQL Server where users specify up to 8000 and then MAX, Oracle allows users to specify exact length of the column, e.g. VARCHAR(25000). Being able to specify the exact maximum size will help optimizer to come up with a better execution plan.

The SQL Server stores in-line within a row up to 8000 bytes and the values larger than 8000 bytes are stored out-of-line. The DB2 stores the variable-length string in-line. Oracle stores in-line up to 4000 bytes by default, and the values larger than 4000 bytes are stored out-of-line. However, the 4000 byte threshold can be changed using a parameter, giving users the maximum flexibility in terms of column placement and storage format for a better performance. As far as we know, none of the vendors offers this capability Oracle provides.

As the size of the string increases, it may have implications on the index key size. The maximum index key size for the SQL Server is 900 bytes and the maximum index key size for DB2 is 1024 bytes. The maximum index key size for Oracle depends on the block size. For 8 K block size, it allows users to build an index of size up to 6398 bytes.

2   Datatype and Storage

The first thing that should be considered in a design of a new feature is the backward compatibility and upgrade/downgrade issues. Changing the maximum capacity of VARCHAR, NVARCHAR, and RAW data types in SQL can cause application-level incompatibilities, even if no data changes. This happens because of cases where these data types are implicitly limited to the maximum capacity of the type, such as:

- A VARCHAR2 column with character length semantics is limited to maximum byte length for VARCHAR, even if this is not sufficient to hold the declared number of arbitrary characters.
- An Oracle PL/SQL function returning VARCHAR2, NVARCHAR2, or RAW will be treated by SQL as a maximum length VARCHAR2, NVARCHAR2, or RAW.

- Some operators such as string concatenation are implicitly limited to maximum length of the data type. For example, if A and B are both VARCHAR2(4000) columns, then A ‖ B will be a VARCHAR2(4000) type if the maximum length of the varchar type is 4000 bytes, and will be a VARCHAR2(8000) if the maximum length of the varchar type is 32767 bytes.

To ensure application compatibility, support of these 32767-byte types is conditional based on a parameter value. For the user, creation of a large string in a table is just a matter of specifying the maximum size they want, but before the user issues such DDL, the user needs to set the MAX_STRING_SIZE parameter to EXTENDED. The EXTENDED setting means that the 32767 byte limits apply, whereas the other valid value for the parameter, STANDARD, means that Oracle's historical length limits apply:

```
CREATE TABLE TAB1(a int, b varchar2(27000));
```

Whether or not the large string is stored as a scalar value or as a BLOB should be based on the workload and users' application objectives. Some users may prefer to store the whole large string as a scalar value to avoid the LOB storage even if they suffer from slow I/O performance due to row chaining.

The storage format is controlled by a parameter, which sets the threshold at which VARCHAR, NVARCHAR, and RAW columns are stored as BLOB. Columns with maximum size larger than this threshold are stored as BLOB, and columns with maximum size equal to or smaller than this threshold are stored as inline scalars. This decision is made at DDL time, so all rows in the table will be stored in the same format, and the length of the actual data in the column does not matter. By default this threshold will be 4000 bytes, however the parameter will allow users to set this threshold anywhere in the range [0, 32767]. With threshold value 0, all string columns will be stored as BLOB whereas threshold value 32767 will have all string columns stored as scalar values possibly with chained rows.

The storage format of a column is set when the column is created as part of a CREATE TABLE or an ALTER TABLE adds a new column. We support ALTER TABLE to modify the capacity of an existing column, but this never changes the storage format of the column. Therefore, even if the parameter is left at its default value, after ALTER TABLE is used to modify the capacity of an existing column we could have a 32767 byte column stored in traditional format, or a 1 byte column stored in BLOB format.

Oracle provides several LOB data types, including BLOB for binary data and CLOB for character data. Large VARCHAR, NVARCHAR and RAW types are all stored internally as BLOB. We did not choose CLOB for VARCHAR or NVARCHAR because CLOB requires additional character set conversions within the LOB layer. For VARCHAR and NVARCHAR all required data type conversions are done as if it were a VARCHAR or NVARCHAR type stored in traditional format. Then, the resulting value is stored in the BLOB type.

However, unlike regular BLOB columns, evaluation of the BLOB for large string storage is optimized for better performance, namely there will be no text-to-binary conversion (i.e. hexadecimal code interpretation), no privilege checking, and no in-memory locator build. In addition, during optimization phase there is no evaluation operator applied to the large string type so that all regular optimizer statistics for the VARCHAR type column is utilized. The evaluation operator is applied when we transform the large string column at runtime. Please see Sect. 4 for details.

Oracle provides two kinds of LOB storage format, BasicFile LOB and Secure-File LOB. SecureFile LOB is more advanced format and it provides compression, encryption, etc. while providing scalability [4]. Our LOB segments are always located in the same tablespace as the base table, and SecureFile LOBs are always used when the tablespace supports SecureFile LOBs.

3 Locator vs. Scalar Value

As described in Sect. 1, the large string stored as BLOB will have two representations to deal with: locator and scalar value. The scalar value is the actual value of the large string and the locator is the pointer into the BLOB segment. The locator is stored in table rows and hence when read from a table, the large string column buffer will contain the locator. Given this locator we need to get the scalar value for subsequent SQL processing. This BLOB to VARCHAR conversion (BL2 V) is done by evaluating an internally generated BL2 V operator.

The scalar values are used for most cases like queries, index keys, constraints, etc., but the locator stored in table rows is required for some DML scenarios such as compare columns for updates, as explained in Sect. 5.

Since SQL gets the scalar value from the BL2V operator buffer, we need to generate the scalar value from the column buffer. Depending upon the value in the column buffer, we need to employ one of two new operators. If the value in the column buffer is a scalar value like in the index scan, we just need to copy the scalar value from the column buffer to operator buffer using LVCPY (long varchar copy) operator. On the other hand, if the value in the column buffer is a locator, we need to read the BLOB value that the locator points to using LVEVA (long varchar evaluate) operator.

Therefore, the LVCPY and LVEVA operators tell us how to compute the result of the BL2 V operator. LVCPY(BL2 V(b)) will copy the scalar value from the column b's buffer to the operator buffer. LVEVA(BL2 V(b)) will evaluate b's BLOB and put the scalar value into the operator buffer. Because these operators are internal to Oracle kernel, the execution plan will show them as INTERNAL_FUNCTION. For example, if a user issues a query `select * from tab1`, the plan will show as follows:

```
----------------------------------------------
| Id  | Operation            | Name          |
----------------------------------------------
|   0 | SELECT STATEMENT     |               |
|*  1 |   TABLE ACCESS FULL  | TAB1          |
----------------------------------------------
Predicate Information (identified by operation id):
----------------------------------------------
   1 - filter(INTERNAL_FUNCTION("TAB1"."B"))
```

As can be seen above in the predicate information, the LVEVA operator is applied to the long varchar column 'B.'

4 Predicate Filter Injection and Operator Evaluation Optimization

A problem with using an operator is that a naïve implementation would evaluate the operator every time it's encountered. Because the operator might need to read the out-of-line LOB it's relatively expensive to evaluate and hence evaluating it unnecessarily will negatively affect performance. To prevent this from happening we do not evaluate the BL2 V operator by default. Instead we implement the special LVEVA and LVCPY operators to evaluate BL2 V and add these special operators to index scan and table scan filter predicates so that the BL2 V operator is evaluated only when needed and only once.

When transforming a compile-time large string column to its runtime representation, the large string column c is transformed into an operator BL2V(c). The evaluation operator of BLOB to get the scalar value (BL2V) is applied to the large string column to convert BLOB to characters. However, the large string may be referenced multiple times in a query, and we do not want to evaluate every time. Instead we want to evaluate once and then use the scalar value anytime we encounter the same column. To accomplish this, we evaluate the large string when we process filter predicates for indexes and tables. If there is no filter predicate, we create the filter and add the BL2V operator to it. As can be seen from the execution plan in Sect. 3 above, even if the query does not contain any filter clauses, the plan shows a filter applied to the long varchar column. The optimization works as follows:

For index scan, the column buffer will contain the scalar value because the index keys are scalar values. If a large string column is referenced in the index filter we add the LVCPY operator for the column to the front of the filter so the scalar value is available when we evaluate the filter. For large string columns that are referenced in the query but are not referenced in the index filter we add the LVCPY operator for the column to the end of the filter so that the scalar value is copied into the BL2V operator buffer. The reason why we put the LVCPY operator to the end is that we do not want to copy the scalar value unnecessarily if the row does not satisfy the filter condition. Once this is done, we set a flag to prevent re-evaluation of the operator for the same column later on.

For table scan, the column buffer will contain the LOB locator. Therefore we add the LVEVA operator to the front of the filter predicates for large string columns included in the filter because we need the scalar values to check the filter condition. For large strings columns that are referenced in the query but not referenced in the filter we add the LVEVA operator to the end of the filter predicates. Again for the same reason as the index scan, we do not evaluate large string columns that do not satisfy the filter condition. As in the index scan, we set the flag to prevent re-evaluation of the operator for the same column.

Both operators always return TRUE and are always added to the filter with an AND conjunction so that they will not affect the outcome of the filter condition. When they are added to the empty filter predicate, the table rows always satisfy the filter condition. For example, in the following scenario:

```
create table t1(a int, b varchar2(5000),
                c varchar2(12000));
create index i1 on t1(b);
select length(b), length(c)
from t1 where b > rpad('aa',20,'bb');
```

The execution plan shows:

```
-----------------------------------------------
| Id  | Operation              | Name          |
-----------------------------------------------
|   0 | SELECT STATEMENT       |               |
| * 1 |   TABLE ACCESS BY      |               |
|     |    INDEX ROWID BATCHED | T1            |
| * 2 |   INDEX RANGE SCAN     | I1            |
-----------------------------------------------

Predicate Information (identified by operation id):
-----------------------------------------------
1 - filter(INTERNAL_FUNCTION("C"))
2 - access("B">'aabbbbbbbbbbbbbbbbbbbb')
filter(INTERNAL_FUNCTION("B"))
```

As can be seen above, the table rows are filtered by the index key first and then LVCPY operator is applied to the column 'b' to copy the value to the operator buffer. On table access we apply LVEVA operator to the column 'c' because column c is fetched from the table. No operator is required for column b in the table access because the column 'b' has already been evaluated as part of the index access.

5 Index and DML Issues

Oracle currently has a limit on building indexes on a large string. Due to B-tree index requirement, it allows users to build an index key of size up to 6398 bytes for 8 K block size. If users want to build an index on a larger size column, they can create a functional index using the substring operator (SUBSTR) or a hash function. We implicitly add the corresponding operator to the query predicates so that the index can be used. For example, if there is an index on SUBSTR(a,1,5) and the predicate contains a > 'abcdefghi', then the system adds the predicate SUBSTR(a,1,5) > SUBSTR ('abcdefghi',1,5) so that the index is used. For example, in the following scenario:

```
create table t1(a int, b varchar2(5000),
                c varchar2(12000));
create index i2 on t1(substr(c,1,5000));
select length(b), length(c)
 from t1 where c > rpad('aa',20,'bb');
```

The execution plan shows:

```
-------------------------------------------------
| Id  | Operation                | Name        |
-------------------------------------------------
|   0 | SELECT STATEMENT         |             |
| * 1 |  TABLE ACCESS BY         |             |
|     |   INDEX ROWID BATCHED|   T1            |
| * 2 |   INDEX RANGE SCAN       | I2          |
-------------------------------------------------

Predicate Information (identified by operation id):
-------------------------------------------------
1 - filter(INTERNAL_FUNCTION("C")
           AND "C">'aabbbbbbbbbbbbbbbbbbb'
           AND INTERNAL_FUNCTION("B"))
2 - access(SUBSTR("C",1,5000)
           >='aabbbbbbbbbbbbbbbbbbb')
```

As can be seen above, the index key access is augmented with SUBSTR operator. Note that evaluation operator for the long varchar column 'b' is appended to the filter whereas that for the column 'c' is prepended because the column 'c' is referenced in the filter condition.

For DMLs, the LOB locator is written to the table row and is used to check consistent read semantics, but the scalar value is used for index keys, constraint checking, partition key evaluation, and materialized view logs.

Oracle's multi-version read consistency semantics ensures that readers never block writers. Readers operate on a snapshot view of the data blocks, and writers modify the current data blocks. An UPDATE or DELETE is a hybrid operation that does both reads and writes. The read phase of the operation finds rows that satisfy the WHERE-clause predicates as of the read snapshot time. The write phase of the operation updates the block to effect the operation. To ensure transaction isolation, the write phase of the operation checks that none of the columns referenced in the WHERE clause have changed since the read phase. For performance reasons we do not read the LOB with current-mode semantics to compare it against the value read from the LOB at the snapshot read time. Instead this check is done based on the LOB locator. As described previously the large VARCHAR column is usually transformed to a BL2 V operator that represents the scalar VARCHAR value. For DELETE and UPDATE we make an exception and introduce an additional operand to represent the LOB locator so the locator value is available when the write phase compares the current value against the value that was read.

During a DML we may need the "old" scalar value (i.e. the pre-DML value for UPDATE or DELETE) or "new" scalar value (i.e. the post-DML value for UPDATE or INSERT) for a column. To find the old value we take the LOB locator from the read phase of the operation and we read from the LOB as of the read snapshot time. We get the new value from the operand in the DML that provides the value for the column.

6 STANDARD_HASH Operator

As we discussed above, a hash function is useful in building an index on a large string. We introduce a new operator that computes hash functions that have been standardized by National Institute of Standards and Technology [3]. It takes two arguments (expression, hash algorithm), like STANDARD_HASH(c, 'SHA512'). The hash algorithm can be one of standard hash algorithms: 'MD5', 'SHA1', 'SHA256', 'SHA384' and 'SHA512'. If the second argument is omitted, it defaults to 'SHA1.' The expression can be of any type except user-defined data types, LOBs, and LONGs.

This hash function is useful for authentication and data integrity as well. It is widely used in security applications, such as digital signatures, checksums, fingerprinting, etc. Since we support strong cryptographic hash functions, such as SHA512, that are collision resistant many applications can use STANDARD_HASH in a primary key constraint to enforce primary key for a large string column.

As we mentioned above in Sect. 5, the operator is implicitly added to the predicates so that the index is used for equality queries. For example, if the index is built on STANDARD_HASH(a), and we have the predicate a = 'xyz', then the system adds the predicate STANDARD_HASH(a) = STANDARD_HASH('xyz'), and this predicate can be used as an index access driver. For null values, we still apply STAN-DARD_HASH operator to them because the STANDARD_HASH(null) does not generate null value:

```
SELECT STANDARD_HASH(NULL,'MD5') FROM DUAL;
STANDARD_HASH(NULL,'MD5')
--------------------------------
D41D8CD98F00B204E9800998ECF8427E
```

For example, in the following scenario:
```
create table t1(a int, b varchar2(5000),
                c varchar2(12000));
create index i3
       on t1(standard_hash(c,'SHA1'));
select length(b), length(c)
from t1 where c = rpad('aa',20,'bb');
```

The execution plan shows:

```
-------------------------------------------------
| Id | Operation                    | Name       |
-------------------------------------------------
|  0 | SELECT STATEMENT             |            |
|* 1 |  TABLE ACCESS BY             |            |
|    |   INDEX ROWID BATCHED        | T1         |
|* 2 |   INDEX RANGE SCAN           | I3         |
-------------------------------------------------

Predicate Information (identified by operation id):
-------------------------------------------------
1 - filter(INTERNAL_FUNCTION("C")
         AND "C"='aabbbbbbbbbbbbbbbbbb'
         AND INTERNAL_FUNCTION("B"))
2 - access(STANDARD_HASH("C",'SHA1')
         =HEXTORAW(XX))
```

As can be seen above, the index key access is augmented with STAN-DARD_HASH operator. Note that evaluation operator for the long varchar column 'b' is appended to the filter whereas that for the column 'c' is prepended because the column 'c' is referenced in the filter condition.

7 Efficient Memory Management

It is obvious that memory requirements will increase with the large string size because more buffer space is needed to accommodate the long varchar columns. Historically Oracle has allocated statically sized buffers for operator result values. However, allocating static buffers for the long varchar column when most values are of small size and there are only a few large size strings is a waste of memory. Instead allocating the memory dynamically as needed is a good way to conserve memory.

For each operator that supports these dynamic buffers implementation changes are required so that the operator computes the size of its result and calls the dynamic buffer layer to ensure that the dynamic buffer is sufficiently sized for the result before writing the result to the buffer.

Our dynamic allocation strategy initially allocates 4096 bytes for each buffer. When a larger buffer is required we release the old buffer and allocate a new buffer of size 8192, 16384, or 32767 bytes. When a larger buffer is required our system always chooses the smallest size that is sufficient to satisfy the operator's request.

In our initial implementation we never reduce the size of a buffer. In combination with the fact that we have a limited number of possible buffer sizes, this guarantees that we never resize a buffer more than four times, thereby limiting the memory management overhead caused by dynamic buffers. Buffers that are released are not returned immediately to the system. Instead they are held in a session-level pool so they can be reused for another allocation.

In the following, we show how much memory we save by using dynamic memory allocation:

7.1 Mutable Memory and 32 K Varchar

We examine runtime memory allocation, measured by a special tracing tool, to quantify how the long varchar feature affects ephemeral frame size for several synthetic examples. We have conducted three experiments to check the efficiency of the dynamic memory allocation: the concatenation operator, SQL compilation, and a pathological case where result buffer allocation is large. We cannot provide details on the experiments due to lack of space.

In this case of concatenation operator, enabling 32 k varchar increases allocations by 93 %. The proposed dynamic allocation for operator result buffers reduces this to a 15 % increase. In the case of SQL compilation, enabling 32 k varchar increases allocations by 426 %. The proposed dynamic allocation for operator result buffers reduces this to a 143 % increase. In the case of a pathological case, enabling 32 k varchar increases allocations by 671 %. The proposed dynamic allocation for operator result buffers reduces this to a 27 % increase.

8   Query and DML Performance

This section describes a performance experiment to measure the effect of storage format (traditional varchar vs. BLOB) for INSERT, SELECT, and DELETE with varchar data of various size. Also, an experiment on the efficiency of the operator evaluation optimization is conducted.

8.1   Experimental Setup

The experiments were conducted using Oracle 12c Release 1 (12.1.0.1.0) on a Red Hat Enterprise Linux Server release 5.6 system with two 2.93 GHz Intel Xeon cores and 6 GB main memory. The database block size is 8 KB and the database is configured with a 256 MB buffer cache. Our experimental data is stored in an *Automatic Segment Space Management* tablespace with uniform extent size of 8 MB.

All experiments are performed with a simple three-column table:

```
create table t(i int,
               vc varchar2(32767),
               j int);
```

We use the parameter described in Sect. 1 to change the storage format of the vc column between traditional varchar storage format and BLOB storage format. All experiments report execution time in seconds.

8.2   Experiment I: INSERT

In this experiment we execute 25,000 single-row INSERT INTO t VALUES(...) operations to load the table. Across the 25,000 rows the integer columns range from 1 to 25,000, and the varchar column is loaded with data that is always N bytes long. We run a total of 12 trials: for each N in { 1024, 2048, 4096, 8192, 16384, 32767 }, we run with both traditional varchar storage format and BLOB storage format. The results are shown in Fig. 1.

For N = 1024 and N = 2048 we see comparable performance for the two storage formats. for BLOB storage values this small will be stored as inline LOB; compared to

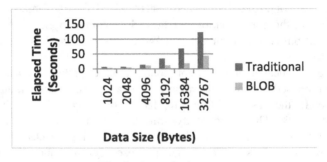

Fig. 1. Insert performance

traditional varchar storage inline LOB has small space overhead for LOB locator data, and small time overhead for code path. For N >= 4096 the LOB will be stored out-of-line and we see that LOB storage outperforms traditional storage format, with a significant difference for N >= 8192. Again this is expected since N >= 8192 will require block chaining for inline storage, and writes to the out-of-line LOB are more efficient than writes to chained data blocks.

8.3   Experiment II: SELECT

For this experiment we consider three SELECT statements, without a WHERE clause, executed as full table scans, to investigate the performance of (1) selecting column I, which precedes the large varchar column in the row, (2) selecting the large varchar column VC, and (3) selecting column J, which succeeds the large varchar column in the row. For each query we select MAX(column), which limits the amount of data returned to the client but ensures that the column value is read for every row. Each case is tested with 50,000 rows for each of the 12 configurations described in Experiment I. Statistics were gathered for the table before running these queries. Each query is run twice; the first run is ignored since it includes query compilation time, and the timing for the second run is reported.

The results are shown in Figs. 2, 3 and 4.

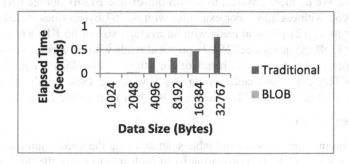

Fig. 2. Select performance (Case 1)

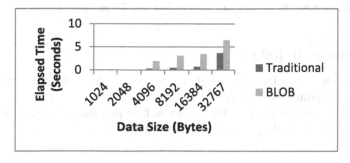

Fig. 3. Select performance (Case 2)

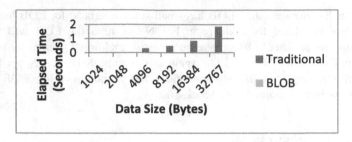

Fig. 4. Select performance (Case 3)

Case 1: select column I

To query column I we perform a full table scan and read the first column in each row. Since we do not traverse the rows to find other columns in the table, the cost of reading all table blocks is the dominating factor here.

When the LOB is stored out-of-line we see large performance gains because the table segment is smaller.

Case 2: select column VC

In this case, traditional storage performs better than LOB storage because the full table scan uses large multi-block I/Os to read all blocks in the table with relatively few physical I/Os. We do more physical I/Os with out-of-line BLOB storage because fewer blocks are read with each I/O. For example, with 32767 byte values stored inline the query performs 17,127 physical reads with an average size of 66 KB; when the values are stored as LOB the query does 25,271 physical reads with an average size of 40 KB. We would not expect to see this behavior for other access plans – e.g. if the query accessed an index and did table lookup by rowid, the table would be read with single-block I/Os and we would do larger I/Os when reading the LOB value.

Case 3: select column J

To query column J we perform a full table scan and read the third column in each row. We must traverse over the first two columns of each row to get to the third column. In this case it is clear that storing large values in traditional format can cause major performance problems, even if the large value is not selected, because we must traverse over the chained rows to find values that follow the large value in the row.

8.4   Experiment III: DELETE

In this experiment we DELETE 50,000 rows for each of the 12 configurations described in Experiment I. The results are shown in Fig. 5.

Deleting an out-of-line LOB is more efficient than processing the delete for chained rows because we are dealing with less number of blocks.

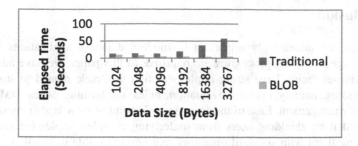

Fig. 5. Delete performance

8.5 Experiment IV: Operator Evaluation Optimization

In this experiment we investigate the performance of the operator evaluation optimization described in Sect. 4. We load 50,000 rows into the table, with a 32,000 byte varchar value stored in BLOB format for each row. Then we run the following query, varying the parameter k to select 1 %, 10 %, 50 %, 90 %, and 100 % of the rows in the table.

```
select max(vc) from t where mod(i, 100) < k;
```

This query is executed using a full table scan, so it will apply the where clause predicate to all 50,000 rows in the table.

We test two configurations: with the optimization enabled we will read the varchar from the BLOB after the filter predicate is applied to reduce the result set; in the naïve configuration we disable the optimization so the BLOB will be read before the filter is applied. Each query is run twice; the first run is ignored since it includes query compilation time, and the timing for the second run is reported. The results are shown in Fig. 6.

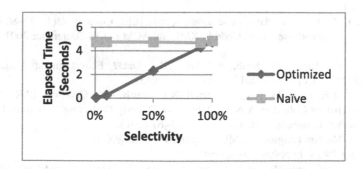

Fig. 6. Operator evaluation optimization performance

The optimization yields large benefits when there is a selective predicate since we avoid reading the BLOB for rows that are filtered out by the predicate. The benefit is smaller for less selective queries, but in no case is there any runtime performance degradation due to the optimization.

9   Conclusion

Large strings are increasingly important in modern database applications and their efficient storage and query processing are required. In this paper, we have addressed a number of issues facing large string implementation in Oracle related to storage and query processing, namely, operator evaluation, index and hashing issues, DML issues, and memory management. Ease of use from a user's point of view is also important and we addressed it by shielding users from underlying complex implementation details and providing them with standard operators that are applicable to small strings. This feature enables users to easily build applications that use XML and JSON documents by allowing them to store the documents directly in large string columns and manipulate them just like regular varchar strings.

Acknowledgements. Many people contributed to the initial discussion of this project. Our thanks to Thierry Cruanes, Cetin Ozbutun, Dmitry Potapov, Kumar Rajamani, Shrikanth Shankar, Sankar Subramanian, Andy Witkowski, and Mohamed Zait. We apologize if we missed anyone.

References

1. IBM DB2 String data types. http://publib.boulder.ibm.com/infocenter/dzichelp/v2r2/index. jsp?topic=%2Fcom.ibm.db2z10.doc.intro%2Fsrc%2Ftpc%2Fdb2z_stringdatatypes.htm
2. Microsoft SQL Server char and varchar. http://technet.microsoft.com/en-us/library/ms 176089.aspx
3. NIST Secure Hashing. http://csrc.nist.gov/groups/ST/toolkit/secure_hashing.html
4. Kunchithapadam, K., Zhang, W., Ganesh, A., Mukherjee, N. Oracle database filesystem. In: ACM SIGMOD Conference 2011, 1149–1160 (2011)
5. Oracle® Database SecureFiles and Large Objects Developer?s Guide, 12c Release 1 (12.1), E17605-10
6. Hendler, J.: Web 3.0: the dawn of semantic search. IEEE Comput. **43**(1), 77–80 (2010)
7. Vianu, V.: A web odyssey: from Codd to XML. In: ACM PODS Conference 2001, pp. 1–15 (2001)
8. Sumbaly, R., Kreps, J., Shah, S.: The ?Big Data? Ecosystem at LinkedIn. In: ACM SIGMOD Conference 2013, 1125–1134 (2013)
9. Guadagno, R.E., Loewald, T.A., Muscanell, N.L., Barth, J.M., Goodwin, M.K., Yang, Y.: Facebook History Collector: A New Method for Directly Collecting Data from Facebook. Int. J. Interact. Commun. Syst. Technol., **3**, 57–67 (2013)
10. Extensible Markup Language (XML). http://www.w3.org/XML/
11. Introducing JSON. http://www.json.org/
12. MySQL 5.7 Reference Manual. http://dev.mysql.com/doc/refman/5.7/en/char.html
13. PostgreSQL 9.3.5 Documentation. http://www.postgresql.org/docs/9.3/static/datatype-character.html

SAM: A Sorting Approach for Optimizing Multijoin Queries

Yong Zeng[1]([⊠]), Amy Nan Lu[1], Lu Xia[1], Chris Xing Tian[2], and Y. C. Tay[1]

[1] National University of Singapore, Singapore, Singapore
matzeng@nus.edu.sg
[2] Beihang University, Beijing, China

Abstract. Finding the optimal join order for a multijoin query is an old, yet very important topic for relational database systems. It has been studied for the last few decades and proven to be NP-hard. The mainstream techniques, first proposed in System R, are based on dynamic programming. These techniques are widely adopted by commercial database systems.

However, it is well known that such approaches suffer from exponential running time in finding the optimal join order for most queries, except simple ones like linear queries. Therefore, a query optimizer must resort to finding a suboptimal join order when the number of tables is large.

This paper proposes SAM, which departs from current practice in two ways: (1) SAM orders the joining attributes before ordering the tables; (2) SAM sorts the tables by comparing selectivities for "table blocks". This approach reduces the exponential time complexity in the optimization; in particular, it can find, in polynomial time, the optimal ordering for clique queries that take exponential time to optimize by dynamic programming.

Experiments comparing SAM to the query optimizers in MySQL and PostgreSQL, using real data, show that its performance is similar for small queries, but much better for large queries.

1 Introduction

Finding the optimal join order for a multijoin query is an old problem, studied since the beginning of relational database systems. Optimality here refers to the total intermediate join sizes, so the topic is very important because query execution time depends crucially on the join order. Unfortunately, it is NP-hard to decide the optimal join order [4].

The mainstream optimization techniques used by commercial database systems are based on dynamic programming [12]. However, it is well known that such traditional approaches suffer from exponential running time in finding the optimal join order for most queries, except simple ones like linear queries. Some widely-adopted heuristics help to reduce the search space: one heuristic is avoiding Cartesian products; another considers only left-deep plans, where the tables join the execution tree one at a time in a linear order. Even so, the worst-case optimization time remains exponential.

© Springer International Publishing Switzerland 2015
Q. Chen et al. (Eds.): DEXA 2015, Part I, LNCS 9261, pp. 367–383, 2015.
DOI: 10.1007/978-3-319-22849-5_25

The issue is growing bigger nowadays, since a database system can have thousands of tables, and queries are often not manually crafted but generated via views and by tools and applications. When the number of tables in a query is large, an exponential-time optimization may become impractical, and the query optimizer must settle for a suboptimal plan.

To illustrate, consider the following two queries:

Example 1. This SQL query is a simple linear query with three tables:

   Select * from A, B, C where A.a = B.a and B.b = C.b

If we draw a graph with A, B and C as nodes, and an edge between two tables that can be joined, then there are edges between A and B, between B and C, but not between A and C — if we avoid Cartesian products. The graph is thus linear.   □

Under the assumption of no Cartesian products, the existing approaches can find the optimal join order for linear queries in time that is polynomial in the number of tables, as the number of possible join orders is largely restricted by the linear topology.

Example 2. This SQL query is a clique query with four tables:

   Select * from A, B, C, D where A.a = B.a and B.a=C.a and C.a=D.a

As we can see, all tables have the same join attribute, so any two tables can be joined together in the join order. Here, drawing an edge between every pair of tables that can be joined will result in a graph that is a clique.   □

Since any two tables can be joined with each other in a clique query, any ordering of the tables is a valid join plan. There are $n!$ such ordering of n tables, so the search space is exponentially large. Therefore, although dynamic programming approaches can improve the efficiency, they still suffer from exponential running time for finding the optimal join order in such a large search space.

This paper introduces the idea of "sorting" as an approach to optimizing left-deep multijoin ordering without Cartesian products. We call the resulting technique **SAM**. Although optimization time remains exponential in the worst case, SAM is able to find, in polynomial time, an optimal join order for clique queries that otherwise require exponential time with traditional dynamic programming approaches.

We begin by surveying related work in Sect. 2, and introduce the sorting approach in Sect. 3. Section 4 presents experiments that compare SAM to other optimization techniques; we use two metrics in this comparison: (i) the **plan time** for finding the join order and (ii) the **execution time** for running the query using the join order. The experiments are run with MySQL and PostgreSQL, using two real datasets. Sect. 5 concludes with a summary and a description of current work.

2   Related Work

Finding the optimal join order for a query is NP-hard in general [4]. The following reviews this important and challenging problem for relational databases.

Finding the Optimal Join Order. The dynamic programming approach is the most popular technique for optimizing the join order for a query. It is now used in many commercial database systems. This use was first proposed for the System R optimizer [12]. The main idea of this approach is: Given an optimal plan of n tables, then for any i ($i < n$), the first i tables in the plan is also the optimal plan for those i tables. Such an approach may generate $O(3^n)$ different plans, so it has exponential worst case plan time.

There are different ways of designing the dynamic programming process, e.g. using subset enumeration [15], or focusing on specific query topologies, like linear queries, stars and cliques [7]. One variation to dynamic programming uses memoization [2], and the technique was later generalized to handle complex join predicates [8].

However, all the existing approaches suffer from exponential worst case running time for handling non-linear queries.

Finding Suboptimal Join Orders. Since plan time for choosing a join order can increase exponentially, many heuristics have been proposed as a compromise between join order search time and join order optimality.

Greedy algorithms form one class of suboptimal algorithms [3,5,6], where the quality of the join order found depends on the heuristics used, and optimality is not guaranteed. Another class of suboptimal algorithm use randomization [14]. This class includes genetic algorithms [13].

Our Approach. There are hybrid approaches that adaptively switch from finding an optimal join order to finding one that is suboptimal, depending on the query complexity [9,11]. SAM, however, takes an entirely different, sorting approach.

It starts with a hypergraph to represent the query, partially orders the joining attributes, constructs a tree for the tables, and uses a comparator and the concept of a table block to sort the tables in the tree.

Intuitively, the effectiveness of this approach lies in the difference in complexity between sorting (SAM) and enumeration (dynamic programming).

3   Sorting Approach to Finding an Optimal Join Order

A **multijoin query** is a set of tables $\{t_1, t_2, ..., t_n\}$ and a set of join conditions $\{t'_1.a = t'_2.a, ...\}$, where t'_i is a table and $t'_i.a$ denotes attribute a of table t'_i.

To simplify the notation, we use the same symbol to denote the join attribute on both sides of $t_i.a = t_j.a$, even if the attribute names may not be the same. For example, if the join condition is table t_1's attribute uid equals to t_2's attribute fid, such a condition can be denoted as $t_1.a = t_2.a$, where $t_1.a$ refers to uid and $t_2.a$ refers to fid.

We follow most of related work in limiting the search space to left-deep join order without Cartesian products.

Given a multijoin query, the **optimal** join order is defined as an order $t'_1 \bowtie t'_2 \bowtie ... \bowtie t'_n$ that minimizes the **cost** that is measured by the total intermediate result size $|t'_1 \bowtie t'_2| + |t'_1 \bowtie t'_2 \bowtie t'_3| + \cdots + |t'_1 \bowtie t'_2 \bowtie \cdots \bowtie t'_n|$, where $\bowtie$ denotes join operation (that is not a Cartesian product) and $|t|$ denotes the number of tuples in table t. We do not consider other operators, like projections and semijoins, which can reduce the cost but increase the search space [10,11].

Given a (nonempty) table t, if it joins a (nonempty) table or intermediate result t', then its **selectivity** is $p_t = \frac{|t' \bowtie t|}{|t'| |t|}$.

Suppose there are N distinct values in the join attribute for $t' \bowtie t$. If the tuples of t' and t are uniformly distributed among the N values, then t' and t have $|t'|/N$ and $—t—/N$ tuples for each value, so the join size is

$$|t' \bowtie t| = N \frac{|t'|}{N} \frac{|t|}{N} = \frac{|t'||t|}{N}, \quad \text{i.e. } p_t = \frac{1}{N}.$$

In fact, this is how current database systems estimate selectivity, and one can interpret it as assuming tuples are uniformly distributed over the join attribute. This assumption can give bad estimates of join sizes.

SAM does not make such a uniformity assumption, and this is partly why SAM's choice of join order is often better than that chosen by current systems.

3.1   A Comparator for Sorting

We first present the sorting algorithm for finding the optimal join order for a set of **table blocks** (instead of tables). This algorithm is the core idea in SAM.

A **table block** $\mathbb{T} = (t_1, t_2, ..., t_n)$ is an ordered set of tables, and $|\mathbb{T}| = |t_1 \bowtie t_2 \bowtie ... \bowtie t_n|$. This concept captures the fact that any optimal plan that includes these tables must join them in this order, *with no other tables interrupting the sequence*. Table blocks arise naturally as we construct the optimal join order.

For $\mathbb{T} = (t_1, t_2, ..., t_n)$ and $\mathbb{T}' = (t'_1, t'_2, ..., t'_m)$, $\mathbb{T} \bowtie \mathbb{T}' = t_1 \bowtie t_2 \bowtie ... \bowtie t_n \bowtie t'_1 \bowtie t'_2 \bowtie ... \bowtie t'_m$, where the join order is as indicated. By **intermediate result** of $\mathbb{T} \bowtie \mathbb{T}'$, we mean an intermediate result of $t_1 \bowtie t_2 \bowtie ... \bowtie t_n \bowtie t'_1 \bowtie t'_2 \bowtie ... \bowtie t'_m$.

Consider this **Table Sorting Problem**: Given table blocks $\{\mathbb{T}_1, ..., \mathbb{T}_n\}$, where each table $t \in \mathbb{T}_i$ is assigned a fixed selectivity p, find a join order $\mathbb{T}'_1 \bowtie \mathbb{T}'_2 \bowtie \cdots \bowtie \mathbb{T}'_n$ that minimizes the cost (i.e. total intermediate result size).

The sorting algorithm orders the table blocks without changing the order within each block. This sorting uses a **comparator** $\preceq$, where $\mathbb{T}_i \preceq \mathbb{T}_j$ means $\mathbb{T} \bowtie \mathbb{T}_i \bowtie \mathbb{T}_j$ will generate a cost smaller than or equal to that of $\mathbb{T} \bowtie \mathbb{T}_j \bowtie \mathbb{T}_i$ for any $\mathbb{T}$. The sorting algorithm uses this comparator to determine if swapping two table blocks will reduce the cost. Since each swap decreases the cost, the swapping eventually terminates.

Suppose $\mathbb{T} = (t_1, t_2, ..., t_n)$ and t_i's selectivity is p_i. Then $\prod(\mathbb{T})$ denotes $|t_1|p_1|t_2|p_2...|t_n|p_n$ and $C(\mathbb{T})$ denotes $|t_1|p_1 + |t_1|p_1|t_2|p_2 + \cdots + |t_1|p_1|t_2|p_2 \cdots |t_n|p_n$. We then define

$$\mathbb{T}_1 \preceq \mathbb{T}_2 \quad \text{iff} \quad C(\mathbb{T}_1) + \prod(\mathbb{T}_1)C(\mathbb{T}_2) \leq C(\mathbb{T}_2) + \prod(\mathbb{T}_2)C(\mathbb{T}_1);$$

$$\text{and} \quad \mathbb{T}_1 \doteq \mathbb{T}_2 \quad \text{iff} \quad C(\mathbb{T}_1) + \prod(\mathbb{T}_1)C(\mathbb{T}_2) = C(\mathbb{T}_2) + \prod(\mathbb{T}_2)C(\mathbb{T}_1).$$

Example 3. Suppose $\mathbb{T}_2 = (t_2)$ and $\mathbb{T}_3 = (t_3, t_4)$. Given the join order $\mathbb{T}_1 \bowtie \mathbb{T}_2 \bowtie \mathbb{T}_3$, the cost is $|\mathbb{T}_1||t_2|p_2 + |\mathbb{T}_1||t_2|p_2|t_3|p_3 + |\mathbb{T}_1||t_2|p_2|t_3|p_3|t_4|p_4 = |\mathbb{T}_1|(C(\mathbb{T}_2) + \prod(\mathbb{T}_2)C(\mathbb{T}_3))$. Similarly, the cost of $\mathbb{T}_1 \bowtie \mathbb{T}_3 \bowtie \mathbb{T}_2$ is $|\mathbb{T}_1|(C(\mathbb{T}_3) + \prod(\mathbb{T}_3)C(\mathbb{T}_2))$. Therefore, if $C(\mathbb{T}_3) + \prod(\mathbb{T}_3)C(\mathbb{T}_2) < C(\mathbb{T}_2) + \prod(\mathbb{T}_2)C(\mathbb{T}_3)$, then we can swap $\mathbb{T}_2$ and $\mathbb{T}_3$ to get the join order $\mathbb{T}_1 \bowtie \mathbb{T}_3 \bowtie \mathbb{T}_2$ with a smaller cost.   □

Lemma 1. *Let $S = \{\mathbb{T}_1, \mathbb{T}_2, ..., \mathbb{T}_n\}$ be a set of table blocks, with each table $t \in \mathbb{T}_i$ assigned a selectivity p. Among all left-deep join orders without Cartesian products, the cost is minimum if the joins follow the total order on S defined by $\preceq$.*

Proof. We first prove the claim that $\preceq$ is a total order. It is easy to see that it is reflexive ($\mathbb{T}_1 \preceq \mathbb{T}_1$), antisymmetric (if $\mathbb{T}_1 \preceq \mathbb{T}_2$ and $\mathbb{T}_2 \preceq \mathbb{T}_1$, then $\mathbb{T}_1 \doteq \mathbb{T}_2$), and total ($\mathbb{T}_1 \preceq \mathbb{T}_2$ or $\mathbb{T}_2 \preceq \mathbb{T}_1$ for any $\mathbb{T}_1$ and $\mathbb{T}_2$). To prove transitivity, suppose $\mathbb{T}_1 \preceq \mathbb{T}_2$ and $\mathbb{T}_2 \preceq \mathbb{T}_3$. Then $C(\mathbb{T}_1) + \prod(\mathbb{T}_1)C(\mathbb{T}_2) \leq C(\mathbb{T}_2) + \prod(\mathbb{T}_2)C(\mathbb{T}_1)$ and $C(\mathbb{T}_2) + \prod(\mathbb{T}_2)C(\mathbb{T}_3) \leq C(\mathbb{T}_3) + \prod(\mathbb{T}_3)C(\mathbb{T}_2)$. This gives

$$\frac{\prod(\mathbb{T}_1) - 1}{C(\mathbb{T}_1)} \leq \frac{\prod(\mathbb{T}_2) - 1}{C(\mathbb{T}_2)} \leq \frac{\prod(\mathbb{T}_3) - 1}{C(\mathbb{T}_3)},$$

so $C(\mathbb{T}_1) + \prod(\mathbb{T}_1)C(\mathbb{T}_3) \leq C(\mathbb{T}_3) + \prod(\mathbb{T}_3)C(\mathbb{T}_1)$, i.e. $\mathbb{T}_1 \preceq \mathbb{T}_3$.

Suppose the joins follow the total order π specified by $\preceq$. If there is an optimum permutation π' of S that has a smaller cost C than π, then π' must have two blocks that violate the $\preceq$ ordering and can be swapped to reduce cost further, thus contradicting the minimality of C.   □

In general, the join attribute used by a table depends on its location in the join order, so its selectivity changes accordingly. However, in the Table Sorting Problem, the selectivities are assigned. This is because, in the step where SAM uses the comparator for sorting, the tables are grouped into blocks for the comparison, and the join attribute used by each table is already fixed.

However, how to apply the sorting algorithm to find the optimal join order for a multijoin query is another challenge. Given a multijoin query Q, we say a table block $(t_1, t_2, ..., t_n)$ is **ideal** if $(t_1, t_2, ..., t_n)$ is part of an optimal join order for Q.

As the sorting algorithm works on a set of table blocks, we need to make sure each table block is ideal; otherwise, no matter how we apply the sorting, it will not find an optimal join order for a query.

Furthermore, the sorting algorithm requires each table to be assigned a selectivity that does not change when table blocks are swapped. We next consider how to construct ideal table blocks with this property.

3.2   SAM's 6-Step Sorting Approach

We now describe how SAM uses sorting to find an optimal join order. As a running example, we assume the database schema is sonSchema [1]. For our purpose, the database has 3 join attributes uid, gid and pid and 6 tables; tables *user*, *post* and *social_product* have attribute uid, tables *group*, *post* and *message* have attribute gid, and tables *message, social_product_relationship* and *social_product* have attribute pid.

The main idea in SAM is to generate a set of execution trees (introduced later) for a given query. Then based on each execution tree, we can construct a set of ideal table blocks and infer the selectivity of each table, on which the sorting algorithm can be applied. The approach has 6 steps.

Step 1: Construct the Join Attributes Diagram. The tables and join attributes in a query can be represented by a hypergraph. A **labelled hypergraph** $\mathcal{H}$ is a 3-tuple, $\mathcal{H} =< V, \xi, f_L >$ where V is a nonempty set of nodes, ξ is a nonempty set of hyperedges, $E \subseteq V$ for each $E \in \xi$, and $f_L : \xi \to L$ is a bijection for a set of labels L.

A multijoin query can be represented by a **join attributes diagram (JAD)** that is a labelled hypergraph $< T, \xi, f_A >$ where T is the set of tables to be joined, A is the set of join attributes and for each $E \in \xi$ and for $y = f_A(E)$, $t \in E$ if and only if y is a join attribute for t. For $E, E' \in \xi$, the **border** is $E \cap E'$. For each $t \in T$, $t.\xi = \{E \in \xi : t \in E\}$. For example, given a query Q_{eg} in SQL form:

```
SELECT *
FROM user u, post p, group g, message msg,
social_product sp, social_product_relation spr
WHERE u.uid = p.uid
AND p.gid = g.gid
AND u.uid = sp.uid
AND sp.pid = spr.pid
AND msg.pid = spr.pid
AND msg.gid = g.gid;
```

the corresponding join attributes diagram is shown in Fig. 1, where:

$T = \{g, u, sp, msg, pst, spr\}$
$\xi = \{\{g, msg, pst\}, \{u, sp, pst\}, \{sp, spr, msg\}\}$
$A = \{\text{gid, pid, uid}\}$
$f_A(\{g, msg, pst\}) = \text{gid}$
$f_A(\{u, sp, pst\}) = \text{uid}$
$f_A(\{sp, spr, msg\}) = \text{pid}$
$sp.\xi = \{\{sp, spr, msg\}, \{u, sp, pst\}\}$ and
$\{msg\}$ is the border for $\{g, msg, pst\}$ and $\{sp, spr, msg\}$

In Fig. 1, each hyperedge is a circle that is labelled by a join attribute. Q_{eg} has 3 join attributes: pid, uid and gid. In each circle, tables indicate that they only use this attribute to join other tables regardless of themselves containing other possible attributes. For example, *msg* may contain 3 attributes, pid, uid and gid, but as the query specifies that *msg* joins using only attributes pid and gid, so it appears only at the intersection of pid and gid circles. There are three borders for the given query.

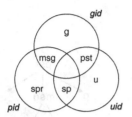

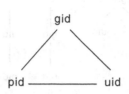

Fig. 1. Join attributes diagram for Q_{eg}.

Fig. 2. Join attributes graph for JAD in Fig. 1

Step 2: Construct the Join Attributes Graph.
We follow common practice in not considering Cartesian products, and this imposes constraints on the join plan.

For example, in Fig. 1, if we start the plan with g, which only has join attribute gid, then it cannot directly join table spr, which only has join attribute pid. Instead, g can first join msg in the border between hyperedges for g and spr, and thus pick up join attribute pid in table msg, so $g \bowtie msg$ can now use pid to join spr. We say that join attribute pid is **included** into the intermediate multijoins.

We use a graph to facilitate the analysis of the above constraint, i.e. how each join attribute is included into the intermediate multijoins.

Given a join attributes diagram $< T, \xi, f_A >$, a **join attributes graph** is an undirected graph $< A, E >$ where A is the set of join attributes and $\{y, z\} \in E$ iff $f_A^{-1}(y) \cap f_A^{-1}(z) \neq \phi$, i.e. the hyperdeges labelled by y and z have a nonempty border.

Figure 2 shows that join attributes graph for the join attributes diagram in Fig. 1. In Fig. 2, the 3 nodes gid, pid and uid represent the 3 hyperedges in the join attributes diagram. Each edge represents the border between two hyperedges. There is no edge between two attributes if their hyperedges have an empty border. As all three borders in Fig. 1 are nonempty, we get 3 edges in Fig. 2.

Step 3: Construct the Spanning Tree.
The next step constructs a spanning tree to order the inclusion of the join attributes into the intermediate multijoins. A **spanning tree** $< A, B, r >$ for a join attributes graph $< A, E >$ is a rooted tree B with nodes (i.e. attributes) A and root r.

Figure 3 shows all possible spanning trees for the join attributes graph in Fig. 2. Each tree specifies a possible way for how the join attributes are to be included into the intermediate multijoins. Take Fig. 3(1) as an example: it means the first join attribute included in the intermediate multijoins is gid, i.e. the first table in the join order must be a table containing attribute gid. Then pid is included into the intermediate multijoins by an edge from gid to pid, which means the intermediate result then joins some table located at the border of $f_A^{-1}(gid)$ and $f_A^{-1}(pid)$ to include join attribute pid. Similarly, uid is included last into the intermediate multijoins by joining some table located at the border of $f_A^{-1}(pid)$ and $f_A^{-1}(uid)$.

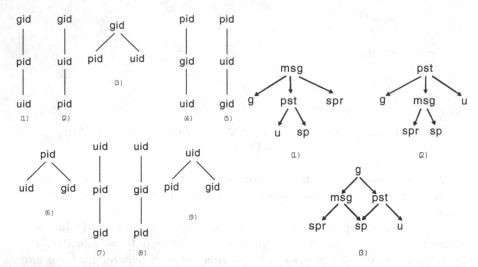

Fig. 3. Spanning trees for join attributes graph of Fig. 2 (the root is the highest in each tree)

Fig. 4. Execution graphs for spanning trees in Fig. 3(3)

Step 4: Construct the Execution Graph. The partial order on join attributes specified by the spanning tree then induces a partial order for sequencing the tables.

Given a set of tables T in a multijoin query, an **execution graph** is a rooted directed graph $< T, L, s >$, where L is the set of directed edges, $s \in T$ is the root, and an edge $< t_1, t_2 >$ indicates table t_1 will appear before table t_2 in the join order. This graph represents the constraints that a join order must obey, in order to follow the chosen ordering of join attributes, and avoid Cartesian products.

For example, the spanning tree in Fig. 3(3) starts with attribute gid, which can be found in tables msg, pst and g. If we start with msg, then it precedes all other tables in the join order. It already includes attribute gid and pid. To include uid, it has to join with a table containing both gid and uid because there is an edge between gid and uid in Fig. 3(3). Such a table is pst, which means uid is included by joining pst. So pst should be joined before all other tables which also contain uid, i.e. u and sp. This gives the execution graph in Fig. 4(1). Figure 4(2) can be derived similarly. If we start with g, then the execution graph is not a tree, as shown in Fig. 4(3).

Algorithm 1 is an algorithm for deriving execution graphs from the spanning trees. Two spanning trees can produce the same execution graph, so this algorithm can be further optimized to generate only distinct execution graphs.

Step 5: Construct the Execution Tree. If the execution graph is not a tree (like in Fig. 4(3)), this next step converts it into a tree. Given an execution graph $EG = < T, L, s >$, an **execution tree** is a directed tree $ET = < T, L', s >$, such

---

Algorithm 1. Construct Execution Graph

---

1 ForEachforeachdoendfch
    input : spanning tree $< A, B, r >$ for join attributes diagram $< T, \xi, f_A >$
    output: a set of execution graphs $\{< T, L, s >\}$
2 **foreach** $s \in f_A^{-1}(r)$ **do**
3      $//s$ as the root; table s contains join attribute r
4      $L \leftarrow \phi$;
5      $C_{traversed} \leftarrow s.\xi$; //use s's join attributes to identify corresponding hyperedges
6      $L \leftarrow L \cup \{< s, t >: t \in H - s \text{ and } H \in s.\xi\}$;
7   **while** $C_{traversed} \neq \xi$ **do**
8   choose $H' \in \xi$ - $C_{traversed}$ such that $< f_A(H), f_A(H') >\in B$ and $H \in C_{traversed}$;
9   // H exists since the hyper graph is connected
10   choose $t' \in H \cap H'$ and $\nexists t''$ such that t'' is an ancestor of t' in L;
11         $L \leftarrow L \cup \{< t', t'' >: t'' \in H' - t'\}$ - $\{< t''', t'' >: < t''', t'' >\in L \text{ and } t'''$ is an ancestor of $t'\}$;
12         $C_{traversed} \leftarrow C_{traversed} \cup \{H'\}$;
13   for each different L generated after line 6-11, output $< T, L, s >$;

---

that if there is a path from table t_i to table t_j in EG, there will also be a path from t_i to t_j in ET.

For example, in Fig. 4(3), there are 2 paths from table g to table sp in the execution graph. These 2 paths generate the 2 execution trees (3) and (4) in Fig. 5. Note that the execution graphs (1) and (2) in Fig. 4 can already be considered as execution trees.

This step is where SAM begins to consider selectivity. The position of a table in an execution tree determines the attributes it uses to join the intermediate result. For example, in Fig. 1, pst has 2 join attributes: **gid** and **uid**. However, pst's ancestors g and msg in Fig. 5(3) only have attributes **gid** and **pid**. Therefore, pst only uses **gid** to join its ancestors. For clarity, we indicate in Fig. 5 the attribute(s) used by each table in joining the intermediate result.

Step 6: Sort the Tables, Bottom-Up, in the Execution Tree. Finally, each execution tree is sorted, bottom up and using the comparator in Sect. 3.1 to give a join order; the one with minimum cost is then selected. In this process, the nodes in the execution tree are thus transformed into table blocks. The pseudocode is shown in Algorithm 2.

We use the execution tree in Fig. 5(3) to illustrate Algorithm 2. Note that the position of table t in the execution tree determines the attributes used by t to join the intermediate result, and thus determines selectivity p_t. Suppose, for each table t, the value $|t|p_t$ is as indicated in Fig. 6.

The only 1-level subtree is the one rooted at pst. Algorithm 2 sorts its children into the order $(sp), (u)$. Since $(pst) \preceq (sp)$, the constraint that (pst) should be before (sp) and (u) will be automatically satisfied if later they are being sorted.

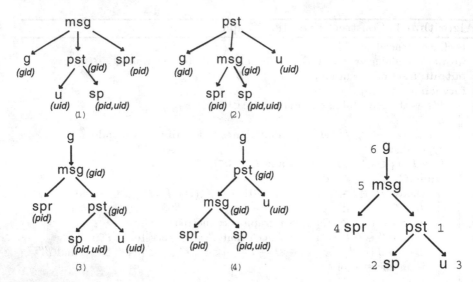

Fig. 5. Execution trees for execution graphs in Fig. 4, with joining attributes indicated in brackets.

Fig. 6. Execution tree in Fig. 5(3) with selectivity*(table size)

So we do not need to do anything. We get three table blocks (pst), (sp) and (u), which are then attached to msg as child nodes.

The subtree rooted at msg now has only 1 level. The children (spr), (pst), (sp) and (u) are sorted in ascending order to be $(pst), (sp), (u), (spr)$. To make sure (msg) precedes (pst), (sp), (u) and (spr) in the join order, a table block of (msg, pst, sp, u, spr) is formed (lines 9–15). Finally, given the constraint that g must be the first table in the order, the algorithm now gives the join order g, msg, pst, sp, u, spr.

Each execution tree in Fig. 5 generates one join order. The ordering with minimum cost among all execution trees is now chosen for query Q_{eg}.

We now return to the clique query discussed in Sect. 1, which makes the traditional dynamic programming approaches suffer from exponential plan time. Example 4 illustrates how SAM handles such a query.

Example 4. For the clique query in Example 2, as there is only one join attribute, its join attributes diagram is shown in Fig. 7(1), where all 4 tables are in 1 hyperedge. The join attributes graph is trivially just 1 attribute, as shown in Fig. 7(2). Hence, the spanning tree is also trivially just 1 node, as shown in Fig. 7(3). According to Algorithm 1, we can construct 4 execution graphs, as shown in Fig. 7(4). Since these graphs are trees, the execution trees are exactly the same, as shown in Fig. 7(5). For each execution tree, there is only one 1-level subtree, so Algorithm 2 just makes the root table as the first table in the join order, and sort the leaf nodes in ascending order. The cost of these 4 join orders are then compared to identify the one with minimum cost for the clique query. □

Algorithm 2. Apply sorting algorithm to execution tree

---

    input : one execution tree ET
    output: optimal join order for ET

1   **foreach** *one-level subtree* $\in ET$ **do**
2      $\mathbb{T} =$ root of the one-level subtree;
3   **if** $\mathbb{T}$ *is the root of* ET **then**
4 | $\mathbb{T}_p = NULL$;
5   **else**
6 | $\mathbb{T}_p =$ parent of $\mathbb{T}$;
7   remove the subtree from ET;
8   sort $\mathbb{T}$'s child nodes, denoted as $S = \{\mathbb{T}_1, \mathbb{T}_2, ..., \mathbb{T}_k\}$;
9   **foreach** *table block* $\mathbb{T}_i \in S$ *(from* $\mathbb{T}_1$ *to* $\mathbb{T}_k$*)* **do**
10   **if** $\mathbb{T}_i \preceq \mathbb{T}$ **then**
11   **foreach** *table* $t \in \mathbb{T}_i$ **do**
12 | append t to the end of $\mathbb{T}$
13   remove $\mathbb{T}_i$ from S;
14   **else**
15 | break;
16   add $\mathbb{T}$ to S as the first table block;
17   **if** $\mathbb{T}_p$ *is not* $NULL$ **then**
18 | attach each $\mathbb{T}_i \in S$ to be a child node of $\mathbb{T}_p$;
19   **else**
20   The order of table blocks in S is the optimal join order.
21   break;

3.3 SAM's Optimality

We now sketch the arguments for proving that SAM constructs the optimal join order. First, note that the table blocks from Algorithm 2 must occur in an optimal join order:

Lemma 2. *The table blocks constructed by Algorithm 2 are ideal.*

Moreover, Algorithm 2 sorts the table blocks:

Lemma 3. *The table blocks constructed by Algorithm 2 are sorted in ascending order.*

Furthermore, the join order from Algorithm 2 obeys all constraints:

Lemma 4. *The join order constructed by Algorithm 2 does not violate any constraint represented by the input execution tree.*

Space constraint prevents us from presenting the proofs for these lemmas, from which we can construct the optimality proof:

Theorem 1. *SAM's 6-step procedure constructs a join order that has minimum total intermediate join sizes among all left-deep join orders that avoid Cartesian products.*

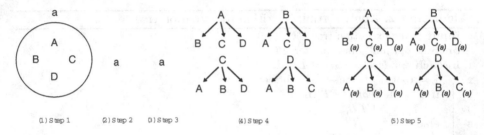

Fig. 7. SAM working on Clique Query in Example 2 (A, B, C, D are the four tables in the query in Example 2)

Proof. Given a multijoin query Q, SAM models Q in Steps 1 and 2. Steps 3, 4 and 5 enumerate all possibilities in constructing the spanning trees, execution graphs and execution trees. Step 6 then runs Algorithm 2 on each of these execution trees.

For a fixed execution tree $\mathcal{T}$, let π_S be the join order constructed by Algorithm 2, and π_o be an optimal join order. If π_S is not optimal, then π_S and π_o are different. By Lemma 2, all table blocks in π_S are ideal, so they occur in π_o.

Compare π_S and π_o from the last table blocks to be joined. Suppose the first different table blocks in π_S and π_o are $\mathbb{T}_S$ in π_S and $\mathbb{T}_o$ in π_o. Therefore, $\mathbb{T}_S$ is joined before $\mathbb{T}_o$ in π_o, and $\mathbb{T}_o$ is joined before $\mathbb{T}_S$ in π_S. It follows from Lemma 3 that $\mathbb{T}_o \preceq \mathbb{T}_S$.

Since all constraints are satisfied in π_S (Lemma 4) and π_o, all tables that must be joined after $\mathbb{T}_S$ or $\mathbb{T}_o$ are joined after $\mathbb{T}_S$ and $\mathbb{T}_o$ in π_S and π_o, respectively.

If there are no tables between $\mathbb{T}_S$ and $\mathbb{T}_o$ in π_o, then we can swap them to get a smaller cost (since $\mathbb{T}_o \preceq \mathbb{T}_S$), thus contradicting the optimality of π_o. Otherwise, consider all tables joined between $\mathbb{T}_S$ and $\mathbb{T}_o$ in π_o as a table block $\mathbb{T}$. Since $\preceq$ is total, either (i) $\mathbb{T}_o \preceq \mathbb{T}$ or (ii) $\mathbb{T} \preceq \mathbb{T}_o$. For (i), we can swap $\mathbb{T}$ and $\mathbb{T}_o$ in π_o to get a smaller cost; for (ii), $\mathbb{T} \preceq \mathbb{T}_o \preceq \mathbb{T}_S$, so $\mathbb{T} \preceq \mathbb{T}_S$ by transitivity (Lemma 1's proof), so we can similarly swap $\mathbb{T}_S$ and $\mathbb{T}_o$ in π_o to get a smaller cost. Either way, we again contradict the optimality of π_o. The contradiction proves the optimality of π_S for $\mathcal{T}$.

The join order that has minimal cost among the π_Ss produced for all the execution trees in Step 6 is thus optimal for Q.   □

3.4 Complexity Analysis

For traditional dynamic programming approaches [7,12,15], the worst case happens when the query is a clique, for which the time complexity is $O(3^{N_t})$, where N_t is the number of tables. For the example in Fig. 7, SAM sorts the leaves of each 1-level execution tree, so its complexity is just $O(N_t^2 \log N_t)$ in that case.

SAM enumerates all execution trees, so its complexity is, in general, exponential. The worst case happens when all tables are in the borders of the join attributes diagram, and the join attributes graph is complete. If the join attributes graph is acyclic, a detailed analysis shows that the complexity is $O(N_a N_t^2 e^{\frac{N_t}{e}})$, where N_a is the number of attributes. Thus, if $N_a = O(N_t)$, as when all joins are on keys, then SAM has complexity $O(1.45^{N_t})$, which is less than $O(3^{N_t})$.

3.5 Selectivity Measurement

When estimating selectivity, current database systems assume tuples are uniformly distributed over the join attribute. (They only resort to more accurate but costly approaches, like histograms, for selectivity estimation if users change the default setting.) This uniformity assumption can give bad estimates of join sizes.

With SAM, the system can record the selectivity p for each table T and its possible join attribute A, whenever T participates in a join using A. This empirical selectivity can be precomputed offline or collected on-the-fly.

4 Experiments

This section presents experiments for comparing SAM's performance to two leading enterprise database systems, both of which have been continuously developed and improved over the last 20 years.

The first system is PostgreSQL[1], which uses dynamic programming to find the optimal join order if the number of tables is small. If there are more than 12 tables, PostgreSQL uses a genetic algorithm[2] to find some approximately optimal ordering.

The other system is MySQL[3]. By default, its query optimizer uses some heuristic rules[4] in a depth-first search to find an approximately optimal ordering (regardless of number of tables). However, there is an option to find the optimal plan.

The performance comparison is in terms of plan time (how long it takes to choose the join order) and execution time (how long it takes to execute the chosen join order).

Experimental Set-Up. All experiments were performed on an Intel Core i5 1.9 GHz machine with 8 GB RAM running 64-bit windows 8.1. Our SAM algorithms are implemented in C++, while PostgreSQL (version 9.1.14) and MySQL (version 5.6.22), both of which are open-source, are also implemented in C++/C.

Datasets. Two real datasets are used in the experiments: (1) Twitter, which contains 5 tables about tweets, users, retweets, etc.; it has 456107 users and 28688584 tweets, and altogether 3.72 GBytes in size. (2) ACMDL[5], which contains 19 tables about publications from 1990 to 2008 indexed by the ACM Digital

---

[1] http://www.postgresql.org/.
[2] http://www.postgresql.org/docs/9.4/static/geqo-intro2.html.
[3] http://www.mysql.com/.
[4] http://dev.mysql.com/doc/internals/en/optimizer-joins-access-methods.html.
[5] Thanks to Craig Rodkin at ACM Headquarters for providing the ACMDL dataset.

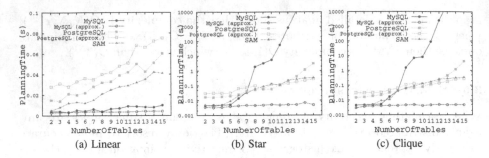

(a) Linear   (b) Star   (c) Clique

Fig. 8. Comparison of plan time for choosing join ordering (ACMDL). Note the log scale vertical axes for star and clique queries. In the legend, MySQL and PostgreSQL are for optimal join orders, MySQL (approx.) and PostgreSQL (approx.) are for join orders found by heuristics. SAM's plan times (in seconds) are 0.005, 0.004, 0.005, 0.009, 0.022, 0.031, 0.061, 0.088, 0.130, 0.110, 0.230, 0.270, 0.320 and 0.340 for the star queries, and 0.003, 0.004, 0.005, 0.009, 0.011, 0.042, 0.061, 0.090, 0.130, 0.172, 0.259, 0.280, 0.320 and 0.350 for the clique queries.

Library; it has 257694 authors and 248185 papers in 4.74 GBytes. Although each dataset is small enough to fit into RAM, the intermediate table sizes can overflow onto disk, thus requiring slow IO that can blow up the execution time. This is exactly why the join order must be carefully chosen even if the tables are small enough to fit into main memory.

4.1 Plan Time for Choosing a Join Order

The plan time complexity analysis in Sect. 3.4 is asymptotic. We now compare the plan times for MySQL and PostgreSQL, with and without heuristics, and SAM (which finds the optimal ordering). We consider linear, star and clique queries of up to 15 tables, so this experiment was done with ACMDL, since the Twitter dataset has only 5 tables.

As shown in Fig. 8(a), all 5 techniques have plan time linear in the number of tables, since the search space is limited when a query is linear.

For star queries, Fig. 8(b) shows the plan times for PostgreSQL and MySQL are consistent with previous analysis [7]: to find the optimal join order, they both take exponential time. In contrast, SAM's plan times for finding the optimal ordering for these queries do not grow exponentially. While MySQL and PostgreSQL have efficient plan times when using their heuristics, the suboptimal ordering that they choose have bad execution times (as we will see).

The results in Fig. 8(c) for clique queries are similar to those for star queries.

4.2 Execution Time for the Chosen Join Order

We next compare SAM, PostgreSQL and MySQL in terms of execution time for the join orders that they choose.

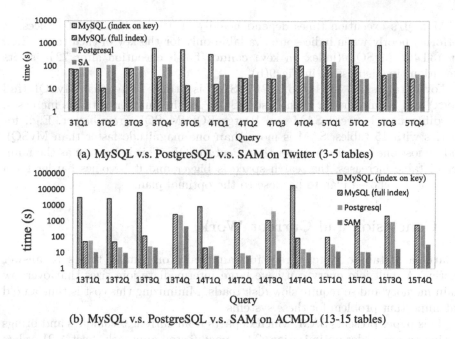

(a) MySQL v.s. PostgreSQL v.s. SAM on Twitter (3-5 tables)

(b) MySQL v.s. PostgreSQL v.s. SAM on ACMDL (13-15 tables)

Fig. 9. Comparison of execution time for each method's chosen join ordering. MySQL uses nested loop joins, while PostgreSQL and SAM use hash joins. Note the log scale on the vertical axes.

There are several algorithms for executing each join in a multijoin order: nested loop, block nested loop, sort-merge join and hash join. MySQL only uses nested loop joins, so we executed each query with two different settings: one with an index on each table's key only, and another with indices on all join attributes. To factor out the effect of indices, we only use hash joins for PostgreSQL, which does not require any index. For SAM, we execute the chosen join order using PostgreSQL (with hash joins).

For small queries of 3–5 tables, we use the Twitter dataset. E.g., 3TQ1 is Query1 of the 3-table queries, which is *"select count(*) from user, follower, tweet where user.uid = follower.uid and user.uid = tweet.uid;"*. Figure 9(a) shows that SAM has the same execution time as PostgreSQL for all queries because they both choose the same join order, except query 5TQ1. For 5TQ1, PostgreSQL's uniform selectivity assumption causes it to choose a suboptimal join order, unlike SAM. In comparison, MySQL is sometimes better, and sometimes worse, depending on how much help it gets from the available indices. The experimental results for ACMDL are similar and hence omitted here.

For larger queries, we use the ACMDL dataset. Both PostgreSQL and MySQL employ heuristics, trading suboptimality for smaller plan times, whereas SAM continues to choose the optimal orderings. Figure 9(b) shows that SAM's plans are always faster than those chosen by PostgreSQL and MySQL.

MySQL's execution times depend heavily on the availability of indices. It performs poorly when indices are available only for the key of each table. E.g., for 13T4Q, MySQL (index on key) cannot finish execution after 72 h, so its execution time is missing in Fig. 9(b).

For 13 tables, i.e. 13T1Q to 13T4Q, SAM is 5 times better than MySQL (full index) and 2 times better than PostgreSQL. As the number of tables increases, the differences between SAM and PostgreSQL/MySQL grow bigger. E.g., for queries with 15 tables, SAM is more than one magnitude faster than MySQL and almost one magnitude faster than PostgreSQL. This is because as the number of table increases, the search space is bigger and it becomes harder for a heuristically chosen plan to be close to the optimal plan.

5   Conclusion and Current Work

Relational database systems rely fundamentally on joining tables to answer queries. A multijoin query can have large intermediate join sizes that overflow main memory and so require slow disk reads. Minimizing this cost is thus an old but important problem for these systems.

This paper presents SAM, which departs from current techniques and brings sorting to join order optimization. This new, 6-step approach (Sect. 3.2) orders the joining attributes (Fig. 3) before ordering the tables (Fig. 5), and introduces the concepts of comparators and table blocks (Sect. 3.1). For clique queries and acyclic join attribute graphs, SAM has smaller time complexity than the dynamic programming techniques (Sect. 3.4). SAM does not make the uniform selectivity assumption (Sect. 3.5) that is used in current systems.

Experiments with Twitter and ACMDL data show that, compared to MySQL and PostgreSQL, SAM has plan time that grows slowly with the number of tables (Fig. 8). They also show that SAM has execution times that are similar to PostgreSQL for small queries, and much better than MySQL and PostgreSQL for large queries (Fig. 9).

This paper introduces SAM for general relational database schemas. Our current work focuses on tailoring SAM for schemas that have a particular form, like sonSchema, which is a conceptual schema for social network datasets [1]. The many elements in a typical social network service — friends, followers, posts, photos, tags, comments, shares, likes — generate many tables, so finding the optimal join order can make a huge difference to the query execution time.

Acknowledgement. We thank Zhifeng Bao for helpful discussions and comments.

References

1. Bao, Z., Tay, Y.C., Zhou, J.: sonSchema: a conceptual schema for social networks. In: Ng, W., Storey, V.C., Trujillo, J.C. (eds.) ER 2013. LNCS, vol. 8217, pp. 197–211. Springer, Heidelberg (2013)

2. DeHaan, D., Tompa, F.W.: Optimal top-down join enumeration. In: Proceedings of the SIGMOD, pp. 785–796 (2007)
3. Fegaras, L.: A new heuristic for optimizing large queries. In: Quirchmayr, G., Bench-Capon, T.J.M., Schweighofer, E. (eds.) DEXA 1998. LNCS, vol. 1460, pp. 726–735. Springer, Heidelberg (1998)
4. Ibaraki, T., Kameda, T.: On the optimal nesting order for computing n-relational joins. ACM Trans. Database Syst. 9(3), 482–502 (1984)
5. Kossmann, D., Stocker, K.: Iterative dynamic programming: a new class of query optimization algorithms. ACM Trans. Database Syst. 25(1), 43–82 (2000)
6. Lee, C., Shih, C.-S., Chen, Y.-H.: Optimizing large join queries using a graph-based approach. IEEE Trans. Knowl. Data Eng. 13(2), 298–315 (2001)
7. Moerkotte, G., Neumann, T.: Analysis of two existing and one new dynamic programming algorithm for the generation of optimal bushy join trees without cross products. In: Proceedings of the VLDB, pp. 930–941 (2006)
8. Moerkotte, G., Neumann, T.: Dynamic programming strikes back. In: Proceedings of the SIGMOD, pp. 539–552 (2008)
9. Neumann, T.: Query simplification: graceful degradation for join-order optimization. In: Proceedings of the SIGMOD, pp. 403–414 (2009)
10. Ngo, H.Q., Porat, E., Ré, C., Rudra, A.: Worst-case optimal join algorithms: [extended abstract]. In: Proceedings of the PODS, pp. 37–48 (2012)
11. Sacco, G.M.: Truly adaptive optimization: the basic ideas. In: Bressan, S., Küng, J., Wagner, R. (eds.) DEXA 2006. LNCS, vol. 4080, pp. 751–760. Springer, Heidelberg (2006)
12. Selinger, P.G., Astrahan, M.M., Chamberlin, D.D., et al.: Access path selection in a relational database management system. In: Proceedings of the SIGMOD, pp. 23–34 (1979)
13. Steinbrunn, M., Moerkotte, G., Kemper, A.: Heuristic and randomized optimization for the join ordering problem. VLDB J. 6(3), 191–208 (1997)
14. Swami, A., Gupta, A.: Optimization of large join queries. In: Proceedings of the SIGMOD, pp. 8–17 (1988)
15. Vance, B., Maier, D.: Rapid bushy join-order optimization with cartesian products. In: Proceedings of the SIGMOD, pp. 35–46 (1996)

GPU Acceleration of Set Similarity Joins

Mateus S.H. Cruz[1]([✉]), Yusuke Kozawa[1], Toshiyuki Amagasa[2],
and Hiroyuki Kitagawa[2]

[1] Graduate School of Systems and Information Engineering,
University of Tsukuba, Tsukuba, Japan
{mshcruz,kyusuke}@kde.cs.tsukuba.ac.jp
[2] Faculty of Engineering, Information and Systems,
University of Tsukuba, Tsukuba, Japan
{amagasa,kitagawa}@cs.tsukuba.ac.jp

Abstract. We propose a scheme of efficient set similarity joins on Graphics Processing Units (GPUs). Due to the rapid growth and diversification of data, there is an increasing demand for fast execution of set similarity joins in applications that vary from data integration to plagiarism detection. To tackle this problem, our solution takes advantage of the massive parallel processing offered by GPUs. Additionally, we employ MinHash to estimate the similarity between two sets in terms of Jaccard similarity. By exploiting the high parallelism of GPUs and the space efficiency provided by MinHash, we can achieve high performance without renouncing accuracy. Experimental results show that our proposed method is more than two orders of magnitude faster than the serial version of CPU implementation, and 25 times faster than the parallel version of CPU implementation, while generating highly precise query results.

Keywords: GPU · Parallel processing · Similarity join · MinHash

1 Introduction

A *similarity join* is an operator that, given two database relations and a similarity threshold, outputs all pairs of records, one from each relation, whose similarity is greater than the specified threshold. It has become a significant class of database operations due to the diversification of data, and it is used in many applications, such as data cleaning, entity recognition and duplicate elimination [3,5]. As an example, for data integration purposes, it might be interesting to detect whether *University of Tsukuba* and *Tsukuba University* refer to the same entity. In this case, the similarity join can identify such a pair of records as being similar.

Set similarity join [10] is a variation of similarity join that works on sets instead of regular records, and it is an important operation in the family of similarity joins due to its applicability on different data (e.g., market basket data, text and images). Regarding the similarity aspect, there is a number of well-known similarity metrics used to compare sets (e.g., Jaccard similarity and cosine similarity).

Q. Chen et al. (Eds.): DEXA 2015, Part I, LNCS 9261, pp. 384–398, 2015.
DOI: 10.1007/978-3-319-22849-5_26

One of the major drawbacks of a set similarity join is that it is a computationally demanding task, especially in the current scenario in which the size of datasets grows rapidly due to the trend of Big Data. For this reason, many researchers have proposed different set similarity join processing schemes [20,22,23]. Among them, it has been shown that parallel computation is a cost-effective option to tackle this problem [15,19], especially with the use of Graphics Processing Units (GPUs), which have been gaining much attention due to their performance in general processing [18].

There are numerous technical challenges when performing set similarity join using GPUs. First, how to deal with large datasets using GPU's memory, which is limited up to a few GBs in size. Second, how to make the best use of the high parallelism of GPUs in different stages of the processing (e.g., similarity computation and the join itself). Third, how to take advantage of the different types of memories on GPUs, such as device memory and shared memory, in order to maximize the performance.

In this research, we propose a new scheme of set similarity join on GPUs. To address the aforementioned technical challenges, we employ MinHash [2] to estimate the similarity between two sets in terms of their Jaccard similarity. MinHash is known to be a space-efficient algorithm to estimate the Jaccard similarity, while making it possible to maintain a good trade-off between accuracy and computation time. Moreover, we carefully design data structures and memory access patterns to exploit the GPU's massive parallelism and achieve high speedups. Experimental results show that our proposed method is more than two orders of magnitude faster than the serial version of CPU implementation, and 25 times faster than the parallel version of CPU implementation. In both cases, we assure the quality of the results by maximizing precision and recall values. We expect that such contributions can be effectively applied to process large datasets in real-world applications.

The remainder of this paper is organized as follows. Section 2 offers an overview of the similarity join operation applied to sets. Section 3 introduces the special hardware used, namely GPU, highlighting its main features and justifying its use in this work. In Sect. 4, we discuss the details of the proposed solution, and in Sect. 5 we present the experiments conducted to evaluate it. Section 6 examines the related work. Finally, Sect. 7 covers the conclusions and the future work.

2   Similarity Joins Over Sets

In a database, given two relations containing many records, it is common to use the join operation to identify the pairs of records that are similar enough to satisfy a predefined similarity condition. Such operation is called *similarity join*. This section introduces the application of similarity joins over sets, as well as the similarity measure used in our work, namely Jaccard similarity. After that, we explain how we take advantage of the MinHash [2] technique to estimate similarities, thus saving space and reducing computation time.

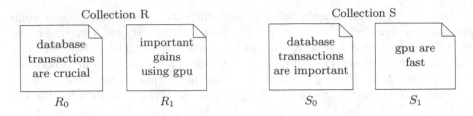

Fig. 1. Two collections of documents (R and S).

2.1   Set Similarity Joins

In many applications, we need to deal with sets (or multisets) of values as a part of data records. Some of the major examples are bag-of-words (documents), bag-of-visual-words (images) and transaction data [1,14]. Given database relations with records containing sets, one may wish to identify pairs of records whose sets are similar; in other words, two sets that share many elements. We refer to this variant of similarity join as a *set similarity join*. Henceforth, we use similarity join to denote set similarity join, if there is no ambiguity.

For example, Fig. 1 presents two collections of documents (R and S) that contain two documents each (R_0, R_1; S_0, S_1). In this scenario, the objective of the similarity join is to retrieve pairs of documents, one from each relation, that have a similarity degree greater than a specified threshold. Although there is a variety of methods to calculate the similarity between two documents, here we represent documents as sets of words (or *tokens*), and apply a set similarity method to determine how similar they are. We choose to use the Jaccard similarity (JS) since it is a well-known and commonly used technique to measure similarity between sets, and its calculation has high affinity with the GPU architecture. One can calculate the Jaccard similarity between two sets, X and Y, in the following way: $JS(X,Y) = |X \cap Y|/|X \cup Y|$. Considering this formula and the documents in Fig. 1, we obtain the following results: $JS(R_0, S_0) = 3/5 = 0.67$, $JS(R_0, S_1) = 1/6 = 0.17$, $JS(R_1, S_0) = 1/7 = 0.14$ and $JS(R_1, S_1) = 1/6 = 0.17$.

The computation of Jaccard similarity requires a number of pairwise comparisons among the elements from different sets to identify common elements, which incurs a long execution time, particularly when the sets being compared are large. In addition, it is necessary to store the whole sets in memory, which can require prohibitive storage [12].

2.2   MinHash

To address the aforementioned problems, Broder et al. proposed a technique called MinHash (Min-wise Hashing) [2]. Its main idea is to create signatures for each set based on its elements and then compare the signatures to estimate their Jaccard similarity. If two sets have many coinciding signature parts, they share some degree of similarity. In this way, it is possible to estimate the Jaccard similarity without conducting costly scans over all elements. In addition one only

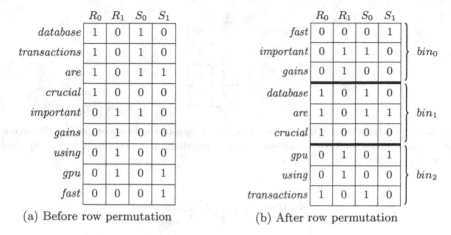

	R_0	R_1	S_0	S_1
database	1	0	1	0
transactions	1	0	1	0
are	1	0	1	1
crucial	1	0	0	0
important	0	1	1	0
gains	0	1	0	0
using	0	1	0	0
gpu	0	1	0	1
fast	0	0	0	1

(a) Before row permutation

	R_0	R_1	S_0	S_1	
fast	0	0	0	1	bin_0
important	0	1	1	0	
gains	0	1	0	0	
database	1	0	1	0	bin_1
are	1	0	1	1	
crucial	1	0	0	0	
gpu	0	1	0	1	bin_2
using	0	1	0	0	
transactions	1	0	1	0	

(b) After row permutation

Fig. 2. Characteristic matrices constructed based on the documents from Fig. 1, before and after a permutation of rows.

needs to store the signatures instead of all the elements of the sets, which greatly contributes to reduce storage space.

After its introduction, Li et al. suggested a series of improvements for the MinHash technique related to memory use and computation performance [11–13]. Our work is based on the latest of those improvements, namely, One Permutation Hashing [13].

In order to estimate the similarity of the documents in Fig. 1 using One Permutation Hashing, first we change their representation to a data structure called *characteristic matrix* (Fig. 2a), which assigns the value *1* when a token represented by a row belongs to a document represented by a column, and *0* when it does not.

After that, in order to obtain an unbiased similarity estimation, a random permutation of rows is applied to the characteristic matrix, followed by a division of the rows into partitions (henceforth called *bins*) of approximate size (Fig. 2b). However, since the actual permutation of rows in a large matrix constitutes an expensive operation, MinHash uses hash functions to simulate such permutation.

For each bin, each document has a value that will compose its signature. This value is the index of the row containing the first *1* (scanning the matrix in a top-down fashion) in the column representing the document. For example, the signature for the document S_0 is *1*, *3* and *8*. It can happen that a bin for a given document does not have any value (e.g., the first bin of set R_0, since it has no *1*), and this case is also taken into consideration during the similarity estimation. Figure 3 shows a data structure called *signature matrix*, which contains the signatures obtained for all the documents.

Finally, the similarity between any two documents is estimated by Eq. 1 (an extensive theoretical analysis is provided by Li et al. [13]), where N_{mat} is the number of matching bins between the signatures of the two documents, b represents the total number of bins composing the signatures, and N_{emp} refers to the number of matching empty bins.

	b_0	b_1	b_2
R_0	*	3	8
R_1	1	*	6
S_0	1	3	8
S_1	0	4	6

Fig. 3. Signature matrix, with columns corresponding to the bins composing the signatures of documents, and rows corresponding to the documents themselves. The symbol * denotes an empty bin.

$$Sim(X, Y) = \frac{N_{mat}}{(b - N_{emp})} \tag{1}$$

The estimated similarities for the given example are $Sim(R_0, S_0) = 2/3 = 0.67$, $Sim(R_0, S_1) = 0/3 = 0$, $Sim(R_1, S_0) = 1/3 = 0.33$ and $Sim(R_1, S_1) = 1/3 = 0.33$. Even though this is a simple example, the estimated values can be considered close to the real Jaccard similarities previously calculated (0.67, 0.17, 0.14 and 0.17). In practical terms, using more bins yields a more accurate estimation, but it also increases the size of the signature matrix.

Let us observe an important characteristic of MinHash. Since the signatures are independent of each other, it presents a good opportunity for parallelization. Indeed, the combination of MinHash and parallel processing using GPUs has been considered by Li et al. [12], as they showed a reduction of the processing time by more than an order of magnitude in online learning applications. While their focus was the MinHash itself, here we use it as a tool in the similarity join processing.

3   General-Purpose Processing on Graphics Processing Units

Despite being originally designed for games and other graphic applications, the applications of Graphics Processing Units (GPUs) have been extended to general computation due to their high computational power [18]. This section presents the features of this hardware and the challenges encountered when using it.

The properties of a modern GPU can be seen from both a computing and a memory-related perspective. In terms of computational components, the GPU's *scalar processors* (SPs) run the primary processing unit, called *thread*. GPU programs (commonly referred to as *kernels*) run in a SPMD (Single Program Multiple Data) fashion on these lightweight threads. Threads form *blocks*, which are scheduled to run on *streaming multiprocessors* (SMs).

The memory hierarchy of a GPU consists of three main elements: *registers*, *shared memory* and *device memory*. Each thread has its own registers (quickly accessible, but small in size) and cannot access the registers of other threads. In order to share data among threads in a block, one can use the shared memory, which is also fast, but still small (16 KB to 96 KB per SM depending on the GPU's capability). Lastly, in order to share data between multiple blocks, the device memory (also called *global memory*) is used. However, it should be noted that the device memory suffers from a long access latency as it resides outside the SMs.

When programming a GPU, one of the greatest challenges is the effective utilization of this hardware's architecture. For example, there are several benefits in exploring the faster memories, as it minimizes the access to the slower device memory and increases the overall performance.

In order to apply a GPU for general processing, it is common to use dedicated libraries that can facilitate such task. Our solution employs NVIDIA's CUDA [16], which provides an extension of the C programming language, by which one can define parts of a program to be executed on the GPU.

In terms of algorithms, a number of data-parallel operations, usually called *primitives*, have been ported to be executed on GPUs in order to facilitate programming tasks. He et al. [7,8] provide details on the design and implementation of many of these primitives.

One primitive particularly useful for our work is *scan* [21]. In this research, we use scan to calculate the positions where each block will write the result of their computation, allowing us to overcome the lack of incremental memory allocation during the execution of kernels and to avoid write conflicts between blocks. We chose to adopt the scan implementation provided by the library Thrust [9] due to its high performance and ease to use.

4   GPU Acceleration of Set Similarity Joins

In the following discussion, we consider the sets to be text documents stored on disk, but the solution can be readily adapted to other types of data. We also assume that techniques to prepare the data for processing (e.g., stopword removal and stemming) are out of our scope, and should take place before the similarity join processing on GPU.

Our proposed scheme comprises three main steps: preprocessing, signature matrix computation, and similarity join.

4.1   Preprocessing

In the preprocessing step, we construct a compact representation of the characteristic matrix, since the original one is usually highly sparse. By doing so, the data to be transferred to the GPU is greatly reduced (more than 95 % for the datasets used in the experimental evaluation in Sect. 5).

This representation is based on the Compressed Row Storage (CRS) format [6], which uses three arrays: *var*, which stores the values of the nonzero elements of the matrix; *col_ind*, that holds the column indexes of the elements in the *var* array; and *row_ptr*, which keeps the locations in the *var* array that start a row in the matrix.

	R_0	R_1	S_0	S_1										
doc_start	0	4	8	12	15									
doc_tok	0	1	2	3	4	5	6	7	0	1	2	4	2	7	8

Fig. 4. Compact representation of the characteristic matrix.

Considering that the nonzero elements of the characteristic matrix have the same value, *1*, there is only need to store their positions. Figure 4 shows such representation for the characteristic matrix of the previous example (Fig. 2a). The array *doc_start* holds the positions in the array *doc_tok* where the documents start, and the array *doc_tok* shows what tokens belong to each document.

After its construction, the characteristic matrix is sent to the GPU, and we assume it fits completely in the device memory. The processing of large datasets, which do not fit into the device memory is part of future work. Nevertheless, the aforementioned method allows us to deal with sufficiently large datasets using current GPUs in many practical applications.

4.2 Signature Matrix Computation on GPU

Once the characteristic matrix is in the GPU's device memory, the next step is to construct the signature matrix. Algorithm 1 shows how we parallelize the MinHash technique, and Fig. 5 illustrates such processing. In practical terms, one block is responsible for computing the signature of one document at a time. Each thread in the block (1) accesses the device memory, (2) retrieves the position of one token of the document, (3) applies a hash function to it to simulate the row permutation, (4) calculates which bin the token will fit into, and (5) updates that bin. If more than one value is assigned to the same bin, the algorithm keeps the minimum value (hence the name MinHash).

During its computation, the signature for the document is stored in the shared memory, which supports fast communication between the threads of a block. This is advantageous in two aspects: (1) it allows fast updates of values when constructing the signature matrix, and (2) since different threads can access sequential memory positions, it favors the coalesced access to the device memory when the signature computation ends. Accessing the device memory in a coalesced manner means that a number of threads will access consecutive memory locations, and such accesses can be grouped into a single transaction. This makes the transfer of data from and to the device memory significantly faster.

The complete signature matrix is laid out in the device memory as a single array of integers. Since the number of bins per signature is known, it is possible to perform direct access to the signature of any given document.

After the signature matrix is constructed, it is kept in the GPU's memory to be used in the next step: the join itself. This also minimizes data transfers between CPU and GPU.

Algorithm 1. Parallel MinHash.

---

input : characteristic matrix $CM_{t \times d}$ (t tokens, d documents), number of bins b
output: signature matrix $SM_{d \times b}$ (d documents, b bins)
1 binSize $\leftarrow \lfloor t/b \rfloor$;
2 **for** $i \leftarrow 0$ *to* d **in parallel do** // executed by blocks
3   **for** $j \leftarrow 0$ *to* t **in parallel do** // executed by threads
4   **if** $CM_{j,i} = 1$ **then**
5               $h \leftarrow hash(CM_{j,i})$;
6               $binIdx \leftarrow \lfloor h/binSize \rfloor$;
7               $SM_{i,binIdx} \leftarrow min(SM_{i,binIdx}, h)$;
8   **end**
9   **end**
10 **end**

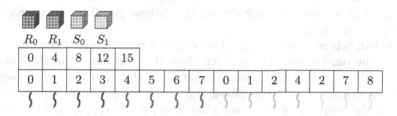

R_0 R_1 S_0 S_1

0	4	8	12	15										
0	1	2	3	4	5	6	7	0	1	2	4	2	7	8

Fig. 5. Computation of the signature matrix based on the characteristic matrix. Each GPU block is responsible for one document, and each thread is assigned to one token.

4.3 Similarity Joins on GPU

The next step is the similarity join, and it utilizes the results obtained in the previous phase, i.e., the signatures generated using MinHash. To address the similarity join problem, we choose to parallelize the nested-loop join (NLJ) algorithm. The nested-loop join algorithm iterates through the two relations being joined and check if the pairs of records, one from each relation, comply with a given predicate. For the similarity join case, this predicate is that the records of the pairs must have a degree of similarity greater than a given threshold.

In practical terms, the parallelization of NLJ consists in executing one of its loops in parallel. For our GPU implementation, this means that each block is responsible for a document of the collection R, and it iterates over documents of the collection S comparing the signatures of the documents (Fig. 6a). Initially, each thread in a block reads one bin of the signature of a document from collection R and copies it to the shared memory. Then, the thread compares the value of that bin to the corresponding bin of a document from collection S, which is accessed in a coalesced manner (Fig. 6b). Using Eq. 1, if the comparison yields a similarity greater than the given threshold, that pair of documents belongs to the final result.

As highlighted by He et al. [8], outputting the result from a join performed in the GPU raises two main problems. Firstly, since the size of the output is

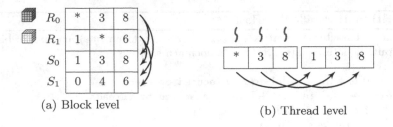

(a) Block level   (b) Thread level

Fig. 6. Parallelization of NLJ.

initially unknown, it is also not possible to know how much memory should be allocated on the GPU to hold the result. In addition, there may be conflicts between blocks when writing on the device memory. For this reason, He et al. [8] proposed a join scheme for result output that allows parallel writing, which we also adopt in this work.

Their join scheme performs the join in three phases. First, the join is run once, and the blocks count the number of similar pairs found in their portion of the execution, writing this amount in an array stored in the device memory. There is no write conflict in this phase, since each block will write in a different position of the array. After that, using the scan primitive, it is possible to know the correct size of memory that should be allocated for the results, as well as where the threads of each block should start writing the similar pairs they found. Finally, the similarity join is run once again, outputting the similar pairs to the proper positions in the allocated space.

After that, depending on the application, the pairs can be transferred back to the CPU and output to the user or kept in the GPU for further processing by other algorithms.

5 Experiments

In this section we present the experiments performed to evaluate our proposal. First, we introduce the used datasets and the environment on which the experiments were conducted. Then we show the results related to performance and accuracy. For all the experiments, unless stated, the similarity threshold was 0.8 and the number of bins composing the documents' signatures was 32.

5.1 Datasets

To demonstrate the range of applicability of our work, we chose datasets from three distinct domains (Table 1). The *Images* dataset, made available at the UCI Machine Learning Repository[1], consists of image features extracted from the Corel image collection. The *Abstracts* dataset, composed by abstracts of

---

[1] http://archive.ics.uci.edu/ml/datasets/.

Table 1. Characteristics of datasets.

Dataset	Cardinality	Size (MB)	# of Tokens per record
Images	68,040	19	32
Abstracts	233,445	255	165
Transactions	1,692,082	1413	177

publications from MEDLINE, were obtained from TREC-9 Filtering Track Collections[2]. Finally, *Transactions* is a transactional dataset available through the FIMI repository[3]. From the original datasets, we randomly chose documents in order to create the collections R and S, whose sizes vary from 1,024 to 524,288 documents.

5.2 Environment

The CPU used in our experiments was an Intel Xeon E5-1650 (6 cores, 12 threads) with 32GB of memory. The GPU was an NVIDIA Tesla K20Xm (2688 scalar processors) with 6 GB of memory. Regarding the compilers, GCC 4.4.7 (with the flag -O3) was used for the part of the code to run on the CPU, and NVCC 6.5 (with the flags -O3 and -use_fast_math) compiled the code for the GPU. For the parallelization of the CPU version, we used OpenMP 4.0 [17].

5.3 Performance Comparison

Figures 7, 8 and 9 present the execution time of our approach for the three implementations (GPU, CPU Parallel and CPU Serial) using the three datasets.

Let us first consider the MinHash part, i.e., the time taken for the construction of the signature matrix. It can be seen from the results (Fig. 7a, b and c) that the GPU version of MinHash is more than 20 times faster than the serial implementation on CPU, and more than 3 times faster than the parallel implementation on CPU. These findings reinforce the idea that MinHash is indeed suitable for parallel processing.

For the join part (Fig. 8a, b and c), the speedups are even higher. The GPU implementation is more than 150 times faster than the CPU Serial implementation, and almost 25 times faster than the CPU Parallel implementation. The speedups of more than two orders of magnitude demonstrate that the NLJ algorithm can benefit from the massive parallelism provided by GPU.

Measurements of the total time of execution (Fig. 9a, b and c) show that the GPU implementation achieves speedups of approximately 120 times when compared to the CPU Serial implementation, and approximately 20 times when compared to the CPU Parallel implementation.

---

[2] http://trec.nist.gov/data/t9_filtering.html.
[3] http://fimi.ua.ac.be/data/.

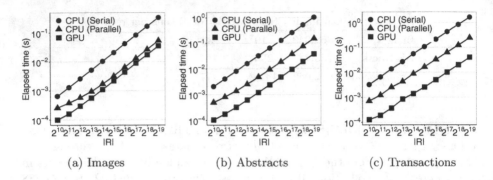

(a) Images   (b) Abstracts   (c) Transactions

Fig. 7. Minhash performance comparison ($|R| = |S|$).

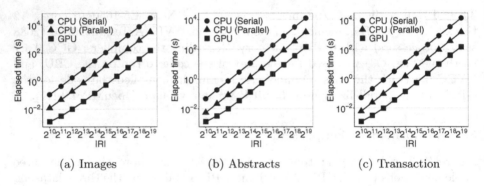

(a) Images   (b) Abstracts   (c) Transaction

Fig. 8. Join performance comparison ($|R| = |S|$).

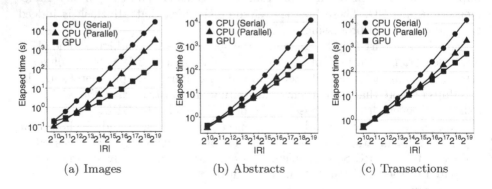

(a) Images   (b) Abstracts   (c) Transactions

Fig. 9. Overall performance comparison ($|R| = |S|$).

Table 2. Breakdown of the execution time in seconds when joining collections of the same size (Abstracts dataset, $|R| = |S| = 524288$).

	GPU	CPU (Parallel)	CPU (Serial)
Read from disk	201.5	200.5	198.1
Preprocessing	9.3	9.4	9.1
MinHash	0.037	0.151	1.033
Join	145	1403	11950
Memory transfer	0.09	0	0
Total	359	1615	12162

Analysis of performance details provides some insights into why the overall speedup is lower than the join speedup. Table 2 presents the breakdown of the execution time for the Abstracts dataset – experiments with the other datasets showed equivalent results. Especially for larger collections, the join step is the most time consuming part for both CPU implementations. For the GPU implementation, reading data from disk constitutes the bottleneck for the execution.

5.4 Accuracy Evaluation

Since our scheme uses the MinHash technique to estimate the similarity between sets, it is also important to evaluate how accurate the results obtained from it are. We evaluated the accuracy of the proposal in terms of precision and recall. Precision relates to the fraction of really similar pairs among all the pairs retrieved by the algorithm, and recall refers to the fraction of really similar pairs that were correctly retrieved.

Table 3 presents the measurements of experiments in which we varied the number of bins composing the signatures of the documents, showing the impact of the number of bins on the number of similar pairs found, as well as on the performance.

As the number of bins increases, the number of similar pairs found nears the number of really similar pairs, thus increasing the values of precision and recall. On the other hand, increasing the number of bins also incurs a longer execution time. Therefore, it is important to achieve a balance between accuracy and execution time. For the used datasets, using 32 bins offered a good trade-off, yielding the lowest execution time without false positive or false negative results.

6   Related Work

A recent survey by Jiang et al. [10] made comparisons between a number of string similarity join approaches. The majority of these works focus on the elimination of unnecessary work and adopt a filter-verification approach [22,23], which initially prunes dissimilar pairs and leaves only candidate pairs that are later verified if they are really similar. The survey highlighted differences concerning the

Table 3. Impact of varying number of bins on precision, recall and execution time (Abstracts dataset, $|R| = |S| = 65536$).

Number of bins	Precision	Recall	Execution time (s)
1	0.0000	0.9999	25.3
2	0.0275	0.9999	25.4
4	0.9733	0.9999	25.6
8	0.9994	0.9999	25.7
16	0.9998	1.0000	26.1
32	1.0000	1.0000	27.4
64	1.0000	1.0000	29.6
128	1.0000	1.0000	34.4
256	1.0000	1.0000	45.8
384	1.0000	1.0000	77.6
512	1.0000	1.0000	133.6
640	1.0000	1.0000	161.5

performance of algorithms based on the size of the dataset and on the length of the joined strings. Jiang et al. [10] also pointed out the necessity for disk-based algorithms to deal with really large datasets that do not fit in memory.

Other works focused on taking advantage of parallel processing to produce more scalable similarity join algorithms. Among these, Vernica et al. [19], Metwally et al. [15] and Deng et al. [4] used MapReduce to distribute the processing among nodes in CPU clusters.

Although the similarity join is a thoroughly discussed topic, works utilizing GPUs for the processing speedup are not numerous. Lieberman et al. [14] mapped the similarity join operation to a sort-and-search problem and used well-known algorithms and primitives for GPUs to perform these tasks. After applying the bitonic sort algorithm to create a set of space-filling curves from one of the relations, they processed each record of the relation set in parallel, executing searches in the space-filling curves. The similarity between the records was calculated using the Minkowski metric.

Böhm et al. [1] proposed two GPU-accelerated nested-loop join (NLJ) algorithms to perform the similarity join operation, and used Euclidean distance to calculate the similarity in both cases. The best of the two methods was the index-supported similarity join, which has a preprocessing phase to create an index structure based on directories. The authors alleged that the GPU version of the indexed-supported similarity join achieved an improvement of 4.6 times when compared to its sequential CPU version.

The main characteristic that discerns our work from the other similarity join schemes is the effective use of MinHash to overcome challenges inherent to the use of GPUs for general-purpose computation, as emphasized in Sect. 2.2

7 Conclusions

We have proposed a GPU-accelerated similarity join scheme that uses Min-Hash in its similarity calculation step and achieved a speedup of more than two orders of magnitude when compared to the sequential version of the algorithm. Moreover, the high levels of precision and recall obtained in the experimental evaluation confirmed the accuracy of our scheme.

The strongest point of GPUs is their superior throughput when compared to CPUs. However, they require special implementation techniques to minimize memory access and data transfer. For this purpose, using MinHash to estimate the similarity of sets is particularly beneficial, since it enables a parallelizable way to represent the sets in a compact manner, thus saving storage and reducing data transfer. Furthermore, our implementation explored the faster memories of GPUs (registers and shared memory) to diminish effects of memory stalls. We believe this solution can aid in the task of processing large datasets in a cost-effective way without ignoring the quality of the results.

Since the join is the most expensive part of the processing, future works will focus on the investigation and implementation of better join techniques on GPUs. For the algorithms developed in a next phase, the main requirements are parallelizable processing-intensive parts and infrequent memory transfers.

Acknowledgments. We thank Neil Millar and the reviewers for their feedback. This research was partly supported by the Grant-in-Aid for Scientific Research (B) (#262 80037) from the Japan Society for the Promotion of Science.

References

1. Böhm, C., Noll, R., Plant, C., Zherdin, A.: Indexsupported similarity join on graphics processors. BTW **144**, 57–66 (2009)
2. Broder, A.Z., Charikar, M., Frieze, A.M., Mitzenmacher, M.: Min-wise independent permutations. J. Comput. Syst. Sci. **60**(3), 630–659 (2000)
3. Chaudhuri, S., Ganti, V., Kaushik, R.: A primitive operator for similarity joins in data cleaning. In: ICDE, p. 5 (2006)
4. Deng, D., Li, G., Hao, S., Wang, J., Feng, J.: Massjoin: a Mapreduce-based method for scalable string similarity joins. In: ICDE, pp. 340–351 (2014)
5. Gravano, L., Ipeirotis, P.G., Jagadish, H.V., Koudas, N., Muthukrishnan, S., Srivastava, D.: Approximate string joins in a database (almost) for free. In: VLDB, pp. 491–500 (2001)
6. Greathouse, J.L., Daga, M.: Efficient sparse matrix-vector multiplication on GPUs using the CSR storage format. In: SC, pp. 769–780 (2014)
7. He, B., Lu, M., Yang, K., Fang, R., Govindaraju, N.K., Luo, Q., Sander, P.V.: Relational query coprocessing on graphics processors. TODS **34**(4), 21:1–21:39 (2009)
8. He, B., Yang, K., Fang, R., Lu, M., Govindaraju, N., Luo, Q., Sander, P.: Relational joins on graphics processors. In: SIGMOD, pp. 511–524 (2008)
9. Hoberock, J., Bell, N.: Thrust: A Productivity-Oriented Library for CUDA (2012)

10. Jiang, Y., Li, G., Feng, J., Li, W.S.: String similarity joins: an experimental evaluation. pvldb **7**(8), 625–636 (2014)
11. Li, P., Knig, A.C.: b-bit Minwise Hashing (2009). CoRR. abs/0910.3349
12. Li, P., Shrivastava, A., König, A.C.: GPU-based minwise hashing. In: WWW, pp. 565–566 (2012)
13. Li, P., Owen, A.B., Zhang, C.H.: One Permutation Hashing for Efficient Search and Learning (2012). CoRR. abs/1208.1259
14. Lieberman, M.D., Sankaranarayanan, J., Samet, H.: A fast similarity join algorithm using graphics processing units. In: ICDE, pp. 1111–1120 (2008)
15. Metwally, A., Faloutsos, C.: V-Smart-Join: a scalable mapreduce framework for all-pair similarity joins of multisets and vectors. PVLDB **5**(8), 704–715 (2012)
16. NVIDIA Corporation: NVIDIA CUDA Compute Unified Device Architecture Programming Guide (2007)
17. OpenMP Architecture Review Board: OpenMP Application Program Interface Version 4.0 (2013)
18. Owens, J.D., Luebke, D., Govindaraju, N., Harris, M., Krger, J., Lefohn, A., Purcell, T.J.: A survey of general-purpose computation on graphics hardware. Computer Graph. Forum **26**(1), 80–113 (2007)
19. Rares, V., Carey, M.J., Chen, L.: Efficient parallel set-similarity joins using Mapreduce. In: SIGMOD, pp. 495–506 (2010)
20. Sarawagi, S., Kirpal, A.: Efficient set joins on similarity predicates. In: SIGMOD, pp. 743–754 (2004)
21. Sengupta, S., Harris, M., Zhang, Y., Owens, J.D.: Scan primitives for GPU computing. In: GH, pp. 97–106 (2007)
22. Wang, J., Li, G., Feng, J.: Can we beat the prefix filtering?: an adaptive framework for similarity join and search. In: SIGMOD, pp. 85–96 (2012)
23. Xiao, C., Wang, W., Lin, X., Yu, J.X.: Efficient similarity joins for near duplicate detection. In: WWW, pp. 131–140 (2008)

Data Mining II

Parallel Eclat for Opportunistic Mining of Frequent Itemsets

Junqiang Liu[1]([✉]), Yongsheng Wu[1], Qingfeng Zhou[1], Benjamin C.M. Fung[2], Fanghui Chen[1], and Binxiao Yu[1]

[1] School of Information and Electronic Engineering,
Zhejiang Gongshang University, Hangzhou 310018, China
jjliu@alumni.sfu.ca
[2] School of Information Studies, McGill University, Montreal, QC, Canada
ben.fung@mcgill.ca

Abstract. Mining frequent itemsets is an essential data mining problem. As the big data era comes, the size of databases is becoming so large that traditional algorithms will not scale well. An approach to the issue is to parallelize the mining algorithm, which however is a challenge that has not been well addressed yet. In this paper, we propose a MapReduce-based algorithm, Peclat, that parallelizes the vertical mining algorithm, Eclat, with three improvements. First, Peclat proposes a hybrid vertical data format to represent the data, which saves both space and time in the mining process. Second, Peclat adopts the pruning technique from the Apriori algorithm to improve efficiency of breadth-first search. Third, Peclat employs an ordering of itemsets that helps balancing the workloads. Extensive experiments demonstrate that Peclat outperforms the existing MapReduce-based algorithms significantly.

Keywords: Big data · Data mining · Frequent itemsets · Vertical format · MapReduce · Parallel algorithms

1 Introduction

Discovering frequent itemsets is always an essential problem in data mining research and database applications. This problem is formulated as follows: Given a transaction database, find all frequent itemsets where a frequent itemset is one that occurs in at least a user-specified number or percentage of transactions, that is, its support is no less than the threshold.

The best-known breadth-first algorithm is Apriori [2] by Agrawal et al. Apriori finds frequent itemsets in a level-wise manner. At each level k, it scans the database once to compute the supports of the k-itemsets whose subsets at level $k-1$ are all frequent. Therefore, Apriori scans the database as many passes as the largest size of frequent itemsets, which incurs high I/O overhead. The FP-growth algorithm [9] by Han et al. is the best-known depth-first algorithm. It uses a novel frequent pattern tree (FP-tree) held in memory to represent the

© Springer International Publishing Switzerland 2015
Q. Chen et al. (Eds.): DEXA 2015, Part I, LNCS 9261, pp. 401–415, 2015.
DOI: 10.1007/978-3-319-22849-5_27

database in a compressed form, and mines all frequent itemsets by recursively projecting the FP-trees. FP-growth as well as any depth-first algorithm does not work as efficiently as supposed when faced with huge and sparse databases.

While most previous works inspired by Apriori [2] and FP-growth [9] employ a horizontal data format, a number of algorithms [6,15,16,19,20] use vertical data format. In a vertical data format, each item is associated with its tidset, the set of identifiers of transactions that contain the item. The best-known vertical mining algorithm is Eclat [19] by Zaki et al., which uses the tidset format. But, when the database is large and dense, Eclat suffers from huge intermediate storage cost. Thus, Zaki proposed the diffset format [20] for dense databases. However, large databases cannot be simply categorized as sparse or dense, with which neither single vertical format can scale well.

From the above analysis, we can observe that sequential mining algorithms do not scale well when databases become huge. Thus, some works [1,3,17,21] proposed parallel algorithms that make use of computing paradigms in the form of multi-processors. But, such parallelization paradigms have issues in balancing workloads and in recovering from hardware or communication failures.

Therefore, Google proposed the MapReduce paradigm [5]. Since then, a few parallel algorithms [4,7,8,10–14,18] based on MapReduce have been proposed for mining frequent itemsets, among which the DPC [13] algorithm is the best derived from Apriori, and PFP [10] is the best derived from FP-growth.

DPC [13] targets the problem of high I/O overhead with Apriori. It generates candidate itemsets and counts supports for as many levels as possible in a MapReduce job, and therefore may reduce the number of database scans. However, DPC still suffers for dense databases as it can only mine one level in one database scan in such a case.

PFP [10] targets a problem with FP-growth, that is, the FP-tree may be too large to be held in memory. It solves the problem by breaking the FP-tree into small subtrees that can be held in memory. But, PFP suffers from data redundancy as the subtrees overlap, which results in a huge amount of redundant data to be sent across the cluster. To alleviate the issue, the PFP implementation by Apache Mahout [22] only outputs the top-k frequent closed itemsets.

In summary, while the best sequential mining algorithms such as Apriori, FP-growth, and Eclat do not work for processing huge databases, the existing parallel algorithms either follow a parallelization paradigm that is not fault-tolerant or do not adequately address the issues with sequential algorithms.

This paper proposes the first algorithm that parallelizes the Eclat algorithm [19] for addressing the shortcomings of the existing MapReduce-based parallel algorithms. We take full advantage of parallelization provided by MapReduce [5], and propose an opportunistic vertical mining approach that improves the Eclat algorithm. Concretely, our contributions are as follows:

- We propose to employ hybrid vertical data formats in the mining process. The novelty is to make a choice on a per itemset basis in selecting a vertical data format that results in a smaller storage footprint. In particular, an itemset can be derived from two itemsets that use different data formats in representing

their transactions. Our approach improves not only scalability but also efficiency as the intermediate storage space, I/O overhead, and computation time
all decrease in the mining process.

- We improve the breadth-first search by adopting the pruning technique from
  the Apriori algorithm [2]. That is, we avoid unnecessary set-intersection operations for computing supports of itmesets whose subsets are infrequent.
- We dynamically rearrange itemsets in the ascending order of supports, which
  results in more groups of itemsets grouped by prefixes and smaller group sizes.
  Such an ordering helps breaking the computation workloads into smaller pieces
  that can be processed in parallel, i.e., balancing the workloads.

The rest of the paper is organized as follows. Section 2 defines the mining problem, Sect. 3 proposes our opportunistic vertical mining approach, Sect. 4 presents
our algorithm, Sect. 5 evaluates our work, and Sect. 6 concludes the paper.

2   Problem Statement and Preliminaries

This section first defines the mining problem, and then analyzes mining strategies
and data formats that will be adapted and improved by our algorithm.

2.1   Problem Statement

The frequent itemset mining problem is stated as follows [2].

Definition 1 (Itemset and Subset). *Let I be a set of items. A set of items
from I is also called an itemset. An itemset consisting of k items is called a
k-itemset. An itemset S is called a subset of an itemset X if all items in S are
also in X, denoted as $S \subseteq X$.*

Definition 2 (Frequent Itemset). *Let D be a transaction database over a
set of items I, where each transaction t in D contains a set of items from I and
has a unique identifier called tid, denoted as t.tid. The support of an itemset
X, denoted as $\sigma(X)$, is the number of transactions that have X as a subset.
The itemset X is frequent or large if its support, $\sigma(X)$, is no less than a user-
specified minimum support threshold, denoted as minsup.*

The problem is to find the set of all frequent or large itemsets, i.e., $L = \{X \subseteq
I \mid \sigma(X) \geq minsup\}$. For example, for the transaction database D in Table 1
and $minsup = 3$, the set L of all frequent itemsets consists of {a}, {b}, {c}, {f},
{m}, {p}, {a,c}, {a,f}, {a,m}, {c,f}, {c,m}, {c,p}, {f,m}, {a,c,f}, {a,c,m},{a,f,m},
{c,f,m}, and {a,c,f,m}.

2.2   Strategies for Mining Frequent Itemsets

Conceptually, all itemsets form a lattice based on a partial order, i.e., the subset
relation on the set of itemsets. For example, Fig. 1 is a lattice on the powerset
of {a,b,c,f,m,p}. To find all frequent itemsets, all algorithms adopt a bottom-up
strategy to search a lattice of itemsets starting with the empty itemset, and
employ the downward closure property partially or fully.

Table 1. Transaction database D

Tid	Items
1	{a, c, d, f, g, i, m, p}
2	{a, b, c, f, l, m, o}
3	{b, f, h, j, o}
4	{b, c, k, p, s}
5	{a, c, e, f, l, m, n, p}

Lemma 1 (Downward Closure Property). *All subsets of frequent itemset must be frequent. An itemset is infrequent if one of its subsets is infrequent.*

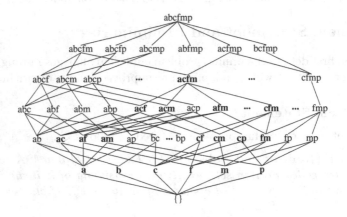

Fig. 1. Lattice of itemsets in the lexicographic order

The Apriori algorithm [2] fully utilizes the downward closure property in its bottom-up, breadth-first search of the lattice. In particular, it generates the set C_k of candidates, i.e., k-itemsets to be search at level k, from the set L_{k-1} of frequent $(k$-1)-itemsets by two functions, join and pruning.

- The join function forms a k-itemset, i.e., a candidate, by joining two $(k$-1)-itemsets that share a $(k$-2)-prefix.
- The pruning function prunes a candidate that has an infrequent subset of size k-1 (which may not be a prefix).

Definition 3 (Prefix). *Let X be a k-itemset, the h-prefix of X is the set of the first h items in X by a given ordering Ω of items, which is denoted as $prefix(X, h, \Omega)$ with $h \leq k$ and abbreviated as $prefix(X, h)$ when the ordering is obvious.*

The FP-growth algorithm [9] partially utilizes the downward closure property in its bottom-up, depth-first search of the lattice. In particular, for each frequent itemset X, it recursively searches the union of X and an item locally frequent in

Table 2. The tidsets and diffsets of frequent items

Frequent item	Tidset	Diffset
a	$\{1, 2, 5\}$	$\{3,4\}$
b	$\{2, 3, 4\}$	$\{1,5\}$
c	$\{1, 2, 4, 5\}$	$\{3\}$
f	$\{1, 2, 3, 5\}$	$\{4\}$
m	$\{1, 2, 5\}$	$\{3,4\}$
p	$\{1, 4, 5\}$	$\{2,3\}$

the conditional database of X, which is similar to the working of the Apriori's join function. The Eclat algorithm [19] also partially utilizes the downward closure property in its bottom-up search of the lattice. In particular, it also employs a join function similar to Apriori's in determining itemsets to be searched.

2.3   Support Counting and Data Formats

When bottom-up searching the lattice of itemsets, algorithms will count the support of each itemset currently being examined, which depends on the data format for representing the database. There are two categories of data formats, horizontal and vertical.

The Apriori algorithm [2] determines the supports of itemsets by directly scanning the database in the horizontal format. The FP-growth algorithm [9] reads off the supports from the FP-tree that compresses the horizontal database.

The Eclat algorithm [19] turns the database into the tidset format as in Definition 4, and calculates the support of an itemset by intersecting the tidsets of the subsets of the itemset. Another vertical format related to tidset, is the diffset format [20].

Definition 4 (Tidset). *Let D be a transaction database over a set of items I. The tidset of an itemset X is denoted by $tidset(X) = \{t.tid \mid t \in D, X \subseteq t\}$. Thus, $\sigma(X) = |tidset(X)|$.*

Definition 5 (Diffset). *The diffset of X is the difference between the tidset of $prefix(X, k\text{-}1)$ and the tidset of X, that is, $diffset(X) = tidset(prefix(X, |X| - 1)) - tidset(X)$. Thus, $\sigma(X) = \sigma(prefix(X, |X| - 1)) - |diffset(X)|$.*

For example, Table 2 shows both tidsets and diffsets of frequent items for D in Table 1 and $minsup = 3$. Moreover, the tidset of itemset $\{a,c\}$ is the intersection of the tidsets of $\{a\}$ and $\{c\}$, that is, $\{1,2,5\} \cap \{1,2,4,5\} = \{1,2,5\}$. Similarly, $tidset(\{c,p\}) = tidset(\{c\}) \cap tidset(\{p\}) = \{1,4,5\}$, and $tidset(\{a,c,p\}) = \{1,2,5\} \cap \{1,4,5\} = \{1,5\}$. Given the lexicographic order, $diffset(\{a,c\}) = \{\}$ and $diffset(\{a,c,p\}) = \{2\}$, and $\sigma(\{a,c,p\}) = \sigma(\{a,c\})$ - $|diffset(\{a,c,p\})| = 2$.

3 Opportunistic Vertical Mining Approach

We propose an opportunistic vertical mining approach that improves the Eclat algorithm [19] by employing a hybrid vertical data format per itemset, by pushing additional pruning into breadth-first search, and by facilitating balanced parallelization of depth-first search.

3.1 Our Hybrid Vertical Format

Traditional vertical mining algorithms [6,15,16,19,20] utilize only one vertical format, either tidset or diffset. When databases are sparse, the tidset is the preferred choice. When databases are dense, the diffset may perform better. But when databases get huge, neither single vertical format scales well. A good approach is to independently choose the vertical format for each individual itemset by considering the resulting storage footprint.

For example, the tidset is good for {a,b} as $tidset(\{a,b\}) = \{2\}$ and $diffset(\{a,b\}) = \{1,5\}$ while the diffset is good for {a,c,p} as $tidset(\{a,c,p\}) = \{1,5\}$ and $diffset(\{a,c,p\}) = \{2\}$, given the lexicographic ordering of items.

To accommodate such a good approach, we present a hybrid vertical format, mixset, as in Definition 6.

Definition 6 (Mixset). *The mixset of an itemset X, denoted as $mixset(X)$ is defined as*

$$mixset(X) = \begin{cases} tidset(X) & if \ |tidset(X)| \leq |diffset(X)| \\ diffset(X) & otherwise \end{cases}$$

For example, $mixset(\{a,b\}) = tidset(\{a,b\}) = \{2\}$, and $mixset(\{a,c,p\}) = diffset(\{a,c,p\}) = \{2\}$.

Notice that mixset is an instrument for facilitating our approach, that is, to automatically select the tidset format or the diffset format on a per itemset basis, which is made possible both in breadth-first search and in depth-first search. Our approach is greatly different from approaches that use only the tidset format [19] or the diffset format for all itemsets, or manually determine at which level to switch from one format to another [20].

3.2 Enabling Our Opportunistic Vertical Mining

To enable our opportunistic vertical mining approach, we need to compute the mixset for any itemset that is a union of two k-itemsets X and Y that share a $(k$-1)-prefix, which we denote as $mixset(X \cup Y) = mixset(X) \cap mixset(Y)$.

Notice that there is no prior work that computes the tidset and diffset of the union of two itemsets when only given the tidset of one itemset and the diffset of the other. Therefore, we investigate in six cases by Theorem 1 how to compute $mixset(X) \cap mixset(Y)$.

Theorem 1. *Given an ordering Ω where itemset X is listed before itemset Y, and $|X| = |Y| = k$ and $prefix(X, k\text{-}1) = prefix(Y, k\text{-}1)$, we have:*

$$
\begin{aligned}
tidset(X \cup Y) &= tidset(X) \cap tidset(Y) && (1)\\
&= tidset(X) - diffset(Y) && (2)\\
&= tidset(Y) - diffset(X) && (3)\\
diffset(X \cup Y) &= tidset(X) - tidset(Y) && (4)\\
&= tidset(X) \cap diffset(Y) && (5)\\
&= diffset(Y) - diffset(X) && (6)
\end{aligned}
$$

For example, for D in Table 1 with the lexicographic ordering, we have $tidset(\{a\}) = \{1, 2, 5\}$, $tidset(\{c\}) = \{1,2,4,5\}$, and $tidset(\{p\}) = \{1,4,5\}$ in Table 2. According to Theorem 1, we get $tidset(\{a,c\}) = \{1,2,5\}$ and $tidset(\{a,p\}) = \{1,5\}$ by (1), and $diffset(\{a,c\}) = \{\}$ and $diffset(\{a,p\}) = \{2\}$ by (4). Furthermore, we can get $tidset(\{a,c,p\}) = \{1,5\}$ by (1), by (2), or by (3), and $diffset(\{a,c,p\}) = \{2\}$ by (4), by (5), or by (6).

3.3   Our Search Strategy

To avoid repetitively enumerating itemsets in bottom-up searching the lattice, all algorithms assume an ordering of itemsets. Thus, all algorithms can also be thought of as top-down searching an itemset enumeration tree, and fall into three categories, breadth-first, depth-first, or the hybrid.

First, we push additional pruning into the breadth-first search. Notice that Eclat [19] determines the support of any k-itemset by directly intersecting the tidsets of any two of its size $(k\text{-}1)$ subsets. In other words, Eclat executes the itemset-union (Apriori's join) and tidset-intersection (support-counting) without Apriori's pruning. We propose to push the Apriori's pruning to avoid unnecessary tidset-intersection operations since all frequent $(k\text{-}1)$-itemsets are available when enumerating k-itemsets in a breadth-first search.

Second, we facilitate parallelization of depth-first search by dynamically arranging sibling itemsets with the same parent itemset from left to right in the ascending order of supports. With such an ordering, more frequent itemsets will root smaller subtrees to be further searched, which results in a smaller and more balanced itemset enumeration tree. Clearly, such an ordering helps balance the workload when depth-first searching the subtrees in parallel, and ultimately improves the efficiency.

4   New Parallel Algorithm Based on MapReduce

This section presents our new parallel algorithm for mining frequent itemsets based on MapReduce [5]. We call our algorithm Parallel Eclat or Peclat for short to note the connection between our algorithm and Eclat [19].

Our algorithm significantly differs from Eclat as discussed in Sect. 3, and parallelizing our approach based on the MapReduce paradigm as presented in this section is nontrivial.

4.1 Our Peclat Algorithm

The pseudo code of our Peclat algorithm is listed in Algorithm 1, which works as follows. First, Peclat invokes a MapReduce job, mrCountingItems, to find all frequent items and to output a list of frequent items together with the mixset for each frequent item (at line 1).

> **Algorithm 1.** Peclat
> **Input:** D, $minsup$
> **Output:** all frequent itemsets
> 1: $L_1 \leftarrow$ mrCountingItems($D, minsup$);
> 2: **for** ($k = 2$; $k \leq m \wedge L_{k-1} \neq \{\}$; k++) **do**
> 3: $L_k \leftarrow$ mrLargeK($L_{k-1}, minsup$);
> 4: **end for**
> 5: $L \leftarrow$ mrMiningSubtrees($L_m, minsup$);

Second, Peclat calls a MapReduce job, mrLargeK, in a maximum m rounds to search the itemset enumeration tree in a breadth-first manner (at lines 2 – 4). The breadth-first search will also terminate if all frequent itemsets are found.

Finally, Peclat switches to a depth-first search by invoking a MapReduce job, mrMiningSubtrees that mines subtrees of the itemset enumeration in parallel (at line 5). The output L_k of the MapReduce jobs is a list of ⟨key = itemset, value = mixset⟩ pairs.

4.2 The mrCountingItems Job

This MapReduce Job first produces the tidset for each item in its map phase, and then outputs each frequent item with its mixset in its reduce phase.

Concretely, in the map phase, the map function is invoked as many times as the number of transactions. Each time the map function reads one transaction and outputs an ⟨item, tid⟩ pair for every item in the transaction.

For example, given D in Table 1 and $minsup = 3$, one invocation of the map function will read the first transaction ($tid =1$) and will output the following ⟨item, tid⟩ pairs: ⟨a, 1⟩, ⟨c, 1⟩, ⟨d, 1⟩, ⟨f, 1⟩, ⟨g,1⟩, ⟨i, 1⟩, ⟨m, 1⟩, and ⟨p, 1⟩, which will be passed to the reduce phase.

In the reduce phase, the reduce function is invoked as many times as the number of distinct items. Each time the reduce function gets a distinct item and a list of $tids$ of the transactions that contain the item. If the item is frequent, it outputs the item with its mixset. For example, the pairs ⟨a, 1⟩, ⟨a, 2⟩, and ⟨a, 5⟩ produced in the map phase will be fed into one invocation of the reduce function, which will first get $tidset(\{a\}) = \{1, 2, 5\}$ and output $\{a\}$ with $mixset(\{a\}) = diffset(\{a\}) = \{3, 4\}$ as diffset is smaller than tidset.

MapReduce Job: mrCountingItems
Input: D, *minsup*
Output: L_1
Map (key = null, value = transaction T_i)
1: **for each** *item* $\in T_i$ **do**
2:     output $\langle item, T_i.tid \rangle$;
3: **end for**
Reduce (key = *item*, value = *list-of-tids*)
1: $tidset(item) \leftarrow \{\}$;
2: **for each** value $t \in$ *list-of-tids* **do**
3:     $tidset(item) \leftarrow tidset(item) \cup \{t\}$;
4: **end for**
5: **if** $\sigma(item) \geq minsup$ **then**
6:     output $\langle item, mixset(item) \rangle$;
7: **end if**

The reduce phase for the previous example eventually outputs the following $\langle$1-itemset, mixset$\rangle$ pairs: $\langle\{a\}, \{3,4\}_{diffset}\rangle$, $\langle\{b\}, \{1,5\}_{diffset}\rangle$, $\langle\{c\}, \{3\}_{diffset}\rangle$, $\langle\{f\}, \{4\}_{diffset}\rangle$, $\langle\{m\}, \{3,4\}_{diffset}\rangle$, and $\langle\{p\}, \{2,3\}_{diffset}\rangle$.

In examples, additional information for each $\langle$itemset, mixset$\rangle$ pair expressed as labels, e.g., $\langle\{a\}, \{3,4\}_{diffset}\rangle$ indicates that $\{a\}$ is with its diffset $\{3,4\}$.

4.3   The mrLargeK Job

This MapReduce job mines all frequent k-itemsets for a given k. First of all, the map phase is initialized by reading the frequent $(k\text{-}1)$-itemsets L_{k-1} and calling the join and pruning functions in Sect. 2.2 to generate the candidate k-itemsets C_k. For example, given D in Table 1 and $minsup = 3$, when calling the job with $k = 3$, we get $C'_3 = \{\{a,c,f\}, \{\{a,c,m\}, \{a,f,m\}, \{c,f,m\}, \{c,f,p\}, \{c,m,p\}\}$ after join, and $C_3 = \{\{a,c,f\}, \{\{a,c,m\}, \{a,f,m\}, \{c,f,m\}\}$ after pruning.

After initialization, each invocation of the map function will reads a $\langle X, mixset(X) \rangle$ pair and match the itemset X with each candidate c in C_k, and will produce a $\langle c, mixset(X) \rangle$ pair if X is a subset of c and shares a common $(k\text{-}2)$-prefix with c, which will be passed to the reduce phase for computing $mixset(c)$.

For example, when calling the job with $k = 3$, one invocation of the reduce function inputs $\langle\{a,c\}, \{\}_{diffset}\rangle$, and outputs $\langle\{a,c,f\}, \langle\{a,c\}, \{\}_{diffset}\rangle\rangle$ and $\langle\{a,c,m\}, \langle\{a,c\}, \{\}_{diffset}\rangle\rangle$. Another invocation inputs $\langle\{a,f\}, \{\}_{diffset}\rangle$, and outputs $\langle\{a,c,f\}, \langle\{a,f\}, \{\}_{diffset}\rangle\rangle$ and $\langle\{a,f,m\}, \langle\{a,f\}, \{\}_{diffset}\rangle\rangle$.

In the reduce phase, each invocation of the reduce function will get a candidate itemset c and the mixsets of two itemsets X_1 and X_2 that share a common prefix with c, and will compute $mixset(c)$ as in Sect. 3.2. If the candidate is frequent, it outputs a $\langle c, mixset(c) \rangle$ pair.

MapReduce Job: mrLargeK
Input: L_{k-1}, $minsup$
Output: L_k
Initialization of Map Phase:
1: read L_{k-1};
2: $C'_k \leftarrow \text{join}(L_{k-1})$;
3: $C_k \leftarrow \text{pruning}(C'_k, L_{k-1})$;
Map (key = itemset X, value = $mixset(X)$)
1: load C_k;
2: **for each** candidate $c \in C_k$ **do**
3:    **if** $X \subset c \wedge prefix(c, k\text{-}2) = prefix(X, k\text{-}2)$
4:    **then** output $\langle c, mixset(X) \rangle$;
5: **end for**
Reduce (key = candidate c, value = $list\text{-}of\text{-}mixsets$)
1: $mixset(X_1) \leftarrow$ the first in $list\text{-}of\text{-}mixsets$;
2: $mixset(X_2) \leftarrow$ the second in $list\text{-}of\text{-}mixsets$;
3: $mixset(c) \leftarrow mixset(X_1) \cap mixset(X_2)$;
4: **if** $\sigma(c) \geq minsup$ **then** output $\langle c, mixset(c) \rangle$;

For example, when calling the job with $k = 3$, one invocation of the reduce function takes $\{a,c,f\}$ and the list of $\langle\{a,c\}, \{\}_{diffset}\rangle$ and $\langle\{a,f\}, \{\}_{diffset}\rangle$ as input, and outputs $\langle\{a,c,f\}, \{\}_{diffset}\rangle$. The reduce phase eventually outputs $\langle\{a,c,f\}, \{\}_{diffset}\rangle$, $\langle\{a,c,m\}, \{\}_{diffset}\rangle$, $\langle\{a,f,m\}, \{\}_{diffset}\rangle$, and $\langle\{c,f,m\}, \{\}_{diffset}\rangle$ for the job with $k = 3$.

MapReduce Job: mrMiningSubtrees
Input: L_k, $minsup$
Output: k'-frequent itemsets for $k' \geq k$
Map (key = k-itemset X, value = $mixset(X)$)
 1: output $\langle prefix(X, k\text{-}1), \langle X, mixset(X)\rangle\rangle$;
Reduce (key = $(k\text{-}1)$-$prefix$, value = $siblings$)
 1: mine($siblings$, $minsup$);
Subroutine: mine($siblings$, $minsup$)
 1: **for each** $\langle X, mixset(X)\rangle \in siblings$ **do**
 2:    $children.clean()$;
 3:    **for each** $\langle Y, mixset(Y)\rangle \in siblings$ after X **do**
 4:       $Z \leftarrow X \cup Y$;
 5:       $mixset(Z) \leftarrow mixset(X) \cap mixset(Y)$;
 6:       **if** $\sigma(Z) \geq minsup$ **then**
 7:          $children.push(\langle Z, mixset(Z)\rangle)$;
 8:          output Z;
 9:       **end if**
10:    **end for**
11:    **if** $children \neq \emptyset$ **then** mine($children$, $minsup$);
12: **end for**

Table 3. Features of datasets used in experiments

Dataset	numTrans	Items	maxLenTrans	avLenTrans	Category
BMS-WebView-1	59,602	497	267	2.5	sparse
BMS-WebView-2	77,512	3,340	161	5.6	sparse
Connect4	67,557	150	43	43.0	dense
Pumsb*	49,045	2,088	63	50.5	dense
T40I10D100k	99,999	150	68	37.0	mixed
T30I10D100k	99,998	120	53	27.8	mixed

4.4   The mrMiningSubtrees Job

This MapReduce job splits the itemset enumeration tree into small subtrees, and then mines the subtrees in parallel.

In the map phase of this job, each invocation of the map function inputs an ⟨itemset, mixset⟩ pair, and outputs the prefix of the itemset as the key and the ⟨itemset, mixset⟩ pair as the value.

In the reduce phase, each invocation of the reduce function can get a list, *siblings*, of all the itemsets with a same prefix together and their mixsets as the input value, and the prefix as the input key. Therefore, the branches of the itemset enumeration tree rooted at *siblings* can be recursively searched by the mine subroutine.

For example, given all the frequent 2-itemsets, {a,c}, {a,f}, {a,m}, {c,f}, {c,m}, {c,p}, and {f,m} together with their mixsets as the input, the map phase will distribute these pairs into three groups with {a}, {c}, and {f} as their prefixes respectively if the lexicographic order is used, or into four groups with {a}, {m}, {p}, and {c} respectively if the ascending order of supports is used. The subtree holding the itemsets in a group, e.g., {a,c}, {a,f}, and {a,m}, will be mined by one invocation of the reduce function.

5   Experimental Evaluation

We evaluate our Peclat by experimentally comparing with DPC [13] and mahout (closed) [22], an implementation of PFP [10], that mines frequent closed itemsets. All algorithms were implemented in Java.

Six datasets[1,2,3] are used in experiments and described in Table 3. All experiments were performed on a Hadoop-0.20.2 cluster of three nodes, 1 master and 2 workers, where each node is equipped with an Intel(R) Core(TM) i5-2400 3.10GHz CPU, 2GB RAM, and a 500GB hard disk running Ubuntu11.10.

---

[1] http://www.sigkdd.org/kdd-cup-2000/.
[2] http://fimi.ua.ac.be/data/.
[3] http://sourceforge.net/projects/ibmquestdatagen/.

5.1 Performance Comparison with Other Algorithms

We compare the running time of DPC [13], mahout(closed) [22], and our Peclat on six datasets. Peclat runs in a breadth-first search for two levels and then switch to a depth-first search. The result is summarized by Fig. 2.

For sparse datasets as in Fig. 2(a)(b), the algorithms from the least efficient to the most efficient are DPC, mahout(closed), and Peclat. For example, when $minsup = 0.1\%$ for BMS_WebView_1, DPC takes 282 s, mahout(closed) 153, and Peclat 97. When $minsup = 0.3\%$ for BMS_WebView_2, DPC takes 313 s, mahout(closed) 141, and Peclat 102.

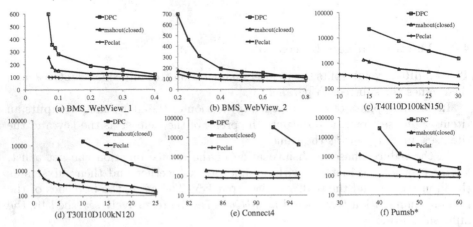

Fig. 2. Running-time (seconds) of three algorithms vs. $minsup(\%)$

For datasets of mixed characteristics as in Fig. 2(c)(d), DPC is the least efficient, and Peclat is the most efficient. For example, when $minsup = 15\%$ for T40I10D100k, DPC takes 22,432 s, mahout(closed) 1,150, and Peclat 254. When $minsup = 10\%$ for T30I10D100k, DPC takes 14,690 s, mahout(closed) 401, and Peclat 227.

For dense datasets as in Fig. 2(e)(f), Peclat beats mahout(closed), and mahout(closed) beats DPC. For example, when $minsup = 92\%$ for Connect4, DPC takes 32,41 s, mahout(closed) 136, and Peclat 78. When $minsup = 40\%$ for Pumsb*, DPC takes 27,548 s, mahout(closed) 450, and Peclat 98.

The observations are as follows.

- Our Peclat is the most efficient algorithm, and DPC [13] is the least efficient for all the six datasets.
- Although the evaluation favours mahout(closed) as it only finds frequent closed itemsets, i.e., it works on a much easier mining problem, our Peclat still outperforms mahout(closed) significantly.

5.2 Anatomy of Opportunistic Vertical Mining Approach

To explain the observed results in the last subsection, this subsection analyzes the three techniques in our opportunistic vertical mining approach.

Selecting Vertical Format per Itemset. We first study the numbers of tidsets and diffsets chosen by our Peclat algorithm in the mining process, denoted as tidset_num and diffset_num respectively.

The aggregate tidset_num and diffset_num with varying *minsup* are shown in Fig. 3(a)(b), and tidset_num and diffset_num at each level of the itemset enumeration tree (i.e. grouped by the size of itemsets) are shown in Fig. 3(c)–(f). For T40I10D100k, our Peclat algorithm selects the tidset format for 25 % of mixsets and the diffset format for 75 %. For T30I10D100k, 45 % of mixsets are in the tidset format, and 55 % are in the diffset format.

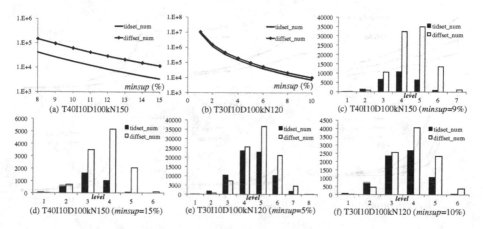

Fig. 3. Numbers of tidsets and diffsets (aggregate and per level)

Our Peclat, depicted as Peclat_mixset in Fig. 4(a)(b) as it employs the mixset format, is compared with variants that employ the tidset and diffset formats. By piecing together the results in Figs. 3 and 4(a)(b), we observe the following.

– Our mixset format is superior to both tidset and diffset since no dataset is purely sparse or purely dense. Selecting a data format on a per itemset basis greatly reduces the computation cost and the space usage.
– Our Peclat algorithm is good for all datasets regardless of data characteristics while other algorithms [20] may need manual fine-tuning by users and cannot take full advantage of a hybrid data format.

Ordering Itemsets in Ascending Supports. We study the improvement by sorting itemsets in the ascending order of supports by comparing our Peclat with its variant, Peclat_Lexi, that sorts itemsets in the lexicographic order.

Such an improvement is significant as shown in Fig. 4(c)(d). For example, when *minsup* = 8 % for T40I10D100k, Peclat takes 443 s while Peclat_Lexi takes 640 s. When *minsup* = 1 % for T30I10D100k, Peclat takes 983 s while Peclat_Lexi takes 4,403 s.

Pushing Additional Pruning. Finally, we analyze the effect of applying the Apriori's pruning in improving the breadth-first search. To do so, we compare two variants of our algorithm, Peclat_BF that integrates the Apriori's pruning in the breadth-first search and Peclat_BF_noPrune that does not.

The result is that the improvement by integrating the Apriori's pruning is significant as shown in Fig. 4(e)(f). For example, when $minsup = 15\%$ for T40I10D100k, Peclat_BF takes 1,168 s while Peclat_BF_noPrune takes 2,551 s. When $minsup = 10\%$ for T30I10D100k, Peclat_BF takes 1,320 s while Peclat_BF_noPrune takes 2,306 s.

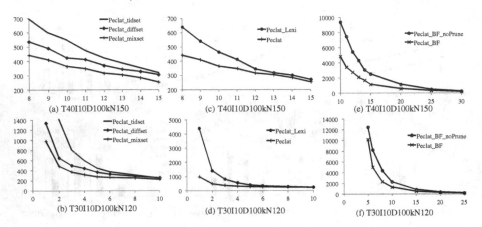

Fig. 4. Running-time of (a)(b) using tidsets, diffsets, mixsets; (c)(d) with/without ascending order of supports; (e)(f) with/without more pruning on 2 datasets

6   Conclusion

This paper has proposed Peclat, a parallel Eclat-like algorithm based on MapReduce. Peclat outperforms the existing algorithms because it improves the Eclat algorithm in three aspects including a hybrid data format that saves both time and space, an ordering of itemsets that helps balancing workloads, and additional pruning for breadth-first search.

Acknowledgements. This work was supported in part by the National Natural Science Foundation of China (61272306), and the Zhejiang Provincial Natural Science Foundation of China (LY12F02024).

References

1. Agrawal, R., Shafer, J.: Parallel mining of association rules. IEEE Trans. Knowl. Data Eng. **8**, 962–969 (1996)
2. Agrawal, R., Srikant, R.: Fast algorithms for mining association rules in large databases. In: 20th VLDB, p. 487 (1994)

3. Chen, X., He, Y., Chen, P., Miao, S., Song, W., Yue, M.: HPFP-Miner: a novel parallel frequent itemset mining algorithm. ICNC **3**, 139–143 (2009)
4. Cyrans, J.-D., Ratt, S., Champagne, R.: Adaptation of apriori to MapReduce to build a warehouse of relations between named entities across the web. In: 2010 DBKDA, pp. 185–189 (2010)
5. Dean, J., Ghemawat, S.: MapReduce: simplified data processing on large clusters. Commun. ACM **51**(1), 107–113 (2008)
6. Dunkel, B., Soparkar, N.: Data organization and access for efficient data mining. In: 15th ICDE, pp. 522–529 (1999)
7. Farzanyar, Z., Cercone, N.: Efficient mining of frequent itemsets in social network data based on MapReduce framework. In: ACM International Conference on Advances in Social Networks Analysis and Mining, pp. 1183–1188 (2013)
8. Hammoud, S.: MapReduce network enabled algorithms for classification based on association rules. Ph.D. Thesis, Brunel University (2011)
9. Han, J., Pei, J., Yin, Y.: Mining frequent patterns without candidate generation. In: 2000 SIGMOD, pp. 1–12 (2000)
10. Li, H., Wang, Y., Zhang, D., Zhang, M., Chang, E.Y.: PFP: parallel FP-growth for query recommendation. In: 2008 ACM Conference on Recommender System (RecSys 2008), pp. 107–114 (2008)
11. Li, L., Zhang, M.: The strategy of mining association rules based on cloud computing. In: BCGIN, pp. 475–478 (2011)
12. Li, N., Zeng, L., He, Q., Shi, Z.: Parallel implementation of apriori algorithm based on MapReduce. In: ACIS International Conference on Software Engineering, Artificial Intelligence, Networking & Parallel/Distributed Computing, pp. 236–241 (2012)
13. Lin, M.-Y., Lee, P.-Y., Hsueh, S.-C.: Apriori-based frequent itemset mining algorithms on MapReduce. In: ICUIMC (2012)
14. Riondato, M., DeBrabant, J.A., Fonseca, R., Upfal, E.: PARMA: a randomized parallel algorithm for approximate association rule mining in MapReduce. In: 21st CIKM, pp. 85–94 (2012)
15. Sarawagi, S., Thomas, S., Agrawal, R.: Integrating association rule mining with databases: alternatives and implications. In: 1998 SIGMOD, pp. 343–354 (1998)
16. Shenoy, P., Haritsa, J.R., Sudarshan, S.: Turbo-charging vertical mining of large databases. In: 2000 SIGMOD, pp. 22–33 (2000)
17. Sohrabi, M.K., Barforoush, A.A.: Parallel frequent itemset mining using systolic arrays. Knowl. Based Syst. **37**, 462–471 (2013)
18. Yang, X.Y., Liu, Z., Fu, Y.: MapReduce as a programming model for association rules algorithm on Hadoop. In: ICIS (2010)
19. Zaki, M.J.: Scalable algorithms for association mining. IEEE Trans. Knowl Data Eng. **12**(3), 372–390 (2000)
20. Zaki, M.J., Gouda, K.: Fast vertical mining using diffsets. In: 9th SIGKDD, pp. 326–335 (2003)
21. Zaki, M.J., Ogihara, M., Parthasarathy, S., Li, W.: Parallel algorithms for discovery of association rules. Data Min. Knowl. Disc. **1**(4), 343–373 (1997)
22. Apache Mahout. http://mahout.apache.org/

Sequential Data Analytics by Means of Seq-SQL Language

Bartosz Bebel[1], Tomasz Cichowicz[1], Tadeusz Morzy[1], Filip Rytwiński[1], Robert Wrembel[1(✉)], and Christian Koncilia[2]

[1] Institute of Computing Science, Poznan University of Technology, Poznań, Poland
{bartosz.bebel,tadeusz.morzy,robert.wrembel}@cs.put.poznan.pl
[2] Institute of Informatics Systems, Klagenfurt University, Klagenfurt, Austria
koncilia@isys.uni-klu.ac.at

Abstract. Ubiquitous devices and applications generate data, whose natural feature is order. Most of the commercial software and research prototypes for data analytics allow to analyze set oriented data, neglecting their order. However, by analyzing both data and their order dependencies, one can discover new business knowledge. Few solutions in this field have been proposed so far, and all of them lack a comprehensive approach to organize and process such data in a data warehouse-like manner. In this paper, we contribute an SQL-like query language for analyzing sequential data in an OLAP-like manner, its prototype implementation and performance evaluation.

1 Introduction

Large volumes of data arrive to analytical systems in order. Some typical examples of applications generating ordered data include: workflow management systems, website monitors for click stream analysis, network packets monitoring, RFID infrastructures, public transportation infrastructures, intelligent buildings, and various smart meters. Some of the data have the character of events that last an instant, whereas others last for a given time period - an interval. In this respect, sequential data can be categorized either as *time point-based* or *interval-based* [20], but for all of them the order in which they were generated is important. This order implies the existence of various patterns, that in turn, provide additional piece of information of business value.

The ability to analyze sequential data and recognize patterns became important for current data analytics. Three classes of such approaches were proposed (cf. Sect. 3), namely: (1) discovering patterns in sequential data, e.g., [10,15–17,19,32,36], (2) complex event processing, e.g., [9,12], and (3) OLAP-like analytics. In the last class, most of the approaches are based on an extended relational data model, e.g., [5,8,18,24,25,30,33] and they focus only on pattern queries, neglecting the analysis of interval data in the context of multiple dimensions. With this respect, there is a need for developing a dedicated data model, suitable for sequential data and a query language capable of analyzing such data in an OLAP-like manner.

© Springer International Publishing Switzerland 2015
Q. Chen et al. (Eds.): DEXA 2015, Part I, LNCS 9261, pp. 416–431, 2015.
DOI: 10.1007/978-3-319-22849-5_28

The goal of this paper is to verify the feasibility of the approach to analyzing time point-based sequential data in an OLAP-like manner, by means of a SQL-like query language. To this end, we contribute the following. First, based on the sequential data model that we developed in [1,2] (cf. Sect. 4), we propose a query language *Seq-SQL* that allows: (1) to analyze sequential data in an OLAP-like manner and (2) to query a database of sequences for various patterns (cf. Sect. 5). Second, we present the implementation of a prototype system based on our data model, show the implementation of *Seq-SQL*, and present the results of its preliminary performance evaluation (cf. Sect. 6).

2   Leading Example

Let us consider data about car failures. Initially, the data are stored in table *Failures*, composed of 8 attributes, as shown in Table 1. A car is characterized by its identifier, model, make, production year, and mileage, whereas a failure is characterized by its date, description, and repair cost. Every record in this table represents an *event* of a car failure.

Table 1. An example table storing events on car failures

Failure date	Car id	Car model	Car make	Production year	Car mileage	Failure description	Repair cost
2012.04.04	BB111	308	Peugeot	2003	145 500	F1	1 500
2012.06.11	BB111	308	Peugeot	2003	160 000	F2	800
2012.06.12	AA222	207	Peugeot	2004	184 000	F3	2 100
2012.06.13	CC333	Golf	VW	2007	80 000	F2	790
2012.07.27	BB111	308	Peugeot	2003	179 000	F3	2 200
2012.12.02	AA222	207	Peugeot	2004	201 123	F4	650
2012.12.08	CC333	Golf	VW	2007	120 000	F4	660
2012.12.13	DD444	Polo	VW	2005	110 000	F1	1 400
2013.01.30	EE555	308	Peugeot	2000	190 000	F3	1 900
2013.02.16	DD444	Polo	VW	2005	121 000	F2	850

Retrieving frequencies of failures and finding a failure that occurred the most frequently is straightforward by means of standard SQL. However, to find cases where one failure frequently precedes other given failures, one has to analyze sequences of failures rather than single, independent failures.

Example Search. Let us assume that for every car someone searches for patterns of failures where the same failure occurred within mileage ≤ 10000 km and at least one other failure occurred in between. For such patterns, the total cost of repair per car, has to be computed but only repairs of the total cost ≥ 5000 are of interest.

2.1   Standard Approach

Let us investigate programming effort to implement the example search using a standard relational DBMS. The first issue is to construct sequences of failures from the set of elementary events. It can be done by means of: (1) the set of SQL DDL commands used for constructing additional tables and populating them with data or (2) materialized views. The second issue is to define dimensions and measures, by means of another set of DDL and DML commands. Finally, to analyze sequences, a complex SQL query is used, composed of multiple joins, advanced filtering, and grouping. Unfortunately, some operations cannot be expressed using only SQL commands. An example is filtering sequences according to a given pattern. The filtering algorithm is very complex (with recursion) and has to be implemented using a procedural language like PL/SQL, Java, C, etc.

The example search implemented in Oracle11g required the creation of three additional materialized views - for storing sequences, the values of a sequence dimension, and the values of a sequence measure (altogether approx. 60 lines of SQL code). The code for analyzing and filtering the sequences according to the pattern was written in Java. The processing required multiple recursions and the program included over 300 lines of code (due to space limitation we do not present its implementation here). For this reason, we argue that sequence analysis is a complex task and needs support from analytical tools and dedicated query languages.

2.2   Our Approach

In our *Seq-SQL* approach the example search is implemented as follows. First, sequences are constructed by partitioning the set of failure events by attribute *car_id*, cf. Command 1.1. For every car, the command creates sequences of failures by ordering the events by mileage they occurred at. Clause *forming attributes* allows to partition the set of events and *ordering attribute* allows to order events within each partition by the values of an indicated attribute.

Command 1.1. Sequence definition

```
create seq failures_seq on failures
forming attributes car_id ordering attributes car_mileage;
```

Second, the constructed sequences with the attributes of interest are retrieved, cf. Command 1.2. Notice that the *from* clause points to the set of already created sequences.

Command 1.2. Sequences retrieval

```
select seq failure_date, car_id, car_model, car_make, failure_description
from failures_seq;
```

Third, one has to define a dimension for the sequences, cf. Command 1.3 and measure, cf. Command 1.4, like in a standard OLAP. Dimension *car_dim* has three levels that form a hierarchy, namely $car \longrightarrow model \longrightarrow make$, each of type varchar2.

Command 1.3. Sequence dimension

```
create seq dimension car_dim on failures_seq
level car as [select distinct(f.car_id) from failures f,
    failures_seq s where s.event=f.rowid and s.seq_id=sequence.seq_id]
       type [varchar2(20)]
is child of level model as [select distinct car_model
   from failures where car_id = event.value] type [varchar2(100)]
is child of level make as [select distinct car_make
   from failures where car_model = event.value] type [varchar2(100)];
```

Measure *total_cost*, for the whole sequence is defined as a total repair cost for this sequence.

Command 1.4. Sequence measure

```
create seq measure total_cost
on failures_seq as
  [select sum(sequence.repair_cost) from dual] type [number(10)];
```

At last, one can analyze the sequences, as expressed by example Command 1.5. First, sequences for which a total cost of repairs ≥5000 are selected (recall that a sequence represents ordered failures of a given car). Second, the sequences are further filtered according to a pattern that evaluates to true if: (1) a sequence consists of two identical failures that both occurred within a mileage ≤10000 km, and (2) at least one other failure took place in between. Third, sequences are grouped according to values of sequence dimension *car_dim* at level *model*. Finally, for each car model the number of sequences that fulfill the aforementioned predicates are counted.

Command 1.5. An example complex analysis

```
select seq seq.dimension.car_dim.model,
       [count(distinct seq.seq_id) as num_cases]
from failures_seq
where seq [seq.measure.total_cost >= 5000]
where pattern[XYX] with mapping [X to failure_description, Y to failure_description]
    with options[X.2.car_mileage − X.1.car_mileage <= 10000]
group by seq.dimension.car_dim.model;
```

3 Related Work

One of the first approaches to analyzing database-stored sequences of data was presented in [28–30]. The authors applied an object-relational data model to define three different types of sequences, namely a base, constant, and derived sequence, with unary and aggregate operators. The unary operators included a selection, projection, and offset within a sequence. The aggregate operators included aggregation (for aggregating in an n-element window) and compose (for combining/joining two sequences). Another SQL-like language, called the Sorted Relational Query Language (SRQL) [24], was based on an extended relational algebra with four operators, namely: Sequence (for creating a sequence out of relational data), Shift (for matching two tuples at a given position in a sequence), ShiftAll (for matching a tuple at a given position in a sequence with tuples within a sequence window), and WindowAggregate (for aggregating sequences within a given window). The extension of SRQL is yet another language - Simple Query Language for Time Series (SQL-TS) described in [25–27]. SQL-TS introduced three operators, namely: Cluster By (for clustering sequences), Sequence By (for sorting sequences), and As (for defining a pattern on sequences).

In [14], the *S-OLAP* concept was proposed. It organizes sequences as Sequence Cuboids that represent answers to the so-called Pattern-Based Aggregate queries. The proposed data model defines a set of operators for the purpose of analyzing patterns. The model was further extended in [5,6] with an algorithm for supporting pattern-based aggregate queries. Again, *S-OLAP* extends the relational data model and storage for sequential data.

[11] proposed a data model based on one dimensional arrays (called arrables) to represent ordered data. With respect to storage, arrables are similar to the columnar storage. The proposed query language - AQuery offers selection, projection, join, grouping, and the 'each' modifier. The latter allows to apply a given function to an array cell. In [35], the authors proposed another array-based model and query language - SciQL that allows to manipulate arrays of sequences and perform SQL-like selects on the arrays. The support for pattern queries and OLAP queries has not been a focus of these two approaches.

[7,8] focused on warehousing RFID data. The authors proposed a few techniques for reducing the size of sequential data, for constructing RFID cuboids, and for computing higher level cuboids from lower level ones. They applied a relational data model to represent RFID data and their order. The proposed approach lacks a formal data model and a query language for analyzing sequences. In [4] the authors focused on constructing the so-called RFID warehouse. Unfortunately, no in-dept discussion on RFID data storage and analysis was provided.

Other research efforts focused on mining sequential patterns either on data stored in a data warehouse, e.g., [10,15,17] or in data streams, e.g., [16,19,36]. The developed algorithms are able to discover patterns in sequences but they do not support typical OLAP-like analysis of sequential data.

In the area of process mining, [32] proposed to store process instances in cells of a *process cube*. The process cube is constructed based on: (1) a process cube structure that defines the "schema" of the cube, (2) a process cube view that is analogous to a relational view, and (3) a materialized process cube view that is the process cube view filled in with data coming from a repository, called event base. The cells of the process cube store process instances that are the subjects of process discovery. The process cube is organized by dimensions that define the context of process mining. The author redefined typical OLAP operators (slice, dice, roll-up, and drill-down) to work on the process cube.

In the area of *Complex Event Processing* [3,34], *Stream Cube* [9] was developed to provide tools for OLAP analysis of stream data within a given time window. [12,13] presented a more advanced concept, called *E-Cube*. It allows to execute OLAP queries on data streams, also within a given time window. *E-Cube* includes: a query language (called SEQ) allowing to query events of a given pattern, a concept hierarchy allowing to compute coarser aggregates based on finer ones, a hierarchical storage with data sharing, and a query optimizer. SEQ enables grouping events by means of attributes and computing aggregate functions like COUNT. Nonetheless, this approach has been developed for the analysis of current data and is unable to perform OLAP-like analysis.

From commercial systems only *Oracle* [33] and *Teradata Aster* [31] support SQL-like pattern analysis in their OLAP engines. *Teradata Aster* supports query language, called SQL-MapReduce (SQL-MR), where the nPath clause is applied to analyze sequences. *Oracle* together with *IBM* defined an ANSI SQL standard for finding patterns within sequences stored in tables [18]. To this end, the MATCH_RECOGNIZE clause was proposed. It allows to search for patterns, define patterns and pattern variables. Nonetheless, these technologies do not offer OLAP functionality on sequential data.

To conclude, (1) [24–30] use an extended relational data model and [11,35] are based on an array model, whereas *Seq-SQL* is based on a dedicated data model to represent and process sequences, (2) [24,28,29] focus on a standard sequence analysis, [5,6,14,25–27] and commercial systems [18,31,33] focus on pattern searching, whereas *Seq-SQL* was developed for the purpose of OLAP-like analysis, where sequences are treated as facts and analyzed in the context of dimensions, which constitute an equivalent of a data cube. With this respect, *Seq-SQL* is similar to a process cube [32], but the model that we proposed is richer as it includes 12 operations on sequences, as compared to 4 proposed in [32]. [4,7,8] focus on storage for RFID data, whereas we proposed a general data model and showed one of its possible implementations in a relational database. The limitations of *Seq-SQL* are as follows: (1) it does not support mining of either patterns or processes. It wasn't developed to analyze sequences in a data stream either.

4   Seq-SQL Data Model

In this section we outline our data model for sequential data (its detailed description can be found in [2]). The model is based on the three fundamental components, namely: an *event*, a *sequence*, and a *dimension*. A sequence is created from events by clustering and ordering them. Sequences and events have distinguished attributes – *measures* that can be analyzed in an OLAP-like manner in contexts set up by *dimensions*. Processing of sequential data is performed by means of *operations*.

Event and its Attributes. An *event* represents an elementary data item, whose duration is a chronon. Formally, event $e_i \in \mathbb{E}$, where $\mathbb{E}$ is the set of events. e_i is n-tuple of attributes' values: $(a_{i1}, a_{i2}, ..., a_{in})$, where a_{ij} is the value of attribute

A_j in event e_i. $A_j \in \mathbb{A}$, where $\mathbb{A}$ is the set of event attributes and $a_{ij} \in dom(A_j)$. $dom(A_j) \subseteq \mathbb{V}$, where $\mathbb{V}$ is the set of atomic values $\cup$ *null*.

Attribute Hierarchy. An event attribute may have associated an explicitly defined hierarchy. Let $\mathbb{L} = \{L_1, L_2, ..., L_k\}$ be the set of levels in the hierarchies of the event attributes. The hierarchy of $A_i \in \mathbb{A}$, is described by $(\mathbb{L}_{A_i}, \triangleright_{A_i})$, where $\mathbb{L}_{A_i} \subseteq \mathbb{L}$ and $\triangleright_{A_i}$ is a partial order on set $\mathbb{L}_{A_i}$. The sets of level values belong to $\mathbb{V}$.

Sequence and its Measures. An ordered set of events constitutes a *sequence*. The order of events is defined by values of selected event attributes, called *ordering attributes*. A sequence is composed of the events that have identical value of distinguished attribute (or attributes), called *forming attributes*. If a forming attribute has a hierarchy associated, then a selected level in the hierarchy can also be a forming attribute. Formally, sequence $s_i \in \mathbb{S}$, where $\mathbb{S}$ is the set of sequences, is pair $(\mathbb{E}_i, \triangleright)$, where $\mathbb{E}_i \subseteq \mathbb{E}$ and $\triangleright$ is a partial order on $\mathbb{E}$.

Creating a Sequence. Sequences are created by means of operator $CreateSequence(\mathbb{E}, \mathbb{F}, \mathbb{A}_o, p) = \mathbb{S}$ where: (1) $\mathbb{F}$ is the set of pairs (A_i, L_j), where A_i is a forming attribute and L_j is the level in the hierarchy of the forming attribute A_i, or (A_i, ϕ) if attribute A_i does not have a hierarchy, (2) $\mathbb{A}_o$ is the set of ordering attributes, (3) $p \in \mathbb{P}$ is a logical predicate which selects events into sequences.

Notice that sequences are not defined statically, but their structure is dynamically constructed based on the features of analyses, for which the sequences are created.

Fact and Measure. Measure m_i belongs to the set of measures $\mathbb{M}$ and $dom(m_i) \subseteq \mathbb{V}$. A measure can be either an attribute of an event or the property of the whole sequence. In order to treat measures uniformly, a measure is defined as function $ComputeMeasure(s, name, p)$ that associates an atomic value with a sequence, where: (1) $s \in \mathbb{S}$ is a sequence, for which the values of the measure are computed, (2) *name* is the name of the measure, (3) $p \in \mathbb{P}$ is an expression that computes the values of the measure for a given sequence.

Dimension. Dimension D_i belongs to the set of dimensions $\mathbb{D}$. A dimension can be either an attribute of an event or the property of the whole sequence. The *CreateContext* operator associates a dimension with either an event attribute or the whole sequence. It also defines a dimension hierarchy, namely the set of levels and a partial order on the set. The syntax of the operator is as follows: $CreateContext(name_{D_i}, A_{D_i}, p_{D_i}, \mathbb{H}_{D_i}) = D_i$ where: (1) $name_{D_i}$ is the name of dimension D_i, (2) A_{D_i} equals to A_j if the dimension is an event attribute A_j or $A_{D_i} = \phi$ if the dimension is the property of the whole sequence, (3) p_{D_i} equals to predicate p if the dimension is the property of the whole sequence (p is an expression that computes the values of the dimension) or $p_{D_i} = \phi$ if the dimension is an event attribute (in this case values of an attribute A_{D_i} are taken as dimension values), (4) $\mathbb{H}_{D_i}$ is the set of hierarchies of dimension D_i or $\mathbb{H}_{D_i} = \phi$ if dimension D_i does not have a hierarchy.

Operations on sequences include:

1. $First(s)$ and $Last(s)$ – return event $e_i \in \mathbb{E}$ from sequence $s \in \mathbb{S}$ by removing from s all events except the first and the last one, respectively.
2. $Subsequence(s, m, n)$ – subtracts from $s \in \mathbb{S}$ a subsequence starting at the m-th and ending at the n-th event.
3. $Split(s, expression)$ – creates the set of new sequences by splitting initial sequence $s \in \mathbb{S}$, where $expression$ defines how to split s by means of predicates on attributes.
4. $Combine(\mathrm{S})$ – creates a new sequence with events of all sequences in $\mathrm{S} \subseteq \mathbb{S}$. The events in the new sequence are ordered by the values of ordering attributes in S.
5. $SelectSequences(\mathrm{S}, expression)$ – filters sequences in $\mathrm{S} \subseteq \mathbb{S}$ that fulfill a given $expression$. The result of the operation is the set of sequences $\subseteq \mathbb{S}$.
6. $SelectEvents(\mathrm{S}, expression)$ – removes from sequences in $\mathrm{S} \subseteq \mathbb{S}$ all events that do not fulfill a given $expression$.
7. $GroupBy(\mathrm{S}, expression \mid D_i)$ – assigns sequences from $\mathrm{S} \subseteq \mathbb{S}$ to groups according to the results of a given grouping $expression$ (case A) or to the values of dimension D_i (case B). Sequences having the same value of the grouping expression or dimension value belong to the same group. The result of the operation is set $\mathbb{G}$ of pairs $(value, s)$, where: (1) $value$ is the value of the grouping expression, (2) $s \in \mathbb{S}$ is a sequence with $value$ of a grouping expression (case A), or set of pairs $(value(D_i), s)$, where $value(D_i)$ is the value of dimension D_i (case B).
8. $Join(\mathrm{S}_{in_1}, \mathrm{S}_{in_2}, join\ condition,\ filtering\ predicate)$ – merges events in sequences from S_{in_1} with events in sequences from S_{in_2} and outputs a new set of sequences $\mathbb{S}_{out}$. The structures of events in sequences from both sets can be different and the output structure of events (set A_{out} of attributes) in the sequences of $\mathbb{S}_{out}$ is the union of events' attributes of sequences from S_{in_1} and S_{in_2}.
9. Set operations: union $\cup$, difference $\setminus$, and intersection $\cap$ – they are standard set operations.
10. $RollUp(D_i, \mathrm{S})$ – navigates one level up in the hierarchy of $D_i \in \mathbb{D}$ for all sequences in $\mathrm{S} \subseteq \mathbb{S}$.
11. $DrillDown(D_i, \mathrm{S})$ – navigates one level down in the hierarchy of $D_i \in \mathbb{D}$ for all sequences in $\mathrm{S} \subseteq \mathbb{S}$.
12. $Aggregate(F, \mathrm{S}, m, p)$ – aggregates the value of measure m. F is an aggregation function, namely Avg, Count, Max, Min, Sum, and S is the set of sequences. Aggregating a measure that is the property of multiple sequences is the same as in traditional OLAP. However, aggregating an event measure within a sequence may require the semantics of the aggregation to be provided by the user, by means of algorithm p (cf. [2]).

5   Seq-SQL Language

Based on the data model we developed a query language. It supports the following DDL commands: `create event measure`, `create event dimension`, create sequence, create sequence measure, create sequence dimension, and their drop counterparts. For the purpose of analyzing sequences, we proposed the SQL-like select seq command. Below we show the syntax diagrams of some of the commands and illustrate them with examples.

Recall the example from Sect. 2 where we created sequence *failures_seq*, sequence dimension *car_dim*, and sequence measure *total_cost*. The commands followed the general syntax, shown in Figs. 1, 2, and 3, respectively.

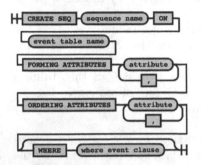

Fig. 1. Create sequence   Fig. 2. Create sequence dimension

Based on the example, let us create event dimension *Time*, called *dim_time* that includes levels day, month, and year. To this end, Command 1.6 is applied and it follows the general syntax shown in Fig. 4.

Command 1.6. Event dimension

```
create event dimension dim_time on failures
level day as [select event.failure_date from dual] type [date]
is child of level month as
    [select to_char(event.value,'yyyy.mm') from dual] type [varchar2(7)]
is child of level year as
    [select substr(event.value,1,4) from dual] type [varchar2(4)];
```

The *level* clause allows to define a level in a dimension hierarchy and the *is child of* clause allows to define the hierarchy of levels, similarly as in a traditional OLAP. Sequence set *failures_seq* (created by Command 1.1, cf. Sect. 2) can now be processed by the *select seq* command (cf. its general syntax shown in Fig. 5).

failures_seq is filtered with respect to different levels of dimension *dim_time*. To this end, we apply the *where event* clause that allows to define a filtering condition applied to events in a sequence, cf. Command 1.7. In this case, sequences whose level *year* in dimension *dim_time* is equal to 2000 are selected. The *split on* clause, allows to divide original sequences into the set of new sequences based on values of a given expression. In this example, sequences originally describing individual car failures within the whole car lifetime, are split into sequences of individual car failures within a year. *e_d_dim_time_year* is the alias of level *year* in the hierarchy of dimension *dim_time*.

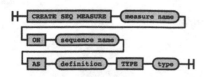

Fig. 3. Create sequence measure

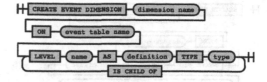

Fig. 4. Create event dimension

Command 1.7. Sequence retrieval and splitting

---

select seq ev.**dimension**.dim_time.**day**, failure_description
from failures_seq
where event [ev.**dimension**.dim_time.**year** > 2000] **split on** e_d_dim_time_year;

A user can also perform an analysis similar to a standard OLAP. Recall Command 1.3 from Sect. 2 that created sequence dimension *car_dim*. Commands 1.5 and 1.8 show scenarios of aggregating along sequence dimension *car_dim* at coarser level *make* (Command 1.5) and at finer level *model* (Command 1.8). Additionally, in the *where seq* clause of Command 1.8: (1) the sequences are filtered based on the value of the sequence measure (*seq.measure.total_cost>=500*) and (2) the sequences are restricted (sliced) according to the value of the indicated dimension level (*seq.dimension.dim_ar.make='VW'*).

Command 1.8. Complex analysis – drilling-down and slicing

---

select seq seq.**dimension**.car_dim.model, [**count**(**distinct seq**.seq_id) **as** num_cases]
from failures_seq
where seq [seq.**measure**.total_cost >= 5000 **and seq.dimension**.dim_car.make = 'VW']
where pattern[XYX] **with mapping** [X **to** failure_description, Y **to** failure_description]
    with options[X.2.car_mileage − X.1.car_mileage < 10000]
group by seq.**dimension**.car_dim.model;

Our language also supports filtering sequences with respect to a given pattern, as shown in Command 1.8. To this end, the *where pattern* clause is applied. Its first element, i.e., *[XYX]* defines the pattern. Its second element maps the pattern elements to the objects of the analysis (e.g., event attribute or event dimension). To this end, the *with mapping* clause is applied. The *with options* clause allows to define conditions to filter elements of sequences during pattern matching. A general syntax for creating an event measure (not illustrated with an example) is shown in Fig. 6.

6 Seq-SQL Prototype

6.1 Architecture

As the proof of concept, we implemented a prototype system, whose architecture is shown in Fig. 7. Its main components include: a data storage, a console, and

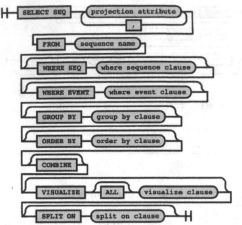

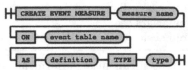

Fig. 5. Select sequence   **Fig. 6.** Create event measure

a *Seq-SQL* engine. In the current implementation, the data storage uses object-relational features of Oracle 11g. The component stores raw data, sequential data (e.g., events, sequences), dimension data, and a metadata dictionary.

The user perceives and processes sequential data in the model that we implemented. The sequential data are made available by means of the *Seq-SQL* engine. The console is used to construct, execute, and view the results of *Seq-SQL* commands (cf. Fig. 8). From the console, a command is sent to the *Seq-SQL* engine that: (1) parses the command, (2) translates it to the set of commands in SQL and PL/SQL, using metadata, and (3) sends the set for execution to the database server. Data retrieved from the database server are processed by the *Seq-SQL* engine, and presented in the console that allows also to visualize the results by means of simple charts. The *Seq-SQL* engine was implemented in Java and is served by Apache Tomcat. The process of parsing and translating *Seq-SQL* commands was implemented with the support of the ANTLR library [21].

6.2   Performance Evaluation

Preliminary experimental evaluations have been run on the presented prototype. We built a data warehouse storing events, which represent sensor measurements. The experiments were run on the data warehouse of the following sizes: 100 MB, 200 MB, 400 MB, 600 MB, and 800 MB. The *Seq-SQL* prototype was installed on a virtual machine (32-bit MS Windows 7, 3 GB RAM, 1 core of Intel Core i7 2.8 GHz, HDD SATA 3.0 TOSHIBA MK7559GSXF with the observed transfer speed between 30 MB/s and 80 MB/s). Notice that no dedicated index structures were used on the sequential data. The results that we discuss represent average values.

We have performed the four following experiments evaluating the performance of:

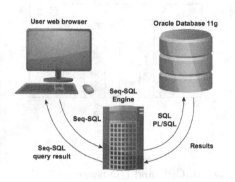

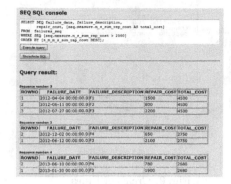

Fig. 7. *Seq-SQL* system architecture   **Fig. 8.** *Seq-SQL* consile

- the creation of sequences, sequence measures, sequence dimensions, and event measures,
- CPU and I/O usage for DDL operations,
- queries of multiple selectivities on multiple data volumes,
- *Group By* and *Combine* operators on multiple data volumes.

The performance of **creating sequences, sequence measures, and sequence dimensions**, for 5 different data volumes, is shown in Fig. 9. As we can observe, for all these objects except event measures the creation time grows steadily (linearly) with the increase of the data volume. The creation time of event measures increases approximately by 2.25–2.29 with the increase of the data volume up to 600 MB, and increases approximately by 1.01 to 1.43 for larger data volumes. This behavior can be explained by some unexpected buffering of the DBMS and the virtual machine.

CPU and I/O usage for DDL Operations is shown in Fig. 10. DDL operations included: (1) creating a sequence - Cr-Seq, (2) creating an event measure - Cr-Ev-M, (3) creating a sequence measure - Cr-Seq-M, and (4) creating a sequence dimension - Cr-Seq-D. Figure 10 clearly shows that the most intensive operations are on sequences.

Figure 11 shows the performance of a **query selecting 1 %** and **10 % of sequences**. Both queries were executed on all the available data volumes. The characteristics show that the execution time grows linearly w.r.t. to the data volume. Figure 12 shows the dependency between the execution time and the selectivity of a query on sequences, for data volumes of 100 MB and 200 MB. The execution time grows from 2.2 to 2.4 times with the increase of a data volume by 2. The execution time growth is close to linear w.r.t. the query selectivity.

Figure 13 shows the execution time of operators *Group By* and *Combine* w.r.t. to a data volume. As we can observe, *Combine* is more expensive than *Group By* as the former creates new sequences and sorts them, whereas the latter sorts the sequences and computes the aggregates. Creating sequences and sorting are I/O intensive operations that contribute to the higher cost of *Combine*.

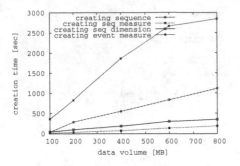

Fig. 9. Creation times

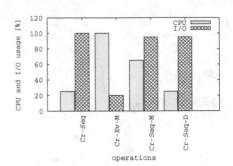

Fig. 10. CPU and I/O usage for DDL

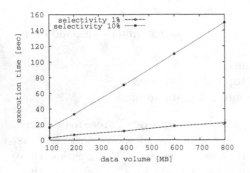

Fig. 11. Exec. times w.r.t. data volume

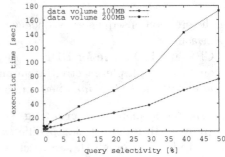

Fig. 12. Exec. times w.r.t. selectivity

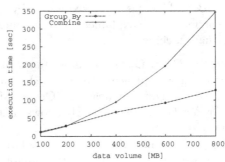

Fig. 13. Execution times of a group by query and a combine query

The execution time of *Group By* grows linearly with the growth of a data volume, whereas for *Combine* the growth is fastest than linear.

7   Conclusions and Future Work

The goal of this paper was to verify the feasibility of the approach to analyzing time point-based sequential data in an OLAP-like manner. Based on the model that we proposed in [2], we developed SQL-like language for manipulating and analyzing sequential data. Next, we outlined a prototype implementation of a sequence data warehouse and the *Seq-SQL* language. Finally, we presented preliminary experimental evaluation of the prototype.

The evaluation that we have conducted clearly showed that in most of the cases processing time grew linearly with a data volume or with a query selectivity. The exception was operator *Combine* whose processing time grew faster than linear with the increase of a data volume.

It must be stressed that the sequence data warehouse that we created was not optimized - neither buffer tuning, nor query optimization techniques, nor special indexes were applied. For this reason, the next steps will address: (1) investigating the most efficient storage model for sequential data (in the prototype we used an object-relational storage for the purpose of fast implementation), (2) developing physical data structures (e.g., indexes) to optimize access to sequential data, (3) query optimization techniques focusing on OLAP-like queries and pattern queries, and (4) comparing the performance of our prototype with commercial systems including Oracle and Teradata Aster w.r.t. pattern queries.

Furthermore, *Seq-SQL* does not support feature extraction techniques to summarize time series, in the spirit of [22,23]. Our future work will focus also on extending *Seq-SQL* with this functionality.

References

1. Bębel, B., Morzy, M., Morzy, T., Królikowski, Z., Wrembel, R.: OLAP-Like analysis of time point-based sequential data. In: Castano, S., Vassiliadis, P., Lakshmanan, L.V.S., Lee, M.L. (eds.) ER 2012 Workshops 2012. LNCS, vol. 7518, pp. 153–161. Springer, Heidelberg (2012)
2. Bebel, B., Morzy, T., Królikowski, Z., Wrembel, R.: Formal model of time point-based sequential data for OLAP-like analysis. Bull. Pol. Acad. Sci. (Tech. Sci.) **62**(2), 331–340 (2014)
3. Buchmann, A.P., Koldehofe, B.: Complex event processing. Inf. Tech. **51**(5), 241–242 (2009)
4. Chawathe, S.S., Krishnamurthy, V., Ramachandran, S., Sarma, S.: Managing RFID data. In: Proceedings of International Conference on Very Large Data Bases (VLDB), pp. 1189–1195. VLDB Endowment (2004)
5. Chui, C.K., Kao, B., Lo, E., Cheung, D.: S-OLAP: an OLAP system for analyzing sequence data. In: Proceedings of ACM SIGMOD International Conference on Management of Data, pp. 1131–1134. ACM (2010)
6. Chui, C.K., Lo, E., Kao, B., Ho, W.-S.: Supporting ranking pattern-based aggregate queries in sequence data cubes. In: Proceedings of ACM Conference on Information and Knowledge Management (CIKM), pp. 997–1006. ACM (2009)
7. Gonzalez, H., Han, J., Li, X.: FlowCube: constructing RFID flowcubes for multidimensional analysis of commodity flows. In: Proceedings of International Conference on Very Large Data Bases (VLDB), pp. 834–845. VLDB Endowment (2006)
8. Gonzalez, H., Han, J., Li, X., Klabjan, D.: Warehousing and analyzing massive RFID data sets. In: Proceedings of International Conference on Data Engineering (ICDE), p. 83 (2006)

9. Han, J., Chen, Y., Dong, G., Pei, J., Wah, B.W., Wang, J., Cai, Y.D.: Stream cube: an architecture for multi-dimensional analysis of data streams. Distrib. Parallel Databases 18(2), 173–197 (2005)
10. Han, J.-W., Pei, J., Yan, X.-F.: From sequential pattern mining to structured pattern mining: a pattern-growth approach. J. Comput. Sci. Technol. 19(3), 257–279 (2004)
11. Lerner, A., Shasha, D.: AQuery: query language for ordered data, optimization techniques, and experiments. In: Proceedings of International Conference on Very Large Data Bases (VLDB), pp. 345–356 (2003)
12. Liu, M., Rundensteiner, E., Greenfield, K., Gupta, C., Wang, S., Ari, I., Mehta, A.: E-Cube: multi-dimensional event sequence analysis using hierarchical pattern query sharing. In: Proceedings of ACM SIGMOD International Conference on Management of Data, pp. 889–900. ACM (2011)
13. Liu, M., Rundensteiner, E.A.: Event sequence processing: new models and optimization techniques. In: Proceedings of SIGMOD Ph.D. Workshop on Innovative Database Research (IDAR), pp. 7–12 (2010)
14. Lo, E., Kao, B., Ho, W.-S., Lee, S.D., Chui, C.K., Cheung, D.W.: OLAP on sequence data. In: Proceedings of ACM SIGMOD International Conference on Management of Data, pp. 649–660 (2008)
15. Mabroukeh, N.R., Ezeife, C.I.: A taxonomy of sequential pattern mining algorithms. ACM Comput. Surv. 43(1), 3:1–3:41 (2010)
16. Marascu, A., Masseglia, F.: Mining sequential patterns from data streams: a centroid approach. J. Intell. Inf. Syst. 27(3), 291–307 (2006)
17. Masseglia, F., Teisseire, M., Poncelet, P.: Sequential pattern mining. In: Wang, J. (ed.) Encyclopedia of Data Warehousing and Mining, pp. 1800–1805. IGI Global (2009)
18. Melton, J. (ed.) Working Draft Database Language SQL - Part 15: Row Pattern Recognition (SQL/RPR). ANSI INCITS DM32.2-2011-00005 (2011)
19. Mendes, L.F., Ding, B., Han, J.: Stream sequential pattern mining with precise error bounds. In: Proceedings of IEEE International Conference on Data Mining (ICDM), pp. 941–946 (2008)
20. Mörchen, F.: Unsupervised pattern mining from symbolic temporal data. SIGKDD Explor. Newsl. 9(1), 41–55 (2007)
21. Parr, T. (ed.) The Definitive ANTLR Reference: Building Domain-Specific Languages. Pragmatic Bookshelf (2007)
22. Perng, C., Wang, H., Zhang, S.R., Jr., D.S.P.: Landmarks: a new model for similarity-based pattern querying in time series databases. In: Proceedings of International Conference on Data Engineering (ICDE), pp. 33–42 (2000)
23. Rafiei, D., Mendelzon, A.O.: Querying time series data based on similarity. IEEE Trans. Knowl. Data Eng. (TKDE) 12(5), 675–693 (2000)
24. Ramakrishnan, R., Donjerkovic, D., Ranganathan, A., Beyer, K.S., Krishnaprasad, M.: SRQL: sorted relational query language. In: Proceedings of International Conference on Scientific and Statistical Database Management (SSDBM), pp. 84–95 (1998)
25. Sadri, R., Zaniolo, C., Zarkesh, A., Adibi, J.: Optimization of sequence queries in database systems. In: Proceedings of ACM SIGMOD-SIGACT-SIGART Symposium on Principles of Database Systems (PODS), pp. 71–81. ACM (2001)
26. Sadri, R., Zaniolo, C., Zarkesh, A., Adibi, J.: Expressing and optimizing sequence queries in database systems. ACM Trans. Database Syst. 29(2), 282–318 (2004)

27. Sadri, R., Zaniolo, C., Zarkesh, A.M., Adibi, J.: A sequential pattern query language for supporting instant data mining for e-services. In: Proceedings of International Conference on Very Large Data Bases (VLDB), pp. 653–656 (2001)

28. Seshadri, P., Livny, M., Ramakrishnan, R.: Sequence query processing. In: SIGMOD Record, vol. 23, no. 2 (1994)

29. Seshadri, P., Livny, M., Ramakrishnan, R.: SEQ: a model for sequence databases. In: Proceedings of International Conference on Data Engineering (ICDE), pp. 232–239 (1995)

30. Seshadri, P., Livny, M., Ramakrishnan, R.: The design and implementation of a sequence database system. In: Proceedings of International Conference on Very Large Data Bases (VLDB), pp. 99–110. Morgan Kaufmann Publishers Inc. (1996)

31. Aster   nPath.   http://developer.teradata.com/aster/articles/aster-npath-guide. Retrived 13 March 2014

32. van der Aalst, W.M.P.: Process cubes: slicing, dicing, rolling up and drilling down event data for process mining. In: Song, M., Wynn, M.T., Liu, J. (eds.) AP-BPM 2013. LNBIP, vol. 159, pp. 1–22. Springer, Heidelberg (2013)

33. Witkowski, A.: Analyze this! Analytical power in SQL, more than you ever dreamt of. Oracle Open World (2012)

34. Wu, E., Diao, Y., Rizvi, S.: High-performance complex event processing over streams. In: Proceedings of ACM SIGMOD International Conference on Management of Data, pp. 407–418. ACM (2006)

35. Zhang, Y., Kersten, M., Manegold, S.: SciQL: array data processing inside an RDBMS. In: Proceedings of ACM SIGMOD International Conference on Management of Data, pp. 1049–1052 (2013)

36. Zheng, Q., Xu, K., Ma, S.: When to update the sequential patterns of stream data? In: Whang, K.-Y., Jeon, J., Shim, K., Srivastava, J. (eds.) Proceedings of Pacific-Asia Confernece on Advances in Knowledge Discovery and Data Mining (PAKDD), vol. 2637, pp. 545–550. Springer, Heidelberg (2003)

Clustering Attributed Multi-graphs
with Information Ranking

Andreas Papadopoulos[✉], Dimitrios Rafailidis, George Pallis,
and Marios D. Dikaiakos

Department of Computer Science, University of Cyprus, Nicosia, Cyprus
{andpapad,drafail,gpallis,mdd}@cs.ucy.ac.cy

Abstract. Attributed multi-graphs are data structures to model real-world networks of objects which have rich properties/attributes and they are connected by multiple types of edges. Clustering attributed multi-graphs has several real-world applications, such as recommendation systems and targeted advertisement. In this paper, we propose an efficient method for Clustering Attributed Multi-graphs with Information Ranking, namely CAMIR. We introduce an iterative algorithm that ranks the different vertex attributes and edge-types according to how well they can separate vertices into clusters. The key idea is to consider the 'agreement' among the attribute- and edge-types, assuming that two vertex properties 'agree' if they produced the same clustering result when used individually. Furthermore, according to the calculated ranks we construct a unified similarity measure, by down-weighting noisy vertex attributes or edge-types that may reduce the clustering accuracy. Finally, to generate the final clusters, we follow a spectral clustering approach, suitable for graph partitioning and detecting arbitrary shaped clusters. In our experiments with synthetic and real-world datasets, we show the superiority of CAMIR over several state-of-the-art clustering methods.

Keywords: Attributed multi-graphs · Spectral clustering · Information ranking

1 Introduction

Information networks in many different domains such as biology, telecommunications, software engineering and social networking, can be modelled as attributed multi-graphs. In an attributed multi-graph, vertices have various attributes and are connected by multiple types of edges, such as user interactions and personal relations. Clustering attributed multi-graph methods aim at grouping related vertices into clusters by optimizing different objective functions, e.g. by minimizing the entropy and/or maximizing the connectivity and the density of the generated clusters. Recent works have shown that taking into account both the network structure and the vertex attributes can lead to improved clustering accuracy [1].

© Springer International Publishing Switzerland 2015
Q. Chen et al. (Eds.): DEXA 2015, Part I, LNCS 9261, pp. 432–446, 2015.
DOI: 10.1007/978-3-319-22849-5_29

In real-world networks, each vertex property, i.e. attribute and edge-type, contains different information, while some of these properties may be irrelevant to the clustering task. For instance, clustering a bibliography network, authors' attribute 'area of interest' is important, while attributes 'name' and 'gender' may introduce noise and thus reduce the clustering accuracy. Hence, the significance of each attribute and edge-type must be identified. One way to achieve this is to rank the vertex attributes and the edge-types, providing thus for each vertex property different importance in the clustering [1,2].

Recently, many methods for clustering attributed graphs have been proposed [1,3–5]. These works use unified distance/similarity measures or models that combine the vertex structural and attribute properties. They aim to partition attributed graphs into dense clusters with vertices having similar attributes, where attribute similarity is modelled by low entropy within each generated cluster. However, using weights to capture the different importance of vertex attributes in attributed graphs has received limited attention [1,6], while several methods ignore the existence of multiple edge-types. HASCOP [2] is one approach that considers the different importance of vertex attributes and multiple edge-types. Nonetheless, similarly to the other methods, the multiple edge-types and the vertex attributes are considered separately to perform the weighting. Therefore, a mechanism to identify the importance of each vertex attribute and edge-type for attributed multi-graphs is required.

In this paper, we consider the aforementioned challenges and we propose *a new approach for Clustering Attributed Multi-graphs with Information Ranking*, namely CAMIR. Specifically, we present *a mechanism to rank and consequently weigh the vertex properties in attributed multi-graphs*, by iteratively co-regularizing the clustering hypotheses across the vertex properties. Co-regularization is a well-known technique [7], which we use to compute the 'agreement' among the vertex attributes and the edge-types in attributed multi-graphs. Two vertex properties 'agree' if they assign vertices the same cluster labels when they are used individually. The vertex property with the highest agreement best separates the vertices into clusters, while the property with the lowest agreement introduces noise and reduces the clustering accuracy. In our method, we rank all vertex properties accordingly; that is, we assign the highest and the lowest ranks to the properties with the highest and the lowest agreements respectively. Taking advantage of our ranking mechanism, we *assign a weight parameter to each vertex property to compute a unified similarity measure for attributed multi-graphs*. Finally, we follow *a spectral clustering approach to partition the attributed multi-graph* and to generate the final clusters. The reason for using spectral clustering is that it identifies clusters of arbitrary shapes and sizes, by performing clustering in the eigenspace. Our experimental evaluation on synthetic and real-world data sets, i.e. two bibliography networks and a network of software packages available on Google code repository, shows that the proposed *CAMIR method outperforms several state-of-the-art clustering algorithms*.

The remainder of this paper is organized as follows, Sect. 2 describes the related work, and in Sect. 3 the problem of clustering attributed multi-graphs is formally defined. Section 4 presents the proposed CAMIR method and Sect. 5 demonstrates our experimental evaluation. Finally, Sect. 6 concludes the paper.

2   Related Work

Several clustering methods have been proposed for plain graphs (with a single edge-type) and multi-graphs [8–10]. However, these methods ignore the case of having vertex attributes and cannot be directly applied to attributed graphs. Representative approaches of attributed graph clustering methods are divided into two main categories: (a) distance-based [1,6] and (b) model-based [5,11]. Also, methods for clustering attributed graphs identify clusters either in the full space of the network [5,6,12] or in multiple sub-spaces [4,13,14]. In this paper, we focus on the case of clustering attributed graphs in the full space.

2.1   Distance-Based Clustering

SACluster [6] is a distance-based clustering method for attributed graphs, which computes a unified distance measure to capture both the structural and attribute similarities of the vertices. The key idea is to build an attribute augmented graph, equal to the initial graph enriched with new vertices each of which represents an attribute value. An edge from a graph vertex to an attribute vertex is added on the condition that the graph and attribute vertices have the same attribute value. The weight of the new edge depends on the importance that each attribute has. SACluster uses random walk on the augmented graph to measure the distance between two vertices, computing thus a unified distance measure. Several extensions of SACluster have been proposed to minimize the computational cost of random walks, such as SA-Cluster-Opt [15] and Inc-Cluster [1]. However, SACluster and its extensions do not work in multi-graphs, but only in attributed graphs.

PICS [3] identifies clusters with high similar connectivity, i.e., a cluster where the neighborhoods of its vertices highly overlap. The goal of PICS is to compress the adjacency, formed by the edges, and the attribute matrices that represent the attributed graph, by minimizing the cost of encoding in bits the clustering result. To minimize the encoding cost, PICS rearranges the rows and the columns of the adjacency and attribute matrices, so that all rows representing vertices in the first cluster are listed first, followed by rows representing vertices in the second cluster, and so on. At each iteration, PICS splits the cluster with the maximum entropy per attribute/edge, until a new splitting increases the total encoding cost. PICS does not consider multiple edge-types and consequently does not work in multi-graphs.

HASCOP [2] is an iterative algorithm for clustering attributed multi-graphs. HASCOP computes a heuristic unified distance function that combines the structural and attribute properties of the vertices based on their importance. The unified similarity of the vertices is defined as the product of their structural and attribute similarities. It tries to optimize similar connectivity and attribute homogeneity. Initially, each vertex is assigned to a singleton cluster. At each iteration, vertices with the highest unified similarities form the new clusters. The algorithm terminates if the number of clusters is not reduced further at the end of an iteration. Also, HASCOP considers the different importance of the attributes and edge-types over the iterations. However, HASCOP has high computational cost for re-evaluating weights at each iteration.

Spectral clustering of attributed graphs constructs a graph, where the edges between the vertices represent the vertex similarities based on their attributes and connections. Then, spectral clustering solves a relaxation of the normalized min-cut problem on the constructed graph to generate the final clusters [7]. Also, spectral clustering on attributed graphs has been used to identify clusters in the subspace projections of the attribute data; e.g. vertices that are close based on a subset of their attributes are grouped together [13,14].

2.2 Model-Based Clustering

Model-based methods enforce the intra-cluster similarity by modelling the attribute values and the connections/edges of vertices in a cluster by various distributions. For instance, BAGC [12] uses Bayesian inference, considering that the vertices in the same cluster should follow a common multinomial distribution for each of their attributes and a common Bernoulli distribution for their connections. BAGC starts with a random assignment of the vertices into clusters. Then, the parameters of all the distributions are iteratively recalculated. Vertices are assigned to a cluster, if vertex attributes and connections follow the respective attributes and connections distributions of the cluster. GBAGC [5] extends BAGC so as to handle weighted attributed graphs. The big advantage of BAGC and GBAGC is their scalability; however, they do not work for multi-graphs. CESNA [11] defines a model on attributed graphs that also enforces the intra-cluster similarities. CESNA models vertex attributes and connections in the same cluster with Bernoulli distributions. Nonetheless, CESNA differs from BAGC and all the aforementioned methods, by identifying overlapping communities.

3   Problem Definition

Our notation is presented in Table 1. Following standard notations, sets are denoted by calligraphic upper case letters, e.g., $\mathcal{A}$; matrices by plain upper case letters, e.g., A; functions by lower case calligraphic letters, e.g. a; numbers and set elements by lower case letters, e.g. a.

An attributed multi-graph G is a set of $|\mathcal{T}|$ different attributed graphs defined over the same set of vertices, where $\mathcal{T} = \{t_i\}$ is the set of edge-types. Formally, $G = \{G_t(\mathcal{V}, \mathcal{E}_t)\}_{t=1}^{\mathcal{T}}$, where $\mathcal{V} = \{v_i\}$ is the set of vertices; $\mathcal{E}_t = \{(v_i, v_j) : v_i, v_j \in \mathcal{V}\}$ is the set of all edges of type t; $\omega_t : \mathcal{E}_t \rightarrow (0, 1]$ returns the weight of the edges of type t, i.e. $\omega_t(v_i, v_j)$ is the weight of the edge of type t from v_i to v_j. Each vertex has $|\mathcal{A}|$ different attribute values, where $\mathcal{A} = \{\alpha_i : 1 \leq i \leq |\mathcal{A}|\}$ is the set of all vertex attributes. The attribute values of a vertex are given by the functions a_α, i.e. $a_\alpha(v_i)$ is the attribute value of vertex v_i on attribute α. Also, we denote as $\mathcal{P} = \{p : p \in \mathcal{T} \vee p \in \mathcal{A}\}$ the set of all vertex properties, i.e. edge-types and attributes.

Table 1. Notations

Symbol	Description
$\mathcal{V}$	Set of vertices $V = \{v_i : 1 \leq i \leq |\mathcal{V}|\}$
$\mathcal{E}_t$	Set of edges of type t. $\mathcal{E}_t = \{(v_i, v_j) : v_i, v_j \in \mathcal{V}\}$
$\omega_t : \mathcal{E}_t \to (0, 1]$	Function that returns the weight of the edges of type t
$\mathcal{A}$	Set of attributes $A = \{\alpha_i : 1 \leq i \leq |\mathcal{A}|\}$
$a_\alpha(v_i)$	Value of vertex v_i on attribute α
$\mathcal{T}$	Set of edge-types, with $\mathcal{T} = \{t_i : 1 \leq i \leq |\mathcal{T}|, \mathcal{E}_t \neq \emptyset\}$
$\mathcal{P} = \{\mathcal{A} \cup \mathcal{T}\}$	Set of all vertex properties (attributes and edge-types)
$w : \mathcal{P} \to (0, 1]$	Function that returns the importance/weight of a vertex property p
k	Number of clusters
$S^p \in \mathbb{R}^{|\mathcal{V}| \times |\mathcal{V}|}$	Similarity matrix for property p
$L^p \in \mathbb{R}^{|\mathcal{V}| \times |\mathcal{V}|}$	Normalized Laplacian matrix of S^p
$U^p \in \mathbb{R}^{|\mathcal{V}| \times k}$	First k eigenvectors of L^p

The problem of clustering attributed multi-graphs is defined as follows: given an attributed multi-graph G with $|\mathcal{P}|$ different vertex properties and a number of clusters k, the goals are (a) to compute a weight $w(p)$ for each vertex property $p \in \mathcal{P}$, in order to construct a unified similarity matrix $S \in \mathbb{R}^{|\mathcal{V}| \times |\mathcal{V}|}$; (b) to generate the k clusters based on S, by maximizing the similarity of vertices within a cluster and minimizing the similarity between vertices in different clusters.

4 The Proposed CAMIR Method

4.1 Overview

The proposed CAMIR method consists of the following steps: (a) the attributed multi-graph is processed to rank the vertex properties and to calculate their weights accordingly; then a unified similarity measure is computed by considering all vertex properties and their importance based on the calculated weights. (b) a spectral clustering approach is adopted to embed the vertices into the respective eigenspace of the unified similarity measure. The reason for selecting spectral clustering is that it identifies clusters of arbitrary shapes and sizes. The key idea in spectral clustering is to achieve graph partitioning by finding the best cut. Spectral clustering finds the best cut by performing eigendecomposition of a graph Laplacian matrix, constructing thus a k-dimensional eigenspace. Spectral clustering methods differ in how they define and construct the Laplacian matrix L of the similarity matrix S and, thus, which eigenvectors are selected to represent the graph, aiming to exploit special properties of different matrix formulations [16]. For the interested reader, Ulrike von Luxburg's tutorial [16] includes examples of different Laplacians' constructions. Moreover, different objective functions are used to derive the best cut. For example, Ratio Cut [17] tries to minimize the total cost of the edges crossing the cluster boundaries, normalized by the size of the k clusters, to encourage balanced cluster sizes. Normalized Cut (NCut) [18] uses the same objective criterion as Ratio Cut, normalized by the total degree of each cluster, making thus the clusters having similar degrees. In our approach we followed the NCut method.[1] In the following Sections we present each step of the proposed CAMIR method in detail.

4.2 Information Ranking

Given an attributed multi-graph G, we compute the affinity matrices $S^p \in \mathbb{R}^{|\mathcal{V}| \times |\mathcal{V}|}$, where $S_{ij}^p \geq 0$ denotes the relationship - similarity between v_i and v_j for property $p \in \mathcal{P} \equiv \{\mathcal{A} \cup \mathcal{T}\}$. For each edge-type $t \in \mathcal{T}$, the respective similarity matrix is calculated as follows:

$$S_{ij}^t = \omega_t(v_i, v_j) \tag{1}$$

For each vertex attribute $a \in \mathcal{A}$ we calculate the respective similarity matrix based on the Gaussian kernel:[2]

$$S_{ij}^\alpha = \exp\left(-\frac{||a_\alpha(v_i) - a_\alpha(v_j)||^2}{2 \cdot \sigma_i \cdot \sigma_j}\right) \tag{2}$$

where σ_i is a scaling parameter to control how rapidly the similarity S_{ij}^α reduces according to the distance between v_i and v_j. For each vertex v_i, the scaling parameter σ_i allows the self-tuning of the vertex-to-vertex distances according to the local statistics of the neighbourhoods surrounding v_i. We followed the self-tuning strategy of [19]. Provided that v_i has ϵ neighbours, σ_i is calculated as the average of the ϵ distances.

According to Eqs. (1) and (2), for each vertex property $p \in \mathcal{P}$ we construct an affinity matrix, denoted as $S^p \in \mathbb{R}^{|\mathcal{V}| \times |\mathcal{V}|}$. We compute the normalized Laplacian $L^p \in \mathbb{R}^{|\mathcal{V}| \times |\mathcal{V}|}$ of S^p as follows:

$$L^p = I - D^{p^{-1/2}} \cdot S^p \cdot D^{p^{-1/2}} \tag{3}$$

where I is the identity matrix and D^p is a diagonal matrix calculated as follows:

$$D_{ii}^p = \sum_{j=1}^{|\mathcal{V}|} S_{ij}^p \tag{4}$$

where $D^{p^{-1/2}}$ indicates the inverse square root of D^p. For any S with $S_{ij} \geq 0$, the Laplacian matrix is symmetric positive semi-definite [16]. After computing

---

[1] Alternatively, several parallel spectral clustering methods could be used in the proposed approach, such as the works of [19,20], to reduce the computational time of spectral clustering.

[2] Also, other types of kernel functions could be used, such as linear and polynomial, thoroughly examined in [21] for machine learning methods.

the $|\mathcal{P}|$ different normalized Laplacians, we perform eigendecomposition on each of the normalized Laplacians to retrieve their top k eigenvectors, denoted by $U^p \in \mathbb{R}^{|\mathcal{V}| \times k}$ for the p-th Laplacian L^p.

According to [7] the vertex property p that best separates the vertices is selected using the following equation:

$$f(\mathcal{P}) = \underset{U^p \in \mathbb{R}^{|\mathcal{V}| \times k}}{\arg\max} \; tr \left[U^{p^T} \cdot \left(L^p + \lambda \cdot \sum_{\substack{i=1 \\ p_i \neq p}}^{|\mathcal{P}|} \left(U^{i^T} \cdot U^i \right) \right) \cdot U^p \right] \tag{5}$$

where $\mathcal{P}$ is the set of all vertex properties, tr is the trace of the matrix, U^{i^T} is the transpose matrix of U^i, and λ is a co-regularization parameter that controls the penalization of a property according to its 'disagreement' with the other properties,[3] denoted by the sum in Eq. (5). According to [7] two different vertex properties 'agree' if they produced the same clustering result when used individually. Equation (5) returns the property p that has the highest 'agreement' with the rest properties, assuming that if we use only the selected property p for clustering we expect to find accurate clusters, independently of the rest of the properties. The mathematical proof can be found in [7].

However, considering just a single property to perform clustering contradicts with works elaborating on the use of all vertex properties to improve clustering accuracy [1,4]. We propose to iteratively apply Eq. (5) $|\mathcal{P}|$ times. At each iteration we exclude the properties that have been already ranked. In particular, given the set of unranked properties, denoted as $\mathcal{P}_u$, we compute the ranking $r(p)$ of a property p by the following equation:

$$r(p) = \begin{cases} |\mathcal{P}_u| : |\mathcal{P}_u| > 1 \wedge f(\mathcal{P}_u) = U^p \\ 1 \quad : |\mathcal{P}_u| = 1 \end{cases} \tag{6}$$

In doing so, the highest rank is given to the property with the highest 'agreement' with the other properties; while the lowest rank is given to the property that is selected last, does not 'agree' with the rest of properties, and consequently introduces noise to the clustering. Equation (6) maps the vertex properties to the ranks $\{|\mathcal{P}|, |\mathcal{P}| - 1, \dots, 1\}$ according to the order the properties are selected. To calculate the weight of each property, we perform a normalization of the properties' ranking as follows:

$$w(p_i) = \frac{r(p_i)}{\sum_{j=1}^{|\mathcal{P}|} r(p_j)} \tag{7}$$

Equation (7) assigns higher weights to more important vertex properties and down-weighs the properties that have lower ranks and introduce noise over the

---

[3] Following [7] we use a common λ for all properties. In practice though, $\lambda = 0.001$ is an appropriate value to control the impact of the other properties, as we observed in our experiments.

Algorithm 1. CAMIR Algorithm

```
Input: Attributed multi-graph G, number of clusters k
Output: Cluster labels
    P_u = P
    while |P_u| ≠ ∅ do
        Rank a property p using Equation (6)
        P_u = P_u \ {p}
    end while
    Compute weights w(p_i) based on Equation (7)
    Compute the unified similarity matrix S using Equation (8)
    Compute the normalized Laplacian L of S according to Equation (3)
    Perform eigendecomposition on L to obtain U - the top k eigenvectors
    Run k-means on U to generate the final cluster labels
```

---

clustering. Finally, according to property weights we compute the unified similarity matrix $S \in \mathbb{R}^{|\mathcal{V}| \times |\mathcal{V}|}$ as the weighted sum of the $|\mathcal{P}|$ similarity matrices S^p:

$$S = \sum_{\forall p \in \mathcal{P}} w(p) \cdot S^p \tag{8}$$

4.3   Generating the Final Clusters

In our method, spectral clustering is formulated as follows: given the $|\mathcal{V}|$ vertices of the attributed multi-graph G and the unified similarity matrix $S \in \mathbb{R}^{|\mathcal{V}| \times |\mathcal{V}|}$, calculated based on Eq. (8), the goal is to find k disjoint vertex subsets, namely clusters, whose union is the whole data set, by solving the following standard eigendecomposition problem:

$$\arg\max_{U \in \mathbb{R}^{|\mathcal{V}| \times k}} tr\left(U^T \cdot L \cdot U\right), \text{ s.t. } U^T \cdot U = I \tag{9}$$

where L is the normalized Laplacian of the unified similarity matrix S given by Eq. (8). The columns of $U \in \mathbb{R}^{|\mathcal{V}| \times k}$ are the top k eigenvectors of the Laplacian matrix L, while its rows are the embeddings of the $|\mathcal{V}|$ vertices in the k-th dimensional eigenspace. The final k clusters are generated by applying k-means algorithm to the $|\mathcal{V}|$ embeddings of the vertices. Algorithm 1 presents the pseudocode of the two-step CAMIR algorithm.

5   Experiments

5.1   Datasets

In our experiments we use synthetic and three real-world datasets, summarized in Table 2.

Table 2. Real-world datasets

Dataset	$|\mathcal{V}|$	$|E|$	$|\mathcal{A}|$	$|\mathcal{T}|$	$|\mathcal{P}|$
DBLP-1K	1000	17128	2	1	3
DBLP-10K	10000	65734	2	1	3
GoogleSP-23	1297	268956	5	2	7

Synthetic datasets have been generated using a modified version of the generator in [4]. The $|\mathcal{V}|$ vertices are divided into k groups. F or each group two parameters specify its similar connectivity and attribute homogeneity. If a vertex in a group connects to a vertex u, then Similar Connectivity parameter specifies the least fraction of vertices in the group that also connect to vertex u. Attribute Homogeneity parameter specifies the least fraction of vertices in a group sharing the same attribute value. The values of the attributes are drawn from a Bernoulli distribution. We set Similar Connectivity and Attribute Homogeneity parameters to 0.8 and, unless stated otherwise, the synthetic datasets have: 1000 vertices, 25 clusters, 1 edge-type and each vertex is characterized by 4 attributes. The reason for selecting 1 edge-type is that many competitors, e.g. SACluster [6], BAGC [12], and PICS [3], do not work on multi-graphs, but only on attributed graphs (Sect. 2).

DBLP-1K and DBLP-10K datasets[4] consist of 1000 and 10000 vertices, respectively. The vertices represent the top authors from the complete DBLP dataset. An author has the following properties: publications and the primary area of interest, e.g. databases, data mining, information retrieval, artificial intelligence, etc. A weighted edge between two authors represents the number of publications they have co-authored. In both datasets the clusters are unknown.

GoogleSP-23 is a dataset constructed by crawling the file system of a virtual machine in which 23 software packages were downloaded from the Google code repository[5]. This dataset was also used in the work of [2]. A vertex represents a software file and has the following attributes: file size, last access time, modification time, creation time and file-type. There are two types of edges in this dataset based on the file name and file path similarities. Given two files we calculate the distance of their file system paths using a string edit distance algorithm. An edge between the two files is added if their paths' distance is less than the average paths distance in the network. Similarly, the edges based on names are added. Thus, GoogleSP-23 is an attributed multi-graph, where each vertex has 5 attributes and there are 2 edge-types. In this dataset the clusters are known, where each cluster corresponds to a software package.

---

[4] The full DBLP dataset is available online at http://kdl.cs.umass.edu/data/dblp/dblp-info.html.

[5] Available online at http://code.google.com.

5.2   Evaluation Protocol

In our experiments we use the (a) Normalized Mutual Information (NMI) and (b) entropy metrics. NMI is in the range of $[0,1]$. High NMI is equivalent to high similarity between the resulted clustering of an examined method and the ground-truth. Given the clustering of an examined method, denoted by $\mathcal{S}_1 = \{\mathcal{B}_1, \ldots, \mathcal{B}_{k1}\}$ and the ground-truth in $\mathcal{S}_2 = \{\mathcal{C}_1, \ldots, \mathcal{C}_{k2}\}$, where it may hold $k1 \neq k2$, NMI is calculated as follows:

$$NMI(\mathcal{S}_1, \mathcal{S}_2) = \frac{H(\mathcal{S}_1) - H(\mathcal{S}_1|\mathcal{S}_2)}{\min(H(\mathcal{S}_1), H(\mathcal{S}_2))} \tag{10}$$

where:

$$H(\mathcal{S}) = -\sum_{i=1}^{|\mathcal{S}|} \frac{|\mathcal{C}_i|}{|\mathcal{V}|} \cdot \log\left(\frac{|\mathcal{C}_i|}{|\mathcal{V}|}\right)$$

$$H(\mathcal{S}_1|\mathcal{S}_2) = -\sum_{i=1}^{|\mathcal{S}_1|}\sum_{j=1}^{|\mathcal{S}_2|} \frac{m_{i,j}}{|\mathcal{V}|} \cdot \log\left(\frac{m_{i,j}/|\mathcal{V}|}{|\mathcal{C}_j|/|\mathcal{V}|}\right)$$

where $m_{i,j}$ is the number of common vertices between clusters $\mathcal{B}_i$ and $\mathcal{C}_j$. In case $NMI(\mathcal{S}_1, \mathcal{S}_2) = 1$, then the two clusterings are identical. At this point we must mention that NMI can be calculated, only if the ground-truth is available. Thus, in the synthetic datasets and GoogleSP-23 we report NMI for each clustering method, while in DBLP-1K and DBLP-10K NMI is not available, since the clusters are unknown.

Entropy ranges in $[0, \infty)$ and is measured for each attribute a as follows:

$$entropy(\alpha) = \sum_{j=1}^{k} \frac{|\mathcal{C}_j|}{|\mathcal{V}|} \cdot entropy(\alpha, \mathcal{C}_j) \tag{11}$$

Low entropy is equivalent to high homogeneity between the attributes of the vertices in the same cluster. The overall entropy for a clustering method is the average entropy for all attributes. Thus, the goal of each clustering method is to achieve low overall entropy.

We evaluate the proposed CAMIR method against HASCOP [2], SA-Cluster [6], BAGC [12], and PICS [3]. All experiments were conducted on a desktop pc equipped with a quad core Intel i7 2.8 GHz processor and 8 GB RAM.

5.3   Evaluation on Synthetic Datasets

In this set of experiments, we vary the number of vertices in $\{100, 500, 1000, 5000, 10000\}$ with 4 attributes. Also, we vary the number of attributes in $\{2, 4, 8, 16, 32\}$ with 1000 vertices. For each variation, we generate 5 random graphs and in Figs. 1(a)–(f) we report the average entropy, NMI and computational time out of the 5 runs. We must mention that for number of vertices equal to 10000, we omit the results of HASCOP, since it does not scale due to its heavy

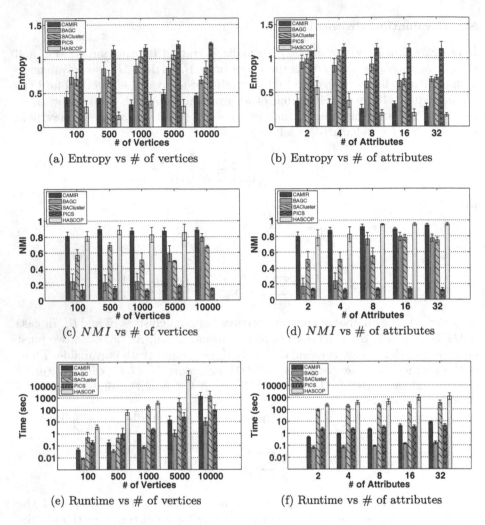

Fig. 1. Clustering performance on synthetic attributed graphs

computational cost. According to Figs. 1(a)–(d) CAMIR outperforms all its competitors, except HASCOP, in terms of entropy and NMI. The high clustering accuracy of CAMIR is based not only on the spectral clustering technique which identifies arbitrary shaped clusters, but also on the proposed weighting mechanism that correctly identifies the importance of the different vertex properties. PICS has high entropy and low NMI, because it converges too early and returns few clusters, by using a self-tuning strategy to determine the number of clusters. HASCOP and CAMIR achieve comparable clustering accuracy, since HASCOP also weighs the vertex properties efficiently. However, HASCOP is much slower than CAMIR, at least two orders of magnitude, as presented in Figs. 1(e)–(f). This happens because HASCOP has a high computational cost of re-evaluating

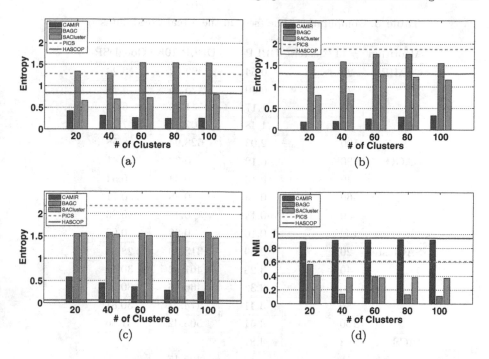

Fig. 2. Performance evaluation on real-world datasets; (a) DBLP-1K (b) DBLP-10K (c)–(d) GoogleSP-23. Since HASCOP and PICS use a self-tuning strategy to determine the number of clusters, both methods are denoted by straight lines

weights at each iteration (Sect. 2). BAGC is the fastest approach; nevertheless, it has limited clustering accuracy as shown, in Figs. 1(a)–(d), since it does not use a weighting scheme to identify the importance of the vertex properties.

5.4   Evaluation on Real-World Datasets

DBLP-1K and DBLP-10K. In this set of experiments we vary the number of clusters from $k = 20$ to 100 by a step of 20. PICS and HASCOP use a self-tuning method to determine the number of clusters, returning 8 and 85 clusters in DBLP-1K and 18 and 763 clusters in DBLP10-K, respectively. In Figs. 2(a) and (b) we present the average entropy on both datasets. NMI is not available, since the clusters in both datasets are unknown. We observe that in both DBLP datasets CAMIR achieves the best results in terms of entropy for all number of clusters variations. The reason is that CAMIR is robust to noise, by incorporating the importance of the vertex properties according the proposed weighting scheme. Besides the high clustering accuracy, CAMIR requires comparable time to SACluster and PICS, as presented in Table 3. BAGC is the fastest approach but has limited clustering accuracy, denoted by high entropy, while HASCOP is the slowest clustering method.

Table 3. Computation time (sec) in the 3 real-world datasets

Algorithm	# of Clusters	DBLP-1K	DBLP-10K	GoogleSP-23
CAMIR	20	0.54	399.13	3.26
	40	0.80	425.15	4.78
	60	1.17	541.39	6.04
	80	1.48	607.81	7.25
	100	2.03	628.94	8.58
BAGC	20	0.12	0.33	0.41
	40	0.13	0.34	0.54
	60	0.14	0.38	0.71
	80	0.14	0.46	0.86
	100	0.16	0.49	0.97
SACluster	20	2.59	313.19	28.84
	40	3.28	407.84	30.11
	60	3.32	466.76	30.57
	80	3.41	474.93	30.81
	100	3.51	503.42	31.86
PICS	8	4.87	-	-
	18	-	495.17	-
	16	-	-	476.49
HASCOP	85	882.18	-	-
	763	-	32957.10	-
	51	-	-	4675.27

GoogleSP-23. In this set of experiments, HASCOP and PICS identify 16 and 51 clusters, respectively. Since GoogleSP-23 is an attributed multi-graph we run PICS, SACluster and BAGC separately with only one edge-type and all the vertex attributes, since these methods do not work in multi-graphs, but only in attributed graphs. In Figs. 2(c) and (d) we report the best results for the three aforementioned methods. CAMIR handles efficiently the noise in the dataset, by identifying the importance of the vertex properties, resulting thus to low entropy and high NMI. In this set of experiments, HASCOP slightly outperforms CAMIR in terms of entropy; however HASCOP requires a significantly higher computational time than CAMIR, almost 3 orders of magnitude, as presented in Table 3.

6   Conclusion

In this paper we proposed CAMIR, a method for clustering attributed multi-graphs. CAMIR ranks vertex properties of an attributed multi-graph by exploiting the information from edge-types and attributes. Based on the identified ranking CAMIR weighs the vertex properties to construct a unified similarity measure and performs spectral clustering to generate the final clusters. Our experimental evaluation on synthetic and real-world graphs against state-of-the-art attributed graph clustering techniques showed the effectiveness of the proposed CAMIR method, in terms clustering accuracy and computational time.

Real-world information networks are continuously updated. Recently, several incremental spectral clustering strategies have been proposed in the literature [22]. These strategies are able to handle not only insertions and deletions of new objects but also similarity changes between them, by efficiently updating the eigenspace. In our future research we plan to examine both the incremental and evolving strategies of the proposed method in the context of spectral clustering in Big Data [23].

Acknowledgments. This work was partially supported by the EU Commission in terms of the PaaSport 605193 FP7 project (FP7-SME-2013).

References

1. Cheng, H., Zhou, Y., Yu, J.X.: Clustering large attributed graphs: a balance between structural and attribute similarities. ACM Trans. Knowl. Discov. Data **5**(2), 12:1–12:33 (2011)
2. Papadopoulos, A., Pallis, G., Dikaiakos, M.D.: Identifying clusters with attribute homogeneity and similar connectivity in information networks. In: Proceedings of the 2013 IEEE/WIC/ACM International Joint Conferences on Web Intelligence (WI) and Intelligent Agent Technologies (IAT), WI-IAT 2013, vol. 01, pp. 343–350. IEEE Computer Society, Washington, DC (2013)
3. Akoglu, L., Tong, H., Meeder, B., Faloutsos, C.: Pics: parameter-free identification of cohesive subgroups in large attributed graphs. In: Proceedings of the 12th SIAM International Conference on Data Mining, SDM 2012, pp. 439–450. SIAM/Omnipress (2012)
4. Perozzi, B., Akoglu, L., Iglesias Sánchez, P., Müller, E.: Focused clustering and outlier detection in large attributed graphs. In: Proceedings of the 20th ACM SIGKDD International Conference on Knowledge Discovery and Data Mining, KDD 2014, pp. 1346–1355. ACM, New York (2014)
5. Xu, Z., Ke, Y., Wang, Y., Cheng, H., Cheng, J.: GBAGC: a general bayesian framework for attributed graph clustering. ACM Trans. Knowl. Discov. Data **9**(1), 5:1–5:43 (2014)
6. Zhou, Y., Cheng, H., Yu, J.X.: Graph clustering based on structural/attribute similarities. Proc. VLDB Endow. **2**(1), 718–729 (2009)
7. Kumar, A., Rai, P., Daume, H.: Co-regularized multi-view spectral clustering. In: Shawe-Taylor, J., Zemel, R., Bartlett, P., Pereira, F., Weinberger, K. (eds.) Advances in Neural Information Processing Systems 24, pp. 1413–1421. Curran Associates, Inc., NY (2011)
8. Karypis, G., Kumar, V.: Multilevel algorithms for multi-constraint graph partitioning. In: Proceedings of the 1998 ACM/IEEE Conference on Supercomputing, SC 1998, pp. 1–13. IEEE Computer Society, Washington, DC (1998)

9. Papalexakis, E., Akoglu, L., Ience, D.: Do more views of a graph help? community detection and clustering in multi-graphs. In: 2013 16th International Conference on Information Fusion (FUSION), pp. 899–905 (2013)
10. Xu, X., Yuruk, N., Feng, Z., Schweiger, T.A.J.: SCAN: a structural clustering algorithm for networks. In: Proceedings of the 13th ACM SIGKDD International Conference on Knowledge Discovery and Data Mining, KDD 2007, pp. 824–833. ACM, New York (2007)
11. Yang, J., McAuley, J.J., Leskovec, J.: Community detection in networks with node attributes. [24], pp. 1151–1156
12. Xu, Z., Ke, Y., Wang, Y., Cheng, H., Cheng, J.: A model-based approach to attributed graph clustering. In: Proceedings of the 2012 ACM SIGMOD International Conference on Management of Data, SIGMOD 2012, pp. 505–516. ACM, New York (2012)
13. Günnemann, S., Färber, I., Raubach, S., Seidl, T.: Spectral subspace clustering for graphs with feature vectors. [24], pp. 231–240
14. Günnemann, S., Boden, B., Färber, I., Seidl, T.: Efficient mining of combined subspace and subgraph clusters in graphs with feature vectors. In: Pei, J., Tseng, V.S., Cao, L., Motoda, H., Xu, G. (eds.) PAKDD 2013, Part I. LNCS, vol. 7818, pp. 261–275. Springer, Heidelberg (2013)
15. Zhou, Y., Cheng, H., Yu, J.X.: Clustering large attributed graphs: an efficient incremental approach. In: Webb, G.I., Liu, B., Zhang, C., Gunopulos, D., Wu, X. (eds.) ICDM 2010, pp. 689–698. IEEE Computer Society, Washington, DC (2010)
16. Luxburg, U.: A tutorial on spectral clustering. Stat. Comput. 17(4), 395–416 (2007)
17. Chan, P.K., Schlag, M.D.F., Zien, J.Y.: Spectral K-way ratio-cut partitioning and clustering. IEEE Trans. CAD Integr. Circuits Syst. 13(9), 1088–1096 (1994)
18. Shi, J., Malik, J.: Normalized cuts and image segmentation. IEEE Trans. Pattern Anal. Mach. Intell. 22(8), 888–905 (2000)
19. Chen, W.Y., Song, Y., Bai, H., Lin, C.J., Chang, E.: Parallel spectral clustering in distributed systems. IEEE Trans. Pattern Anal. Mach. Intell. 33(3), 568–586 (2011)
20. Kang, U., Meeder, B., Papalexakis, E.E., Faloutsos, C.: Heigen: spectral analysis for billion-scale graphs. IEEE Trans. Knowl. Data Eng. 26(2), 350–362 (2014)
21. Hofmann, T., Schölkopf, B., Smola, A.J.: Kernel methods in machine learning. Ann. Stat. 36(3), 1171–1220 (2008)
22. Ning, H., Xu, W., Chi, Y., Gong, Y., Huang, T.S.: Incremental spectral clustering by efficiently updating the eigen-system. Pattern Recogn. 43(1), 113–127 (2010)
23. Mall, R., Langone, R., Suykens, J.A.K.: Kernel spectral clustering for big data networks. Entropy 15(5), 1567–1586 (2013)
24. Xiong, H., Karypis, G., Thuraisingham, B.M., Cook, D.J., Wu, X. (eds.): 2013 IEEE 13th International Conference on Data Mining. IEEE Computer Society, Washington, DC (2013)

Indexing and Decision Support Systems

Building Space-Efficient Inverted Indexes
on Low-Cardinality Dimensions

Vasilis Spyropoulos[(✉)] and Yannis Kotidis

Athens University of Economics and Business, 76 Patission Street, Athens, Greece
{vasspyrop,kotidis}@aueb.gr

Abstract. Many modern applications naturally lead to the implementation of inverted indexes for effectively managing large collections of data items. Creating an inverted index on a low cardinality data domain results in replication of data descriptors, leading to increased storage overhead. For example, the use of RFID or similar sensing devices in supply-chains results in massive tracking datasets that need effective spatial or spatio-temporal indexes on them. As the volume of data grows proportionally larger than the number of spatial locations or time epochs, it is unavoidable that many of the resulting lists share large subsets of common items. In this paper we present techniques that exploit this characteristic of modern big-data applications in order to loselessly compress the resulting inverted indexes by discovering large common item sets and adapting the index so as to store just one copy of them. We apply our method in the supply chain domain using modern big-data tools and show that our techniques in many cases achieve compression ratios that exceed 50 %.

1 Introduction

Many applications implement auxiliary indexes of the form $id \leftarrow list(itemId)$, often referred to as inverted indexes. For example, such indexes are used in performing keyword search in a collection of documents ($word \leftarrow list(documentId)$). In supply-chain applications equipped with RFID tracking technology tagged objects are recorded while passing through a check point (reader's location) in the supply-chain [1,2]. Inverted indexes can be used to implement spatial ($locationId \leftarrow list(tagId)$), temporal ($epochId \leftarrow list(tagId)$) or spatio-temporal (($locationId, epochId) \leftarrow list(tagId)$) indexes. When inverted indexes are build on a low cardinality data domain (e.g. $locationId$, $epochId$), the resulting lists tend to share a possibly large number of common item references (e.g. $tagIds$). A straightforward implementation of these lists will result in excessive replication of item references, increasing the space overhead of the inverted

This research has been co-financed by the European Union (European Social Fund ESF) and Greek national funds through the Operational Program "Education and Lifelong Learning" of the National Strategic Reference Framework (NSRF) - Research Funding Program: RECOST.

© Springer International Publishing Switzerland 2015
Q. Chen et al. (Eds.): DEXA 2015, Part I, LNCS 9261, pp. 449–459, 2015.
DOI: 10.1007/978-3-319-22849-5_30

index. While there is a wealth of techniques that compress inverted indexes at binary level, there is an opportunity to achieve significant compression ratios by looking at the actual content of the lists. The existing methods usually represent the index lists as a sorted sequence of integers and transform it into more compressible form by using the distances between the integers (d-gaps) and applying integer coding techniques such as *Variable Byte*, *Golomb* or *Rice* codes. Our technique takes place at the information level and, thus, can be safely applied before any low level compression of the index. In brief, our method is aiming at finding intersections of the lists and materialize them as new lists, which we call *derived lists*. When a large portion of the original lists are replaced by references to derived lists, we are able to reduce the space of the index since a single copy of item ids is referred by many lists. We identify the following challenges:

Performing Intersections of Large Lists: While checking combinations of lists in order to generate the candidate derived lists we need to have a way so as to efficiently compute the size of their intersection. Since the actual computation of a large number of intersections of lists, each containing millions of records, is a costly operation we utilize approximation techniques. In a first phase we reduce the number of intersections to compute by testing only combinations of similar lists stored in the same bucket among buckets populated by the use of the minhashing/LSH technique. Then, instead of computing the actual intersections, we estimate their size reusing the already computed minhash signatures of the lists and a novel adaptation of the inclusion-exclusion principle for the Jaccard similarity measure, as we discuss in Sect. 2.

Frequent Itemset Mining for Candidate Generation: A derived list is actually a set of frequent items in the index. Unfortunately, a straightforward implementation of the well known Frequent Itemset Mining *A-priori* algorithm [3] will not work in our setting. In applications like the ones we described the goal is to find frequent itemsets (derived lists) of very large cardinality which can be in the order of thousands or millions, a number that corresponds to the number of iterations the *A-priori* algorithm would have to perform. Clearly, this process will not terminate in our setting. Instead we propose a novel adaptation that seeks to find "frequent" itemsets (derived lists) that are contained in the intersections of lists. In this new setting, the algorithm iterates over the number of intersecting lists and not their sizes. Thus, while *A-priori* in the kth iteration finds frequent itemsets of size k, our adaptation finds frequent itemsets produced by intersecting k lists. Furthermore, unlike the original *A-priori* we are not given any monotonicity guarantees, that is if the intersection of k lists qualifies for use in the compression of the index then we cannot say that any subset of them also qualifies. The reason behind this is that in contrary to the *support* of an itemset which is just a count of many times the itemset exists in the dataset, the support (or gain in our terminology) of a derived list is a function of two variables (number of lists, number of items in the intersection of these lists), one increasing and one decreasing, while the algorithm augments the existing candidates. We discuss the adaptation of the *A-priori* algorithm in order to address these shortcomings in Sect. 3.

Conflicting Derived Lists: In Sect. 2 we explain that the candidate derived lists from a dataset cannot all be used for compression at the same time due to conflicts among them. Briefly, a conflict is present among a pair of derived lists when they have been constructed by using at least one common list and they also include at least one common item. Then using both of these lists for the compression of the index would result in duplicate items. The challenge is to select a subset of the candidate derived lists that presents no conflict and at the same time maximizes the compression ratio. In order to address this challenge we propose a heuristic based on an Integer Linear Programming modelling of the problem, which is presented in Sect. 4.

2 The Derived List

Given a set of lists $\mathcal{L} = \{l_1, l_2, \ldots, l_m\}$, a derived list dl is constructed by taking the items in the intersection of a subset $\mathcal{L}'$ of $\mathcal{L}$. We refer to the set $\mathcal{L}'$ as the base of dl and denote it as $dl.base$. We call $dl.items$ the set of items in the intersection of the lists in $dl.base$, and we refer to the number of items as $dl.size$. Last, we define a metric $dl.gain$, which is the gain (reduction in number of items to store) that the use of this derived list adds to the compression of the index. Assume a derived list dl and it's n base lists $l_1, l_2, \ldots, l_n$, where $n \geq 2$ ($n = |dl.base|$). If we select to materialize dl (e.g. write it to disk), then we can use it in each of it's base lists in the place of the original items, substituting them by a reference to dl. That essentially means that we are using only one copy of these items instead of n copies. Thus, when used in all of it's base lists, the gain of the dl (reduction of index size) is:

$$dl.gain = (|dl.base| - 1) \times dl.size \tag{1}$$

In order to compute the gain of a derived list we need to know the number of its base lists and the number of items in their intersection. Due to the way that we construct the derived lists, we know beforehand the number of base lists. That means that we just need to estimate the size of their intersection so as to estimate the derived list's gain. As we show next this can be done by calculating the Jaccard similarity of the lists using their already computed minhash signatures and a result based on the inclusion-exclusion principle.

Jaccard Similarity: Jaccard similarity is a measure of sets similarity, defined as the ratio of the size of their intersection over the size of their union. Given the sets $A_1, A_2, \ldots, A_n$, we compute their Jaccard similarity $JS(A_1, A_2, \ldots, A_n)$ as:

$$JS(A_1, A_2, \ldots, A_n) = \frac{|A_1 \cap A_2 \cap \cdots \cap A_n|}{|A_1 \cup A_2 \cup \cdots \cup A_n|} \tag{2}$$

Minhashing: Given a large collection of sets with values from a domain, min-hashing [4] is the process of constructing a small signature for each one of them by applying a series of hash functions to each of the set elements. The min-hash signatures present a very interesting property: given two sets A and B and their respective minhash signatures $mh(A)$ and $mh(B)$, the probability that $mh(A) = mh(B)$ equals to the Jaccard similarity between A and B.

Inclusion-Exclusion Principle: From the inclusion-exclusion principle [5] we know that we can calculate the union of a number of finite sets $A_1, \ldots, A_n$ as

$$\left| \bigcup_{i=1}^{n} A_i \right| = \sum_{k=1}^{n} (-1)^{k+1} \left(\sum_{1 \le i_1 < \cdots < i_k \le n} |A_{i1} \cap \cdots \cap A_{ik}| \right) \tag{3}$$

Size Estimation of Set Expressions: Assume n finite sets $A_1, \ldots, A_n$. We want to estimate the size of their intersection $|A_1 \cap A_2 \cap \cdots \cap A_n|$, given their respective minhashes $mh(A_1), \ldots, mh(A_n)$ and sizes $|A_1|, \ldots, |A_n|$. Let $S(A_1, \ldots, A_n)$ be the Jaccard similarity of sets $A_1, \ldots, A_n$:

$$S(A_1, \ldots, A_n) = \frac{|A_1 \cap A_2 \cap \cdots \cap A_n|}{|A_1 \cup A_2 \cup \cdots \cup A_n|} \tag{4}$$

The Jaccard similarity can also be estimated using the set's minhashes with strong guarantees as:

$$S(A_1, \ldots, A_n) = \frac{mh(A_1) \cap mh(A_2) \cap \cdots \cap mh(A_n)}{mh(A_1) \cup mh(A_2) \cup \cdots \cup mh(A_n)} \tag{5}$$

We continue by using the inclusion-exclusion principle (Eq. 3) so as to recursively compute the intersection of the given sets. For $n = 2 * m$, the union of the n sets can be written as $U^- - a$, where U^- is the size of the union of the n sets minus the last term of the sum:

$$U^- = \sum_{k=1}^{n-1} (-1)^{k+1} \left(\sum_{1 \le i_1 < \cdots < i_k \le n-1} |A_{i1} \cap \cdots \cap A_{ik}| \right) \tag{6}$$

This last term, which is not present in U^-, is represented by a and is actually the size of the intersection of the n sets ($a = |A_1 \cap A_2 \cap \cdots \cap A_n|$). Then, the Jaccard similarity of the n sets can be expressed as:

$$S(A_1, \ldots, A_n) = \frac{a}{U^- - a} \tag{7}$$

and, thus, the intersection a is derived from the following formula:

$$a = \frac{S(A_1, \ldots, A_n)}{1 + S(A_1, \ldots, A_n)} \times U^- \tag{8}$$

In the case that $n = 2 * m + 1$, the respective formula is:

$$a = \frac{S(A_1, \ldots, A_n)}{1 - S(A_1, \ldots, A_n)} \times U^- \tag{9}$$

We can easily compute the Jaccard Similarity S for any combination of sets from their minhashes using Eq. 5. We compute U^- recursively using Eq. 6 and estimate the intersection size of the sets using Eq. 8 (or 9).

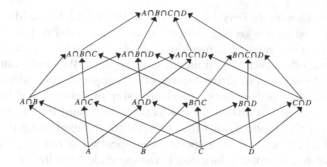

Fig. 1. The derived lists Lattice for lists A, B, C, D

Conflicting Derived Lists: In Fig. 1 we depict the derived lists lattice that can be constructed from four example base lists A, B, C and D. The derived lists are connected with directed edges that form paths from the lower level (trivial derived lists with one base list) to the higher level (a single derived list constructed by intersecting all base lists). When there is a path connecting two derived lists these derived lists conflict, meaning that the use of one of them for compression suggests that we cannot use the other. This is explained from the fact that since they are on the same path they share at least one base list and also that the lower level one contains all of the items in the higher level one since the latter is constructed by further intersecting the former. However, this is not the only case where two derived lists conflict. Any pair of derived lists that share at least a base list and at least one item also conflicts. The important distinction between the two cases is that in the first case we can infer that two derived lists conflict without computing their intersection, while in the second case we have to compute their intersection in order to decide whether they conflict. We collectively refer to all the conflicting derived lists of a derived list dl by $dl.relatives$.

3   Candidates Generation

Candidates generation aims at discovering a set of candidate derived lists. Given a set of lists $\mathcal{L}$, at the end of the candidate generation phase we have populated a set of candidate derived lists $\mathcal{DL}$ described not by the actual items they include but by their base lists and their estimated gain.

Preprocess - Minhashing - LSH: The input data is the set of lists $\mathcal{L}$ in the inverted index. For each $l \in \mathcal{L}$ we compute it's minhash signature, and count it's size (number of items in l). At the same time we apply the LSH technique for minhashes [6] so as to populate buckets of similar lists. The buckets are populated with triplets of the form $(l.id, l.minhash, l.size)$. All these tasks are parallelizable and can be efficiently computed for large datasets using modern distributed processing frameworks such as Spark or MapReduce.

Generating the Candidate Derived Lists: For a derived list to qualify we demand that it's computed gain is greater than or equal to a user-defined *mingain* value. The default *mingain* value is 1, meaning that we add to the candidates set any derived list that offers even the minimum gain possible. We model our candidate generation problem as an intuitive variant of a frequent itemset mining problem, where we seek to find "itemsets" (derived lists) by intersecting base lists as we move upwards in the lattice. In this setting we would like to find itemsets (derived lists) with support (gain) that exceeds the value of mingain. While this adaptation is intuitive, there is a specific complication in our setting. The gain function is not a monotonous function because, while "moving" from the lower levels of the lattice to higher ones, it is computed as the product of an increasing function ($|dl.base| - 1$) by a decreasing one ($dl.size$), as intersecting more lists results in a new derived list of smaller or equal size. What this actually means is that the A-priori principle does not hold in our case. In order to overcome this problem, we are using a user-defined upper bound in the number of base lists that a derived list can have which we call k_{max}. That way we set a bound to the first term of Eq. 1 (number of base lists). Parameters k_{max} and *mingain* can be tuned so that the system will favor either speed of execution, or compression ratio. Then, we solve Eq. 1 for the second term, which is the size of the derived list. It follows that for a qualifying derived list dl of kth order (having k base lists) it must hold that:

$$dl.size \geq \frac{mingain}{k-1} \tag{10}$$

Then, for $k = k_{max}$ it must hold that:

$$dl.size \geq \frac{mingain}{k_{max} - 1} = minsize \tag{11}$$

where *minsize* is the minimum size a derived list can be of so as to benefit us with gain larger than *mingain*. This conclusion is justified by the fact that the size of a derived list dl_1 that is the intersection of $n \geq 2$ lists is less or equal to the size of a derived list dl_2 that is the intersection of any combination of m of those lists ($1 \leq m < n$). With this modification, our candidate generation problem reduces to the problem of frequent itemset mining and can be solved using e.g. the A-priori algorithm within each LSH bucket. An important difference that we must point out is that *minsize* is the minimum size for a derived list of order k_{max} to qualify, but since k_{max} is the highest order we use it as an overall minimum gain. The reason we use *minsize* for all orders in a first pass is that a derived list always reduces in size (or stays the same) when adding a new base list. In a second pass over the set of produced candidates within the bucket, we calculate the actual minimum size for each order and keep only the truly qualifying lists (e.g. those with $dl.gain \geq mingain$). As has been discussed, an important property of our technique is that all calculations are performed on the compact statistics that we maintain for each derived list, without looking at the original dataset. An optimization that we apply during candidate generation is the early pruning of candidates within a bucket by removing any of them that can be inferred to be a relative of another one with higher gain.

4   Derived Lists Selection

The input to this stage is the set of candidate derived lists generated in the previous phase. Each candidate derived list dl is described as a tuple ($dl.base,dl.gain$) where $dl.gain$ is an estimation of the gain as computed previously. Our goal is to select a non-conflicting subset of the candidate derived lists which maximizes the gain and then use it to compress the index. A natural choice for this kind of problems (selection-maximization) is ILP (Integer Linear Programming) [7].

ILP Formulation of the Problem: Let A be a matrix having as many rows as many conflicts exist among the candidates and one column for each candidate. For each conflict we add to A a row with the value 1 at the columns of the two conflicting candidates and 0 everywhere else. Also, let w be a vector of length equal to the number of candidates where $w[i]$ is the gain of the ith derived list candidate. Finally, assume a solution vector x of length equal to the number of candidates with each position i taking a value of 1 or 0, respectively meaning that the ith candidate is part or no of the solution represented by x. We want to find the vector x that maximizes the gain $x : max(\sum w \cdot x)$ restricted by the rule $A \cdot x \leq 1$ so that there are no conflicts in the solution. If we had the processing power required to solve the problem for all the candidates at once, it would suffice to feed A and w to an ILP solver and get the optimal solution in one step. In most practical cases, this would not be feasible when the number of candidates is in the order of thousands or millions. In such cases, we propose a greedy heuristic that uses the ILP approach, as described next.

Greedy Solver Algorithm: We load the candidate derived lists in decreasing order of their gains in a priority queue PQ, and we initialize an empty set *globalSolution* to store the qualifying candidates. Then, depending on the ILP solver capabilities, we select a value N which is the number of candidates we shall use at each iteration. We get the top-N candidates out from the PQ, construct the corresponding matrix A and vector w, and pass them to the solver. The solution returned by the solver is a non-conflicting set of candidates that locally maximizes the gain for the candidates that it examined. We add this local solution to *globalSolution* and remove from PQ any candidate that conflicts with any of the derived lists in the solution. Then, we start over using the next top-N candidates until PQ is empty or a desired compression ratio has been achieved.

5   Experimental Evaluation

We implemented our method using Apache Spark, Hadoop and HBase running on a small test cluster consisting of nine Linux virtual machines. The data used in our experiments were synthetically generated based on 20 publicly available real supply chain networks selected from [8]. Each supply chain is a directed acyclic graph and its nodes can be distinguished in source, inner and sink nodes. We created RFID data for each supply chain using a custom data generator, which loads the topology graph and the available average

Table 1. Supply chains statistics

Id	#nodes	#source nodes	#inner nodes	#sink nodes	#edges	#paths	Max path length	#rfid records
1	40	12	26	2	48	22	8	7636760
2	152	21	33	98	211	1157	5	3037082
3	154	49	77	28	224	172	8	5122505
4	156	44	97	15	263	528	9	3956007
5	156	74	80	2	169	282	10	5132532
6	186	76	76	34	359	772	7	4224544
7	271	198	48	25	524	486	3	2533934
8	334	209	83	42	1245	4055	6	4119465
9	409	94	142	173	853	1158	3	2999721
10	468	401	65	2	605	579	6	2132407
11	482	418	52	12	941	889	5	3222313
12	577	398	89	90	2262	15181	8	3430016
13	617	128	124	365	753	3789	5	3490857
14	626	1	405	220	632	227	5	3553438
15	844	309	313	222	1685	2814	5	3283238
16	976	119	525	332	1009	1688	8	5799807
17	1206	1148	5	53	4063	4320	3	2001061
18	1386	619	731	36	1857	1140	6	5041132
19	1479	274	646	559	2069	6062	4	3998868
20	2025	820	646	559	16225	97085	4	3998940

demand values for the terminal nodes, and then, using this information, performs a number of random graph walks simulating tagged objects moving through the network. The generated walks are used to produce RFID records of format $(tagId, locationId, timestamp)$. Then, using these data, we create location-based indexes as inverted lists of the form $locationId \leftarrow list(tagId)$. Table 1 describes the supply chains and the dataset produced by performing one million graph walks on each of them. Because of the variable number of nodes and paths in each of the supply chains, the number of RFID records is not the same for all of them, as is depicted in the table.

Comparison with Frequent Itemset Mining (FIM) Algorithms: Our technique is aiming at the efficient discovery of common itemsets in a collection of lists. The most well known algorithms for such tasks are A-priori [3], FP-growth [9] and Eclat [10]. While these algorithms focus on the discovery of frequent itemsets in a large number of relatively small transactions, we are instead mostly interested in discovering large itemsets which are not necessarily very frequent. These algorithms are not optimized for computing very large itemsets that often arise by

Table 2. Comparison of FIM algorithms and our Approx-A-priori

Number of records		Time (ms)			
# graph walks	# rfid records	A-priori	FP-growth	Eclat	LSH+Candidate generation
10	13	4	5	2	628
100	429	24	404,784	405,150	1,242
1,000	5,050	NA	NA	NA	1,253
10,000	51,270	NA	NA	NA	1,538
100,000	512,100	NA	NA	NA	3,867
1,000,000	5,122,505	NA	NA	NA	33,254

the intersections of very long lists. We used the implementations of A-priori, FP-growth and Eclat from [11] to demonstrate their inefficiency in such a setting, and the results are depicted in Table 2 (NA means that the respective algorithm either exited due to insufficient memory or space, or that it did not finish after running for twenty four hours). For fairness, all algorithms were executed in a single machine. Our method, by utilizing approximation and our novel adaptation of the A priori algorithm within each LSH bucket, manages to scale to datasets orders of magnitude larger than the other algorithms do.

Compression Ratio Achieved: The compression ratio indicates the reduction in the size of the index due to the use of our techniques. Let $\mathcal{DL}$ be the set of the selected derived lists and $\mathcal{L}$ the lists in the original index. We define compression ratio as $\sum_{dl \in \mathcal{DL}} dl.gain / \sum_{list \in \mathcal{L}} |list.items|$. For these experiments we set $k_{max}=5$ and *mingain*=1. The value of the compression ratio achieved in each supply chain is presented in Fig. 2(a). We observe an average compression ratio of 49 % and we can see compression ratios well above 50 % (maximum is 74 %).

Execution Time: We measured the execution time for the three discrete stages of the method (LSH, candidate generation, compression). In Fig. 2(b) we present the execution times for the LSH and the candidate generations stages. The time required for LSH is related to the number of RFID records generated for each supply chain, while the time required for the candidate generation phase is proportional to the number of total nodes contained in each supply chain. The execution times for running *Greedy Solver* to compress the indexes are depicted in Fig. 2(c). LSH and candidate generation are parallelizable tasks and this results in fast execution times. Selection of derived lists and compression of the index is partly parallelizable and also performs heavy read operations while retrieving the original lists, so these operations are slower. We observe that the time needed to perform the compression of the index is proportional to the number of paths that exist in the network. A higher number of paths results in the creation of a larger number of candidates with more complicated relations.

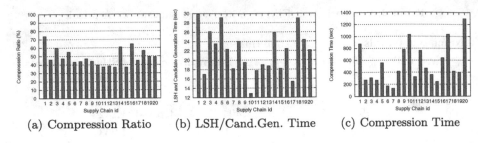

|(a) Compression Ratio | (b) LSH/Cand.Gen. Time | (c) Compression Time |

Fig. 2. Compression ratio and execution time for 20 supply chain networks

Effect of Parameters k_{max} and Mingain: In order to examine how different values of k_{max} affect performance and outcome we ran a new set of experiments. We selected four out of the twenty supply chains and varied k_{max} from 2 to 7 ($mingain=1$). We observed that lower values of k_{max} are not desired since in some cases result in lower compression ratio with no gain in execution time. For higher values of k_{max} the required time decreases or remains on the same levels while someone would instead expect an increase. This behavior is explained by the early pruning optimization discussed at the end of Sect. 3. Next we varied *mingain* value from 1 to 2000, keeping $k_{max} = 5$. As *mingain* increased, the fewer the generated candidates were. For most of the datasets, as *mingain* increased, the overall execution time decreased, but compression ratio remained at levels near the ones achieved by a value of 1. This is a consequence of the fact that the derived lists selected by *Greedy Solver* are the ones with the more gain and, thus, the ones with small gains are not likely to be selected anyway.

6   Conclusions

In this paper we presented a framework for compressing large inverted indexes built on low cardinality domains. Such indexes are common in many applications of interest and tend to produce lists that share large sets of common item references. Our techniques are able to discover the most promising of these common sets in a single pass over the original lists by utilizing novel dimensionality reduction and approximation techniques. We complemented our method with a greedy heuristic that uses an intuitive ILP formulation in order to select a subset of these common sets so as to construct a solution that maximizes the compression ratio. Finally, we implemented our framework using modern big data tools and used it for compressing data generated on real supply chain networks. As future work, we plan to explore ways to adopt our method so as to support incremental updates of the compressed indexes on streaming data.

References

1. Bleco, D., Kotidis, Y.: RFID Data Aggregation. In: Trigoni, N., Markham, A., Nawaz, S. (eds.) GSN 2009. LNCS, vol. 5659, pp. 87–101. Springer, Heidelberg (2009)

2. Bleco, D., Kotidis, Y.: Business intelligence on complex graph data. In: Proceedings of the 2012 Joint EDBT/ICDT Workshops, EDBT-ICDT 2012, pp. 13–20. ACM, New York (2012)
3. Agrawal, R., Srikant, R.: Fast algorithms for mining association rules in large databases. In: Proceedings of the 20th International Conference on Very Large Data Bases, VLDB 1994, pp. 487–499. Morgan Kaufmann Publishers Inc., San Francisco (1994)
4. Broder, A.Z., Charikar, M., Frieze, A.M., Mitzenmacher, M.: Min-wise independent permutations. J. Comput. Syst. Sci. **60**(3), 630–659 (2000)
5. Knuth, D.E.: The Art of Computer Programming, vol. 1 (3rd Ed.): Fundamental Algorithms. Addison Wesley Longman Publishing Company Inc, Redwood City (1997)
6. Rajaraman, A., Ullman, J.D.: Mining of massive datasets. Cambridge University Press, Cambridge (2012)
7. Papadimitriou, C.H., Steiglitz, K.: Combinatorial Optimization: Algorithms and Complexity. Prentice-Hall Inc, Upper Saddle River (1982)
8. Willems, S.P.: Data set–real-world multiechelon supply chains used for inventory optimization. Manufact. Serv. Oper. Manage. **10**(1), 19–23 (2008)
9. Han, J., Pei, J., Yin, Y.: Mining frequent patterns without candidate generation. In: Proceedings of the 2000 ACM SIGMOD International Conference on Management of Data, SIGMOD 2000, pp. 1–12. ACM, New York (2000)
10. Zaki, M.J.: Scalable algorithms for association mining. IEEE Trans. Knowl. Data Eng. **12**(3), 372–390 (2000)
11. Viger, P.F., Gomariz, A., Gueniche, T., Soltani, A., Wu, C.W., Tseng, V.S.: SPMF: a java open-source pattern mining library. J. Mach. Learn. Res. **15**, 3389–3393 (2014)

A Decision Support System for Hotel Facilities Inventory Management

Giuseppe Monteleone, Raffaele Di Natale$^{(\boxtimes)}$, Piero Conca,
Salvatore Michele Biondi, Antonio Rosario Intilisano,
Vincenzo Catania, and Daniela Panno

Dipartimento di Ingegneria Elettrica Elettronica Informatica,
Università degli Studi di Catania, Catania, Italy
{gmonth,piero.conca}@gmail.com, {raffaele.dinatale,
sbiondi,aintilis,vincenzo.catania,
daniela.panno}@dieei.unict.it

Abstract. A major goal of a tourism supply chain is a profitable collaboration between actors involved. Small hotel facilities tend to order small amounts of each good. The unit cost is generally unfavorable compared to that of large hotel facilities. To overcome this disadvantage, small facilities can collaborate by placing aggregate orders to a single vendor. Consequently, the increased quantity ordered can afford a unit cost reduction. This paper investigates the effectiveness of a set of novel demand forecast techniques for supporting this order aggregation. We describe four different algorithms: they all use the orders' history; in addition, two of them forecast the numbers of guests. The performed tests use large amount of anonymous real-world data and show that the algorithms that also use the numbers of guests performs better than those based only on the orders' history.

Keywords: Decision support system · Tourism supply chain · Inventory management · Forecasting

1 Introduction

The purpose of this work is to define a Decision Support System (DSS) [1] for hotel facilities. The DSS is designed to be part of a web application (running in the cloud) that will implement a volume discount environment [2].

Tourism supply chain requires hotel facilities to stock up goods needed for carrying out the usual treatment for guests [3]. Usually, small facilities tend to order small amounts of each good; consequently, the unit cost is generally unfavorable if compared to that of larger orders placed by big hotel facilities. In order to overcome this disadvantage, many small facilities could join to place aggregate orders for the same item to a unique vendor [4, 5] (Fig. 1). In such a way, a vendor, thanks to an increased total order quantity, may offer a cheaper unit cost. In this way, the described marketplace could allow small hotel facilities to compete with the larger ones. On the other side, supply providers could offer lower prices by selling large item quantities and potentially enlarge their number of customers. In order to carry out the selection of a vendor

© Springer International Publishing Switzerland 2015
Q. Chen et al. (Eds.): DEXA 2015, Part I, LNCS 9261, pp. 460–470, 2015.
DOI: 10.1007/978-3-319-22849-5_31

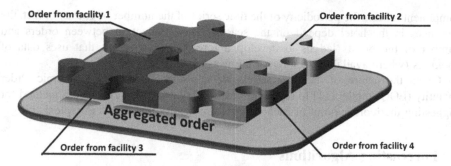

Fig. 1. Aggregated order from different facilities.

and the creation of groups of multiple-buyers, the system has to know the needs of the facilities concerning the goods. The DSS is designed to accomplish this task, by forecasting new orders of a given good and advertising them to the facilities. The hotel facilities can then accept or modify the suggested new orders.

For the design and the test of the DSS, we used anonymous data of real hotel facilities. The data analysis is discussed in Chapter 2. The DSS uses demand forecasting techniques [6] to estimate the time point in which a facility may need to replenish the stock of a good. These are forecasting-based techniques [7, 8], of which two are novel algorithms specifically developed for the context of the tourism supply chain. In particular, these novel algorithms forecast the number of the guests that a hotel facility will host. The algorithms are presented in Chapter 3. Their novelty consists in the correlation between the orders quantity and the number of the guests. No other similar works were found in literature review, because there has been a lack of research done for hotel supply chain management [9]. The availability of different algorithms allows the system to employ the most fitting for a given item. In order to evaluate our work, we present, in Chapter 4, the comparison between the performances of these different types of algorithms.

2   Data Analysis

The whole system design and the testing of the forecasting algorithms were performed by analyzing and processing data consisting of time series of orders placed by anonymous hotel facilities. Every order placed by the facilities was tracked by an online system. Each order contains information such as item description, quantity, price and date. In addition, data of the number of guests hosted each day were also available for most of the facilities. The data does not contain any kinds of information concerning lead times and safety stocks, neither presents a history of inventory quantity.

Most of the goods ordered by the facilities were perishable, in particular food for catering service. The trend of the orders for this kind of goods is different from the trend of non-perishable ones [10]. In fact, the former is mostly irregular in frequency and quantity while the latter is more regular. Nonetheless, a seasonal component is present in the majority of the time series of orders. In particular, we noticed that such component reflects the periodicity of the time series of the number of guests. In fact, the presences in the hotel depend on the season. This connection between orders and number of the guests led us to develop a forecasting algorithm that uses data of bookings (where available) to forecast the usage rate of a good.

Given the nature of the time series, we could not apply the Economic Order Quantity (EOQ) model [11] to represent orders. However, we developed a method for suggesting the reorder point of each item to any facility that needs to replenish it.

3 Forecasting Algorithms

The aim of this work is to deliver a flexible system to suggest reorder points in different conditions, with or without an historical track of the orders and the number of the guests. In particular, the DSS uses four algorithms, which differ for the type of information they use (Table 1). All four algorithms perform the estimation of the *daily usage rate* of a given item in the period that follows a new order. In order to provide a forecast when the stock limit occurs and a new order must be placed (to avoid stock out), the daily usage rate is evaluated. Because of missing information from the available data (lead times, safety stocks and inventory reviews), we had to make some assumptions. In particular, we consider that, between an order and the next one, the item quantity decreases linearly until total depletion. Therefore, a new order replenishes the inventory to its maximum quantity.

Table 1. Forecasting algorithms characteristics summary

Algorithm	Last two order quantities	Last year order quantity history	Booking track record
I Simple moving average	✓		
II Order history based	✓	✓	
III Simple moving average with the number of guests	✓		✓
IV Order history based with number of guests	✓	✓	✓

3.1   First Algorithm (Simple Moving Average)

The simplest of all four algorithms only uses information about the last two orders. In general, for a given item, the daily usage rate of the ith order and the following order can be computed as:

$$C_i = \frac{Q_i}{(T_{i+1} - T_i)}. \tag{1}$$

where:

- C_i is the daily usage rate for the given item after the ith order
- Q_i is the order quantity for the ith order

- T_i and T_{i+1} represent, respectively, the days in which the ith order and the next one are placed;

When a new order is placed, the forecast of the daily usage rate is computed as the average of the two previous daily usage rates [7, 8]. The estimation of the new usage rate, for the kth order, is calculated as follows:

$$C_k = 0.5 * (C_{k-1} + C_{k-2}). \tag{2}$$

where C_{k-1} and C_{k-2} are the daily usage rate for the $(k-1)^{th}$ and $(k-2)^{th}$ orders computed by means of Eq. (1). This estimation requires a history containing at least two orders. Once the daily usage rate is estimated, the maximum number of days before the depletion of the last order quantity can be forecasted as:

$$FT_{k+1} = \frac{Q_k}{C_k}. \tag{3}$$

Then, depending on lead-time and safety stock, FT_{k+i} can be used to forecast the next reorder point for a given item.

3.2   Second Algorithm (Order History Based)

The second algorithm can be used when an history of orders of at least one year is available. Based on the order history of the previous year, this technique accounts for the seasonal components of the time series of orders.

Given a period of days T, the daily usage rate for a new order is computed as follows [7, 8]:

$$C_i = C_i' * \left(\frac{C_{i-1}}{C_{i-1}'}\right). \tag{4}$$

where:

- $T = 30$ (days)
- $C_i' = \frac{\sum Q_k'}{T}$
  - C_i' is the estimated daily usage rate related to the previous year of the T days following the day of the ith order;
  - Q_k' are the order quantities placed in the next T days after the ith order during the previous year (obtained from order history);
- $C_{i-1} = \frac{\sum Q_j}{T}$
  - C_{i-1} is the estimated daily usage rate of the T days immediately before the day of the ith order
  - Q_j are the order quantities placed in the T days before the ith order

- $C'_{i-1} = \frac{\Sigma(Q'_j)}{T}$

  - C'_{i-1} is the estimated daily usage rate related to the previous year of the T days immediately before the day of the ith order
  - Q'_j are the order quantities for the previous T days before i^{th} order during the previous year (obtained from order history)

Once C_i is computed, the maximum number of days before stock depletion is forecasted based on Eq. (3).

3.3 Third Algorithm (Simple Moving Average with Number of the Guests)

The available data also contained the track record of reservations and booking confirmations for many of the anonymous facilities. This led us to design a novel algorithm that exploits the number of guests present in the hotel to estimate the daily usage rate. By correlating the usage rate to the presences in the facility, the system can compute the *daily usage rate per person*. For past orders, the daily usage rate per person is obtained by dividing the order quantity by the total presences in the facility during the days included between the ith order and the next one.

$$CP_i = \frac{Q_i}{\sum_{y=T_i}^{T_{i+1}} P_y}. \tag{5}$$

where:

- CP_i is the daily usage rate per person
- Q_i is the ith order quantity
- P_y is the total number of guests for *day y* (y is a day included between T_i and T_{i+1})
- $\left(\sum_{y=T_i}^{T_{i+1}} P_y\right)$ represents the total guests, summed up day by day, during the period between the ith order and the next one $i + 1$

The estimation of the daily usage rate per person for a new order k is achieved by means of simple moving average (2), but in this case C_{k-1} and C_{k-2} represent the daily usage rate per person and they are computed with (5).

$$CP_k = 0.5 * (CP_{k-1} + CP_{k-2}). \tag{6}$$

Once the daily usage rate per person is estimated, the daily forecasted presences in the hotel facilities are needed. We developed a dedicated algorithm for the forecasts of the numbers of the guests in the facility (**see paragraph 3.5, Forecasting pick-up**). For a given day, by multiplying the daily usage rate per person by the daily number of guests, we obtain the daily usage rate. This daily usage rate differs from a day to another.

$$C_y = CP_k * FP_y. \tag{7}$$

where:

- C_y is daily usage rate for the given y *day*
- CP_k is the daily usage rate per person
- FP_y is the forecasted number of guests for the y *day*

Finally, in order to estimate the number of days before a new order quantity Q_i is ordered, on day T_i, the following rule is used:

$$FT_{i+1} = \max\{t \mid t > T_i\}: \quad Q_i - \sum_{y=T_i}^{t} (CP_i * FP_y) \geq 0. \tag{8}$$

where:

- FT_{i+1} is the number of days forecasted before the depletion of the last ordered quantity Q_i
- CP_i is the daily usage rate per person computed by means of (6)
- $CP_i * FP_y$ represents the daily usage rate for day y (7); these products are summed up day-by-day, as long as the sum is less than the order quantity.

3.4   Fourth Algorithm (Order History Based with Number of the Guests)

The fourth algorithm consists of a mixture of the second and third ones. It uses all the information available, namely, the booking track record and the previous year order history. In this case, the daily usage rate per person is computed considering the order history. Daily usage rate per person of previous periods are computed on intervals of T days.

$$CP_i = CP'_i * \left(\frac{CP_{i-1}}{CP'_{i-1}}\right). \tag{9}$$

where:

- $T = 30$ (days)
- $CP'_i = \dfrac{\sum Q'_k}{\sum_{y=T_i}^{T} P'_y}$

  - CP'_i is the estimated daily usage rate per person during the previous year in the T days immediately following the day of the ith order
  - Q'_k are the order quantities placed in the next T days after the ith order during the previous year (obtained from order history)
  - P'_y are the daily presences in the facility for the T days flowing the ith order during the previous year (obtained from booking track record history)

- $CP_{i-1} = \dfrac{\sum Q_j}{\sum_{y=T_i-T}^{T_i} P_y}$

  - CP_{i-1} is the estimated daily usage rate per person during the previous T days immediately before the day of the order
  - Qj are the order quantities placed in the T days before the ith order
  - P_y are the daily presences in the facility for the T days before the ith order

- $CP'_{i-1} = \dfrac{\sum (Q'_j)}{\sum_{y=T_i-T}^{T_i} P'_y}$

  - CP'_{i-1} is the estimated daily usage rate per person during previous year in the T days immediately before the day of the ith order
  - Q'_j are the order quantities for the previous T days before i[th] order during the previous year (obtained from order history)
  - P'_y are the daily presences in the facility for the T days before the ith order during the previous year (obtained from booking track record history)

Once CP_i is computed, the maximum number of days before stock depletion is forecasted with (8).

3.5   Forecasting Pick-up

The following algorithm, designed to perform the pick-up forecasting in a hotel facility, uses the clustering technique of K-means [13]. Many factors may influence the booking trends of a hotel facility: one of the most important is the seasonality but also the external events such as conferences, ceremonies, etc. The forecasting of this kind of event is very complex due to its sporadic nature.

In this study, we used the analysis of the time series of the booking records of the same anonymous hotel facilities described before: it shows the importance of the seasonal component that highly influences the booking distribution along the seasons. Seasonality is a characteristic of time series, we model it and we use it as the base of the algorithm being proposed. The algorithm uses the K-Means Clustering, whose initial goal is to define four clusters: they represent the four different seasonal periods by which we classify the booking curves.

For the forecasting algorithm, a booking curve a fixed day is a row containing the sum of the booking record starting from 365 before this date. This row describes also other information like the month, the day and the week of the month, useful to define the classification of the booking curves (Fig. 2).

The algorithm forecasts the trend of bookings using the 4 groups created by K-means by the subdivision of the Dataset. For a not-reached Date, we obtain the cumulative bookings curve still incomplete (Fig. 3).

This curve will be used as input to the K-means clustering algorithm, which will generate a cluster to which the curve belong. After the selection of a cluster, all its rows (curves) are retrieved by getting the ith (days before of date for making forecasting) element for each row that will be saved in an array.

1 Day Before	..	28 Days before	29 Days before	..	364 Days before	365 Days before	Day of week	Week of Month	Month
25	..	10	5	..	0	0	7	2	8
30		15	7		0	0	1	3	12
20		10	7		1	0	3	4	1

Fig. 2. Dataset example (forecasting curve/row)

1 Day Before	..	28 Days before	29 Days before	..	364 Days before	365 Days before	Day of week	Week of Month	Month
?	..	?	4	..	0	0	3	4	5

Fig. 3. Incomplete cumulative bookings curve

The array is used to retrieve the maximum value (*ValMax*) and the minimum value (*ValMin*) in order to provide a range of values in which the curves of the cluster, selected by the algorithm, are distributed for the *I* days before of forecasting.

The forecasted value, that fills one missing point in the incomplete cumulative booking curve *R*, will be chosen between the values contained in the range with the restriction that the value must be greater than or equal to the previous value because the curve is cumulative.

The following algorithm defines the criteria for the forecasting:

```
If(ValMin > R[I+1])
    Prev = ValMin
else if((ValMin+ValMax)/2) > R[I+1])
    Prev =((ValMin+ValMax)/2)
else if(ValMax > R[I+1])
    Prev = ValMax
else
    Prev= R[i+1]
```

The choice of the forecasting value can be further improved, dividing the interval into several parts, by comparing the previous value (*R [I + 1]*) with each individual part. This way, you will avoid carry over estimates that would determine the value assignment *ValMax* to forecasted value.

4 Results

The performances of the four algorithms were tested using all the available order records. The booking data was not available for all the hotel facilities and not all the order time series contained more than one year of order history. For this reason, we selected data from nine facilities, which contain the booking track record; each set is composed of a different number of items and orders. We tried to forecast the reorder point for each order in the available history and compared it to the actual day of the next order (if available). The metric chosen for the evaluation of the algorithms is the Mean Absolute Percentage Error (MAPE) [12]. This mean is computed across the results of every item for each hotel facility. Figure 4 sums up the results.

The test results show that the algorithms III and IV, which use the information about the daily presences in the hotel facilities, perform slightly better than algorithm II in most cases (order lines) we have considered. Algorithm I has the worst results; it does not use any data regarding previous year history and number of guests, hence it does not possess any information regarding the seasonal components of the orders time series. In fact, almost all order lines presents an increased number of orders during peak season. By considering previous year history and the presences in the facilities, the other three algorithms have additional information (related to season) regarding the demand of goods for the considered next period.

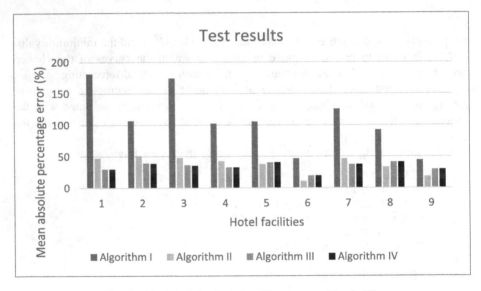

Fig. 4. MAPE of the four algorithms grouped by facility

5 Conclusion

This paper has presented a DSS conceived to help small hotel facilities to group together in order to place larger orders for a given good with the goal of decreasing unit cost. In particular, such DSS helps facilities to determine an appropriate reorder point for each good and this information can be used to aggregate orders of different facilities. The aggregation process could be facilitated by the implementation of an interoperable cloud platform with the purpose of connecting different hotel facilities and vendors.

Four different algorithms for forecasting reorder points have been presented. Preliminary tests show that the error of such a prediction can reach 100 %. This result is related to the dynamicity of the time series employed, which makes their forecasting a complex task [14]. In order to improve the performance of forecasts, we have used booking data in order to predict the occupancy and therefore the rate of consumption. This approach has led to satisfactory results. Moreover, we must remark that the results we obtained consider the discrepancy between the actual reorder point and its forecast. Moreover, we expect that the users of such a DSS, by using the system, will learn from it and, therefore, this should produce a further reduction of errors.

6 Future Works

The design of a model for orders aggregation and vendor selection is the next step of our work. This model will use the reorder point suggested by the DSS for finding optimal combination, in terms of unit cost, of multiple orders from a selected vendor.

Acknowledgments. Authors were in part supported by the Sicily region grants PROG-ETTO POR 4.1: "CLOUTER A Tourism Terminal in the Cloud".

References

1. Power, D.J.: A brief history of decision support systems. DSSResources. COM, World Wide Web (2007). http://DSSResources.COM/history/dsshistory.html.version 4
2. Dahel, N.E.: Vendor selection and order quantity allocation in volume discount environments. Supply Chain Manage.: Int. J. **8**(4), 335–342 (2003)
3. Zhang, X., Song, H., Huang, G.Q.: Tourism supply chain management: a new research agenda. Tour. Manag. **30**(3), 345–358 (2009)
4. Drezner, Z., Wesolowsky, G.O.: Multi-buyer discount pricing. Europ. J. Oper. Res. **40**(1), 38–42 (1989)
5. Kothari, T., Hu, C., Roehl, W.S.: e-Procurement: an emerging tool for the hotel supply chain management. Int. J. Hospitality Manage. **24**, 369–389 (2005)
6. Archer, B.H.: Forecasting demand: quantitative and intuitive techniques. Int. J. Tourism Manage. **1**(1), 5–12 (1980)
7. Smith, B.T., Wight, O.W.: Focus forecasting: computer techniques for inventory control. Monochrome Press, Incorporated (1994)
8. Smith, B.T.: Focus Forecasting and DRP: Logistics Tools of the Twenty-First Century. BT Smith & Assoc. (1991)
9. Odoom, C.K.: Logistics and supply chain management in the hotel industry: Impact on hotel performance in service delivery (2012)
10. Nahmias, S.: Perishable inventory theory: A review. Oper. Res. **30**(4), 680–708 (1982)
11. Harris, F.W.: How many parts to make at once. Oper. Res. **38**(6), 947–950 (1990)
12. Shcherbakov, M.V., et al.: A Survey of Forecast Error Measures (2013)

13. MacQueen, J.: Some methods for classification and analysis of multivariate observations. In: Proceedings of the Fifth Berkeley Symposium on Mathematical Statistics and Probability, vol. 1. no. 14 (1967)
14. Plossl, G.: Production and Inventory Control: Techniques and Principles, 2nd edn. Prentice Hall, Englewood Cliffs (1985)

TopCom: Index for Shortest Distance Query in Directed Graph

Vachik S. Dave[✉] and Mohammad Al Hasan

Department of Computer and Information Science, Indiana University Purdue
University Indianapolis, Indianapolis, USA
vsdave@iupui.edu, alhasan@cs.iupui.edu

Abstract. Finding shortest distance between two vertices in a graph
is an important problem due to its numerous applications in diverse
domains, including geo-spatial databases, social network analysis, and
information retrieval. Classical algorithms (such as, Dijkstra) solve this
problem in polynomial time, but these algorithms cannot provide real-
time response for a large number of bursty queries on a large graph.
So, indexing based solutions that pre-process the graph for efficiently
answering (exactly or approximately) a large number of distance queries
in real-time are becoming increasingly popular. Existing solutions have
varying performance in terms of index size, index building time, query
time, and accuracy. In this work, we propose TopCom, a novel indexing-
based solution for exactly answering distance queries in a directed acyclic
graph (DAG). Our experiments with two of the existing state-of-the-art
methods (IS-Label and TreeMap) show the superiority of TopCom over
these two methods considering scalability and query time.

Keywords: Shortest distance query · Indexing method for distance
query · Directed acyclic graph

1 Introduction

Finding shortest distance between two nodes in a graph (*distance query*) is one of
the most useful operations in graph analysis. Besides the application that stands
for its literal meaning, i.e. finding the shortest distance between two places in
a road network, this operation is useful in many other applications in social
and information networks. For instance, in social networks, the shortest path
distance is used for computing various centrality metrics [10], link prediction [6],
and community detection [2]. In information networks, shortest path distance
is used for keyword search and relevance ranking [8]. Researchers have been
studying this problem from the ancient time, and several classical algorithms
(Dijkstra, Bellman-Ford, Floyd-Warshall) exist for this problem, which run in
polynomial time over the number of vertices and the number of edges of the
network. However, as the real-life graphs grow in the order of thousands or
millions of vertices, classical algorithms deem inefficient for providing real-time

© Springer International Publishing Switzerland 2015
Q. Chen et al. (Eds.): DEXA 2015, Part I, LNCS 9261, pp. 471–480, 2015.
DOI: 10.1007/978-3-319-22849-5_32

answers for a large number of distance queries on such graphs. So, there is a growing interest for the discovery of more efficient methods for solving this task.

Various approaches are considered to obtain an efficient distance query method for large graphs. One of them is to exploit topological properties of real-life networks; examples include the spatial and planar properties of road networks [13]. The second approach is to perform pre-processing on the host graph and build an index data structure which can be used at runtime to answer the distance query between an arbitrary pair of nodes more efficiently. Several indexing ideas are used, but two are the most common, landmark-based indexing [1,11] and two-hop cover indexing [4,7]. Methods adopting the landmark-based idea identify a set of landmark nodes and pre-compute all-single source shortest paths from these landmark nodes. During query time, distances between a pair of arbitrary nodes are answered from their distances to their respective closest landmark nodes. Methods adopting the two-hop cover indexing store a collection of hops for each node such that the shortest path between a pair of arbitrary vertices can be obtained from the intersection of the hops of those vertices.

A related work to the shortest path problem is the reachability problem. Given a directed graph $G(V, E)$, and a pair of vertices u and v, the reachability problem answers whether a path exists from u to v. This problem can be solved in $O(|V| + |E|)$ time using graph traversal. However, using a reachability index, a better runtime can be obtained in practice. All the existing solutions [14] of the reachability problem solve it for a directed acyclic graph (DAG) as any directed graph can be converted to a DAG such that a DAG node is a strongly connected component (SCC) of the original graph; since all nodes in an SCC are reachable to each other, the reachability solution in the DAG easily answers the reachability question in the original graph. The indexing idea that we propose also exploits the SCC, but unlike existing works, we solve the distance query problem instead of reachability.

In this work, we propose TOPCOM[1], an indexing based method for obtaining exact solution of a distance query in a directed acyclic graph. In principle, TOPCOM uses a 2-hop-cover solution, but its indexing is different from other existing indexing methods. Specifically, the basic indexing scheme of TOPCOM is designed for a DAG and it has some similarities with the indexing solution of the reachability queries proposed in [3]. We show experiment results on DAG that validate TOPCOM's superior performance over IS-Label [5] and TreeMap [12] which are two of the fastest known methods in the recent years. Following other recent works, we also compare our method with bi-directional Dijkstra, which is a well-accepted baseline method for distance query solutions in directed graph.

2   Method

2.1   Topological Compression

The main idea of TOPCOM is based on topological compression of DAG, which is performed during the pre-processing step. During the compression, additional

---

[1] TOPCOM stands for **Top**ological **Com**pression which is the fundamental operation that is used to create the index data structure of this method.

distance information is preserved in a data structure which TOPCOM uses for answering a distance query efficiently.

Topological Level. Given a DAG G, we use $V(G)$ and $E(G)$ to represent the set of vertices and the set of edges of G. The topological level of any vertex $v \in V(G)$, named $topo(v)$, is one if v has no incoming edge, otherwise it is at least one higher than the topological level of any of v's parents. Mathematically,

$$topo(v) = \begin{cases} \max_{(u,v) \in E(G)} \ topo(u) + 1, & \text{if } v \text{ has incoming edges} \\ 1, & \text{otherwise} \end{cases}$$

If the topological level number of a vertex is even, we call it an even-topology vertex, otherwise it is an odd-topology vertex. For any edge $e = (u, v) \in E(G)$, it is called a single-level edge if $topo(v) - topo(u) = 1$, otherwise it is called a multi-level edge. For the DAG G, its topological level number is defined as: $topo(G) = \max_{v \in V(G)} \ topo(v)$

Topological compression of a DAG G halves the $topo(G)$ value by removing all odd-topology vertices from G. Compression also removes all the edges that are incident to the removed vertices. Thus, all single-level edges are removed, as one of the adjacent vertices of these edges is removed during compression. Multi-level edges are also removed if at least one of the endpoints of these edges is an odd-topology vertex. The compression process is applied iteratively to generate a sequence of DAGs $G^1, G^2, \cdots, G^t$ such that the topological level number of each successive DAG is half of that of the previous DAG, and the topological level number of the final DAG in this sequence is 1; i.e., $topo(G^i) = \lfloor topo(G^{i+1})/2 \rfloor$, and $topo(G^t) = 1$, where $t = \lfloor \log_2 topo(G) \rfloor$.

To preserve the information that is lost due to the removal of multi-level edges, during topological compression we insert additional even-topology vertices, together with additional edges between the even-topology vertices to prepare the DAG for the compression. The insertion of additional vertices and edges for preserving the information of a removed DAG multi-edge $e = (u, v)$ is discussed below along with an example given in Fig. 1. In this figure, the topological levels are mentioned in rectangle boxes. On the left side we show the original graph, and on the right side we show the modified graph which preserves information that is lost due to compression. There are four possible cases:

Case 1: ($topo(u)$ is odd and $topo(v)$ is even). Compression removes the vertex u, so we add a **fictitious** vertex u' such that $topo(u') = topo(u) + 1$, and remove the multi-level edge (u, v) and replace it with with two edges (u, u') and (u', v). Since topological level number of both u' and v are even, the topological compression will not delete the edge (u', v). For example, consider the multi-level edge (b, l) in Fig. 1(a), $topo(b) = 1$ (odd), and $topo(l) = 4$ (even). In Fig. 1(b) this edge is replaced by two edges (b, b') and (b', l), where b' is the fictitious node.

Case 2: ($topo(u)$ is even and $topo(v)$ is odd). This case is symmetric to Case 1, where the compression removes the vertex v instead of u. We create v_1, called a **copied** vertex of v such that $topo(v_1) = topo(v) - 1$ and replace the multi-level edge (u, v) with two edges (u, v_1) and (v_1, v). Note that, to distinguish the

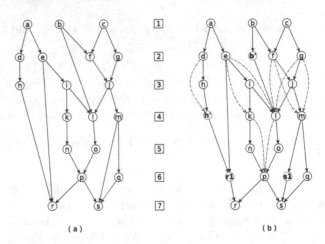

Fig. 1. (a) Original DAG G and (b) Modified DAG G_m

vertices added in Case 1 and Case 2, we give them different names, **fictitious** and **copied**; the justification of different naming will be clarified in latter part of the text. Example of Case 2 in Fig. 1(a) is edge (m, s), which is replaced by edges $(m, s1)$ and $(s1, s)$.

Case 3: $(topo(u)$ is odd and $topo(v)$ is odd). In this case we use a combination of above two methods and add two new vertices u' and v_1. We set topological level numbering of new vertices as mentioned above. Also we replace multi level edge (u, v) with three different edges (u, u'), (u', v_1), and (v_1, v). Multi level edge (h, r) in Fig. 1(a) is an example of this case. As shown in Fig. 1(b), we add two new vertices h' and r_1 and three new edges, (h, h'), (h', r_1), and (r_1, r) after deleting the original edge (h, r). Note that, if $topo(u) = topo(v) - 2$, $topo(u') = topo(v_1)$. In this case, we treat it as Case 1 by adding only u' (but not v_1) and follow the Case 1. It generates a single-level edge (u', v), which we do not need to handle explicitly.

Case 4: $(topo(u)$ is even and $topo(v)$ is even). Neither u nor v is removed by the compression process. In fact, the changes in the above three cases convert those cases into Case 4.

As we described earlier, we don't need to handle single-level edges separately. However if two continuous single-level edges are removed, we still need to maintain the logical connection between the even-topology vertices. For example, in Fig. 1(a) edges (e, i), (i, k), and (i, l) are single-level edges which will be deleted after the first compression iteration because $topo(i) = 3$. Now, information of a logical (indirect) connection between e to k and l needs to be maintained, because all three vertices will exist after the compression. To handle this, we add new **dummy edges** (e, k) and (e, l); dummy edges are shown as dotted lines in Fig. 1(b). For the same start and the end nodes, if there are multiple dummy edges, TOPCOM considers the edge with the smallest distance. To find

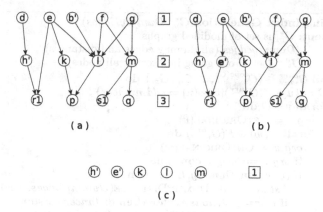

Fig. 2. (a) 1-Compressed Graph G^1, (b) Modified 1-compressed Graph G_m^1, (c) 2-Compressed Graph G^2

the dummy edges, we scan through all odd-topology vertices and find their single-level incoming and outgoing edges. We store all these dummy edges along with their corresponding distance values in a list called *DummyEdges*, which we use during the index generation step. For example dummy edge (d, h') has distance 2 in Fig. 1, then $\langle (d, h'), 2 \rangle$ is stored in *DummyEdges*.

At each compression iteration, we first obtain a modified graph, with fictitious vertices, copied vertices, and dummy edges and then apply compression to obtain the compressed graph of the subsequent iteration. The fictitious vertices, copied vertices, and dummy edges of the modified graph in the earlier iteration become regular vertices and edges of the compressed graph in the subsequent iteration. The above modification and compression proceed iteratively until we reach t-compressed graph, G^t, for which the topological level number is 1. We use G_m to denote the modified uncompressed graph, G_m^1 to denote the modified 1-compressed graph, G_m^2 to denote the modified 2-compressed graph and so on. For example, Fig. 2(a) shows G^1 which is obtained by compressing the modified graph G_m in Fig. 1(b). Figure 2(b) shows G_m^1, the modified 1-compressed graph, and Fig. 2(c) shows G^2, the 2-compressed graph. We refer set of all modified compressed graphs as G_m^*, i.e. $\{G_m^t, ..., G_m^2, G_m^1\}$.

2.2   Index Generation

For each vertex v of the input graph, TopCom's index data structure contains two lists: (i) outgoing index I_v^{out}, which stores shortest distances from v to a set of vertices reachable from v; and (ii) incoming index I_v^{in}, which stores shortest distances between v and a set of vertices that can reach v. Both the lists contain a collection of tuples, $\langle vertex\_id, distance \rangle$, where $vertex\_id$ is the id of a vertex other than v, and *distance* is the corresponding shortest path distance between v and that vertex. For the sake of simplicity, in subsequent discussion we will

---

Outgoing Index Generation(G_m^*, $DummyEdges$)
Comment: G_m^* is set of modified graphs
                $Dummyedges$ (all dummy edges and distances)
                I_*^{out} (set of out going indexes for all nodes)

1. **forall** $G_m^{curr} \in \{G_m^{topo(G)}, ..., G_m^1, G_m\}$ **do**
2. $O = \{u \in V(G_m^{curr}) | topo(u) = odd\_number\}$
3.   **forall** $v \in O$ **do**
4. $org_v = $ GETORIGINAL(v)
5.   **forall** $(v, w) \in E(G_m^{curr})$ **do**
6. $org_w = $ GETORIGINAL(w)
7.   **if** $org\_v == org\_w$ **continue**
8.   **if** (v, w) **isa** $Dummy\_Edge$ **then**
9. $distance = $GETDUMMYDISTANCE$(DummyEdges, v, w)$
10.   **if** $w == fictitious\_vertex$ **then** $distance = distance - 1$
11.   RECURSIVEINSERT$(I_{org\_v}^{out}, org\_w, distance, out)$

RecursiveInsert(I_v^{io}, a, $distance$, in_or_out)
1. **if** $in\_or\_out == in$
2. $I_a^{io} = I_a^{in}$
3. **else**
4. $I_a^{io} = I_a^{out}$
5. **if** $I_a^{io} == \emptyset$
6. $add\_tuple(I_v^{io}, a, distance)$
7. **else**
8. $add\_tuple(I_v^{io}, a, distance)$
9.   **forall** $(x, dist) \in I_a^{io}$
10.   RECURSIVEINSERT$(I_v^{io}, x, distance + dist, in\_or\_out)$

Fig. 3. Outgoing index generation pseudo-code

assume that the given graph is unweighted for which the weight of each edge is 1 and distance between two vertices is the minimum hop count between them.

At the beginning, for each vertex v, TOPCOM generates index from G_m^t and repeats the process in reverse order of graph compression i.e. from graph G_m^t to G_m. In i'th iteration of index building, it uses G_m^{t-i+1} and inserts a set of tuples in I_v^{out} and I_v^{in}, only if v is an odd-topology vertex in G_m^{t-i+1}. Thus, during the first iteration, for every odd-topology vertex v of G_m^t, for an incoming edge (u, v) TOPCOM inserts $\langle u, 1 \rangle$ in I_v^{in} and for an outgoing edge (v, w) TOPCOM inserts $\langle w, 1 \rangle$ in I_v^{out}. TOPCOM also inserts elements of I_u^{in} and I_w^{out} into I_v^{in} and I_v^{out}, respectively, as needed.

Figure 3 shows the pseudo-code of the index generation procedure for outgoing indexes only. An identical piece of code can be used for generating incoming indexes also, but for that we need to exchange the roles of fictitious and copied vertices, and change the I_*^{out} with I_*^{in} in Line 5–11. TOPCOM collects all odd-topology vertices in variable O (Line 2) and builds out-index for each of these vertices using all single level edges from these vertices to other vertices in the graph (Line 5). If v and w are fictitious or copied vertex, TOPCOM uses the subroutine GETORIGINAL() to obtain the corresponding original vertex (Line 4

Table 1. Index for DAG in Fig. 1

Node	Out index	In index	Node	Out index	In index
a	$\{\langle d,1\rangle, \langle e,1\rangle, \langle h,2\rangle, \langle k,3\rangle, \langle l,3\rangle\}$	$\emptyset$	i	$\{\langle k,1\rangle, \langle l,1\rangle\}$	$\{\langle e,1\rangle\}$
b	$\{\langle l,1\rangle, \langle f,1\rangle, \langle m,3\rangle\}$	$\emptyset$	j	$\{\langle l,1\rangle, \langle m,1\rangle\}$	$\{\langle f,1\rangle, \langle g,1\rangle\}$
c	$\{\langle f,1\rangle, \langle g,1\rangle, \langle l,3\rangle, \langle m,3\rangle\}$	$\emptyset$	n	$\{\langle p,1\rangle\}$	$\{\langle k,1\rangle\}$
d	$\{\langle h,1\rangle\}$	$\emptyset$	o	$\{\langle p,1\rangle\}$	$\{\langle l,1\rangle\}$
e	$\{\langle k,2\rangle, \langle l,2\rangle\}$	$\emptyset$	p	$\emptyset$	$\{\langle k,2\rangle, \langle l,2\rangle\}$
f	$\{\langle l,2\rangle, \langle m,2\rangle\}$	$\emptyset$	q	$\emptyset$	$\{\langle m,1\rangle\}$
g	$\{\langle l,2\rangle, \langle m,2\rangle\}$	$\emptyset$	r	$\emptyset$	$\{\langle h,1\rangle, \langle e,1\rangle, \langle p,1\rangle,$ $\langle k,3\rangle, \langle l,3\rangle\}$
h	$\emptyset$	$\{\langle d,1\rangle\}$	s	$\emptyset$	$\{\langle m,1\rangle, \langle p,1\rangle, \langle k,3\rangle,$ $\langle l,3\rangle, \langle q,1\rangle\}$

and 6). Similarly, if (v,w) is a dummy edge, TOPCOM obtains the actual distance using *DummyEdges* data structure. In case the end-vertex w is a fictitious vertex, TOPCOM decrements the distance value by 1 (Line 10), because for each fictitious vertex, an extra edge with distance 1 is added from the original vertex to the fictitious vertex which has increased the distance value by one. On the other hand, if w is a copied vertex, TOPCOM does not make this subtraction. This is the reason why we have to make a distinction between the fictitious vertices and the copied vertices. Also note that, after generating indexes for each vertex, there may be multiple entries for some vertices; from these multiple entries we need to get the smallest value (entry) and remove others. For building in-indexes, TOPCOM subtracts 1 for a copied vertex but does nothing for a fictitious vertex, as the roles of the start and end vertices are opposite for the in-indexes.

Say, we want to find the outgoing index for node a of Fig. 1. We start with G_m^2 which is shown in Fig. 2(c). It has h', e', k, l and m nodes but no edges. Here, for k, l and m outgoing indexes (incoming also) will be empty, but h' and e' are fictitious nodes, indexes for corresponding nodes h and e will be built. In the next iteration from graph G_m^1, we build $I_d^{out} = \{\langle h,1\rangle\}$, which uses the distance of dummy edge $(d,h') = 2$ and then replaces the fictitious vertex h' with h and obtains a distance of 1 by subtracting 1 from 2. Similarly $I_e^{out} = \{\langle k,2\rangle, \langle l,2\rangle\}$. In the next iteration, considering G_m, we insert $\langle d,1\rangle$ in I_a^{out} using **RecursiveInsert** subroutine (Line 11 in Fig. 3); this function also inserts $\langle h,2\rangle$ in I_a^{out}; the recursion stops at h because I_h^{out} is empty. Similarly, $\langle e,1\rangle$ and $\langle k,3\rangle, \langle l,3\rangle$ are inserted in I_a^{out} recursively from I_e^{out}. At the end of the algorithm, we remove duplicate entries from indexes. For example, for the incoming index of s, we have two entries for vertex m i.e. $\langle m,1\rangle$ and $\langle m,2\rangle$, one corresponding to edge $(m,s1)$ in G_m^1 and the other is a recursive result from q to s in G_m. We only consider $\langle m,1\rangle$ and discard the other entry from I_s^{in}. For the DAG in Fig. 1, the final index data structure is shown in Table 1.

For a weighted graph, TOPCOM makes a few changes. Note that distance values are stored either in the indexes or in the *DummyEdge* data structure. Many of these distances are implicitly 1 for an unweighted graph which is not correct for a weighted graph, so, for the latter TOPCOM stores the distance explicitly, ensuring that the distance value between fictitious (or copied) vertices and the original vertex is zero. Thus, for weighted graphs, no special treatment is needed for fictitious and copied vertices.

2.3 Query Processing

For query processing, TOPCOM uses the distance indexes that it has built earlier. To find $\delta(u,v)$, TOPCOM intersects outgoing Index of u i.e. I_u^{out} and incoming Index of v i.e. I_v^{in} and finds common *vertex_id* in I_u^{out} and I_v^{in}, along with the distance values. Here tuples $\langle u,0 \rangle$ and $\langle v,0 \rangle$ are added in I_u^{out} and I_v^{in}, respectively and then intersection of indexes is found. If the intersection set size is 0, there is no path from u to v and hence the distance is infinity. Otherwise, the distance is simply the sum of distances from u to *vertex_id* and *vertex_id* to v. If multiple paths exist, we take the one that has the smallest distance value.

Example: We want to find $\delta(a,s)$ in Fig. 1. From Table 1, $I_a^{out} \cap I_s^{in} = \{k,l\}$. Now, we need to sum up the corresponding *distance* values, that gives $\{\langle k,6 \rangle, \langle l,6 \rangle\}$. Now we need to find smallest distance value; in this case both the values are same, hence we can provide any one as a result.

3 Experimental Evaluation

We implement TOPCOM in C++ and compare its performance with two of the recent methods (IS-Label and TreeMap) using the code provided by the respective authors. We also compare TOPCOM with a baseline method, Bidirectional Dijkstra, which is one of the fastest online methods for single source shortest distance queries. We perform all experiments with a machine with Intel 2.4 GHz processor, 8 GB RAM running Ubuntu 14.04 OS. For our experiments, we use seven real world datasets from different domains (See, Table 2), which we have obtained from SNAP (Stanford Network Analysis Project) web page[2] except the *Epinion* trust network dataset, which we collected from [9]. For all graphs, we add distance values randomly over the edges of the DAG. In Table 2, $|V|$ and $|E|$ are the number of vertices and the number of edges respectively. Similarly $|V_{DAG}|$ and $|E_{DAG}|$ are the number of vertices and the edges in the DAG of the corresponding graph. AD and AD_{DAG} are average degree values for the graph and its DAG counterpart, respectively.

For DAG, TOPCOM outperforms IS-Label and TreeMap methods for all datasets as shown in Table 3. Here the results are average query time over 10 executions, where each execution calculates 10 K random queries. We can see that for datasets *Epinion, Gnutella09* and *Gnutella31* TOPCOM outperforms

---

[2] http://snap.stanford.edu/data/index.html.

Table 2. Real world datasets and basic information

Name	| V |	| E |	AD	| V_{DAG} |	| E_{DAG} |	AD_{DAG}
AS_Caida	26,374	2,304,095	87.36	26,358	48,958	1.86
Email_Eu	265,214	420,045	1.58	231,000	223,004	0.97
Epinion	49,289	487,183	9.88	16,264	16,497	1.01
Gnutella09	8,114	26,013	3.21	5,491	6,495	1.18
Gnutella31	62,586	147,892	2.36	48,438	55,349	1.14
WikiTalk	2,394,385	5,021,410	2.1	2,281,879	2,311,570	1.01
WikiVote	7,116	103,689	14.57	5,817	19,540	3.36

Table 3. Average query time for DAG

Name	TopCom	IS-Label	Bi-Djk	TreeMap[a]
AS_Caida	0.1036	0.2237	24.75	0.2471
Email_Eu	0.1059	0.3865	1657.46	0.2674
Epinion	0.0360	0.2388	14.83	0.1722
Gnutella09	0.0345	0.3292	7.27	0.115
Gnutella31	0.0752	0.2095	50.74	0.254
WikiTalk	0.1551	0.3494	43.11	-[b]
WikiVote	0.1551	0.3494	43.11	0.2131

[a]Unweighted DAG results
[b]Unable to execute the dataset on the machine due to memory issue

both IS-Label and TreeMap by an order of magnitude. For other datasets also TopCom performs 2 to 3 times better than both the methods. Also, TopCom performs multiple order of magnitudes better than Bi-Dijkstra for all the datasets.

4 Conclusions

In this paper, we propose TopCom, a novel indexing method to answer distance query for directed acyclic graphs. This method uses topological ordering of DAG for achieving a distance preserving compression of DAG, which can be used as a 2-hop cover index. We compare TopCom with two recently proposed methods and show that it performs better than both IS-Label and TreeMap for DAG.

Acknowledgment. This research is supported by M. Hasan's NSF CAREER award (IIS-1149851)

References

1. Akiba, T., Iwata, Y., Yoshida, Y.: Fast exact shortest-path distance queries on large networks by pruned landmark labeling. In: ACM SIGMOD, pp. 349–360 (2013)
2. Backstrom, L., Huttenlocher, D., Kleinberg, J., Lan, X.: Group formation in large social networks: membership, growth, and evolution. In: SIGMOD, pp. 44–54 (2006)
3. Cheng, J., Huang, S., Wu, H., Fu, A.W.C.: TF-Label: a topological-folding labeling scheme for reachability querying in a large graph. In: SIGMOD, pp. 193–204 (2013)
4. Cohen, E., Halperin, E., Kaplan, H., Zwick, U.: Reachability and distance queries via 2-hop labels. In: SODA, pp. 937–946 (2002)
5. Fu, A.W.C., Wu, H., Cheng, J., Wong, R.C.W.: IS-Label: an independent-set based labeling scheme for point-to-point distance querying. VLDB **6**, 457–468 (2013)
6. Hasan, M., Zaki, M.: A survey of link prediction in social networks. In: Aggarwal, C.C. (ed.) Social Network Data Analytics, pp. 243–275. Springer, US (2011)
7. Jin, R., Ruan, N., Xiang, Y., Lee, V.: A highway-centric labeling approach for answering distance queries on large sparse graphs. In: SIGMOD, pp. 445–456 (2012)
8. Kargar, M., An, A.: Keyword search in graphs: Finding r-cliques. Proc. VLDB Endow. **4**(10), 681–692 (2011)
9. Massa, P., Avesani, P.: Trust-aware bootstrapping of recommender systems. In: ECAI Workshop on Recommender Systems, pp. 29–33 (2006)
10. Okamoto, K., Chen, W., Li, X.-Y.: Ranking of closeness centrality for large-scale social networks. In: Preparata, F.P., Wu, X., Yin, J. (eds.) FAW 2008. LNCS, vol. 5059, pp. 186–195. Springer, Heidelberg (2008)
11. Qiao, M., Cheng, H., Chang, L., Yu, J.: Approximate shortest distance computing: a query-dependent local landmark scheme. IEEE Trans. Knowl. Data Eng. **26**(1), 55–68 (2014)
12. Xiang, Y.: Answering exact distance queries on real-world graphs with bounded performance guarantees. VLDB J. **23**(5), 677–695 (2014)
13. Yan, D., Cheng, J., Ng, W., Liu, S.: Finding distance-preserving subgraphs in large road networks. In: ICDE, pp. 625–636 (2013)
14. Yildirim, H., Chaoji, V., Zaki, M.: Grail: a scalable index for reachability queries in very large graphs. VLDB J. **21**(4), 509–534 (2012)

A Universal Distributed Indexing Scheme for Data Centers with Tree-Like Topologies

Yuang Liu, Xiaofeng Gao$^{(\boxtimes)}$, and Guihai Chen

Shanghai Key Laboratory of Scalable Computing and Systems,
Department of Computer Science and Engineering,
Shanghai Jiao Tong University, Shanghai 200240, China
liuyuang2012@sjtu.edu.cn, {gchen,gao-xf}@cs.sjtu.edu.cn

Abstract. The indices in the distributed storage systems manage the stored data and support diverse queries efficiently. Secondary index, the index built on the attributes other than the primary key, facilitates a variety of queries for different purposes. In this paper, we propose U^2-Tree, a universal distributed secondary indexing scheme built on cloud storage systems with tree-like topologies. U^2-Tree is composed of two layers, the global index and the local index. We build the local index according to the local data features, and then assign the potential indexing range of the global index for each host. After that, we use several techniques to publish the meta-data about local index to the global index host. The global index is then constructed based on the collected intervals. We take advantage of the topological properties of tree-like topologies, introduce and compare the detailed optimization techniques in the construction of two-layer indexing scheme. Furthermore, we discuss the index updating, index tuning, and the fault tolerance of U^2-Tree. Finally, we propose numerical experiments to evaluate the performance of U^2-Tree. The universal indexing scheme provides a general approach for secondary index on data centers with tree-like topologies. Moreover, many techniques and conclusions can be applied to other DCN topologies.

Keywords: Two-Layer index · Cloud storage system · Data center network

1 Introduction

Nowadays, the unprecedented development of cloud storage systems is drawing attentions from both academia and industry. The efficient queries in distributed

This work has been supported in part by the National Natural Science Foundation of China (Grant number 61202024, 61472252, 61133006, 61422208), China 973 project (2014CB340303), the Natural Science Foundation of Shanghai (Grant No.12ZR1445000), Shanghai Educational Development Foundation (Chenguang Grant No.12CG09), Shanghai Pujiang Program 13PJ1403900, and in part by Jiangsu Future Network Research Project No. BY2013095-1-10 and CCF-Tencent Open Fund.

Q. Chen et al. (Eds.): DEXA 2015, Part I, LNCS 9261, pp. 481–496, 2015.
DOI: 10.1007/978-3-319-22849-5_33

cloud systems require the construction of indices. Other than the index based on the primary key in the key-value storage, we need the *secondary indices* on other attributes for various applications. However, a centralized secondary index both consumes huge volume of storage space and causes the issue of access congestion. Therefore, a common design is to distribute the index on servers, and organize them as a *two-layer indexing scheme*.

The two-layer index consists of a global layer and a local layer. The global index collects the information published from the local indices as an overlay. When processing queries, the host will first request the information on the global index hosts, and further forward the request to the corresponding local index host. The two-layer indexing scheme efficiently solve the problem of the secondary index construction. Nonetheless, current researches on the two-layer indexing [4,15,21–23] are mainly concentrated on the *Peer-to-Peer (P2P) network*, whereas nowadays a typical cloud storage system is organized as a data center. *Data Center Networks (DCNs)* [1,9–12,18–20], the backbone of data centers, have the feature of scalability, reliability, and energy efficiency. The DCNs differ from P2P networks due to their specific physical topologies. Hence, we should take advantage of DCN topologies to design an efficient two-layer indexing scheme. However, there are few researches [7,8,21] regarding such design. Besides, each of them focuses on only one specific DCN topology. Consequently, we are motivated to design a general two-layer indexing scheme based on the properties of DCNs.

In this paper, we consider a series of DCNs with *tree-like topologies*. Nowadays, all commercial DCNs adopt the tree-like topologies. Moreover, they provide high bandwidth, satisfying fault tolerance, and regular structures. Therefore, we take advantage of the tree-like topologies to build an efficient secondary index. We first introduce some representative tree-like topologies including Fat Tree [1], Aspen Tree [20], and VL2 [9], and then extend our discussion to general tree-like topologies. Finally, we construct a Universal TWO-layer indexing built on TREE-like topologies named U^2-*Tree*.

We divide the construction of U^2-Tree into 4 steps. The first step is to build the local index on each host depending on the local data features. Next, we assign the potential indexing range of each host based on the characteristic of data distribution. We then publish the information about local index to the corresponding global index host. Finally, the global index is constructed according to the collected information. We explain the process of each step in detail, and introduce several optimization techniques for them. We further compare the performance and applicable conditions of techniques.

Moreover, we discuss the effects of two index update types, and explain the situations of lazy and eager updates. We also compare and analyze different index tuning schemes, and give the applicable scenarios. We then introduce the re-convergence problem after the link failure based on the fault tolerance of topologies. Finally, we validate our universal indexing design and compare different techniques by simulation results.

The contributions of this paper are: We propose a universal indexing scheme U^2-Tree utilizing the advantages of tree-like DCN topologies. We explain and compare the detailed implementations in index construction and maintenance. More importantly, many techniques and conclusions can also be extended to provide a general platform for secondary index construction on DCNs.

The rest of this paper is organized as follows. Section 2 briefly introduces the relate work. Section 3 is an illustration and comparison of tree-like topologies. Section 4 thoroughly explains the construction processes of U^2-Tree. Section 5 is an discussion about index updating, tuning, and fault tolerance. Section 6 proposes the processing approaches for various types of queries. Section 7 evaluates the performance of techniques in U^2-Tree. Finally, Sect. 8 summarizes and concludes the previous contents.

2   Related Work

The cloud storage systems are developing rapidly with the explosive growth of data. Massive data sets are distributed on several nodes, and nodes are connected to supply a fast access to non-relational databases. Google's Bigtable [3], Amazon's Dynamo [5], Apache Cassandra [13] are well-known examples of commercial distributed systems.

Typically, a cloud storage system is organized as a data center. Data center network (DCN) interconnects all the resources, such as storage and computational data, of a data center. Therefore, DCN architecture plays a significant role. DCN topologies can be categorized into switch-centric DCNs and server-centric DCNs. High bandwidth and better fault tolerance are the main features of switch-centric DCNs including Fat Tree [1], Aspen Tree [20], and Jellyfish [19], etc. Server-centric DCNs, such as Bcube [10], Dcell [11], and HCN [12], are of high scalability and relatively lower cost.

The design of two-layer index maximizes the topological benefits. Previous two-layer indexing designs [4,15,21–23] are mainly built on P2P networks. The indexing schemes regarding the features of DCNs are rarely referred. The multi-dimensional indices RT-HCN [14] and RB-Index [7] integrated HCN and Bcube topologies and R-tree index. The FT-Index [8] leveraged interval tree and B^+-tree on Fat Tree topology.

3   Data Centers with Tree-Like Topologies

Numerous kinds of DCN architecture designs are proposed in the last ten years. The connection of switches and servers varies among different architectures, and thus they have distinct properties, such as scalability, fault tolerance, energy efficiency, etc.

We mainly focus on tree-like topologies, a category of switch-centric DCNs. In the traditional DCN architecture, the widely used three-tier, multi-rooted tree is an example of tree-like topologies. Usually, the tree-like topologies adopt such multi-rooted structure, and divide the switches into multiple layers. The servers are connected to the bottom layer of switches. The disadvantages of traditional three-tier DCNs include limited bandwidth, poor scalability, and high

Table 1. List of notations

Notation	Definition	Notation	Definition
n	Number of switch layers	c_i	Fault tolerance parameter of L_i
k	Port number of switches	$C = \langle c_2, \cdots c_n \rangle$	Fault tolerance vector
L_i	Tree layer i	$[L, U)$	Boundary of data range
h_i	Host i	pr_i	Potential indexing range of h_i
H	Number of hosts	K	Total number of keys
S	Total number of switches	β	Data density

Table 2. Tree-like topologies

Topology	Structure and features
Fat tree [1]	**Layer:** Three layers of identical k-port switches
	Connection: Half ports of an edge switch connect to servers, and others to aggregation switches. Remaining ports of an aggregation switch connect to core switches. All k ports of a core switch connect to aggregation switches
	Expansion: Can be extended to arbitrary levels adopting such connection rule
	Example: Fig. 1 shows a 3-layer, 4-port Fat Tree topology
Aspen Tree [20]	**Layer:** Arbitrary layers of switches
	Connection: Based on the connection rule of Fat Tree. Adding redundant links between layers to reduce the re-convergence time after link failures
	Diversity: A vector C is used to identify different n-level, k-port Aspen Trees
	Example: Fig. 2 shows two 4-layer, 6-port Aspen Trees with different C
Virtual Layer 2 (VL2) [9]	**Layer:** Three layers of switches
	Connection: Based on the connection rule of Fat Tree
	Specification: A clos topology between D_I-port intermediate switches and D_A-port aggregate switches enables an 1:1 over-subscription
	Example: Fig. 3 shows an example of VL2 architecture
Portland [18]	**Connection:** Similar to Fat Tree
	Specification: A fabric manager for better fault tolerance and multicast

cost. Therefore, there are several novel tree-like topologies presented in recent years aiming to solve these problems. On one hand, the individual specifications enable different topologies to have their own characteristics. On the other hand, we can utilize the regular structure these topologies shared to build our universal two-layer indexing scheme. In this section, we will introduce some representa-

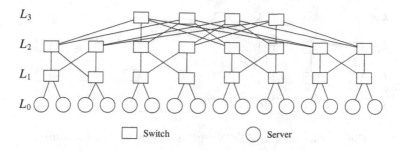

Fig. 1. An example of DCN with a 3-layer, 4-port Fat Tree topology

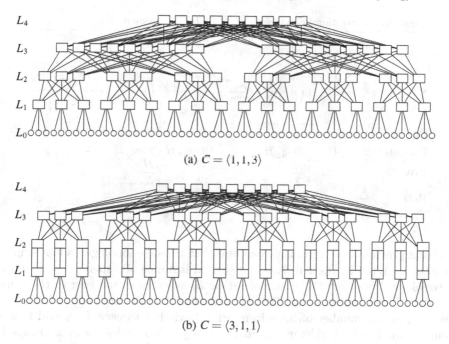

(a) $C = \langle 1, 1, 3 \rangle$

(b) $C = \langle 3, 1, 1 \rangle$

Fig. 2. Examples of DCNs with 4-layer, 6-port Aspen Tree topologies

tive tree-like topologies, including Fat Tree [1], Aspen Tree [20], VL2 [9], and Portland [18], and make a brief comparison among them.

We use k to denote the port number of switches, and n to denote the number of switch layers in tree-like topologies. In this paper, we refer to the switch layers in the three-layer trees as edge layer, aggregate layer, and core layer from bottom to top. Alternatively, we also define them as $L_1, L_2, \cdots, L_n$ in trees with arbitrary layers. Table 1 lists the symbols and their definitions. We will introduce more notations in the following texts. We then give a brief description of popular tree-like topologies in Table 2.

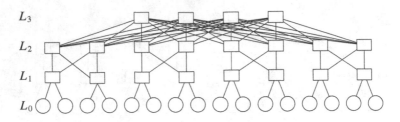

Fig. 3. An example of DCN with a 3-layer VL2 architecture

Table 3. Comparison of tree-like topologies

Topology	Three-tier	Fat tree	VL2	Aspen tree
H	k^n	$\dfrac{k^n}{2^{n-1}}$	$5k^2$	$\dfrac{k^n}{2^{n-1} \cdot \prod\limits_{j=2}^{n} c_j}$
S	$\dfrac{k^n - 1}{k - 1}$	$\left(n - \dfrac{1}{2}\right) \cdot \dfrac{k^{n-1}}{2^{n-2}}$	$\dfrac{k^2 + 6k}{4}$	$\left(n - \dfrac{1}{2}\right) \cdot \dfrac{k^{n-1}}{2^{n-2} \cdot \prod\limits_{j=2}^{n} c_j}$
Degree	1	1	1	1
Diameter	$2\log_2 H$	$2\log_2 H$	$2\log_2 H$	$2\log_2 H$
BiW	1	$\dfrac{H}{2}$	$\dfrac{H}{2}$	$\dfrac{H}{2}$
BoD	$\dfrac{k-1}{k^2} \cdot H^2$	H	H	H

From previous descriptions we can find that tree-like topologies adopt similar structures. Actually, the Fat Tree is a special instance of Aspen Trees in terms of topology. We define the maximal set of L_i switches connecting to the same set of L_{i-1} switches as a pod [20]. Then we can quantify the fault tolerance of level i c_i as the number of links from an L_i switch s to each L_{i-1} pod that s connects to. The fault tolerance vector $C = \langle c_2, c_3, \cdots, c_n \rangle$ since c_1 is always 1. $C = \langle 1, 1, \cdots, 1 \rangle$ for Fat Tree, while the vector is arbitrary for Aspen Tree.

C affects the fault tolerance of the tree, which can be reflected in the density of links. Fat Tree has a poor fault tolerance compared to Aspen Tree and VL2, while Fat Tree supports the most number of hosts when n and k are determined. However, the topology of VL2 is a "best" choice for both better fault tolerance and more supported hosts [20].

Other than the fault tolerance, we can compare tree-like topologies from several aspects. We first compare the maximum number of supported hosts H, and the total number of switches S in the same scale among topologies. We also compare the degree, the number of links from each host to switches, and the diameter, the longest path length of any pairs of hosts, of each topology. Finally, we compare the bisection bandwidth (BiW), the minimal possible bandwidth between the segmented pair of the network, and the bottleneck degree (BoD), the maximal number of flows over a single link under an all-to-all communication model. The comparison results are shown in Table 3.

4   The U²-Tree

The universal two-layer indexing scheme U²-Tree consists of two main parts, the local index and the global index. In the first stage of the index construction, each host builds the local index based on its local data. Then, the whole data range is partitioned into small ranges that each global index is responsible for. After that, the local host publishes some information about the data and local index to the corresponding global index host. Finally, the global index host will collect the received information and construct the global index. Each host in the U²-Tree will maintain a portion of local index and global index. The outline of the construction process is shown in Algorithm 1. Note that for each stage, the construction of index on all hosts can be executed in parallel.

Algorithm 1. U²-TREE CONSTRUCTION (at h_i side)

Input: The data on local host
Output: The universal two-layer index U²-Tree

1 Construct local index on the given key attribute // Local index construction
2 Scan data and calculate $pr_i.n, pr_i.c, pr_i.l, pr_i.u$ // Potential indexing range
3 Assign $pr_i' \subseteq [L, U)$ by Eqs. 2 and 4
4 Select nodes in the local index   // Publishing to global index
5 **foreach** *selected nodes* **do**
6 $\quad$| Publish $(range, ip, pos)$ to the corresponding global index host
7 Collect all information published from h_j // Global index construction
8 Construct the global index based on the information

4.1   Local Index Construction

In a cloud system, the data stored on each local host is of extremely large size. Thus, it is necessary to build a local index on each host so as to reduce the searching time and I/O cost. Typically, B-tree and B⁺-tree, or other search tree structures based on the storage format of data, can be used to build the local index. The nodes in these trees normally have two or more children and thus support efficient query, insertion, and deletion.

The keys we store in the index are non-negative integers, and the integers are unique among hosts, i.e. for each key, it is stored in at most one host. If the keys are not numerical, we can use hash or other techniques to transfer them into integers.

4.2   Potential Indexing Range Assignment

After the construction of local indices, we will determine the *potential indexing range* of each host. The potential indexing range pr_i is the data range that each global index host h_i represents. The pr_i does not intersect with other pr_j ($i \neq j$), while the union of all the potential indexing ranges $\bigcup_i pr_i$ will be the whole data range $[L, U)$. In this way, each point in the range will correspond to exactly one responsible global index host.

In the index publishing stage, each local host will publish some information about a certain data range to the corresponding global index according to the potential indexing range assignment. Besides, we will also query the data using a portion of global index based on the potential indexing range. Therefore, it is beneficial to assign the potential indexing range in a way that the keys are partitioned evenly on each global index host.

It is quite easy if the data are distributed uniformly on the range. If so, the approach proposed in [8] can be used. We just partition the whole range into H subranges with equal length, and assign them to the corresponding hosts in ascending order, i.e.

$$pr_i = \left[L + i \cdot \frac{U-L}{H}, L + (i+1) \cdot \frac{U-L}{H} \right), \quad 0 \leq i \leq H - 1. \tag{1}$$

Otherwise, we need to handle the skewed data distributed on the range. The goal of the assignment is to balance the number of keys published to each global index host. Assume the j-th key distributed on the local host i in ascending order is denoted as k_i^j. Then the keys on host h_i are $K_i = \{ k_i^j \mid k_i^0 < k_i^1 < \ldots < k_i^{|K_i|} \}$. If we sort the keys on all hosts and get $k_{i_0}^{j_0} < k_{i_1}^{j_1} < \ldots k_{i_{K-1}}^{j_{K-1}}$, where $K = |\bigcup_i K_i|$, we can assign

$$pr_i' = \begin{cases} [L, b_1), & i = 0, \\ [b_i, b_{i+1}), & 1 \leq i \leq H - 2, \\ [b_{H-1}, U), & i = H - 1, \end{cases} \tag{2}$$

where $b_i = k_{i \lfloor iK/H \rfloor}^{j \lfloor iK/H \rfloor}$. In this way, the number of keys published to each host is almost the same. However, the ideal method is not practical in that we must collect all the keys on hosts and sort them in ascending order. Now that the data stored in a single host are already massive, it is both storage and time intolerable to achieve such kind of task.

Nevertheless, we can use an approximate method to balance the keys. Zhang et al. [24] offered a *Piecewise Mapping Function (PMF)* that can solve the problem. We first use potential indexing range described in Eq. (1) to divide the whole range $[L, U)$ into H subranges, and count the number of keys distributed on each subranges. Moreover, we record the maximum key $pr_i.u$ and the minimum key $pr_i.l$ in each subrange. Note that this step can be done in parallel, and costs only linear time. We denote the count on each subrange as $pr_i.n$, and the cumulative count, i.e. the number of keys smaller than the right boundary of each subrange, as $pr_i.c$. Clearly, $pr_i.c = \sum_{j \leq i} pr_j.n$. Thus we can get an approximate mapping from $x \in pr_i \subseteq [L, U)$ to the key $k_{i_m}^{j_m}$ where

$$m = \begin{cases} pr_{i-1}.c, & x < pr_i.l, \\ \dfrac{pr_i.n}{pr_i.u - pr_i.l}(x - pr_i.l) + pr_{i-1}.c, & pr_i.l \leq x \leq pr_i.u, \\ pr_i.c, & x > pr_i.u. \end{cases} \tag{3}$$

Conversely, the potential indexing range assignment can be given by Eq. (2), where

$$b_i = \begin{cases} \dfrac{pr_j.u - pr_j.l}{pr_j.n}\left(\dfrac{iK}{H} - pr_{j-1}.c\right) + pr_j.l, & pr_{j-1}.c < \dfrac{iK}{H} < pr_j.c, \\ \dfrac{1}{2}\left(pr_j.u + pr_{j+n+1}.l\right), & \dfrac{i}{H} = pr_j.c = \ldots = pr_{j+n}.c. \end{cases} \tag{4}$$

As an example, we consider the 15 prime keys $\{2, 3, 5, \ldots, 43, 47\}$ distributed on $[0, 50]$. If we divide them into 5 subranges using Eq. (1), the count of keys in each subranges will be $4, 4, 2, 2, 3$, with a variance of 0.8. However, Eqs. (2) and (4) can balance the counts perfectly, as shown in Table 4.

Table 4. An example of potential indexing range assignment

i	pr_i	$pr_i.n$	$pr_i.c$	$pr_i.l$	$pr_i.u$	b_i	pr_i'	$pr_i'.c$
0	$[0,10)$	4	4	2	7	-	$[0,5.75)$	3
1	$[10,20)$	4	8	11	19	5.75	$[5.75,15)$	3
2	$[20,30)$	2	10	23	29	15	$[15,26)$	3
3	$[30,40)$	2	12	31	37	26	$[26,39)$	3
4	$[40,50)$	3	15	41	47	39	$[39,50)$	3

In practice, we assign range for several times to achieve a better performance. Algorithm 2 shows the procedure of potential indexing range assignment for multiple times.

Algorithm 2. POTENTIAL INDEXING RANGE ASSIGNMENT

Input: Original potential indexing range assignment pr_i and assignment *times*
Output: New assignment pr_i'

```
1 foreach t ∈ [1, times] do
2     foreach i ∈ [0, H) do               // Estimate data distribution
3         Calculate pr_i.n, pr_i.c, pr_i.l, pr_i.u
4     foreach i ∈ [1, H) do               // Calculate range boundaries
5         Calculate b_i by Eq. 4
6     foreach i ∈ [0, H) do               // Update potential indexing range
7         Calculate pr_i' by Eq. 2
8         pr_i ← pr_i'
9 return pr_i'
```

4.3 Publishing Scheme

After assigning the potential indexing range, each host will publish some information about the local index to corresponding global index hosts. Since the nodes in the local index typically store keys as the subtree separation values, we will publish the intervals of keys which indicate the possible keys that exist in a subtree of the node. Moreover, we will also publish the ip address of the host and the position of the node stored in the host. Therefore, a tuple of $(range, ip, pos)$ can be published to the global index host to locate the data on the local host if the range of query intersects with the range of node.

However, publishing in such way is plausible but not efficient enough because of the false positives. On one hand, there is no false negative for query, since all possible keys in the subtrees of the nodes are included in the published intervals. On the other hand, there may be many false positives for query since we only publish the minimum and maximum boundaries of the keys. The publishing scheme significantly reduces the size of published information, while it causes the problem of false positives. The false positives will directly increase the hops needed in queries.

Gap Elimination (FT-Gap) and *Bloom Filter (FT-Bloom)*, two methods proposed in [8], can efficiently solve the problem. The false positives can be considered as the gaps in the intervals. For example, if the keys stored in a node are $\{4, 7, 9, 15, 17\}$, we will publish an interval $[4, 17]$. The queries in the "gaps", for instance, $[10, 14]$ will cause false positives. Gap elimination will remove several biggest gaps in the interval and publish the remaining segments. Bloom filter uses hash functions to map the keys in the intervals into a bit array with all 0's and set the corresponding positions to 1. In the query phase, we will hash the query key to check whether the bits are 1. Both methods guarantee no false negative and moderate false positives with tolerable additional space.

4.4   Global Index Construction

The global index is on top of the local index logically. The global index collects the meta-data information published by the local hosts, and will arrange them sensibly to facilitate efficient query. Wu et al. [22] used conventional table or list to store the information. Instead, we can also construct more efficient tree data structures to index the interval ranges. Interval tree, segment tree, and priority search tree are common data structures for storing intervals and supporting various queries.

Interval tree [6, 16] uses the median of the endpoints of the intervals to separate the intervals into three sets: intervals intersecting with the median, those lying in the left of it, and those lying in the right. The subtrees are built recursively on the last two sets.

Segment tree [2] decides the atomic intervals based on the endpoints of the intervals. Each node corresponds to an atomic interval or the union of some atomic intervals. An interval is stored in the nodes whose union is exactly the range of the interval.

Priority search tree [17] stores two-dimension data in the nodes. The tree is a heap for one dimension and also a binary search tree for the other dimension. The two dimensions correspond to the lower bound and the upper bound for storing intervals.

For n intervals in total, interval tree and priority search tree will cost $O(n)$ storage, while segment tree will cost $O(n \log n)$ storage. The construction times of the three trees are all $O(n \log n)$. Moreover, the point query times of them are all in $O(\log n + k)$, where k is the number of reported intervals. Compared with scanning the lists, these structures can significantly reduce the searching time on global index. Typically, the interval tree and priority search tree are preferred to segment tree, while the segment tree can be modified to support multi-dimensional query which is not applicable to the others.

5   Update and Maintenance

5.1   Index Updating

The insertion and deletion of data after the index construction will cause the updating of index. The updates in the local index should be executed immediately in order to guarantee the correctness of index. However, since sending the updates to the global index causes additional network cost, a common method is to divide the updates into lazy ones and eager ones [22]. The lazy updates are those that may increase false positives while not affect the correctness. The eager updates will cause false negatives and thus the index fail to provide correct results. Therefore, we forward the eager updates immediately to the global index host while do a batch update for lazy updates.

The merges and splits of local index nodes that change previous *pos* of the nodes are considered to be eager updates. Any insertions that enlarge the ranges of local index nodes or lie in the eliminated gaps for FT-Gap will also trigger the eager updates. Moreover, for FT-Bloom, the insertions that change the bit array of bloom filter are also eager updates. On the contrary, the changes that do not affect the published information are lazy updates that can be issued in a batch way.

Frequent updates in trees for global index will cause the unbalance problem. Therefore, we can maintain an additional *updates index* to store the updated intervals. After a certain time period or when the size of the updates index is too large, we destroy the original tree and reconstruct a new one.

5.2   Index Tuning

When we select the intervals (nodes) in the local index to build the global index, we have multiple choices. For each leaf node, we should select at least one node in the path from the root to it. Yet it is enough that we select only one node in this path. Moreover, we can reduce the update cost by selecting only one node. The two properties, index completeness and unique index, guarantee the correctness of the index selection. However, even among these right selections, we still have different choices. For example, we can select the nodes near the leaves in order to reduce false positives. On the contrary, the nodes near the root are unlikely to be split or merged, and the update cost is reduced. There are some index tuning approaches to balance the false positives and update cost.

Top-Down Approach. The top-down approaches focus on false positives of index. For example, we only select the nodes that are under a certain level (namely l) to publish. For sibling nodes under l, we select a portion to publish directly, while recursively select the descendants of other s frequently accessed nodes to publish. The access times can be counted incrementally, while the tuning is updated in batch. The top-down approaches can effectively reduce false positives, with a limited size of published nodes. It is especially suitable for the indices that support skewed query, but are seldom updated.

Bottom-Up Approach. The more sophisticated bottom-up approaches consider cost models. The cost model for each node consists of the cost of query processing and index maintenance. Wang and Wu et al. [21–23] gave different cost models based on the type of data and the network feature. The goal of the index tuning is to select a set of nodes with the least total cost. Therefore, we calculate the cost of a node and its children, and select the one(s) with less cost. Recursively doing the same job until we meet the root in a bottom-up approach, we can get the optimal indexing set. The calculation of the cost and the selection of nodes are both executed in a batch way.

5.3   Fault Tolerance

In Sect. 3, we mentioned that VL2 and Aspen Tree introduce additional links between some layers. These links can help the system react to link failures more conveniently and rapidly. Walraed-Sullivan et al. [20] proposed the Aspen Reaction and Notification Protocol (ANP). The notifications are sent upwards to ancestors located near to a failure with ANP, rather than a global re-convergence. Suppose there is a failed link between L_i and L_{i+1}, the global re-convergence can be avoided so that the convergence time is shortened as long as the failed link occurs along the upward segment, or $c_j > 1 (j \geq i)$ along the downward segment. For an n-level tree, the supported hosts are decreased by half if $c_n = 2$ while the convergence time can speed up $70 \sim 80\%$ compared to Fat Tree. VL2 just utilizes the property to gain a better fault tolerance and scalability.

6   Query Processing

U^2-Tree can process kinds of queries, such as point query, range query, and k-NN query.

The range query in U^2-Tree is similar to most other two-layer indexing schemes. The query host first find out the global index hosts with the potential indexing range intersecting with the queried range. The query is then forwarded to search on the global indices. After that, possible hosts storing the data are returned. Since the hosts with more similar ip are connected within less hops, we sort the possible local index hosts by ip and visit them one by one in sequence. Finally, we collect all the data intersecting with the range on these local indices and forward the results to the query host.

The point query is a special case of range query when the bounds of range are equal. The difference are that (1) we only search on one global index host, (2) we can halt the search on the local index hosts if we have retrieved the data.

The k-nearest neighbours (k-NN) query returns the top-k nearest results to the key given the query (key, k). We define the density of data β as

$$\beta = \frac{K}{U - L},\qquad(5)$$

where K is the total number of keys on all hosts. Actually, we have already counted K in the potential indexing range assignment and $K = pr_{H-1}.c$. Therefore, we can use β to estimate the ranges of k nearest results and do the range query. We will first query the range $[key - \gamma k/\beta, key + \gamma k/\beta)$, where γ is a scaling parameter typically slightly large than 0.5. If we can find out more than k results in this range, we simply select the k nearest ones and return. Otherwise, we do range query on $[key - (i + 1)\gamma k/\beta, key - i\gamma k/\beta)$ and $[key + i\gamma k/\beta, key + (i + 1)\gamma k/\beta)$ continuously with increasing i by one for each iteration until the number of results is greater or equal to k.

7   Performance Evaluation

We simulated the U²-Tree on different tree-like topologies in C++. We generated the non-negative integers as keys randomly among the range $[0, U)$. The keys in the range are unique, while each possible key in the range does not necessarily exist due to the data density (or existence probability) $\beta = K/U$. Table 5 shows the experiment settings.

Table 5. Experiment settings

Parameter	Value
Data density (β)	0.3, 0.8, 1
Upper bound of data (U)	500 K, 1 M, 2 M, 3 M
Host number (H)	Depends on topology
Data distribution	Uniform, Zipfian

We first evaluate the performance of the potential indexing range assignment algorithm. The inconsistency of keys distributed on global index host is reflected by the variance of the counts. The variance is defined as

$$V = \frac{1}{H} \sum_i pr_i.n.\qquad(6)$$

Therefore, we use Zipfian distribution data and compare the variance of counts. In Fig. 4, the number of rounds means the result after such times of assignment. From the figure we know that the variance significantly reduces after the assignment. Moreover, the performance is better after more times of assignment.

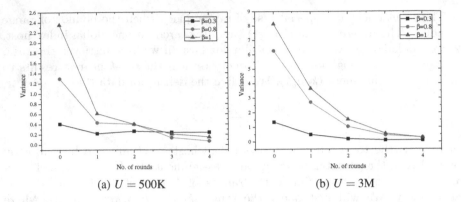

(a) $U = 500\text{K}$ (b) $U = 3\text{M}$

Fig. 4. Performance of potential indexing range assignment algorithm

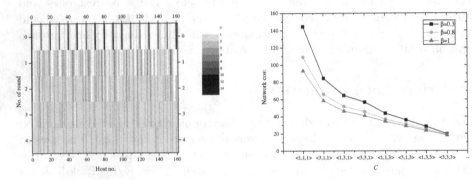

Fig. 5. Heatmap of keys counts on hosts

Fig. 6. Network cost on tree-like topologies

The data with different data densities will result in similar variances eventually. Figure 5 shows the key counts on each host after each time of assignment when $U = 3\text{M}$. The darkness of color shows the number of keys on corresponding hosts. We can find that initially the distribution is extremely unbalanced, while the counts vary a little among hosts after the assignments.

Figure 6 shows the average network cost on different 4-level, 6-port tree-like topologies. The network cost is defined as the hop counts on networks for each point query. The topology with $C = \langle 1, 1, 1 \rangle$ is a Fat Tree, while others are various Aspen Trees. We can learn from the figure that the average hops decreases as the fault tolerance increases. However, we must point out that it does not mean that Aspen Trees are superior, since the number of hosts also reduces as C increases. The data density β also has an effect on the network cost, because if the queries key does not exist in the data range, the query will be forwarded to all possible local index host, instead of halting halfway.

Finally, we compare the performance of different global indices. We record the number of stored and visited intervals in index construction and query processing. The result is shown in Fig. 7. From Fig. 7(a), the number of stored intervals

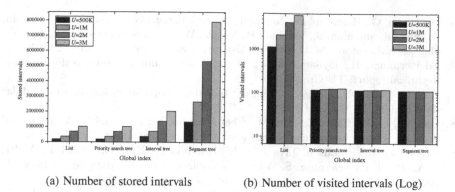

(a) Number of stored intervals   (b) Number of visited intervals (Log)

Fig. 7. Comparison of global index structures

for priority search tree and interval tree is similar to the traditional list, while segment tree consumes much more space. However, the segment tree can be built faster in practice. Figure 7(b) shows the average number of visited intervals in logarithm scale. The visited intervals of list grow linearly with the total number of keys. On the contrary, the performances of three trees are similarly fine and quite stable with the increasing of stored intervals.

8   Conclusion

In this paper, we proposed the U^2-Tree, a universal distributed secondary index scheme with tree-like DCN topologies. We took the topological benefits to build a two-layer indexing. We explained and compared the techniques in the index construction in detail. We also discussed the index maintenance problems including updating, tuning, and fault tolerance. The U^2-Tree can support several types of query processing efficiently. The experiment evaluated the performance of U^2-Tree, and provided further conclusions about different techniques. In a broad sense, the universal indexing scheme can even be applied to other DCN topologies with proper modifications.

References

1. Al-Fares, M., Loukissas, A., Vahdat, A.: A scalable, commodity data center network architecture. ACM SIGCOMM Comput. Commun. Rev. **38**(4), 63–74 (2008)
2. Bentley, J.L.: Solutions to klee's rectangle problems. Technical report, Carnegie-Mellon University, Pittsburgh (1977)
3. Chang, F., Dean, J., Ghemawat, S., Hsieh, W.C., Wallach, D.A., Burrows, M., Chandra, T., Fikes, A., Gruber, R.E.: Bigtable: a distributed storage system for structured data. ACM Trans. Comput. Syst. **26**(2), 4 (2008)
4. Chen, G., Vo, H.T., Wu, S., Ooi, B.C., Özsu, M.T.: A framework for supporting DBMS-like indexes in the cloud. VLDB. **4**, 702–713 (2011)

5. DeCandia, G., Hastorun, D., Jampani, M., Kakulapati, G., Lakshman, A., Pilchin, A., Sivasubramanian, S., Vosshall, P., Vogels, W.: Dynamo: amazon's highly available key-value store. ACM SIGOPS Operating Syst. Rev. **41**(6), 205–220 (2007)
6. Edelsbrunner, H.: Dynamic data structures for orthogonal intersection queries. Technical report, TU Graz (1980)
7. Gao, L., Zhang, Y., Gao, X., Chen, G.: Indexing multi-dimension data in modular data centers. In: DEXA (2015)
8. Gao, X., Li, B., Chen, Z., Yin, M., Chen, G., Jin, Y.: FT-INDEX: A distributed indexing scheme for switch-centric cloud storage system. In: ICC (2015)
9. Greenberg, A., Hamilton, J.R., Jain, N., Kandula, S., Kim, C., Lahiri, P., Maltz, D.A., Patel, P., Sengupta, S.: VL2: a scalable and flexible data center network. ACM SIGCOMM Comput. Commun. Rev. **39**(4), 51–62 (2009)
10. Guo, C., Lu, G., Li, D., Wu, H., Zhang, X., Shi, Y., Tian, C., Zhang, Y., Lu, S.: Bcube: a high performance, server-centric network architecture for modular data centers. ACM SIGCOMM Comput. Commun. Rev. **39**(4), 63–74 (2009)
11. Guo, C., Wu, H., Tan, K., Shi, L., Zhang, Y., Lu, S.: Dcell: a scalable and fault-tolerant network structure for data centers. ACM SIGCOMM Comput. Commun. Rev. **38**(4), 75–86 (2008)
12. Guo, D., Chen, T., Li, D., Liu, Y., Liu, X., Chen, G.: BCN: Expansible network structures for data centers using hierarchical compound graphs. In: INFOCOM, pp. 61–65. IEEE (2011)
13. Lakshman, A., Malik, P.: Cassandra: a decentralized structured storage system. ACM SIGOPS Operating Syst. Rev. **44**(2), 35–40 (2010)
14. Li, F., Liang, W., Gao, X., Yao, B., Chen, G.: Efficient R-tree based indexing for cloud storage system with dual-port servers. In: DEXA, pp. 375–391 (2014)
15. Lu, P., Wu, S., Shou, L., Tan, K.L.: An efficient and compact indexing scheme for large-scale data store. In: ICDE, pp. 326–337 (2013)
16. McCreight, E.M.: Efficient algorithms for enumerating intersection intervals and rectangles. Technical report, Xerox Paolo Alto Reserach Center (1980)
17. McCreight, E.M.: Priority search trees. SIAM J. Comput. **14**(2), 257–276 (1985)
18. Mysore, R.N., Pamboris, A., Farrington, N., Huang, N., Miri, P., Radhakrishnan, S., Subramanya, V., Vahdat, A.: Portland: a scalable fault-tolerant layer 2 data center network fabric. ACM SIGCOMM Comput. Commun. Rev. **39**(4), 39–50 (2009)
19. Singla, A., Hong, C.Y., Popa, L., Godfrey, P.B.: Jellyfish: networking data centers randomly. In: NSDI. vol. 12, p. 17 (2012)
20. Walraed-Sullivan, M., Vahdat, A., Marzullo, K.: Aspen trees: balancing data center fault tolerance, scalability and cost. In: CoNEXT, pp. 85–96 (2013)
21. Wang, J., Wu, S., Gao, H., Li, J., Ooi, B.C.: Indexing multi-dimensional data in a cloud system. In: SIGMOD, pp. 591–602 (2010)
22. Wu, S., Jiang, D., Ooi, B.C., Wu, K.L.: Efficient B-tree based indexing for cloud data processing. VLDB **3**, 1207–1218 (2010)
23. Wu, S., Wu, K.L.: An indexing framework for efficient retrieval on the cloud. IEEE Data Eng. Bull. **32**(1), 75–82 (2009)
24. Zhang, R., Qi, J., Stradling, M., Huang, J.: Towards a painless index for spatial objects. ACM Trans. Database Syst. **39**(3), 19 (2014)

Data Mining III

Improving Diversity Performance of Association Rule Based Recommender Systems

M. Kumara Swamy$^{(\boxtimes)}$ and P. Krishna Reddy

Center for Data Engineering, International Institute of Information
Technology-Hyderabad (IIIT-H), Gachibowli, Hyderabad 500032, India
kumaraswamy@research.iiit.ac.in, pkreddy@iiit.ac.in

Abstract. In recommender systems (RSs), getting high accuracy alone
does not improve the user satisfaction. In the literature, efforts are being
made to improve the variety/diversity of recommendations for higher
user satisfaction. In these efforts, the accuracy performance is reduced
at the cost of improving the variety of recommendations. In this paper,
we propose an approach to improve the diversity as well as accuracy
of association rule based RSs. We propose a ranking mechanism, called
diverse rank, to rank association rules based on the diversity of the items
in the pattern. The recommendations are made based on association rules
with high confidence and diversity. The experimental results on the real-
world MovieLens data set show that the proposed approach improves the
performance of association rule based RSs with better diversity without
compromising the accuracy.

Keywords: Data mining · Association rules · Patterns · Diversity ·
Concept hierarchy · Recommender systems · Electronic commerce

1 Introduction

The recommender system (RS) technology has become a key component of modern e-Commerce applications. Collaborative filtering (CF) [1,10,11] is a popular
recommendation approach, and several variations have been proposed in the literature. Association rule based RSs [1,8] have received a great deal of attention
to address the scalability problem. One of the research issues is to improve the
variety of recommendations for better user satisfaction. The RS should recommend items, which give high user satisfaction [9,12]. Research efforts are on to
improve the variety/diversity of the recommendations [9,12].

The importance of diverse recommendations was proposed in [4] and showed
that the diverse recommendations provide interesting recommendations over the
similarity based recommendations. It was mentioned that traditional RSs recommend items with poor diversity. Ziegler et al. [12] proposed an approach to return
a list of recommendations that better cater to the users' full range of interest
by selecting lists that have been low "inter-list similarity." It was demonstrated
experimentally that real users prefer more diversified results.

© Springer International Publishing Switzerland 2015
Q. Chen et al. (Eds.): DEXA 2015, Part I, LNCS 9261, pp. 499–508, 2015.
DOI: 10.1007/978-3-319-22849-5_34

In this paper, we incorporate the aspect of diversity to recommend the items in association rule based RS. In the literature, an effort [7] was made to rank the pattern (set of items) with the notion of *diverse rank* (*drank*) by analyzing the extent the items of the pattern belong to multiple categories in the corresponding concept hierarchy. The notion of *drank* indicates that the extent the items of the rule belong to multiple categories in the corresponding concept hierarchy. In this paper, we refine the methodology to assign diverse rank to the patterns, explored how the notion of *drank* could improve the diversity of association rule based RSs and proposed a recommendation framework to recommend the diverse as well as relevant items. The experimental results on the real-world MovieLens data set show that the proposed approach provides the recommendations with better diversity without compromising the accuracy as compared to the existing association rule based RS approaches.

The rest of the paper is organized as follows. In the next section, we explain the overview to compute diversity of patterns. In Sect. 3, we explain the proposed approach for diverse recommenations. In Sect. 4, we present experimental results. The last section contains the summary and conclusions.

2   Overview to Compute Diversity of Patterns

The concept of diversity, and framework to assign the diverse rank of a pattern are given in [7]. In this section, we present the overview of diversity. Next, we explain about the rened methodology to assign the diverse rank to the patterns.

2.1   Overview of Diversity

A pattern contains set of items also called as data items. We organize the data items in hierarchical manner in the concept hierarchy to compute the diversity of a pattern. The concept hierarchy is a tree in which all the leaf nodes represent the *items*, the internal nodes represent the *categories* and the top node represent the *root*. The *root* could be a virtual node. In this paper, we consider the concept hierarchy in which a lower-level node is mapped to only one higher-level node.

Figure 1 represents a concept hierarchy. In this, the items *orange*, *apple* and *cherry* are mapped to the category *fruits*. Similarly, the categories *drinks* and *fruits* are mapped to the category *fresh food*. Finally, the categories *fresh food* and *house hold* are mapped to *root*.

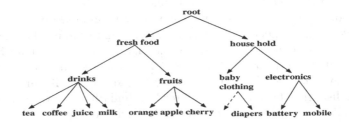

Fig. 1. An example of concept hierarchy

The diversity of a pattern can be computed based on the number of items in the pattern and the dynamics of merging to high-level categories. If the items of a pattern are mapped to the same/few categories in a concept hierarchy, we consider that the pattern has low diversity. Relatively, if the items are mapped to multiple categories, we consider that the pattern has more diversity.

Example 1. Consider the concept hierarchy in Fig. 1, and patterns {tea, juice}, {coffee, orange} and {milk, mobile}. For the pattern {tea, juice}, the items *tea* and *juice* are mapped to the next level category *drinks*. From there the category *drinks* mapped to root crossing the category *fresh food*. For the pattern {coffee, orange}, the item *coffee* mapped to category *drinks* and item *orange* mapped to the category *fruits*. Further, both the categories *drinks* and *fruits* are mapped to the category *fresh food*, and *fresh food* in turn map to *root*. Similarly, the pattern {milk, mobile} merges at the *root* by crossing several internal nodes. The pattern {tea, juice} quickly merges to parent, the diversity of the pattern is low. The pattern {tea, juice} merges slowly as compared to the pattern {tea, juice}, the diversity of the pattern is medium. Similarly, the pattern {milk, mobile} merges slowly at root crossing several internal nodes, the diversity of the pattern is high.

2.2   Approach to Compute the Diversity of Patterns

An approach was present in [7] by capturing the merging behavior of the pattern. Here, we propose a refined methodology using the notion of *projection of a pattern.*

The issue is to compute the merging behavior of the pattern. Given a concept hierarchy and a pattern, we assign the diverse rank to the pattern by analyzing the portion of the concept hierarchy formed by the items in the pattern. Given the pattern, the number of nodes in the corresponding projection of the concept hierarchy vary based on the patterns, if items in the pattern merge quickly, the corresponding projection will have less number of nodes as compared to the projection of a pattern in which the items merge slowly.

The value of *diverse rank* (*drank*) ranges between *0* (lowest) and *1* (highest). Let a pattern be Y, $Y = \{i_1, i_2, \ldots, i_n\}$ with n items and a concept hierarchy C of height h. We first define the notions of projection of pattern Y, maximal projection of Y, and minimal projection of Y.

Definition 1 *Projection of Y ($\Pi(Y/C)$): Let Y be a pattern and C be a concept hierarchy. The $\Pi(Y/C)$ is the projection of C for Y. It is equal to the portion of C with all nodes and edges of the paths of the items of Y to the 'root', including the items of a pattern and the root. In effect, $\Pi(Y/C)$ is the sub-tree which represents a concept hierarchy concerning to the pattern Y.*

The issue is how to extract the $\Pi(Y/C)$ from the C. We use the item-encoding technique [5] to extract the projection of the concept hierarchy. For a given pattern, we number all the nodes of a parent node from left to right, in the concept hierarchy. The encoded representation of an item is the traversal path from *root* to respective item.

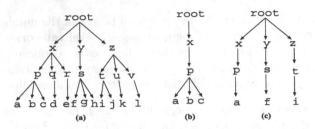

Fig. 2. (a) Sample concept hierarchy (b) $min\Pi(Y/C)$ for $Y = \{a, b, c\}$ (c) $max\Pi(Y/C)$ for $Y = \{a, f, i\}$

Now, we introduce two terms, *maximal projection of Y* ($max\Pi(Y/C)$) and *minimal projection of Y* ($min\Pi(Y/C)$) to explain the computation of $drank(Y)$.

Definition 2 *Maximal projection of Y* ($max\Pi(Y/C)$): *The $max\Pi(Y/C)$ is a projection of C for Y by considering all the leaf-level items merge at the root through distinct intermediate nodes.*

The number of nodes in $max\Pi(Y/C)$ is equal to $(|Y| \times h)$, where h is the height of C and $|Y|$ is the number of items in Y.

Definition 3 *Minimal projection of Y* ($min\Pi(Y/C)$): *The $min\Pi(Y/C)$ is a projection of C for Y by considering all the items merge at the immediate parent node.*

The number of nodes in $min\Pi(Y/C)$ is equal to $(|Y| + h - 1)$, where h is the height of C and $|Y|$ is the number of items Y.

Example 2. Consider the concept hierarchy in Fig. 2, for $min\Pi(\{a, b, c\}/C) = 5$ (Fig. 2(b)) and for $max\Pi(\{a, f, i\}/C) = 9$ (Fig. 2(c)).

Computing the *drank.*

We define the diversity of a pattern as the ratio of $|\Pi(Y/C)|$ to $|max\Pi(Y/C)|$ which is equal to $\frac{|\Pi(Y/C)|}{|max\Pi(Y/C)|}$. The minimum value of this ratio is equal to $\frac{|min\Pi(Y/C)|}{|max\Pi(Y/C)|}$ and the maximum value of this ratio is equal to $\frac{|max\Pi(Y/C)|}{|max\Pi(Y/C)|}$ (which is equal to 1). We call this ratio as diverse rank (*drank*). After applying min-max normalization, the final formula for *drank*, given Y and C, is as follows.

$$drank(Y) = \left(\frac{\left(\frac{|\Pi(Y/C)|}{|Y| \times h} \right) - \left(\frac{|Y| + h - 1}{|Y| \times h} \right)}{1 - \left(\frac{|Y| + h - 1}{|Y| \times h} \right)} \right) (1 - 0) + 0 \tag{1}$$

$$drank(Y) = \frac{(|\Pi(Y/C)|) - (|Y| + h - 1)}{(h - 1)(|Y| - 1)} \tag{2}$$

where, $|\Pi(Y/C)|$ is the number of nodes in $\Pi(Y/C)$, $|Y|$ is the number of items in a pattern, and h is the height of $\Pi(Y/C)$.

Example 3. In continuation of Example 2 (Fig. 2), the $drank(\{a, b, c\}) = \frac{5-5}{2 \times 2} =$ 0.0. Similarly, the $drank(\{a, b, e\}) = 0.25$, $drank(\{a, b, f\}) = 0.5$, $drank(\{a, e, f\})$ = 0.75, and $drank(\{a, f, i\}) = 1.0$.

3   Proposed Approach for Diverse Recommendations

In this section, we explain the association rule based RS and proposed approach.

3.1   Association Rule Based RS Approach

An association rule [2] is in $X \rightarrow Y$ the form, where X and Y are sets of items (or itemsets). The set X is called as *body* and Y is called as *head* of the rule. The meaning of rule $X \rightarrow Y$ is that X and Y are present in the transactional database together with some *minSup* and *minConf*.

The steps for association rule based RS [8] is as follows.

i. **Data set preparation**: The data needs to be converted into the transactional data to execute the frequent pattern mining algorithms. Each user wise, the items purchased/rated by a user forms a transaction. The set of such transactions form transactional data.

ii. **Generation of association rules**: The frequent patterns [2] are generated with *minSup* using the transactional database. The association rules are generated from frequent patterns with the user specified *minConf*. The top-N high confidence rules are selected for recommendations.

iii. **Recommendations**: The top-N association rules are used for recommendation. For a user, the items that appear in the left side (body) of the rule is matched, if they match, the right side (head) item is recommended.

3.2   Proposed Approach

We extract the frequent patterns at certain *minSup*. Next, we generate the association rules using *minConf*. For each pattern, diverse rank is computed using the Eq. 2.

The following options are possible to selection different set of association rules based on confidence and *drank* values.

– HC-LD: Top-N rules with high *confidence* and low *drank*.
– HC-HD: Top-N rules with high *confidence* and high *drank*.
– LC-HD: Top-N rules with low *confidence* and high *drank*.
– LC-LD: Top-N rules with low *confidence* and low *drank*.

In addition, we can consider the following two options.

– HC: Ignore the *drank* and select top-N rules with high *confidence*.
– HD: Ignore the *confidence* and select top-N rules with high *drank*.

In the preceding alternatives, HC is the existing approach. The alternatives HD and HC-HD are the proposed approaches. It is expected that HC provides better accuracy without considering the diversity aspects, and HD provides better diversity performance without considering accuracy. The HC-HD approach is expected to give better accuracy and diversity. The framework is as follows.

i. **Data set preparation**
   (a) **Preparation of user rating data:** This step is the same as the existing approach.
   (b) **Formation of concept hierarchy:** All the items in the transactional database are organized in hierarchical manner.

ii. (a) **Generation of association rules:** Same as the existing approach.
   (b) **Computing the** *drank*: For each set of items in the rules, we compute *drank* using the Eq. 2.

iii. **Recommendations:** Top-N rules are selected using the one of the following approaches.
   (a) HC: Improving the accuracy for RS.
   (b) HC-HD: Improving the accuracy as well as diversity for RS.
   (c) HD: Improving diversity for RS.

4   Experimental Results

In the section, we explain about the data set and methodology. Next, we present the results.

4.1   Preparation of Data Set and Methodology

We conducted experiments on the data set provided by MovieLens project (www.grouplens.org/). The data set contain 100,000 ratings of 943 users on 1682 movies. All ratings follow the 1 to 5 numerical scale indicate bad to excellent. The movies liked by the user are taken as unique transaction. All the movies are included in the transactional database having rated greater than 2. The data

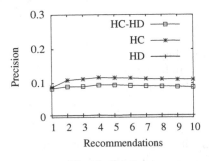

Fig. 3. Precision

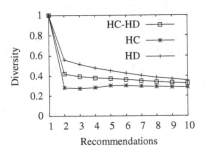

Fig. 4. Diversity

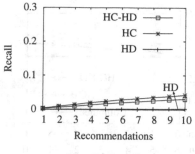

Fig. 5. Recall

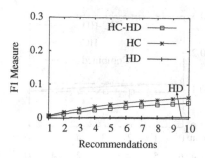

Fig. 6. F1-Score

set has been converted into a user-movie ratings' matrix that had 943 rows (i.e., 943 users) and 1682 columns (i.e., 1682 movies).

Concept Hierarchy: For the experiments, we have taken the taxonomy provided in Internet Movie Database (IMDb) [6] for MovieLens data set. In the concept hierarchy, the *root* is the dummy node. We have included the genre type at first level, genre at second level, sub-genres at third level and the movies at the leaf-level of the concept hierarchy. In the MovieLens data set, there are some movies that are belong to two or more genres. To accommodate these kinds of movies a new genre is added on employing the names of both genres as internal-nodes, and the movie is added to the newly created internal node.

Methodology: We employ Apriori [2] algorithm to generate the association rules with $minSup = 0.10$ and $minConf = 0.05$. By considering the number of items in the rule as a pattern, we compute the *drank* for all association rules.

The data set has been divided into five distinct splits of training and test data. The training set is used to generate the association rules, and the test set is used to predict the recommendation. Experiments have been conducted on the training set, and extracted the association rules, called *top-N* rules. We then look into the test set and match items with *top-N* rules. The items that appear in the left side (body) of the rule is matched to the test set, if they match, the right side (head) item is included in a special set, called *Hit* set, and each match is known as a *Hit*. The *Hit* set is employed for testing the accuracy. All the experiments are performed on each of the five splits, and average is reported. We employ Precision, Recall, F1-metric [3], and Diversity as performance metrics.

In all the experiments, we evaluate the performance of three approaches for 1 to 10 recommendations.

4.2 Results

We show the results on accuracy of HC, HC-HD, HD, combined (HC and HC-HD) and usefulness of diverse recommendations.

(i) Performance

Figure 3 shows the precision performance of HC, HD, and HC-HD. The association rules with high support and confidence indicate that these movies are

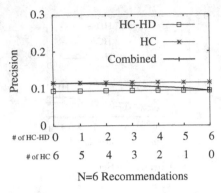

Fig. 7. Precision of HC and HC-HD   **Fig. 8.** Diversity of HC and HC-HD

liked large number of users. As a result, HC exhibits good performance. It can be observed that the performance of HD is significantly low as compared to HC. This is due to the fact that HD does not consider the support and confidence values to rank association rules. Still, the performance of HD does not reach zero. This is due to the appearance of a few high confidence rules. It can be observed that the performance of HC-HD is close to the performance of HC. This is due to the fact that HC-HD contains the rules with both high confidence as well as high drank. Due to high confidence rules, the performance is near to HC. Similar trend can be observed in recall and F1 metric performances as shown in Figs. 5 and 6 respectively.

In Fig. 4, we show the diversity performance. The results show that HC-HD is returning the results with high drank as compared to HC. As expected, the performance of HD is significantly high as compared to HC and HC-HD. The performance of HC-HD is nearer to actual HC and HD.

From the results, it can be concluded that the performance of the HC-HD approach gives recommendations with high diversity and without signifiable compromise the precision performance.

(ii) Combined performances

We conducted another experiment by combining the rules of HC and HC-HD. Selecting x rules from HC and $(6 - x)$ rules from HC-HD, where x is varied from 1 to 6. From both Figs. 7 and 8, the following observations can be made.

- By selecting 2 rules from HC-HD and 4 rules from HC high precision which is equal to HC could be achieved by improving diversity considerably.
- By selecting 4 rules from HC-HD and 2 rules from HC high diversity which is equal to HC-HD could be archived by improving the precision considerably.

From the results, it can be consider that the performance of the accuracy and diversity can be adjusted based on the application demand.

(iii) Usefulness of Diverse Recommendations

In Table 1, recommendations of sample of two users are provided to understand the usefulness of approach. In this table, we have shown userId, movies in

Table 1. Hit sets of HC and HC-HD for sample users

UserId	Movies in test set	Hit set of HC	Hit set of HC-HD
1	The Saint, L.A. Confidential, Twelve Monkeys, Independence Day	*The Saint* (Fusion/FusionAction/ RomanceThriller), *L.A.Confidential* (Fusion/FusionCrime/ Film-Noir- MysteryThriller), *Twelve Monkeys* (Fusion/FusionDrama/ Sci-Fi)	*The Saint* (Fusion/FusionAction/ RomanceThriller), *L.A.Confidential* (Fusion/FusionCrime/ Film-Noir- MysteryThriller), *Independence Day* (**Dark/Violence/ ActionWar**)
2	Aliens, The Princess Bride, Braveheart, Forrest Gump	*Aliens* (Dark/ Violence/ ActionWar), *The Princess Bride* (Dark/ Violence /ActionAdventure), *Braveheart* (Dark/ Violence/ ActionWar)	*Aliens* (Dark/ Violence/ ActionWar), *The Princess Bride* (Dark/ Violence /ActionAdventure), *Forrest Gump* (**Fuslon/ FusionComedy/ RomanceWar**)

test set, hit set of HC, and hit set of HC-HD. From the table, it can be observed that HC-HD is providing at least one diverse item (higher-level category in the concept hierarchy) in the set as compared to HC.

5   Summary and Conclusions

The recommender systems are important in e-Commerce. It has been viewed that the similar recommendations alone do not result in high user satisfaction. In this paper, we have proposed an association rule based recommendation approach to provide the diverse recommendations. After organizing the items in the form of concept hierarchy, the diverse rank of a pattern can be computed based on category of items. The experimental results on the real-world MovieLens data set show that the proposed approach improve the performance of association rule based RSs with better diversity without compromising the accuracy. As a part of future work, we are planning to extend the notion of diversity in other types of recommendation approaches.

References

1. Adomavicius, G., Tuzhilin, A.: Toward the next generation of recommender systems: a survey of the state-of-the-art and possible extensions. TKDE **17**(6), 734–749 (2005)
2. Agrawal, R., Imieliński, T., Swami, A.: Mining association rules between sets of items in large databases. In: Proceedings of the ACM SIGMOD, pp. 207–216. ACM (1993)
3. Basu Roy, S., Amer-Yahia, S., Chawla, A., Das, G., Yu, C.: Constructing and exploring composite items. In: Proceedings of the SIGMOD, pp. 843–854. ACM (2010)
4. Bradley, K., Smyth, B.: Improving recommendation diversity. In: Proceedings of the 12th AICS, pp. 75–84 (2001)
5. Han, J., Fu, Y.: Discovery of multiple-level association rules from large databases. In: Proceedings of the 21th VLDB, pp. 420–431. Morgan Kaufmann Publishers Inc. (1995)
6. IMDb: Internet movie database. http://www.imdb.com/genre/ (2014), [Online; accessed October-2014]
7. Kumara Swamy, M., Reddy, P.K., Srivastava, S.: Extracting diverse patterns with unbalanced concept hierarchy. In: Tseng, V.S., Ho, T.B., Zhou, Z.-H., Chen, A.L.P., Kao, H.-Y. (eds.) PAKDD 2014, Part I. LNCS, vol. 8443, pp. 15–27. Springer, Heidelberg (2014)
8. Lin, W., Alvarez, S.A., Ruiz, C.: Efficient adaptive-support association rule mining for recommender systems. Data Min. Knowl. Discov. **6**(1), 83–105 (2002)
9. McNee, S.M., Riedl, J., Konstan, J.A.: Being accurate is not enough: how accuracy metrics have hurt recommender systems. In: CHI 2006, pp. 1097–1101. ACM (2006)
10. Sarwar, B., Karypis, G., Konstan, J., Riedl, J.: Analysis of recommendation algorithms for e-commerce. In: 2nd Conference on e-Commerce, pp. 158–167. ACM (2000)
11. Su, X., Khoshgoftaar, T.M.: A survey of collaborative filtering techniques. Adv. Artif. Intell. **2009**, 4:2–4:2 (2009)
12. Ziegler, C.N., McNee, S.M., Konstan, J.A., Lausen, G.: Improving recommendation lists through topic diversification. In: Proceedings of the 14th WWW, pp. 22–32. ACM (2005)

A Prime Number Based Approach for Closed Frequent Itemset Mining in Big Data

Mehdi Zitouni[1,2]($\boxtimes$), Reza Akbarinia[2], Sadok Ben Yahia[1,3], and Florent Masseglia[2]

[1] University of Tunis ElManar, Faculty of Sciences of Tunis,
LIPAH-LR 11ES14, 2092 Tunis, Tunisia
`Mehdi.Zitouni@inria.fr`, `Sadok.Benyahia@fst.rnu.tn`
[2] INRIA and LIRMM, Montpellier, France
`{Reza.Akbarinia,Florent.Masseglia}@inria.fr`
[3] Institut Telecom, Telecom SudParis, umr 5157, Cnrs Samovar, Evry, France

Abstract. Mining big datasets poses a number of challenges which are not easily addressed by traditional mining methods, since both memory and computational requirements are hard to satisfy. One solution for dealing with such requirements is to take advantage of parallel frameworks, such as MapReduce, that allow to make powerful computing and storage units on top of ordinary machines. In this paper, we address the issue of mining closed frequent itemsets (CFI) from big datasets in such environments. We introduce a new parallel algorithm, called CLoPN, for CFI mining. One of the worth of cite features of CLoPN is that it uses a prime number based approach to transform the data into numerical form, and then to mine closed frequent itemsets by using only multiplication and division operations. We carried out exhaustive experiments over big real world datasets to assess the performance of CLoPN. The obtained results highlight that our algorithm is very efficient in CFI mining from large real world datasets with up to 53 million articles.

Keywords: Data mining · Closed frequent itemset · MapReduce · Big data · Parallel algorithm · CLoPN

1 Introduction

In the past few years, advances in hardware and software technologies have made it possible for the users of information systems to produce large amounts of transactional data. Such data, which rapidly grows over time and needs more and more space and computing, is referred to as big data. Unfortunately, mining only frequent itemsets in big data generates an overwhelming large number of itemsets. The latter has the disadvantage to make their interpretation almost impossible and affect the reliability of the expected results.

Several studies were conducted to define and generate condensed representations of frequent itemsets. In particular, closed frequent itemsets have received much attention with very general proposals. Since its introduction in [1], closed

© Springer International Publishing Switzerland 2015
Q. Chen et al. (Eds.): DEXA 2015, Part I, LNCS 9261, pp. 509–516, 2015.
DOI: 10.1007/978-3-319-22849-5_35

frequent itemsets (CFI) have been actively studied. However, existing algorithms for mining CFI flag out good performance as far as dealing with small datasets or when the support threshold is high. But, when the scale of dataset grows or the support threshold turns to be low, both memory usage and communication costs become hard to bear. Some early efforts tried to speed up the mining algorithms by running them in parallel [2,3]. Even though, they could improve the mining performance, but promote other issues such as data partitioning, minimization of communication costs, and potential errors caused by node failures.

To deal with the above mentioned issues, we can take advantage of parallel frameworks, such as MapReduce or Spark, that allow to make powerful computing and storage units on top of ordinary machines. In this paper, we introduce a parallel algorithm, called CLoPN (CLOsed itemset as Prime Numbers), in the aim of an efficient CFI mining by using such frameworks. In CloPN, we develop a new approach based on mathematical techniques. The items from the dataset are transformed into prime numbers, and closed frequent itemsets are generated by using only division and multiplication operations in the MapReduce framework. The main contributions of this paper are as follows:

- We propose a numerical representation of transactional datasets using a new transformation technique.
- We design an efficient MapReduce-based parallel algorithm for CFI mining by taking advantage of the mathematical properties of our numerical representation.
- We carried out exhaustive experiments on real world datasets to evaluate the performance of CloPN. The results show that our algorithm sharply outperforms the pioneering algorithms in CFI mining of large real world datasets with up to 53 millions articles.

2   Related Work

Many research efforts [4,5] have been introduced to design parallel FIM algorithms capable of working under multiple threads under a shared memory environment. Unfortunately, these approaches do not address the problem of heavy memory requirement when processing large scale databases.

Since the introduction of closed frequent itemset in [1], numerous algorithms for mining it were proposed [6]. In fact, these algorithms tried to reduce the problem of finding frequent itemsets to the problem of mining closed frequent itemsets by limiting the search space to only closed frequent itemsets rather than the the whole powerset lattice. Furthermore, they have good performance whenever the size of dataset is small or the support threshold is high. However, as far as the size of the datasets become large, both memory use and communication cost are unacceptable. Thus, parallel solutions are of a compelling need. But, research works on parallel mining of closed frequent itemset are few. Authors in [7] introduce a new algorithm based on the parallel FP-Growth algorithm PFP [8] that divides an entire mining task into independent parallel subtasks and achieves quasi-linear speedups. The algorithm mines closed frequent itemsets in four MapReduce jobs and introduces a redundancy filtering approach to deal with the problem of generating redundant itemsets. However, experiments on algorithm were on a small-scale dataset.

3 Preliminary Notions

In this section, we give the definition of some concepts used throughout the paper.

Definition 1 (EXTRACTION CONTEXT). *An extraction context is a triplet $\mathcal{K} = (\mathcal{O}, \mathcal{I}, \mathcal{R})$, where $\mathcal{O}$ represents a finite set of objects, $\mathcal{I}$ is a finite set of items and $\mathcal{R}$ is a binary (incidence)relation (i.e., $\mathcal{R} \subseteq \mathcal{O} \times \mathcal{I}$). Each couple $(o, i) \in \mathcal{R}$ expresses that the object $o \in \mathcal{O}$ contains the item $i \in \mathcal{I}$.*

The closure operator $('')$ denotes the closure operator $\phi \circ \psi$ s.t. (ϕ, ψ) represents a couple of operators defined by $\psi : \mathcal{P}(\mathcal{I}) \rightarrow \mathcal{P}(\mathcal{O})$ s.t. $\psi(I) = \{o \in \mathcal{O} | \forall i \in I, (o, i) \in \mathcal{R}\}$ and $\phi : \mathcal{P}(\mathcal{O}) \rightarrow \mathcal{P}(\mathcal{I})$ s.t. $\phi(O) = \{i \in \mathcal{I} | \forall o \in O, (o, i) \in \mathcal{R}\}$. It is worth of cite, the closure operator induces an equivalence relation on the power set of items partitioning it into disjoint subsets called *equivalence classes* [9]. The largest element (*w.r.t.* the number of items) in each equivalence class is called a *closed itemset* (CI) and is defined as follows:

Definition 2 (CLOSED ITEMSET). *An itemset $I \subseteq \mathcal{I}$ is said to be closed if and only if $I'' = I$ [1]. The support of I, denoted by $Supp(I)$, is equal to the number of objects in $\mathcal{K}$ that contain I. I is said to be frequent if $Supp(I)$ is greater than minsupp.*

4 Frequent Closed Itemset Mining

In this section, we introduce our CLoPN algorithm for CFI mining. It operates in two steps. The first step is dedicated to transform the data and generates a list of transformed frequent items sorted in a descending order of their supports using a MapReduce pass. We denote this list as *F-List*. Using a second MapReduce pass, CLoPN mines the complete set of closed frequent itemsets. Both steps are developed in the following.

4.1 First Step: Prime Number Transformation

In this subsection, we present some definitions and theorems used to transform and manipulate the input data by our algorithm.

Definition 3 (PRIME NUMBER). *An integer $X > 1$ is a prime number, if it is divided evenly only by 1, and itself, e.g., $X = 193$ is a prime number.*

Definition 4 (PRIME FACTORS). *A non-prime integer $X > 1$ is a composed integer. So, X can be presented by a set of prime numbers multiplied together to make the original number. Formally, X can be represented as $X = p_1^{m1} p_2^{m2} ... p_r^{mr}$, having p_i as a prime number, and m_i a positive integer called multiplicity[1] of*

---

[1] There is no duplicated items in a transaction from transactional dataset, so we will suppose that the multiplicity will be $m_i = 1$, without any loss of information.

	A	B	C	D	E
1	×		×	×	
2		×	×		×
3	×	×	×		×
4		×			×
5	×	×	×		×

Item	PN	Supp
B	2	4
C	3	4
E	5	4
A	7	3

ID	$\mathcal{T}$	Primes	$V_{\mathcal{T}}$
1	C, A	3,7	21
2	B, C, E	2,3,5	30
3	B, C, E, A	2,3,5,7	210
4	B, E	2,5	10
5	B, C, E, A	2,3,5,7	210

Fig. 1. (**Left**) An extraction context $\mathcal{K}$. (**Middle**) The Items from $\mathcal{K}$ sorted in descending order with their corresponding prime numbers and supports. (**Right**) The context $\mathcal{K}$ transformed and reduced.

p_i, e.g., The prime factors of $X = 330$ are 2, 3, 5 and 11, since we have $330 = 2 \times 3 \times 5 \times 11$.

Let us now define our data transformation technique.

Definition 5 (TRANSACTION TRANSFORMATION). *Let $\mathcal{T} = \{i_j...i_k\}$ be a transaction, where i_r items from $\mathcal{T}$. The transformation process assign a prime number p_r to each item $i_r \in \mathcal{T}$. Then, the new numeric value of the transaction, denoted as $V_{\mathcal{T}}$, is computed by applying the following equation:*

$$V_{\mathcal{T}} = \prod_{j}^{k} p_r \tag{1}$$

Figure 1 illustrates the transformation phase for a sample dataset.

4.2 Second Step: Closed Frequent Itemset Mining

After the transformation step, CLoPN starts the second MapReduce pass to extract the complete set of frequent closed itemsets. Below, we give a description of the Map and Reduce phases for mining CFIs of our algorithm.

Map Phase: During this step, the CloPN algorithm splits the context into sets of minimized contexts, denoted as Conditional-context. Bellow we give the definition of a conditional-context cited in [10].

Definition 6 (CONDITIONAL-CONTEXT). *[11] Given an extraction context $\mathcal{K}$. Let i be a frequent item in $\mathcal{K}$. The i-Conditional-context is the subset of transactions containing i, while all infrequent items, item i and items following i in the F-list are omitted.*
Let j be a frequent item in X-conditional-context, where X is a frequent itemset. The jX-conditional-context is the subset of transactions in X-conditional-context containing j, while all infrequent items, item j, and items following j in local F-list are omitted.

Having frequent items sorted in descending order of their respective supports, for each item i the map function of CLOPN checks its conditional-context from transactions containing only items that collocate with i. To facilitate the exploration of these sub contexts, we could use the technique proposed in [10] that defines a header table which is associated to each context. This table lists items contained in the corresponding conditional-context, sorted in descending order of their supports. However, in our approach, extracting closed frequent itemset wont need the use of this table (cf following section). Each of our map tasks emits a key-value pair, where the key is an item from the F-list generated in step one, and the value is a transaction which will be part of the conditional-context of the item.

Reduce Phase: Before describing the Reduce phase of our approach, for an itemset X, the following properties holds.

1. Closed itemset X extracted from a conditional-context is discovered by concatenating the items which are as frequent as X (in the conditional-context).
2. There is no need to develop a conditional-context of an itemset Y included in a closed frequent itemset already discovered X such that $\text{Supp}(X)$ is equal to $\text{Supp}(Y)$.

Now, having the following definition,

Definition 7 (GREATEST COMMON DIVISOR (GCD)). *The greatest common divisor of two or more integers, when at least one of them is not zero, is the largest positive integer that divides the numbers without a remainder. e.g. the GCD of 8 and 12 is 4.*

We introduce the following theorem to extract frequent closed itemsets.

Theorem 1 (GCD-CLOSURE). *Let X-conditional-context be the subset of transactions contained X. The greatest common divisor in X-conditional-context represents the closure between all transactions in the conditional-context.*

Proof. As shown in Definition 2, the closure of an itemset X is produced from the intersection between all transactions containing X. With Definition 7 and manipulating the prime numbers, the GCD between integers is unique. Thus, having all VT_{ID}, extracting the closure from a set of transactions amounts to calculate the GCD between them. Thus, the GCD in X-Conditional-context is the closure between transactions composing X−Conditional-context.

After the shuffling process, each reducer will receive an item with its conditional-context as input. In fact, having the prime number representing the item and its transactions as a set of V_Ts, computing the closure from the conditional-context is straightforward by computing the GCD of all transactions of the conditional-context. Doing so, there is no further need to store supports of items contained in the conditional-context. Indeed, if the closure exists, then will have inevitably the same support as that of the item. By concatenating the closure to the candidate item multiplying the prime number of the item and the number representing the closure, the result will be a closed frequent itemset that is represented as a number which is added to the set of final results. The steps of CLoPN algorithm are as follows.

- **Step 0**: A preprocessing step to transform our data into numerical form.
- **Step 1**: Parallel Counting F-list: a MapReduce job is executed in order to count the frequency of all items that appear in $\mathcal{K}$. Each mapper is given one part of $\mathcal{K}$. The result is stored in one F-list for each shard of $\mathcal{K}$ in mappers.
- **Step 2**: Parallel mining closed frequent itemset: in this step CloPN mines the locally closed frequent itemset using the GCD approach in parallel.
- **Step 3**: Collecting Results.

With the above four steps, we can mine the complete set of closed frequent itemsets correctly. Indeed, CLoPN starts after the preprocessing phase, to count supports of items of the ground set. For that, it uses a classical MapReduce counting job, where Map is fed with fragments from $\mathcal{K}$, and the output of Reduce would be the list of items (primes) with their supports sorted in descending order and pruned from those who are not frequent. In the second step, CLoPN starts to mine the set of closed frequent itemsets. CLoPN algorithm uses a second MapReduce pass where it splits the context into a new conditional-contexts, having the dataset as a Map input. It tests the inclusion of the item by dividing the product of the transaction in dataset by the prime number representing the item. This latter is token as key and its conditional-context as value to start the reduce phase. At this point, the CLoPN algorithm tries to extract the closure by applying our new closure operation that calculates the GCD between all transactions in the conditional-context.

5   Experimental Evaluation

In order to assess effectiveness of the proposed approach, we carried out thorough tests on two real-life datasets. The first one, called "Wikipedia Articles", represents a transformed set of Wikipedia articles into a transactional dataset, each line mimics a research article. It contains 7,892,123 transactions with 6,853,616 distinct items, in which the maximal length of a transaction is 153,953, and the size of the whole dataset is 4.7 Gigabytes. The second dataset, called "ClueWeb", consists of about one billion web pages in ten languages that were collected in January and February 2009 and it is used by several tracks of the TREC conference. During our experiments, we used a part of "Clue Web" dataset that contains 53,268,952 transactions including 11,153,752 items with a maximal length of a transaction equal to 689,153. The size of the considered "Clue Web" is 24.9 Gigabytes.

To perform our experiments, we used one of the clusters of Grid5000 [12] which is a large-scale and versatile test-bed for experiment-driven research on parallel and distributed computing. Our experiments were performed on a cluster with 10 nodes equipped with Hadoop 1.3.0 version. One node was designed as

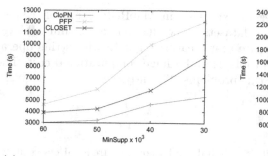
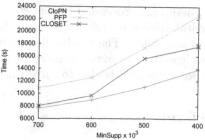

(a) The results on "Wikipedia Articles"   (b) The results on "Clue Web"

Fig. 2. Experimental results of CLoPN

master, which was responsible for scheduling tasks execution among different nodes, and other nodes were set as workers. Our algorithm was implemented using java and the JDK version is openjdk-7-jdk. We compared our algorithm to a basic implementation of CLOSET algorithm in MapReduce and the parallel FP-Growth algorithm (PFP in short), which is described briefly in the Related Work section.

Figure 2(a) illustrates the experimental results tested on "Wikipedia Articles" of the three algorithms under different values of minimum support (Minsupp) less then 1 % of the overall size of the dataset. CLoPN sharply outperforms both other algorithms. The reason is that, "Wikipedia articles" presents a large number of items which is almost equal to transactions numbers. Thus, as far as the Minsupp value is low, PFP and CLOSET generate too many candidates, and a lot of long conditional-contexts for each one. So, the inclusion tests and evaluations under the pruning methods used in these two algorithms causes lead as expected to poor performances. CLoPN overcomes these problems by using prime numbers to generate the conditional-contexts through division operations. Furthermore, the GCD in each conditional-context has eliminated the check of supports between the candidate and its deduced closure. Thus, the phase of computing the frequency of the closure is no longer of need.

Figure 2(b) shows the experimental results tested on "CloueWeb". By reducing Minsupp, the number of frequent closed items does not change significantly. However, the number of matching transactions for candidate itemsets increases quickly, and this has the disadvantage of generating big conditional-contexts for itemset candidates. Interestingly enough, our algorithm performs much better, thanks to the efficient strategies which we developed, particularly the GCD approach that avoids redundant computations for producing the closure from each conditional-context.

6   Conclusion

In this paper, we revisited the closed frequent itemset mining problem and proposed a new algorithm for mining closed frequent itemsets by using prime numbers and processing big integers expressed in MapReduce model. Experimental results on two large scale datasets show that our algorithm achieves good scalability. The results illustrate that treating big data with string operations could cause many memory problems and a huge computational cost. The proposed method is able to solve this problems efficiently.

References

1. Pasquier, N., Bastide, Y., Taouil, R., Lakhal, L.: Discovering frequent closed itemsets for association rules. In: Beeri, C., Bruneman, P. (eds.) ICDT 1999. LNCS, vol. 1540, pp. 398–416. Springer, Heidelberg (1998)
2. El-hajj, M., Zaïane, O.R.: Parallel leap: large-scale maximal pattern mining in a distributed environment. In: Conference on Parallel and Distributed Systems, pp. 135–142 (2006)
3. Chen, K., Zhang, L., Li, S., Ke, W.: Research on association rules parallel algorithm based on FP-growth. In: Liu, C., Chang, J., Yang, A. (eds.) ICICA 2011, Part II. CCIS, vol. 244, pp. 249–256. Springer, Heidelberg (2011)
4. Zaïane, O.R., El-Hajj, M., Lu, P.: Fast parallel association rule mining without candidacy generation. In: ICDM Conference, pp. 665–668 (2001)
5. Liu, L., Li, E., Zhang, Y., Tang, Z.: Optimization of frequent itemset mining on multiple-core processor. In: VLDB Conference, pp. 1275–1285 (2007)
6. Wang, J., Han, J., Pei, J.: Closet+: searching for the best strategies for mining frequent closed itemsets. In: KDD Conference, pp. 236–245 (2003)
7. Wang, S.-Q., Yang, Y.-B., Gao, Y., Chen, G.-P., Zhang, Y.: Mapreduce-based closed frequent itemset mining with efficient redundancy filtering. In: ICDM Workshops, pp. 449–453. IEEE Computer Society (2012)
8. Li, H., Wang, Y., Zhang, D., Zhang, M., Chang, E.Y.: Pfp: parallel fp-growth for query recommendation. In: ACM Conference on Recommender Systems (RecSys), pp. 107–114 (2008)
9. Bastide, Y., Taouil, R., Pasquier, N., Stumme, G., Lakhal, L.: Mining frequent patterns with counting inference. In: KDD Conference, pp. 66–75 (2000)
10. Borgelt, C.: An implementation of the fp-growth algorithm. In: OSDM Workshop, pp. 1–5 (2005)
11. Pei, J., Han, J., Mao, R.: Closet: an efficient algorithm for mining frequent closed itemsets. In: SIGMOD Workshop on Research Issues in, Data Mining and Knowledge Discovery, pp. 21–30 (2000)
12. https://www.grid5000.fr/

Multilingual Documents Clustering Based on Closed Concepts Mining

Mohamed Chebel[1]([✉]), Chiraz Latiri[1], and Eric Gaussier[2]

[1] Research Laboratory LIPAH, Faculty of Sciences of Tunis,
University Tunis El Manar, Tunis, Tunisia
mohammedchebel@gmail.com, chiraz.latiri@gnet.tn
[2] Research Laboratory LIG, AMA Group,
University Joseph Fourier (Grenoble I), Grenoble, France
eric.gaussier@imag.fr

Abstract. The scarcity of bilingual and multilingual parallel corpora has prompted many researchers to accentuate the need for new methods to enhance the quality of comparable corpora. In this paper, we highlight the interest and usefulness of Formal Concept Analysis in multiligual document clustering to improve corpora comparability. We propose a statistical approach for clustering multiligual documents based on multilingual Closed Concepts Mining to partition the documents belonging to one or more collections, writing in more than one language, in a set of classes. Experimental evaluation was conducted on two collections and showed a significant improvement of comparability of the generated classes.

1 Introduction

In this paper, we propose a new multilingual document clustering method of a noisy comparable corpora to improve corpora comparability. We rely on the approach of [5] in IR who took advantage of a coupling between the formal and the relational aspect to take into account the relationships between objects of the same context. We chose to make a coupling between Formal Concepts Analysis (FCA) [3] to formalize the content of one or more document collections of the same language, and a vector model [9], in order to take into account the relations similarity between the documents of various concepts.

Formal Concepts Analysis applied in a text mining context, allows to retrieve document classes in the form of Closed Concepts, denoted CCs. Furthermore, a vector space model based on the vectors extensions of the extracted CCs, allows to align CCs in different languages by calculating the degree of similarity of the extracted monolingual Closed Concepts, with the aim of generating multilingual CCs.

Our motivation is double, namely: first of all, it comes to apply FCA techniques for the extraction of the CCs from comparable corpora; and, second, to exploit these CCs and vector models in the clustering and alignment of multilingual documents. We propose in the following paper to study the contribution

© Springer International Publishing Switzerland 2015
Q. Chen et al. (Eds.): DEXA 2015, Part I, LNCS 9261, pp. 517–524, 2015.
DOI: 10.1007/978-3-319-22849-5_36

of multilingual *CCs* and vector models in multilingual documents clustering and in the improvement of the corpora comparability.

The remainder of the paper is organized as follows: Sect. 2 presents a brief review of the literature on bilingual and multilingual document classification. Section 3 is dedicated to the presentation of the new method and the definition of its steps. An experimental evaluation is presented in Sect. 4. A conclusion and future work are given in Sect. 5.

2   Literature Review

Several approaches have been addressed to bilingual and multilingual documents classification in the literature and could be divided in categories, among which we can mention: *Dependent approaches to language:* These approaches are mainly based on linguistic features of documents language to study [6]. The main drawback of this approach is its specificity to certain languages only. *Independent approaches to language:* More close to our work, these approaches are mainly based on the presentation of multilingual and bilingual documents in an independent representation of the language [4, 10, 11]. *Approaches based on machine translation:* These approaches rely on the automatic translation of all documents in the collection [2] or parts of the documents [1] to allow the mapping of bilingual and multilingual documents in the same representation space.

We propose in this work a new framework of multilingual documents classification, based on the extracted multilingual Closed Concepts, which is independent of the documents language and using a technique of translation and a statistical disambiguation. In our context of research, the independent representation of the documents language is modeled by Closed Concepts extracted from a comparable corpora. In the next section, we introduce the proposed multilingual document clustering approach and its overall process.

3   Multilingual Documents Clustering Approach

The purpose of our contribution is to disclose how that can be achieved when FCA is used within clustering of multilingual documents. Our goal is to improve the comparability of comparable corpora. The different steps of the clustering process are detailed in the rest of the paper.

3.1   Mathematical Foundations: Key FCA Settings

In this paper, we shall use in text mining field, the theoretical framework of Formal Concept Analysis (FCA) presented in [3]. First, we formalize an extraction context made up of documents and index terms, called *textual context*.

Definition 1. *A **textual context** is a triplet $\mathfrak{M} := (\mathcal{C}, \mathcal{T}, \mathcal{I})$ where:*

- $\mathcal{C} := \{d_1, d_2, \ldots, d_n\}$ *is a finite set of n documents of a collection.*
- $\mathcal{T} := \{t_1, t_2, \cdots, t_m\}$ *is a finite set of m distinct terms in the collection. The set $\mathcal{T}$ then gathers without duplication the terms of the different documents which constitute the collection.*
- $\mathcal{I} \subseteq \mathcal{C} \times \mathcal{T}$ *is a binary (incidence) relation. Each couple $(d, t) \in \mathcal{I}$ indicates that the document $d \in \mathcal{C}$ has the term $t \in \mathcal{T}$.*

Definition 2. *A **concept**: A concept $C = (T, D)$ can be defined by its intension and extension where the extension is all documents in $\mathcal{C}$ that belong to the concept while the intension is the set of terms in $\mathcal{T}$ shared by these documents. The support of C in $\mathfrak{M}$ is equal to the number of documents in $\mathcal{C}$ containing all the term of T. The support is formally defined as follows[1]:*

$$Supp(C) = |\{d | d \in \mathcal{C} \wedge \forall t \in T : (d, t) \in \mathcal{I}\}| \tag{1}$$

$Supp(C)$ is called the absolute support of C in $\mathfrak{M}$. The relative support (aka frequency) of $C \in \mathfrak{M}$ is equal to $\dfrac{Supp(C)}{|\mathcal{C}|}$.

A concept is said *frequent* if its terms co-occur in the collection a number of times greater than or equal to a user-defined support threshold, denoted *minsupp*. Otherwise, it is said *unfrequent*.

Definition 3. ***Galois Closure Operator***: *Let a concept $C = (D, T)$. Two functions are defined in order to map sets of documents to sets of terms and vice versa. Thus, for $T \subseteq \mathcal{T}$, we define:*

$$\Psi(T) := \{d | d \in \mathcal{C} \wedge \forall t \in T : (d, t) \in \mathcal{I}\} \tag{2}$$

$\Psi(T)$ is equal to the set of documents containing all the terms of T. Its cardinality is then equal to $Supp(T)$.

For a set $D \subseteq \mathcal{C}$, we define:

$$\Phi(D) := \{t | t \in \mathcal{T} \wedge \forall d \in D : (d, t) \in \mathcal{I}\} \tag{3}$$

$\Phi(D)$ is equal to the set of terms appearing in all the documents of D.

Both functions Ψ and Φ constitute Galois operators. Consequently, the compound operator $\Omega := \Phi \circ \Psi$ is a Galois closure operator which associates to a termset T the whole set of terms which appear in all documents where the terms of T co-occur.

Definition 4. *A concept $C = (D, T)$ is said to be closed, denoted in the rest of the paper CC, if $\Omega(C) = C$. A closed concept is then the maximal set of terms common to a given set of document. A closed concept is said to be frequent w.r.t. the minsupp threshold if $Supp(C) = |\Psi(T)| \geq minsupp$ [7]. Hereafter, we denote by CC a closed concept.*

---

[1] In this paper, we denote by $|X|$ the cardinality of the set X.

It is worth noting that in our work, a Closed Concept represents a class of documents grouped by a set of representative terms.

3.2 Multilingual Closed Concepts Extraction

Extracting Closed Concepts CCs from a French-English comparable corpora requires a linguistic preprocessing step. In order to extract the most representative terms, a linguistic preprocessing is performed on the comparable corpora by using the French morpho-syntactic tagger TREETAGGER[2]. In this application, we focus only on the common nouns, the proper nouns and the adjectives.

In order to extract CCs from the comparable corpora, we adapted the CHARM-L algorithm [12] to our textual context $\mathfrak{M}$. The algorithm extracts all the frequent closed concepts. It also adds the possibility to construct the entire concept lattice, as well as the links between all sub/super-concepts (*i.e.*, frequent closed concepts). The minimal threshold of the minimum and maximum threshold of the support, i.e., *minsupp*, w.r.t. the document collection size and the term distributions. While considering the *Zipf* distribution of every collection, the maximum threshold of the support values is experimentally set in order to spread trivial terms which occur in the most of the documents, and are then related to too many terms. On the other hand, the minimal threshold allows eliminating marginal terms which occur in few documents, and are then not statistically important when occurring in a closed concept.

The algorithm CHARM-L generates iteratively, for each language separately the whole set of frequent concepts and then derives only closed concepts CCs respectively to each language. It is worth recalling that the extracted closed concepts are modeled as pairs : $CC = < \{t_1, t_2, \ldots, t_n\}, \{d_1, d_2, \ldots, d_m\} >$. Each pair represents a set of documents $\{d_1, d_2, \ldots, d_m\}$ sharing a set of terms $\{t_1, t_2, \ldots, t_n\}$ with a support greater than or equal to *minsupp*. The algorithm stops when no more frequent closed concept to handle. So, this algorithm allows to generate as result all Closed Concepts CC_{fr} French and English Closed Concepts CC_{en}, separately.

In the next step, we propose to deploy all the English and French CCs extracted to study their contribution in clustering multilingual documents (Table 1).

Table 1. Examples of CC_{fr} and CC_{en} extracted from the comparable corpora French-English CLEF'2003.

CC_{fr}	CC_{en}
$\{bank, gouvernement\}, \{618, \ldots, 41881\}$	$\{bank, government\}, \{1200, \ldots, 39216\}$
$\{bilan, entreprise, Europe\}, \{230, \ldots, 32556\}$	$\{business, Europe\}, \{16, \ldots, 44112\}$

---

[2] http://www.ims.uni-stuttgart.de/projekte/corplex/TreeTagger/.

3.3 Closed Concepts Translation and Disambiguation

The translation of CCs extensions consists in enriching each term of the extension of a CC in its source language L_s with terms in the target language L_c. Indeed, the extensions are translated from French to English and from English to French. All possible translations of a term is retained. Extensions translation was carried out using an ONLINE MT SYSTEM[3].

To disambiguate the translated terms of CCs extensions, we implemented a statistical method of disambiguation. So, we start with a set of CCs expressed in a source language L_c and their extensions which are translated to the target language L_c and a set of CCs expressed in the target language L_c.

For each CC_{source}, we derive all possible combinations of translated terms of extensions, we then calculate the Euclidean distance from every combination compared to all CC_{target}.

For each CC_{source}, the combination of translated terms that gives the smallest Euclidean distance average of all CC_{target} will be considered as the most appropriate translation of the CC_{source}. The purpose of this step (i.e., translation and disambiguation) is to correspond the terms t_i in CC_{source} extension with the CC_{target} intension (i.e., documents), in order to construct the vectors corresponding to CCs of each language. These vectors will be used in the alignment of CCs (cf Sect. 3.4).

3.4 Multilingual Closed Concepts Alignment

The alignment of the Closed Concepts aims to gather the most similar CC extracted from each language, to generate multilingual Closed Concepts CC_{fr-en} (i.e., multilingual classes) French-English. The idea that we propose is based on the use of a vector space model based on unsupervised classification algorithm, i.e., K-MEANS. In this regard, the K-MEANS algorithm allows to assess the similarity degree between CCs and to combine them based on the vectors related to the translated and disambiguated CCs. Thus, a Closed Concept is represented by a vector in the vector space. The vector corresponding to each CC is composed of representative terms in the Closed Concept (i.e., the terms of the extensions in the CCs).

Our approach starts with a set of CCs vectors expressed in a source language L_s and in a target language L_c. As Output, we obtain multilingual Closed Concepts CC_{fr-en} (i.e., clusters of multilingual documents) and monolingual Closed Concepts CC_{fr}/CC_{en} (i.e., clusters of monolingual documents) as depicted in Table 2.

4 Experiments and Results

4.1 Description of the Comparable Corpora

We thoroughly evaluate our multilingual document clustering approach on a collection of noisy comparable documents. To create this comparable corpora,

---

[3] https://translate.google.com/.

we used a subset of multilingual collection from the evaluation campaign of cross-lingual information retrieval systems CLEF'2003 [8]. We consider two collections namely SDA95 (French) et GlasgowHerald95 (English). The resulted corpora is considered as a noisy comparable corpora. Some statistical features of the used comparable corpora extracted from CLEF'2003 after preprocessing are given in Table 3.

4.2   Evaluation Framework

Frequency based methods are the best alternative to mesure the comparability between the bilingual corpora collections, because it generally focus on the quantity of vocabulary in common between documents, which is the main feature of bilingual lexicons extraction. So, for the purpose of evaluating the new comparability of French and English documents gathered in a same Closed Concept (*i.e.*, cluster), we propose to use standard information retrieval measures, considering two well known models, namely : binary model and vector space model.

– **Binary Measure of Comparability:** We calculated the degree of comparability of our corpora with a binary measure as follows:
  Given a corpora with a source language L_s and target language L_c, the binary $trans(W_s, d_t)$ returns 1 if the translation of a word vocabulary W_s source was found in the target vocabulary d_t and 0 in other cases.
  So $binDC$ for source and target documents is calculated as follows:

$$binDC(d_s, d_t) = \frac{\sum_{W_s \in d_s} trans(W_s, d_t)}{|d_s|} \qquad (4)$$

  We note that $binDC(d_s, d_t)$ and $binDC(d_t, d_s))$ are not symmetrical So the total comparability of our comparable corpora is calculated as the average of the two degrees of comparability.

– **Vector Measure of Comparability:** In the vector information retrieval model, each document vector is composed of indexing terms. The similarity measure is generally the cosine of the angle between the two vectors. For representing documents in the vector space model, we built the sources and targets vectors with the following method: we extracted indexes with LEMUR[4]. The resulting index (*i.e.*, in the source language) has been translated, with the online MT system.

Table 2. Statistical characteristics of bilingual (*i.e.*, French and English) and monolingual CCs resulting from the alignment step for $k = 300$.

Language of CC	CC_{fr-en}	CC_{fr}	CC_{en}
Number of CC	42	78	180

---

[4] http://www.lemurproject.org/.

Table 3. Statistical features of the comparable corpora CLEF'2003 after preprocessing

Language of the corpora	SDA95 (Frensh)	GlasgowHerald95 (English)
Number of documents	42615	56472
Vocabulary size	105010	227301

4.3 Experimental Results and Discussion

To evaluate the evolution of the comparability degree of the initial English-French corpora regarding bilingual classes (*i.e.*, Closed Concepts) generated by our clustering approach, we calculated the comparability degrees of documents within each bilingual class through both the Boolean model $(bin - DC)$ and the vector model from French to English and from English to French. Then we derive an average comparability degree.

Table 4. Comparability results

Measures	$bin - DC$	$cosine - DC$
Average comparability degree of CC_{fr-en}	0.48	0.39
Comparability degree of the initial corpora	0.37	0.32

Table 4 summarizes the results of two comparability measures: $bin - DC$ for the Boolean method and $cosine - DC$ for the vector method. The results show that the $bin - DC$ comparability is better than $cosine - DC$. This result was expected because the measure based on the vector model includes the term weighting unlike the Boolean model that uses a binary weighting.

Our experimental results of comparability measures are promising and yield a significant improvement. Indeed, the average of comparability degree of bilingual classes obtained is interesting (0.48 for the binary measure and 0.39 for vector measurement).

Moreover, if we compare the comparability degree of the original corpora (0.37 for the binary measure and 0.32 for Vector measure) with the average comparability degree of the obtained bilingual classes, we highlight that our FCA-based multilingual documents clustering approach allows achieving better results than state-of-the-art word-based IR models.

Our statistics shows that only 15 % of the obtained classes are bilingual classes, which is expected because only 32 % of the closed concepts which are aligned was French concepts beside 78 % for English concepts. It shows also highlights interesting comparability degrees. In fact, 84 % of the bilingual classes have a comparability degree higher than the initial corpora and 17 % of the bilingual classes have a comparability degree higher or equal to 0.6 which is a fairly significant degree compared to the original corpora.

5 Conclusion

This paper highlights the deployment of Closed Concepts in the multilingual documents clustering in order to improve the comparability of the noisy comparable corpora. Experimental evaluation was conducted on two collections from CLEF'2003 and showed a significant improvement of comparability. As future work we are currently implementing an algorithm in order to have a greater number of multilingual clusters and that are more balanced in terms of documents number. Further work includes evaluation of recall and precision of the extracted multilingual concepts.

Acknowledgements. This work is partially funded by the DGRST-CNRS no 14/R 1401 Franco-Tunisian project, entitled "Text mining for construction of bilingual lexicons and multilingual information retrieval"

References

1. Chen, H.-H., Lin, M.-S., Wei, Y.-C.: Novel association measures using web search with double checking. ACL-44: Proceedings of the 21st International Conference on Computational Linguistics and the 44th Annual Meeting of the Association for Computational Linguistics, pp. 1009–1016 (2006)
2. Evans, D., Klavans, J.: A platform for multilingual news summarization. Technical Report, Department of Computer Science, Columbia University (2003)
3. Ganter, B., Wille, R.: Formal Concept Analysis. Springer, Heidelberg (1999)
4. Gliozzo A., Strapparava C.: Cross language text categorization by acquiring multilingual domain models from comparable corpora. ParaText 2005: Proceedings of the ACL Workshop on Building and Using Parallel Texts (2005)
5. Mimouni, N., Nazarenko, A., S. Salotti: Classification conceptuelle d'une collection documentaire, intertextualité et recherche d'information. CORIA 2012: 9th French Information Retrieval Conference. Bordeaux, France (2012)
6. Montalvo, S., Martínez, R., Casillas, A., Fresno, V.: Multilingual news document clustering: two algorithms based on cognate named entities. In: Sojka, P., Kopeček, I., Pala, K. (eds.) TSD 2006. LNCS (LNAI), vol. 4188, pp. 165–172. Springer, Heidelberg (2006)
7. Pasquier, N., Bastide, Y., Taouil, R., Stumme, G., Lakhal, L.: Generating a condensed representation for association rules. J. Intell. Inf. Syst. **24**(1), 2560 (2005)
8. Peters C.: Result of the CLEF 2003 cross-language system evaluation campaign. In: Notes for the CLEF 2003 Workshop, 21–22 August, Trondheim, Norway (2003)
9. Salton, G., Buckely, C.: Term weighting approaches in automatic text retrieval. Inf. Process. Manage. **24**(5), 513–523 (1988)
10. Romeo, S., Ienco, D., Tagarelli, A.: Knowledge-based representation for transductive multilingual document classification. In: Hanbury, A., Kazai, G., Rauber, A., Fuhr, N. (eds.) ECIR 2015. LNCS, vol. 9022, pp. 92–103. Springer, Heidelberg (2015)
11. Wei, C.-P., Yang, C.-C., Lin, C.-M.: A latent semantic indexing-based approach to multilingual document clustering. Decis. Support. Syst. **45**(3), 606–620 (2008)
12. Zaki, M.-J., Hsiao, C.-J.: Efficient algorithms for mining closed itemsets and their lattice structure. IEEE Trans. Knowl. Data Eng. **17**(4), 462–478 (2005)

Modeling, Extraction, Social Networks

Analyzing the Strength of Co-authorship Ties
with Neighborhood Overlap

Michele A. Brandão and Mirella M. Moro[✉]

Universidade Federal de Minas Gerais, Belo Horizonte, Brazil
{micheleabrandao,mirella}@dcc.ufmg.br

Abstract. Evaluating researchers' scientific productivity usually relies on bibliometry only, which may not be always fair. Here, we take a step forward on analyzing such data by exploring the strength of co-authorship ties in social networks. Specifically, we build co-authorship social networks by extracting the datasets of three research areas (sociology, medicine and computer science) from a real digital library and analyze how topological properties relate to the strength of ties. Our results show that different topological properties explain variations in the strength of co-authorship ties, depending on the research area. Also, we show that neighborhood overlap can be applied to scientific productivity evaluation and analysis beyond bibliometry.

Keywords: Research evaluation · Social networks analysis · Interdisciplinary co-authorship behavior

1   Introduction

With so much research being published, a fundamental question for government decision makers and university administrators is: how to evaluate scientific productivity and quality? To help with such analysis, there is a current trend on designing systems for aiding the evaluation (and ranking) of researchers and their groups, graduate programs and even conferences (e.g. ArnetMiner). Such evaluation depends on various aspects (academic, economic, etc.), but the most common one is bibliometry. Indeed, designing such systems and their evaluation methods have considered the degree of interaction among researchers (i.e., the strength of the relationship between researchers), publications patterns of research areas and the degree of endogamy [16,17,22].

Another current trend is to use social network analysis for considering not only one single researcher's productivity but also the scientific outcomes from a research group or lab, department, graduate program and institution. Considering a social network formed by researchers as nodes and their co-authorships as edges, one central aspect is the *strength of the ties* among researchers. According to Granovetter's theory, the ties are *weak* when they serve as bridges in the network by connecting users from different groups, and *strong* when they link individuals in the same group [11]. Going back to the original question

© Springer International Publishing Switzerland 2015
Q. Chen et al. (Eds.): DEXA 2015, Part I, LNCS 9261, pp. 527–542, 2015.
DOI: 10.1007/978-3-319-22849-5_37

(evaluating research productivity), the strength of ties provides a completely different way to answer it that goes beyond the "simply count publications and their citations". Specifically, recent studies have considered the strength of ties for assessing the quality of conferences or evaluating research teams [22], and evaluating and predicting the influence of a publication [25], among others.

Tie strength may be measured by a combination of the amount of time, the cooperation intensity and the reciprocal services that characterize the tie [11,21]. Such strength may also be measured by using the *neighborhood overlap* metric [8], a numerical quantity that captures the total number of co-authorships between the two ends of each edge. This metric has been used for uncovering the community structure [15], and analyzing structural properties of a large network of mobile phone users [2]. However, it has *not* been well explored to measure the strength of ties in the academic context.

Nonetheless, as we will show in this paper, the strength of ties is just one part of a complex equation when evaluating research networks. Specifically, studying how other *topological properties* of social networks relate to the strength of ties may improve the accuracy and quality of existing methods that combine bibliometry and academic social analysis. For example, Silva et al. [22] show that using (only) weak ties is not suitable for assessing the quality of Computer Science journals. Also, our study helps to understand how topological properties explain variations in the strength of ties, which is important to understand the dynamics of co-authorships among researchers.

Considering the aforementioned scenario of evaluating research outcome that considers bibliometry and social network concepts, this work aims to improve the design of such systems by showing a detailed analysis of how the strength of ties is influenced. Hence, the contributions of this paper are summarized as:

- We build co-authorship social networks considering real datasets of publications from three different areas (computer science, medicine, sociology), which are also quantitatively compared (Sect. 3).
- We then characterize the strength of ties measured by neighborhood overlap and define a nominal scale to classify the ties as weak or strong. Also, we verify if the Granovetter's theory governs the three networks (Sect. 4).
- We analyze how nine topological properties impact on the strength of ties. Initially, we study the correlation between each property and neighborhood overlap. Then, our analysis takes one step forward and considers a regression model to quantify how the combination of each property to neighborhood overlap may improve even further the evaluation results (Sect. 5).

2 Related Work

There is a new trend on designing systems and online applications for aiding the evaluation of researchers productivity as well as the quality of conferences, graduate programs and other sets of researchers. For example, Klink et al. [14] have designed *DBL-Browser*, a system that allows the search of authors/publications

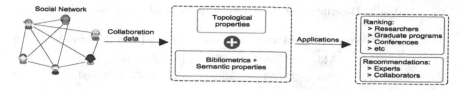

Fig. 1. Architecture of a general research evaluation-oriented system.

and analysis of social networks on the basis of bibliographical data. Such systems are usually designed by relying on specific indicators and quality measures. Quantitative measures of research productivity and quality usually consider total number of citations, number of distinct publication venues, and so on, (i.e., they focus on the *bibliometric* aspects of research productivity) [6,16,17].

Considering each researcher individually has its merits, but taking the evaluation to the *social* level may provide brand new perspectives. For example, the set of publications from a research area, conference, journal or institution may be used to build a co-authorship network. Specifically, co-authorship networks have been used for evaluating collaborations for expert recommendation [4,18], assessing graduate programs [17], and motivating the design of disambiguation frameworks [24].

Here, we focus on co-authorship social networks for understanding the topological and dynamics laws that govern complex networks. Likewise, Silva et al. [22] consider research endogamy as a metric for assessing the quality of conferences based on co-authorship networks, and Gonçalves et al. [10] consider the individual importance in a co-authorship network to estimate a scholar popularity. Overall, they consider topological properties of the networks in order to investigate productivity and quality patterns in the academic context.

Topological properties capture the characteristics of the graph that represents a social network [26]. Here, we consider topological properties that have been applied for analyzing the importance of researchers [3,10,25], the distance among researchers [5,19], and the density of connections in the network [3]. One special property is the strength or weakness of the ties (edges, relationships or links) in the network. Now, we build upon it and propose to analyze how some topological properties relate to the strength of ties in co-authorship social networks. We experimentally analyze such relations and dynamics by considering real datasets and co-authorship networks. To our knowledge, we are the first to show the relevance of other topological properties to the design of research evaluation-oriented systems and applications, and to explore how such properties explain variations in the strength of ties.

3   Datasets Main Features

We consider systems and database applications that measure research productivity by evaluating the social aspects of researchers. Examples of such applications include (but are not limited to) ranking researchers, graduate programs

Table 1. Description of the datasets for building social networks.

Area	#Insts	#Res	#Publs	AvgPubA	#Pairs (#dist)	#SubA
CS	111	543	48,706	89.69	16,312 (1,563)	884
Med	114	368	75,553	205.30	16,089 (778)	664
Soc	43	96	7,195	74.95	322 (39)	68

Note: CS = Computer Science, Med = Medicine, Soc = Sociology

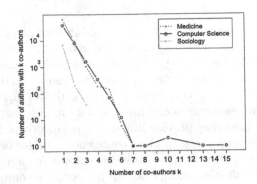

Fig. 2. Distribution of numbers of co-authors for researchers in each area.

and conferences, and recommending experts and collaborators, as presented in [4,16–18,22]. Figure 1 illustrates a general architecture of such systems: from a social network (e.g., co-authorship network), the collaboration data is extracted; then the system applies bibliometrics and analyzes semantic properties, whose results are sent to the applications. Here, we show the importance of analyzing the topological properties as well, as also done in [3].

For characterizing the importance of topological properties, we build three co-authorship social networks using the *CiênciaBrasil* datasets[1]. The publications available in *CiênciaBrasil* are from Brazilian researchers and have been collected from Lattes[2], an online platform for archiving researchers' curriculum vitae, in November 2013. Each network represents the co-authorships among researchers from three areas: computer science, medicine and sociology.

Figure 2 presents the distributions of the number of co-authors for the three areas. We have chosen these areas because there are clearly three different degrees of collaboration: low for sociology (up to three co-authors), medium for medicine (up to seven co-authors) and high for computer science (up to 15 co-authors). Note that for the authors in computer science and medicine, publishing together is a common practice, which is not for sociology (as presented by Simon [23]). For instance, from 7,195 publications in sociology, 83.96 % have only one author. Also, although the three networks are from Brazilian researchers, there are other studies that corroborate such behavior on different datasets [1,9,12]. Finally,

---

[1] Datasets available at http://www.dcc.ufmg.br/~mirella/Tools/DEXA2015/.

[2] Lattes: http://lattes.cnpq.br.

Table 1 summarizes the datasets and shows the number of institutions, number of researchers (authors of papers), number of publications, average number of publications per author, number of pairs of co-authors (and number of distinct pairs of co-authors) and number of subareas.

4 Characterizing the Strength of Ties

In this paper, we measure the strength of co-authorship ties using a metric called *neighborhood overlap*. We now describe and characterize such metric in the three co-authorship social networks (Sect. 4.1). Also, as we are studying the strength of ties, we have to verify if the ties follow Granovetter's theory, i.e., if they govern the co-authorship networks when the strength of ties is measured by neighborhood overlap. In the case of such theory be not valid in the studied co-authorship networks, we would have to consider other theories to characterize the ties as weak or strong. Hence, we analyze the topological properties when weak and strong ties are removed (Sect. 4.2).

4.1 Neighborhood Overlap Characterization

We consider the topological properties of three co-authorship social networks to investigate their impact to the ties strength. Here, the strength of a tie (relationship between a pair of researchers) is estimated by the neighborhood overlap metric of an edge connecting researchers A_1 and A_2 [8]. The metric is given by the equation: $|A_{c_1} \cap A_{c_2}| / (|A_{c_1} \cup A_{c_2}| - (A_1, A_2 \ themselves))$, where A_{c_1} represents the co-authors of researcher A_1, and A_{c_2} the co-authors of A_2.

Using neighborhood overlap to measure the strength of co-authorship ties is not well defined in the literature yet. Hence, such investigation is important to discover whether such metric captures the real importance of the tie to a researcher. We emphasize that neighborhood overlap does not consider the semantic of the relationship between pairs of researchers (for example, the period or the asymmetric importance of the co-authorship to a researcher) because it is not the focus of this paper. However, neighborhood overlap captures the *densities of co-authorship ties among researchers*. Such density is important because it measures the strength of the ties considering the *neighborhood effect*, which means the larger the number of common co-authors that a pair of researchers have, the more such pair tends to initiate a collaboration. This indicates that such tie will be stronger considering the neighborhood overlap. In academic context, such effect is a variable that explains the tendency of a researcher to collaborate with others based on the relational effects of the researcher working in the neighborhood. For instance, given a pair of researchers A_1 and A_2 with a tie in the network where each researcher has six co-authors (including A_2 as co-author of A_1, and contrariwise) with only one in common. Such tie is weak, because there is only one co-author in common from 10 possibilities. In such case, the tie being weak means that the neighbors do not have much effect on the co-authorship, i.e. the intensity of co-authorship is small.

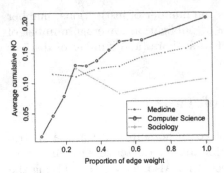
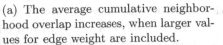
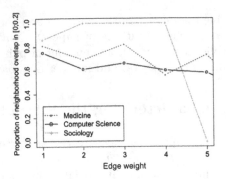

(a) The average cumulative neighborhood overlap increases, when larger values for edge weight are included.

(b) Proportion of weak ties in co-
authorships of same edge weight.

Fig. 3. Analyzing the neighborhood overlap versus edges weight.

Conceptually, the tie strength grows as the neighborhood overlap increases.
A tie is considered *weak* when the neighborhood overlap is very small [8], and the
problem becomes what "very small" means (0.01? 0.1? 0.2?). Note that ties are
strong when their neighborhood overlap has opposite values to those defined for
weak ties. According to Granovetter [11], the tie strength is based on properties
associated to the individuals' relationship (e.g. the intensity or the age). Hence,
in order to define a nominal scale for tie strength, we compare two properties:
neighborhood overlap and edge weight (i.e., an edge is the link between two
researchers and the weight is the number of their co-authored publications).

Considering the three areas, Figs. 3(a) and (b) show the relationship between
their neighborhood overlap and edge weight. Figure 3(a) shows the average cumulative neighborhood overlap for a fraction of edges weight. To compute such
average, we sort all edges in increasing order by weight. Then, we take the top
k $(0 \leq k \leq 1)$ fraction of edges from the sorted list and calculate the average neighborhood overlap for those edges. Observe that when edges with larger
weights are included, the average cumulative neighborhood overlap increases
in the three areas on average. Also, including all weighted edges, the average
cumulative neighborhood overlap does not reach much more than 0.2 (0.21 to
computer science, 0.17 to medicine and 0.11 to sociology). Such average represents the typical value of neighborhood overlap in each network.

Then, Fig. 3(b) presents the proportion of ties with edge weight varying from
one to five when the neighborhood overlap ranges from zero to 0.2. In computer
science and medicine, 55 % of the ties with weights in the range [1;5] have neighborhood overlap in the range [0;0.2]. In sociology, most ties with small weight
have neighborhood overlap in the range [0;0.2]. Such analysis suggests that ties
with small weight also have high proportion of ties with neighborhood overlap
between [0;0.2]. Therefore, we define that *a tie is weak* when the neighborhood
overlap is within [0;0.2] as well. We also note that, in practice, the values of the

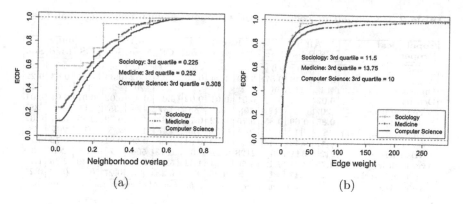

Fig. 4. Empirical CDF of neighborhood overlap and edge weight computed by the co-authorship between pairs of researchers.

nominal scale vary slightly depending on the research area, but to simplify the analysis, we have standardized the values in only one scale, which does not result in loss of information.

Now, the aim is to analyze the distribution of neighborhood overlap and edge weight in the three networks and compare them. Such analysis contributes to understanding the presence of weak and strong ties in the networks. Figure 4 presents the ECDF[3] (Empirical Cumulative Distribution Function) of neighborhood overlap and edge weight for the three co-authorship networks. The third quartile values in the graphic indicate that 75 % of the data is less than that number in each area. It shows that only 25 % of the co-authorships have neighborhood overlap equal or higher than 0.308 in computer science, 0.252 in medicine and 0.225 in sociology. The number of co-authorships among researchers (weight) is also small for the three networks; i.e., 25 % of the pairs of researchers have weight equal or higher than 10 in computer science, 13.75 in medicine and 11.5 in sociology. Hence a direct conclusion is that weak ties are strongly present in co-authorship networks independently from the research area.

Also, the analysis of neighborhood overlap and edge weight distributions indicates that computer science has more ties with co-authors in common than medicine and sociology. Hence, the neighborhood overlap does indeed capture the real co-authorship among researchers in the three networks, because the results concur with the distribution of co-authors from Fig. 2.

4.2   Granovetter's Theory Analysis

Granovetter's theory [11] raises the hypothesis about the importance of weak ties in same situations. Following such theory in the academic context, the weak ties

---

[3] ECDF assigns a probability of $1/n$ to each value of neighborhood overlap and edge weight, sorts the data in increasing order, and calculates the sum of the assigned probabilities up to and including each value.

Table 2. Co-authorship social networks properties when removing weak ties.

Topological Property	All Ties			Equal 0			[0;0.1]			[0;0.2]		
	CS	Med	Soc	CS	Med	Soc	CS	Med	Soc	CS	Med	Soc
# removed ties	0	0	0	199	168	23	481	302	24	896	516	29
Avg. degree	3.44	2.48	0.87	3.00	1.94	0.36	2.38	1.52	0.33	1.47	0.83	0.22
Diameter	14	11	6	12	9	2	14	20	2	13	6	2
Density	0.02	0.02	0.04	0.01	0.01	0.02	0.01	0.01	0.02	0.01	0.01	0.01
# communities	20	23	12	108	118	35	137	128	35	215	205	38
Modularity	0.85	0.69	0.85	0.75	0.72	0.73	0.83	0.8	0.72	0.89	0.87	0.62
Connect. component	8	11	12	98	110	35	127	119	35	212	201	38
Avg. clust. coef.	0.47	0.42	0.43	0.63	0.69	0.91	0.68	0.66	0.81	0.71	0.78	0.79
# triangles	2125	641	6	2128	641	6	1813	506	5	1190	268	3
Avg. path length	4.75	3.31	2.12	4.56	4.21	1.11	5.20	7.14	1.17	4.49	2.36	1.17
Avg. neig. overlap	0.21	0.17	0.11	0.26	0.25	0.31	0.33	0.31	0.27	0.43	0.39	0.25

Note: Avg = average, CS = Computer Science, Med = Medicine, Soc = Sociology

connect researchers from different communities, for instance, different research groups or teams. Considering the case where the ties are strong between two individuals, such theory suggests the existence of a triad which claims that if A and B are connected, and A and C are connected, then B and C will also be connected. In other words, the strong ties link researchers within the researchers groups and teams. In order to better understand the strength of ties behavior when it is measured by neighborhood overlap and verify whether Granovetter's theory governs the networks, we present how removing the weak ties and then the strong ties affects the topological properties.

Table 2 presents the properties of the social networks with all ties, when removing ties with neighborhood overlap equal to zero, and in the ranges [0;0.1] and [0;0.2]. As expected, all three networks are topologically affected by removing weak ties. In general, the average degree, diameter, density and total number of triangles decrease, whereas the total number of communities, total number of connected components and average clustering coefficient increase. Such increases indicate that the weak ties really connect researchers from different communities and validate Granovetter' theory. Note that changes in the topological properties when removing ties are common. However, if such ties are *not* important and/or influential in the network, such changes are *not* significant. Specifically, when all weak ties are removed, most properties have an increased or decreased value of at least 50 % (i.e., weak ties are *essential* in those networks).

Regarding the removal of strong ties, we discuss them briefly without showing their values for space constraints. The ties are strong when their neighborhood overlap is higher than 0.2. We remove the strong ties starting from the range 1.0 to 0.8, then 1.0 to 0.5, and 1.0 to 0.2 (not including 0.2). Most properties change their value (i.e. have greater impact) more when weak ties are removed than strong ties. For example, when weak ties are removed, the diameter varies more than when strong ties are removed. Second, the properties for modularity, average clustering coefficient, average neighborhood overlap and average path length change differently when weak and strong ties are removed. For instance, the average clustering coefficient increases 50 % for computer science when removing all weak ties and decreases 81.22 % when removing strong ties. These results

Table 3. Topological properties (see [8] for formal definitions).

Notation	Description
eBetweenness	edge betweenness of each edge in the network
Weight	frequency of researchers published a work together
ClosenessA1/ClosenessA2	closeness of each researcher in a pair of researchers
AvgCloseness	average closeness of each pair of researchers
EccentricityA1/EccentricityA2	eccentricity of each researcher in a pair of researchers
AvgEccentricity	average eccentricity of each pair of researchers
ClusterCoefA1/ClusterCoefA2	clustering coefficient of each researcher in a pair of researchers
AvgClusterCoef	average clustering coefficient of each pair of researchers
nTrianglesA1/nTrianglesA2	number of triangles of each researcher in a pair of researchers
AvgNTriangles	average number of triangles of each pair of researchers
wDegreeA1/wDegreeA2	weight degree of each researcher in a pair of researchers
AvgWDegree	average weight degree of each pair of researchers
EigenvecA1/EigenvecA2	eigenvector value of each researcher in a pair of researchers
AvgEigenvec	average eigenvector of each pair of researchers
PageRankA1/PageRankA2	page rank of each researcher in a pair of researchers
AvgPageRank	average page rank of each pair of researchers

agree with Granovetter's theory, because the clustering coefficient measures the trend of nodes in a network to form clusters. By definition, the greater the clustering coefficient, the greater the number of closed triads. Additionally, we have found that the giant component breaks more rapidly when a critical number of weak ties is removed. All these results show that Granovetter's theory is *valid to the three networks*: weak ties link researchers from different groups or teams, whereas strong ties connect researchers from the same one.

5   The Impact of the Properties on the Strength of Ties

So far, we have measured the tie strength by the neighborhood overlap of two researchers. However, as our evaluation will show, understanding how such

Table 4. Pearson correlation coefficients between topological properties and neighborhood overlap. Values lower than 0.1 are insubstantial.

Properties	Computer Science	Medicine	Sociology
ClusterCoefA1/	0.47	0.62	0.81
ClusterCoefA2	0.53	0.56	0.79
AvgClusterCoef	0.61	0.73	0.88
nTrianglesA1/	0.37	0.32	0.86
nTrianglesA2	0.28	0.15	0.82
AvgNTriangles	0.38	0.26	0.93
Weight	0.31	0.3	0.34
EigenvecA1/	0.3	0.2	0.65
EigenvecA2	0.23	0.11	0.62
AvgEigenvec	0.3	0.17	0.68
wDegreeA1/	0.24	0.17	0.54
wDegreeA2	0.17	0.03	0.34
AvgWDegree	0.27	0.11	0.6
PageRankA1/	0.2	-0.04	-0.17
PageRankA2	0.09	-0.02	-0.1
AvgPageRank	0.19	-0.03	-0.26
EccentricityA1/	0.11	0.2	0.086
EccentricityA2	0.13	0.2	0.23
AvgEccentricity	0.13	0.21	-0.034
ClosenessA1/	-0.06	-0.14	0.013
ClosenessA2	-0.1	-0.15	-0.2
AvgCloseness	-0.09	-0.15	-0.067
eBetweenness	-0.4	-0.5	-0.26

metric is influenced and using other properties for defining the strength of ties may both contribute considerably to evaluating research productivity. Indeed, considering co-authorship networks, the actual strength of a tie may depend on characteristics of the network graph by itself. Therefore, we relate the strength of ties to topological properties that capture: *(i)* the importance of researchers (weight degree, eigenvector and pageRank) and pairs of researchers (edge betweenness and weight) within a network; *(ii)* the distance of a researcher from the furthest other (closeness, eccentricity); and *(iii)* the degree to which researchers tend to cluster together (clustering coefficient and number of triangles), in Table 3.

Analyzing the relative importance of each property to neighborhood overlap and quantifying their strength may reveal knowledge to improve systems that combine bibliometry and social network analysis (e.g. algorithms for ranking graduate programs [17]). Also, relative measures (consider and compare more than one aspect) can better represent the reality than absolute counts (represents only one aspect) [20]. Hence, the combination of neighborhood overlap with other metrics can generate better results. Here, we analyze the correlations between each property (individually) and the neighborhood overlap (Sect. 5.1), and use a regression model to quantify the importance of sets of properties to neighborhood overlap (Sect. 5.2) as well.

5.1 Correlation Analysis

We quantify the correlations between each property and the strength of ties by using the Pearson linear correlation coefficient and the Spearman's rank correlation coefficient [13]. The Pearson coefficient measures the linear relationship between two variables. When such relationship is not linear, the Spearman's rank correlation coefficient is more appropriate. The Pearson coefficient is presented in Table 4, whereas Spearman results are very similar and thus omitted.

We follow the conventions to interpret correlation coefficient from [7]: greater than 0.7 is *very large*, within [0.5;0.7] is *large*, within [0.3;0.5] is *moderate*; within [0.1;0.3] is *small*, and anything smaller than 0.1 is *insubstantial* (note that the same ranges are valid for negative correlation). Table 4 shows that, for most properties, the correlations with neighborhood overlap tend to be *small* or *moderate*. In computer science and medicine, the neighborhood overlap varies slightly when the other topological properties change. However, in the sociology network, these correlations are balanced (there are 11 large correlations against 12 *small* and *moderate* ones). Such situation may be explained by the smaller size of the sociology co-authorship network and/or limited number of co-authorships (the maximum number of co-authors in a paper is three for sociology network).

As for specific differences among the networks, number of triangles (nTrianglesA1, nTrianglesA2 and AvgNTriangles), weighted degree (wDegreeA1, wDegreeA2 and AvgWDegree) and eigenvector (EigenvecA1, EigenvecA2 and AvgEigenvec) have small or moderate correlations in computer science and medicine, but large in sociology. Such large linear correlation in sociology is expected, because, conceptually, the existence of more triangles indicates more neighbors in common. A direct question is: why the number of triangles is not linearly correlated with neighborhood overlap in computer science and medicine? We may only speculate that such correlation is non linear or there is no correlation due to factors not considered in this work. Furthermore, the large linear correlation between neighborhood overlap and metrics that capture the importance of a researcher in a network (weighted degree and eigenvector) indicates that in sociology, important researchers in the network have co-authorship ties stronger than others. Up to now, we cannot claim such behavior for computer science and medicine.

Furthermore, for computer science and medicine, clustering coefficient is the property most linearly correlated with neighborhood overlap: the correlation between AvgClusterCoef and the neighborhood overlap reaches 0.61 in computer science and 0.73 in medicine. For sociology, the most correlated property is number of triangles (0.93), although the correlation between clustering coefficient and neighborhood overlap is also large (0.88).

These correlations provide evidence of properties that are strongly related to the strength of ties in co-authorship networks, and thus can help to explain such strength. They may also provide insights to improve methods that consider the strength of ties concept. For instance, based on the observed patterns and the research area, the design of an assessment system for conferences or teams evaluation that considers the strength of ties may also consider clustering coefficient, number of triangles, edge betweenness, weighted degree and/or eigenvector to improve its accuracy and overall quality. In addition, the correlated topological properties can be used to answer the question: why a certain tie has a particular strength? Finally, topological properties may also be combined to better explain the strength of ties or have a non linear correlation (greater than linear correlation) with neighborhood overlap. We investigate these issues next.

5.2   Regression Analysis

Another way to further assess the relative importance of each topological property to the neighborhood overlap is to use *regression models*. Such model is obtained from a statistical process called regression analysis that estimates the relationships among variables [13]. The quality of the regression model is estimated by the coefficient of determination R^2, which represents the fraction of the variation in the response variable y that is explained by other variables. Here, the response variable y is neighborhood overlap, and the variables are the nine topological properties described in Table 3. Overall, the goals in this section are: to *identify* which of the topological properties are necessary to build a model that efficiently characterizes the neighborhood overlap in each research area and to *quantify* the relative importance of each property to neighborhood overlap.

To define an appropriate regression model, we apply *linear regression models* without and with logarithm and exponential transformations. First, we use simple linear regression model (without and with transformations) considering each topological property as the variable (factor) and the neighborhood overlap as the response variable (or estimated variable). However, for most topological properties, the quality of regression is poor. For instance, the R^2 value for the linear regression between edge betweenness and neighborhood overlap is 0.161 in computer science (0.068 in sociology and 0.25 in medicine) and with logarithm transformation is 0.329 (−0.22 in sociology and −0.96 in medicine). Nonetheless, using *simple exponential regression model* has improved the quality of regression. For example, the R^2 value for the exponential regression between edge betweennness and neighborhood overlap is 0.966 in computer science (0.746 in sociology and 0.971 in medicine). Also, the results show that most properties are statistically significant (non-zero), with 95 % confidence level, for all three areas.

Therefore, we apply a *multiple exponential regression model* to estimate a response variable y as exponential function of k variables (i.e., topological properties) $x_1, x_2, ..., x_n$ using the following equation: $y = \beta_0 * \beta_1^{x_1} * \beta_2^{x_2} * ... * \beta_k^{x_k}$, or $ln(y) = ln(\beta_0) + ln(\beta_1)x_1 + ln(\beta_2)x_2 + ... + ln(\beta_k)x_k$. We build one model for each social network by determining parameters $\beta_0, \beta_1, ..., \beta_k$ in order to minimize the least squared error for all researchers in the co-authorship network.

The first line of Table 5 shows the R^2 values for the models built considering *all* topological properties ($k = 9$). We have only considered the average values of node properties in the multiple regression models (e.g., for ClosenessA1, ClosenessA2 and AvgCloseness, we use AvgCloseness, where average represents the property value of two researchers in the co-authorship). Then, each subsequent line presents the cumulative results after removing the properties – i.e., the second line is after removing Closeness, the third is after removing Eccentricity from the previous model without Closeness, and so on. The properties

Table 5. Results with all properties and removing one property at a time.

Regression Model	Model Quality(R^2)		
	CS	Med	Soc
All properties	0.953	0.917	-0.103
AvgCloseness (-)	0.954	0.823	-0.105
AvgEccentricity (-)	0.954	0.884	-0.09
AvgClusterCoef (-)	0.969	0.935	-0.04
AvgNTriangles (-)	0.964	0.942	-0.01
AvgWDegree (-)	0.964	0.959	0.009
AvgEigenvec (-)	0.963	0.967	0.453
AvgPageRank (-)	0.967	0.967	0.546
Weight (-)	0.965	0.971	0.747

Table 6. Properties correlation to the strength of ties.

Property	CS	Med	Soc
Clustering coefficient	SLC	SLC	SLC
Edge Betweeness	EC	EC	EC
Number of Triangles			SLC
Weight			SLC
Eigenvector			SLC
Closeness		EC	
Ecentricity		EC	

SLC = Strong linearly correlated,
EC = Exponentially correlated

are removed from the least to the most important. Specifically, the first two to be removed are closeness and eccentricity because their definitions consider the smallest path in the graph, which in theory has no relation to neighborhood overlap. Then, clustering coefficient and number of triangles come next because they are metrics that consider the neighbors information. The last ones to be removed are the most important because they consider the number of relationships of a node, which directly associated to weak ties. After removing the weight, the only variable in the regression model is edge betweenness.

Considering the results, the models can well explain the strength of ties between researchers in computer science and medicine, with R^2 reaching 0.96 for computer science and 0.97 for medicine. For sociology, the R^2 is smaller: -0.105, which indicates a completely inappropriate model. However, after removing some topological properties from the model, the R^2 values start to increase and reach 0.747, which indicates a reasonably good model. The worst R^2 value for each co-authorship network can be explained by applying an exponential regression model. As shown in Table 4, there is a large linear correlation between some topological properties and the neighborhood overlap. Also, there is a noticeable increase of the models accuracy when the clustering coefficient (large correlation with neighborhood overlap in the three networks) is removed.

Analyzing the removal of each property enables to identify which ones are important to the quality of the regression model. Specifically, for computer science, removing clustering coefficient and pageRank increases the R^2 value, which indicates that such properties can *not* be used to explain variations of neighborhood overlap values. Likewise, removing other properties (such as number of triangles, eigenvector and weight) decreases the quality of the model, but not significantly (around -0.5%, -0.1% and -0.2%, respectively). For medicine, removing closeness and eccentricity reduces the regression model's quality, which reinforces the importance of such properties to the model. Thus, the two metrics that measure the distance of a researcher from the furthest other (closeness and eccentricity) non linearly explain variations in the strength of the ties. Likewise, removing other properties increases the R^2 value, showing that they are not very important to the model. Then for sociology, keeping only eBetweenness is enough to have a good model. Also, for the three networks, the quality of the model is good when keeping only edge betweenness.

Such results show the relevance of a metric that represents the importance of a tie between researchers to explain the non linear variation in neighborhood overlap. Note that the computation of neighborhood overlap uses the betweenness concept. Thus, it is expected the correlation between the two metrics. The novelty here is that such correlation is non linear and is a pattern for the three areas. At the end, such results indicate that each area network has the strength of ties related to other topological properties in *different* ways.

Finally, regarding the statistical significance of each model parameter, we set up a series of hypothesis tests, one for each parameter β_i, specified by a null hypothesis $H_0 : \beta_i = 0$. We found that most parameters are statistically significant (i.e., non-zero), with 99 % confidence level for the models with the highest R^2 value. The only exceptions (p-value > 0.05) are the coefficients associated with the average eigenvector and the average page rank for computer science.

6   Conclusions

In this paper, we build co-authorship networks from three areas to quantify the impact of properties on strength of ties (neighborhood overlap). The characterization of neighborhood overlap in the three networks shows that the average value of this metric is around 0.2, i.e. the networks are formed more by weak ties. Additionally, our results showed that the Granovetter's theory governs the three networks, and how topological properties are affected by removing weak and strong ties. Also, the correlation among topological properties and neighborhood overlap was different in each research area, as summarized by Table 6.

We have also evaluated each property for increasing the quality of the regression model. Out of them, the clustering coefficient and edge betweenness were related to neighborhood overlap in the three networks. Such result is trivial, because of the definition of neighborhood overlap. However, the most important contribution is the discovery of other properties related to the strength of ties, and whether the relations are linear or not. Such properties should be considered to improve the quality of systems whose design considers the strength of ties, and to better understand the reasons for a tie being strong or weak.

In the three networks, the neighborhood overlap is an appropriate metric to measure their strength of ties. In this work, we observe that neighborhood overlap captures not only the neighbors of a tie, but also the real intensity of co-authorship between pairs of researchers. Additionally, such metric is easy to compute because it requires only the topological structure of the networks. Future work includes studying other ways to measure the strength of ties between researchers considering semantic aspects.

Acknowledgments. The authors thank CAPES, CNPq and Fapemig - Brazil.

References

1. Acedo, F.J., Barroso, C., Casanueva, B., Galán, J.L.: Co-authorship in management and organizational studies: an empirical and network analysis. J. Manag. Stud. **43**(5), 957–983 (2006)
2. Akoglu, L., Dalvi, B.: Structure, tie persistence and event detection in large phone and sms networks. In: Proceedings of MLG, pp. 10–17 (2010)
3. Barabasi, A.L., Jeong, H., Néda, Z., Ravasz, E., Shubert, A., Vicsek, T.: Evolution of the social network of scientific collaborations. Physica A: Stat. Mech. Appl. **311**(3), 590–614 (2001)
4. Brandão, M.A., Moro, M.M., Lopes, G.R., Oliveira, J.P.M.: Using link semantics to recommend collaborations in academic social networks. In: Proceedings of WWW (2013)
5. Burt, R.S.: Structural holes and good ideas. Am. J. Sociol. **110**(2), 349–399 (2004)
6. Cavero, J.M., Vela, B., Cáceres, P.: Computer science research: more production, less productivity. Scientometrics **98**(3), 2103–2111 (2014)
7. Cohen, J.: Statistical Power Analysis for the Behavioral Sciences, 2nd edn. Lawrence Erlbaum Associates, Hillsdale (1988)
8. Easley, D., Kleinberg, J.: Networks, Crowds, and Markets: Reasoning About a Highly Connected World. Cambridge University Press, Cambridge (2010)
9. Glänzel, W., Schubert, A.: Analysing scientific networks through co-authorship. In: Moed, H.F., Glänzel, W., Schmoch, U. (eds.) Handbook of Quantitative Science and Technology Research, pp. 257–276. Springer, The Netherlands (2005)
10. Gonçalves, G.D., Figueiredo, F., Almeida, J.M., Gonçalves, M.A.: Characterizing scholar popularity: a case study in the computer science research community. In: Proceedings of JCDL (2014)
11. Granovetter, M.S.: The strength of weak ties. Am. J. Sociol. **78**(6), 1360–1380 (1973)
12. Huang, T.H., Huang, M.L.: Analysis and visualization of co-authorship networks for understanding academic collaboration and knowledge domain of individual researchers. In: Proceedings of CGIV, pp. 18–23 (2006)
13. Jain, R.: The Art of Computer Systems Performance Analysis: Techniques for Experimental Design, Measurement, Simulation, and Modeling. Wiley, New York (1991)
14. Klink, S., Reuther, P., Weber, A., Walter, B., Ley, M.: Analysing social networks within bibliographical data. In: Bressan, S., Küng, J., Wagner, R. (eds.) DEXA 2006. LNCS, vol. 4080, pp. 234–243. Springer, Heidelberg (2006)
15. Li, K., et al.: Efficient algorithm based on neighborhood overlap for community identification in complex networks. Physica A: Stat. Mech. Appl. **391**(4), 1788–1796 (2012)
16. Lima, H., Silva, T.H.P., Moro, M.M., Santos, R.L.T., Meira Jr., W., Laender, A.H.F.: Aggregating productivity indices for ranking researchers across multiple areas. In: Proceedings of JCDL (2013)
17. Lopes, G.R., Moro, M.M., Silva, R., Barbosa, E.M., Oliveira, J.P.M.: Ranking strategy for graduate programs evaluation. In: Proceedings of ICITA (2011)
18. Lopes, G.R., Moro, M.M., Wives, L.K., de Oliveira, J.P.M.: Collaboration recommendation on academic social networks. In: Trujillo, J., et al. (eds.) ER 2010. LNCS, vol. 6413, pp. 190–199. Springer, Heidelberg (2010)
19. Newman, M.E.J.: The structure of scientific collaboration networks. In: Proceedings of NAS (2001)

20. Pendlebury, D.A.: The use and misuse of journal metrics and other citation indicators. Arch. Immunol. Ther. Exp. **57**(1), 1–11 (2009)
21. Rana, J., et al.: The strength of social strength: an evaluation study of algorithmic versus user-defined ranking. In: Proceedings of ACM SAC (2014)
22. Silva, T.H.P., et al.: Community-based endogamy as an influence indicator. In: Proceedings of JCDL (2014)
23. Simon, R.J.: The work habits of eminent scholars. Work Occupations **1**(3), 327–335 (1974)
24. Wang, P., Zhao, J., Huang, K., Xu, B.: A unified semi-supervised framework for author disambiguation in academic social network. In: Decker, H., Lhotská, L., Link, S., Spies, M., Wagner, R.R. (eds.) DEXA 2014, Part II. LNCS, vol. 8645, pp. 1–16. Springer, Heidelberg (2014)
25. Yan, R., et al.: To better stand on the shoulder of giants. In: Proceedings of JCDL (2012)
26. Zaki, M.J., Meira Jr., W.: Data Mining and Analysis: Fundamental Concepts and Algorithms. Cambridge University Press, Cambridge (2014)

Event Extraction from Unstructured Text Data

Chao Shang, Anand Panangadan(✉), and Viktor K. Prasanna

University of Southern California, Los Angeles, CA 90089, USA
chaoshangcs@gmail.com, {anandvp,prasanna}@usc.edu

Abstract. We extend a bootstrapping method that was initially developed for extracting relations from webpages to the problem of extracting content from large collections of short unstructured text. Such data appear as field notes in enterprise applications and as messages in social media services. The method iteratively learns sentence patterns that match a set of representative event mentions and then extracts different mentions using the learnt patterns. At every step, the semantic similarity between the text and set of patterns is used to determine if the pattern was matched. Semantic similarity is calculated using the WordNet lexical database. Local structure features such as bigrams are extracted where possible from the data to improve the accuracy of pattern matching. We rank and filter the learnt patterns to balance the precision and recall of the approach with respect to extracted events. We demonstrate this approach on two different datasets. One is a collection of field notes from an enterprise dataset. The other is a collection of "tweets" collected from the Twitter social network. We evaluate the accuracy of the extracted events when method parameters are varied.

1   Introduction

The amount of *unstructured text* data collected in real-world enterprise applications is increasing with the easy availability of portable computers enabling operators to enter notes in the field. For instance, field engineers record their observations into a Computerized maintenance management system (CMMS) using a mix of natural language and domain-specific terms. Such text instances can number in the hundreds of thousands in an enterprise. In this context, "unstructured" refers to natural language that does not strictly follow the rules of the language. Content extraction from such unstructured text collections is the process of identifying only the key terms in an entry that are relevant to a specific application, i.e., the terms that share a particular relation with each other. Content extraction thus transforms unstructured text to a structured representation. For example, in a maintenance application, the content extraction task could be to identify terms describing repair actions and parts that were repaired.

The problem of content extraction has been extensively studied for the case of natural language [8]. These methods rely on natural language parsing and discourse structure analysis as fundamental steps in the analysis of text. Natural language processing methods, such as named entity segmenting and part

© Springer International Publishing Switzerland 2015
Q. Chen et al. (Eds.): DEXA 2015, Part I, LNCS 9261, pp. 543–557, 2015.
DOI: 10.1007/978-3-319-22849-5_38

of speech tagging, assume that the sentences are grammatically correct. Short unstructured text, such as field notes, are comparatively noisy and have a unique style. For instance, capitalization is a key feature for named entity extraction, but this feature is used inconsistently in Twitter where words are often capitalized for emphasis, and named entities are often left all lowercase [11]. Unreliable language parsing also makes using part-of-speech tags unsuitable for identifying actions. Therefore, these methods do not perform well on the type of short unstructured text described above [11]. Content extraction methods are being developed for unstructured text data collected from social networking services such as Twitter. The text messages in these datasets are typically short and use language in idiosyncratic ways. Event extraction methods for these datasets are designed for a particular application with the use of domain-related features [12]. For domain-independent content extraction, the redundancy from the large number of messages enables latent variable modeling methods to discover topics in the dataset [4, 11].

In this work, we apply another method to extract content from short unstructured text. This method iteratively learns sentence patterns that match a set of event mentions and then extracts new mentions using the learnt patterns. This is a bootstrapping approach since the method begins with a small set of representative event mentions. This method was demonstrated for extracting relations from webpages in the DIPRE [6] and Snowball [1] systems. However, the method is accurate only when the relevant event mentions appear within uniform contexts as patterns are matched to sentences using exact or approximate string matching (DIPRE and Snowball, respectively). Thus, directly applying this method to short unstructured text messages with widely varying contexts does not extract events accurately. In information extraction from the web, the goal is to retrieve relevant content from *some* webpage, i.e., the same information is available in multiple webpages and therefore this information should be extracted only from webpages where the retrieval step is expected to be most accurate. Our goal is to extract any relevant content contained within *every* instance of short text in the given collection. Here, we extend the DIPRE approach to make it suitable for content extraction from unstructured text. Specifically, we compute the semantic similarity between the context and set of patterns. We use the WordNet lexical database [10] to compute this semantic similarity. Moreover, it is possible to compute *local structure* features such as frequently occurring bigrams. Defining the context using local structure where possible enables more accurate matching between patterns and sentences. We rank and filter the learnt patterns to balance the precision and recall of the approach with respect to extracted events. The bootstrapping approach of beginning with a few examples of representative content also simplifies the task of specification of the desired content in different domains.

We demonstrate this approach on two different datasets. One is a collection of field notes from an enterprise dataset. The other is a collection of "tweets" collected from the Twitter social network. We evaluate the accuracy of the extracted events when method parameters are varied. The main contributions of this paper are (1) extension of the DIPRE relation extraction approach to unstructured text data, (2) use of semantic similarity in matching event patterns to messages, and (3) demonstration of the framework on two unstructured datasets from different applications.

The rest of the article is organized as follows. Section 2 gives an overview of related work. In Sect. 3, we present background concepts and define the content extraction problem for unstructured data followed by our proposed approach to this problem. In Sect. 4, we present the results of our experimental evaluation of the method. We conclude in Sect. 5.

2   Related Work

Content extraction methods have been extensively studied for natural language text datasets, especially for identifying all instances of a pre-defined set of relations [8]. These methods make extensive use of natural language parsing and also relation-specific text features.

Lin and Pantel [9] describe an unsupervised algorithm to generate a more general set of inference rules from a collection of text documents. Brin [6] introduced the DIPRE method to extract instances of a specific type of relation (such as author-book) from a collection of HTML documents based only on a representative set of instances. Snowball is an extension of this method which uses approximate matching [1]. Many early algorithms for relation extraction, including DIPRE, used little or no syntactic information. Recent approaches have used syntactic information based on parsing the input sentences. Banko et al. [3] describe TEXTRUNNER, an open domain relation extraction system for the Web, that makes use of a natural language parser in the training stage. The above approaches work well when the input data is either well-formed natural language sentences or when the context surrounding the instances of the relation of interest is uniform (as in the case of HTML documents). We focus on unstructured text that uses natural language in idiosyncratic ways and with high variability.

More recently, several works have been published on information extraction from social media, in particular the tweets from the Twitter network. Sakaki et al. [12] describe a method to detect earthquake-related tweets. The method uses featues specific to this application. Benson et al. [4] trained a relation extractor to identify artists and venues from tweets. It develops a graphical model by learning records and records-message alignment. Ritter et al. [11] describe a method based on latent variable modeling to extract event types described in tweets. Features such as tweet popularity and the times of events referred to in the tweets are also used. Zhao et al. [13] describe a method to extract only the most "topical" keywords from tweets. Similar approaches have also been applied to mine relevant information from non-text sources. For instance, Zong et al. [14] describe approximation algorithms to identify critical alerts from a large set of alert sequences. Our approach differs from these methods in that the system is provided a representation of the events of interest only in the form of a set of terms describing representative events. Thus, arbitrary types of events can be discovered without utilizing event type-specific features such as time-of-event.

3   Problem Definition and Approach

We use *unstructured data* to mean a collection of text sentences, where a sentence does not have to follow the rules of a natural language. In particular, sentences can include terms of non-alphabet characters (for instance, alphanumeric sequences representing part identifiers in an industrial application).

We denote by $U = \{S_1, S_2, \ldots, S_N\}$ the unstructured dataset containing N sentences. A sentence S is a sequence of $l(S)$ *words*, where a word is a sequence of alphanumeric characters: $S = (w_1, w_2, \ldots, w_{l(S)})$. To simplify the notation, we also denote a sequence as $\mathbf{w}_{p:q} \equiv (w_p, w_{p+1}, \ldots, w_q)$.

We define the problem of content extraction as that of extracting sub-sequences of words from each sentence that represent an *event* and any associated *event parameters*. Events have been defined elsewhere as an occurrence in place and time [2]. In our work, we use "event" to refer to any action of interest in the domain. In an industrial application, an event can denote actions taken by an operator while the event parameters can denote the equipment which is involved (e.g., "tighten-bolt123" where "bolt1234" is the event parameter referring to a specific piece of equipment). In social media messaging, events could denote an action described by the message sender and the event parameters its associated object (e.g., "watch-game show" where "game show" is the specific event parameter).

Our approach is iterative: event patterns are used to extract matching sentences and these matches are subsequently used to refine the patterns. The approach is semi-supervised since the process has to be bootstrapped with an appropriate seed set containing some of the desired events and this has to be provided by the user. Note that only a set of events are provided at the beginning; the corresponding patterns are generated in the approach. This approach is adapted from DIPRE [1] where patterns are used to extract author-book pairs. While DIPRE relied on regular expressions of characters for patterns, we use approximate matching based on the semantic distance between words. We rank patterns in order to ensure that only high precision patterns are retained. We also tokenize the input sentences to extract local features before searching for patterns. The overview of our approach is shown in Fig. 1. The steps are described in detail next.

3.1   Data Pre-processing

This step transforms the input strings to a sequence of tokens where each token captures any correlation between adjacent words. Other pre-processing steps remove insignificant variations between sentences such as stop words, case, and word endings and variations from non-alphabetic characters (such variations were relevant for finding author-book pairs from webpage, the application for DIPRE). The steps are described below.

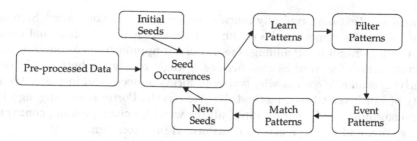

Fig. 1. Approach for event extraction from unstructured data

Tokenization with Bigrams. Since we will use semantic distance to match patterns and sentences, it is important that a sentence is split into semantically coherent tokens. We use Pointwise Mutual Information (PMI) between individual words as they appear in the input corpus to determine if two consecutive words should be considered as one token. PMI is a measure of association and quantifies how much the probability of a particular co-occurrence of events $(p(x, y))$ differs from what is expected based on the probabilities of the individual events and the assumption of independence $(p(x)p(y))$ [5]:

$$PMI(x; y) = log(p(x, y)/p(x)p(y))$$

PMI can be normalized (NPMI) to values in $[-1, 1]$. Then, 0 corresponds to independence, and $+1$ for perfect co-occurrence.

$$NPMI(x; y) = PMI(x; y)/ - log(p(x, y))$$

In our approach, word counts are used to compute the occurrence probabilities, i.e., we count the occurrences of every word $(p(x)$, $p(y))$ and every two adjacent words $(p(x, y))$. Then we use the frequency $p(x)$, $p(y)$ and p(x,y) to calculate the value of Pointwise mutual information(PMI) and Normalized pointwise mutual information (NPMI). The NPMI-based bigram calculation is evaluated in Sect. 4.2.

Removal of Stop Words. Stop words are words that have the same likelihood of occurring in all data classes of interest. For the problem of extracting events, stop words are those that are equally likely to appear in sentences, whether they match an event pattern or not. We use a stop words filter to remove common terms that could skew the calculation of semantic similarity. Stop words are typically removed based on looking up pre-defined lists of words, such as articles and prepositions. However, some classes of words in stopword lists have relevance for event extraction. We evaluated different classes of words to be considered as stop words for event extraction and identified prepositions to be useful. In Sect. 4.3, we quantitatively evaluate this pre-processing step and show the usefulness of removing only a subset of commonly used stop words for purposes of event extraction.

Stemming. Text data typically contains different variants of a word. Stemming is the process for removing the common morphological and inflexional endings from words in English. Stemming is useful in our approach since counting all the inflected forms of a word as instances of a single root word better reflects the underlying semantic reasons why two word forms co-occur and increases the reliability of word-count based methods [7]. We use the Porter stemming algorithm for transforming words. For example, all the words in `consignment`, `consigned`, `consigning`, and `consign` can be transformed into `consign`.

3.2   Event Extraction

Event extraction is an iterative process. The approach maintains a set of event patterns – matching the event patterns to a sentence results in identifying an event from that sentence. At each iteration, the extracted events is used to learn an expanded set of patterns. These patterns are ranked and only the most reliable ones are retained for the next iteration. The process is initiated with a set of representative events, the seed, provided by the user. These steps are listed in Algorithm 1 and are described next.

Algorithm 1. ExtractEvents (initial seed events, E; set of sentences, S)

1: Find all occurrences of known events in data: $O \leftarrow$ findOccurrence(E, S);
2: Generate patterns from occurrences: $P \leftarrow$ generatePatterns(O, S);
3: Rank patterns: $R \leftarrow$ rankPatterns(P, E, S);
4: Retain only high ranking patterns: $P^* \subset P, R(p) > \tau, p \in P^*$;
5: Match patterns to sentences to get new events: $E \leftarrow$ matchPatterns(P^*, S);
6: **if** E has expanded, **then** goto Step 1;
7: **return** E;

Find Occurrences. Given a specific (Event, EventParameter) pair, we extract the set of *occurrences* of this event from the data. Each sentence in the dataset, Event, and EventParameter are treated as sequences of words. An event occurrence is found when all words of the Event and EventParameter are matched to a sentence. An occurrence of an (Event,Event Parameter) pair is then defined as a 7-tuple:

$$(\texttt{Eventsequence}, \texttt{EventParametersequence}, \texttt{Prefix}, \texttt{Middle}, \texttt{Suffix}, \texttt{Order})$$

This is illustrated in Fig. 2. Here, `Prefix`, `Middle`, and `Suffix` represent the context of the occurrence. `order` is a binary variable indicating the relative position of Event and EventParameter (i.e., which appears first in the sentence). These will be used to generate patterns.

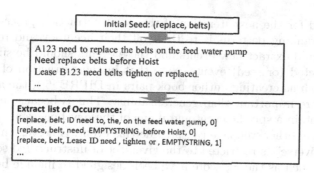

Fig. 2. Finding event occurrences given an initial seed.

Generating Patterns. We define a pattern as a 4-tuple, (`prefix`, `middle`, `suffix`, `order`), where `prefix`, `middle`, and `suffix` are sequences of words and `order` $\in \{0, 1\}$. We use a simple rule for generating patterns from the set of all occurrences. For each occurrence, we create a patterns for every combination of prefix and suffix sub-sequence with length less than 5 words. Specifically, let $w_{1:N}$ denote the sequence of words in a sentence from the dataset that matches a given `Event` and `EventParameter`. Let $w_{E_1:E_2}$ denote the sub-sequence of words that match the words in `Event`. Similarly, let $w_{E_3:E_4}$ denote the sub-sequence of words that match the words in `EventParameter`. Multiple patterns can match such a given sentence. We generate sequences of words that are candidates for the prefix, middle, and suffix portions of a pattern. A candidate prefix is any sub-sequence of words of maximum length M_{prefix} that ends at w_{E_1-1}. Similarly, a candidate suffix is any sub-sequence of words of maximum length M_{suffix} that begins at w_{E_4+1}. There is only one candidate middle: the sub-sequence of words between w_{E_2} and w_{E_3} (i.e., $w_{E_2+1:w_{E_3}-1}$). An example of the generation of patterns from an occurrence is shown in Fig. 3.

Pattern Ranking. While a large number of patterns can explain a given occurrence, only a few are likely to generalize to other sentences. Ranking of patterns according to how well they are expected to generalize to sentences from which events have not been extracted will enable only the most generalizable patterns

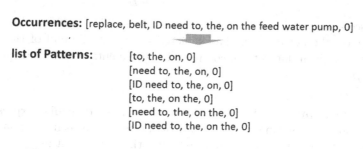

Fig. 3. Generating a set of patterns from one occurrence.

to be retained for the next iteration of the event extraction process. The main challenge in ranking patterns is to trade-off their accuracy and recall: ideally, a pattern should extract new events and these events should be similar to previously extracted (or seed) events. Note that in the extraction of events from webpages (such as creating author-book pairs in DIPRE [6]), the accuracy of a pattern is more important than generalizability. This is because multiple webpages can contain a specific event and if one of these webpages are not matched by a pattern, another webpage can provide this event. However, in our application, the "universe" is restricted to the given set of unstructured sentences.

Measures such as the F_β score have been designed to include both precision and recall into a single numeric value that can then be used to rank patterns. The F_β score is defined as follows.

$$F_\beta = (1 + \beta^2) \frac{\text{Precision} \times \text{Recall}}{(\beta^2 \times \text{Precision}) + \text{Recall}}$$

The smaller the value of β, the more the emphasis that is placed on precision as compared to recall. In typical applications, the calculation of precision and recall is performed by comparison with a ground truth dataset. In our approach, pattern ranks have to be calculated in every iteration and hence precision and recall have to be calculated without access to any ground truth. Instead, we calculate precision of a pattern by comparing matched events with the set of seed events in that iteration and its recall by measuring the number of sentences that it matched. Specifically, let A_{P_i} represent the number of sentences in the complete data set that match pattern P_i and the extracted event is one of the seed events in that iteration. Let B_{P_i} denote the number of sentences that contain a seed event but P_i did not extract an event from them. C_{P_i} is the number of sentences that match P_i but the extracted event is not one of the seeds. Then, $A_{P_i} + B_{P_i}$ is the number of sentences that contain a seed event and $A_{P_i} + C_{P_i}$ is the total number of sentences that the pattern P_i matched. The precision and recall of P_i is then calculated as

$$\text{Precision}(P_i) = \frac{A_{P_i}}{A_{P_i} + C_{P_i}}$$

$$\text{Recall}(P_i) = \frac{A_{P_i}}{A_{P_i} + B_{P_i}}$$

The F_β score is calculated from the above values. The specific value of β has an impact on the accuracy of the extracted events at the end of the iterative process. For a given dataset, we conducted experiments to determine a suitable value for β.

Pattern Matching. DIPRE [6] and Snowball [1] used regular expressions to represent patterns and match sentences. String matching using regular expressions is less effective in unstructured text since the same event may be expressed in a sentence with slight variations in syntax and word order while their semantic similarity is high. In our work, we compute an approximate match between a pattern and a sentence by using the semantic similarity between pairs of words in the given pattern and sentence. In particular, we compute the semantic similarity between pairs of words without considering the order of the words within a sentence or pattern.

We use WordNet as the basis for computing semantic similarity. WordNet is an online lexical database for English that groups together related words into *synsets*. Synsets are connected to each other according to different relation types and these relations can be used to compute the semantic relatedness between two words in the WordNet corpus [10]. Different algorithms to compute relatedness are also available in a freely available software package [10].

The match score is calculated separately for each part of the context of the match — prefix, middle, and suffix — and then combined. This requires that the sequence of words comprising the prefix, middle, and suffix of a match has to be extracted from a sentence for a given pattern. To extract these sub-sequences, the last word of the prefix, the first word of the middle, and the first word of the suffix of a pattern are used to indicate the sub-sequence start/end points. Let $S = w_{1:l(S)}$ denote the sentence and let

$$P = (P^{\text{pre}} = v_{1:p}^{\text{pre}}, P^{\text{mid}} = v_{1:m}^{\text{mid}}, P^{\text{suf}} = v_{1:s}^{\text{suf}}, o)$$

denote the pattern components. Then, the prefix sub-sequence, $S^{\text{pre}} = v_{i-p+1:i}$; the middle sub-sequence, $S^{\text{mid}} = v_{j:j+m-1}$; and the suffix sub-sequence, $S^{\text{suf}} = v_{k:k+s-1}$. The start/end points, i, j, and k, are calculated such that

$$w_i = v_p^{\text{pre}}, w_j = v_1^{\text{mid}}, w_k = v_1^{\text{suf}}, i+1 < j, j+m < k$$

We then compute the degree of semantic matching between the prefix match of the sentence and pattern prefix as

$$\text{Match}(S^{\text{pre}}, P^{\text{pre}}) = \sum_{w \in S^{\text{pre}}} \max_{v \in P^{\text{pre}}} S_{\text{WN}}(w, v)$$

where $S_{\text{WN}}(w, v)$ is the WordNet similarity between two words w and v. In this calculation, if a pair of words does not appear in the WordNet database (as happens when one of the "words" is a bigram extracted during the tokenization step), then the result of the exact match between them is used as their similarity. The semantic similarity between the middle and suffix matches of the sentence and pattern is computed in a similar manner. The match score for the complete sentence and the pattern is then computed as

$$\text{Match}(S, P) = \text{Match}(S^{\text{pre}}, P^{\text{pre}}) + \text{Match}(S^{\text{mid}}, P^{\text{mid}}) + \text{Match}(S^{\text{suf}}, P^{\text{suf}})$$

Note that in this approach to computing a match score, the order of words in the sentence and pattern is not considered. In particular, a word in the sentence can match multiple words in the pattern. The match can be made more restrictive by enforcing the constraint that every word is matched exactly once. With

> Welder Repair *PART* on Section `DIGIT` spent *PART* pump `ID DIGIT` suction line
> install motor valve @ `ID DIGIT`, on `ID NUM`, for well `ID DIGIT`

Fig. 4. Examples from enterprise dataset. `ID` and `DIGIT` represent alphanumeric characters and numbers respectively that are replaced with type tags before event extraction. *PART* represents words not shown for privacy concerns.

this restriction, computing the match becomes equivalent to finding a maximum weight matching in a weighted bipartite graph where the graph nodes correspond to sentence and pattern words, edges correspond to similarity between sentence and pattern words, and the maximum weight matching identifies word pairs that together represent maximum total similarity.

4   Experimental Evaluation

We evaluated the bootstrap-based event extraction algorithm on data from two different sources. The first is text comments in an enterprise dataset. The second is a collection of "tweets" from the Twitter social media service.

4.1   Datasets

Enterprise Dataset. The enterprise-scale dataset contained approximately 400,000 lines and contains field notes. Events in this application correspond to actions taken by engineers and the event parameter represents the object involved in an action. The average length of a line is 13 words. The lines are a mixture of English words, numbers, and alphanumeric terms. Since variations in numbers and alphanumeric terms between otherwise similar sentences do not typically represent different events, we replaced instances of such terms with a generic tag. Two examples of lines in this dataset are shown in Fig. 4.

Twitter Dataset. We accessed approximately 1 million tweets sent between April 6, 2009 to May 29, 2009. These tweets represent a wide range of events. As is typical with messages on Twitter, this dataset includes extensive use of special characters (such as in hashtags). Example tweets are shown in Fig. 5.

> @xxxx I wish I had know about the anchor hunt before! !I'd love show
> @xxxx hi, also. lol!
> http://twitpic.com/xxxx - One Of The Last Group Photos To Be Taken At School
> @xxxx Care to tell what movie u r watching

Fig. 5. Examples of tweets in Twitter dataset. Hashtags and URLs have been redacted for privacy.

4.2   Evaluation of Bigrams

We first report on the accuracy of the extracted bigrams based on normalized PMI (NPMI), i.e., what proportion of the bigrams represent semantically related words. For the enterprise dataset, we first discarded words with frequency less than 1000 occurrences. For the Twitter dataset, we discarded words with frequency less than 10 occurrences. For evaluation, we chose the 20 bigrams with the highest NPMI and for each bigram we determined if it represented a semantically sensible pairing. These are shown in Figs. 6(a) and (b) for the enterprise and Twitter dataset respectively. The accuracy of bigrams in the enterprise dataset is 75 % and for the Twitter dataset is 80 %. (Note that these bigrams are not Event-Event parameter pairs; in our work, bigrams are used to extend the pattern representation from single words to frequently occurring two-word pairs.)

Phrases	Frequency	Phrases	Frequency
positive displacement	2291	fire extinguisher	1073
provide supporting	17797	polish rod	6513
supporting activities	17790	heat exchanger	2025
operational troubleshoot	1091	reclosure thermography	1027
route bulk	3313	producer lubrication	1863
fill poly	4573	lube route	3890
stuffing box	11990	staging facility	1204
see comments	9291	vacuum truck	12361
hydrotest leadline	1083	rig activity	8504
load cell	2826	heater treater	1632

(a) Enterprise dataset

Phrases	Frequency	Phrases	Frequency
rajeev motwani	21	milton keynes	22
elora danan	12	carne asada	19
abu dhabi	17	roland garros	37
porto alegre	25	hoedown throwdown	48
wal mart	122	rascal flatts	32
sri lanka	34	krispy kreme	80
feas feas	18	cli gs	105
clonk clonk	13	susan boyle	356
twurl nl	253	haagen dazs	15
sml vg	29	hong kong	81

(b) Twitter dataset

Fig. 6. Correctness of extracted bigrams. The bigrams are arranged in decreasing order of NPMI. Bigrams marked in red are those marked incorrect based on evaluation by the authors.

4.3 Evaluation of Removing Stop Words

We compared the relative accuracy of events extracted when different types of stop words are removed during the pre-processing step. In particular, we consider retaining prepositions instead of removing all stop words. This evaluation was performed on the enterprise dataset. The results are shown in Fig. 7. These indicate that retaining prepositions gives more accurate events (70 %) while removing all stop words is still better than retaining all stop words (60 % as compared to 50 %).

Retain stop words		Remove all stop words		Retain prepositions	
Events and *Parameters*	*Frequency*	Events and *Parameters*	*Frequency*	Events and *Parameters*	*Frequency*
replace belts	769	repair *PART*	558	need belt	678
repair *PART*	704	need belt	464	repair *PART*	613
replace the	10	replace belt	57	replace belt	182
well is	7	need set	56	repair brake	80
be replace	7	instal belt	11	need set	56
backhoe and	6	belt ID	8	to belt	36
replace air	5	c DIGIT	8	tighten belt	22
repair water	5	ID belt	7	install belt	14
belts on	4	gwa DIGIT	6	gwa DIGIT	12
check belts	4	put belt	6	pipe at	11

Fig. 7. Accuracy of extracted event-event parameter pairs after removing stop words. Event pairs in red were judged to be incorrect by the authors.

4.4 Event Extraction from the Enterprise Dataset

We selected the following seed Event-Event parameter word pairs for extracting new events from the enterprise dataset: "replace belts", "repair *PART*", "need belts", and "repair break." We included multiple pairs in the seed set since the desired events can be of multiple types. For the F_β-based pattern ranking step, the value of $\beta = 0.2$, the threshold for filtering pattern is F_β score > 0.3, frequency of pattern ≥ 20, precision > 0.6, and recall > 0.1. With these parameters and seed pairs, the patterns found after the first iteration are shown in Table 1.

To compare the impact of approximate matching between patterns and data with exact word matching, we extract events using both methods. Figure 8 shows ten events and event parameter pairs extracted using exact pattern matching. Each event was marked as correct or otherwise by the authors to quantify the accuracy of event extraction. This indicates that the method achieves a precision of 70 %. Note that 30 % of the extracted event-event parameter contain prepositions and DIGIT. A filtering step to remove events containing such terms can improve the accuracy of event extraction.

Table 1. Patterns found after first iteration in enterprise dataset

Prefix	Middle	Suffix	Order	F Score	Frequency	Precision	Recall
EMPTYSTRING	ID,ppm, DIGIT,day	EMPTYSTRING	1	0.318	279	1	0.017
fugem	ID, ppm, DIGIT, day	EMPTYSTRING	1	0.318	279	1	0.017
DIGIT	EMPTYSTRING	well	0	0.315	438	0.74	0.02
EMPTYSTRING	new	EMPTYSTRING	0	0.268	428	0.638	0.017
Digit	EMPTYSTRING	well, is	0	0.241	225	0.884	0.012
to	EMPTYSTRING	on	0	0.227	279	0.713	0.013

Events	need	repair	replace	repair	need	to	tighten	install	gwa	pipe
Event Parameters	belt	PART	belt	brake	set	belt	belt	belt	DIGIT	at
Frequency	678	613	182	80	56	36	22	14	12	11

Fig. 8. Exact pattern matching: Subset of event and event parameter pairs extracted after first iteration. Pairs in red are considered incorrect by the authors.

Figure 9 shows events and event parameter pairs extracted with approximate pattern matching using WordNet similarity. This result shows that a larger number of event-event parameter pairs was extracted as compared to exact matching and with greater accuracy.

4.5   Event Extraction from the Twitter Dataset

We initialized the event extraction process with the event-event parameter pair "watch movie". The patterns found after the first iteration is shown in Table 2. The extracted event-event parameter pairs after the first iteration and their frequency are shown in Fig. 10. We labeled each extracted event as correct or not based on its context — events are expected to describe an action described by the message sender and the event parameters its associated object. As compared to event extraction from the enterprise dataset, fewer events are identified and with less accuracy.

In Table 3, we compare the precision of event extraction of our proposed methods with the DIPRE approach on both the enterprise and Twitter datasets. Precision is calculated as the frequency of correctly identified event and event

Event and Event parameters	Frequency	Event and Event parameters	Frequency
repair brake	1626	repair *PART*	1024
replace belt	89	new belt	61
repair ppm	56	adjust and	40
brake repair	37	need belt	33
repair water	28	change belt	26
repair belt	20	install ladder	15

Fig. 9. Approximate pattern matching: Subset of event and event parameter pairs after first iteration. Pairs in red are considered incorrect by the authors.

Table 2. Patterns found after first iteration in Twitter dataset

Prefix	Middle	Suffix	Order	F Score	Frequency	Precision	Recall
to	a	EMPTYSTRING	0	0.346	118	0.618	0.029
to	the, mtv	awards	0	0.213	26	1	0.010

Events	watch	be	get	take	take	buy
Event Parameters	movie	good	new	nap	shower	new
Frequency	284	216	208	183	149	145

Fig. 10. Subset of event and event parameter pairs extracted after first iteration from Twitter dataset. Pairs in red are considered incorrect by the authors.

Table 3. Precision of event extraction methods on enterprise and Twitter datasets

Method	Enterprise dataset	Twitter dataset
DIPRE	73.1 %	27.7 %
Pre-processing and exact matching	90.3 %	36.5 %
No pre-processing and approximate matching	85.8 %	40.1 %
Pre-processing and approximate matching	91.7 %	48.9 %

parameters (as labeled by the authors) in the 20 most frequently extracted events. DIPRE has significantly lower accuracy for event extraction from unstructured data. The pre-processing steps have a large impact on event extraction. Approximate matching between patterns and sentences using WordNet similarity improves the accuracy of event extraction as compared to exact matching. The event extraction methods have much lower accuracy for the Twitter dataset as compared to the enterprise dataset.

5   Conclusion

We described how a bootstrapping method that was developed for extracting relations from webpages can be extended to the problem of extracting content from large collections of short unstructured text. The method iteratively learns sentence patterns that match a set of representative event mentions and then extracts different mentions using the learnt patterns. In contrast to the original method, we used WordNet-based semantic similarity between the text and patterns to determine if a pattern was matched. Learnt patterns were ranked to balance the precision and recall of the approach with respect to extracted events. The results on two different datasets demonstrate that the original relation extraction method has significantly lower accuracy when applied to these datasets. Our pre-processing steps have a large impact on the accuracy of event extraction. Approximate matching using WordNet similarity improves the accuracy of event extraction as compared to exact word-to-word matching. The method has much lower accuracy for the Twitter dataset as compared to the enterprise dataset indicating the need for designing domain-specific local features.

Acknowledgment. This work is supported by Chevron U.S.A. Inc. under the joint project, Center for Interactive Smart Oilfield Technologies (CiSoft), at the University of Southern California.

References

1. Agichtein, E., Gravano, L.: Snowball: extracting relations from large plain-text collections. In: ACM Conference on Digital Libraries, pp. 85–94 (2000)
2. Allan, J., Papka, R., Lavrenko, V.: On-line new event detection and tracking. In: Proceedings of the 21st Annual International ACM SIGIR Conference on Research and Development in Information Retrieval, pp. 37–45. ACM (1998)
3. Banko, M., Cafarella, M.J., Soderland, S., Broadhead, M., Etzioni, O.: Open information extraction for the web. In: IJCAI, vol. 7, pp. 2670–2676 (2007)
4. Benson, E., Haghighi, A., Barzilay, R.: Event discovery in social media feeds. In: Proceedings of the 49th Annual Meeting of the Association for Computational Linguistics: Human Language Technologies, pp. 389–398 (2011)
5. Bouma, G.: Normalized (pointwise) mutual information in collocation extraction. In: Proceedings of the Biennial GSCL Conference, pp. 31–40 (2009)
6. Brin, S.: Extracting patterns and relations from the world wide web. In: Atzeni, P., Mendelzon, A.O., Mecca, G. (eds.) WebDB 1998. LNCS, vol. 1590, pp. 172–183. Springer, Heidelberg (1999)
7. Bullinaria, J.A., Levy, J.P.: Extracting semantic representations from word co-occurrence statistics: stop-lists, stemming, and svd. Behav. Res. Methods **44**(3), 890–907 (2012)
8. Doddington, G.R., Mitchell, A., Przybocki, M.A., Ramshaw, L.A., Strassel, S., Weischedel, R.M.: The automatic content extraction (ace) program-tasks, data, and evaluation. In: LREC (2004)
9. Lin, D., Pantel, P.: Discovery of inference rules for question-answering. Nat. Lang. Eng. **7**(04), 343–360 (2001)
10. Pedersen, T., Patwardhan, S., Michelizzi, J.: Wordnet:: Similarity: measuring the relatedness of concepts. In: Demonstration Papers at HLT-NAACL 2004, pp. 38–41. Association for Computational Linguistics (2004)
11. Ritter, A., Mausam, E.O., Clark, S.: Open domain event extraction from twitter. In: Proceedings of the 18th ACM SIGKDD International Conference on Knowledge Discovery and Data Mining, pp. 1104–1112. ACM (2012)
12. Sakaki, T., Okazaki, M., Matsuo, Y.: Earthquake shakes twitter users: real-time event detection by social sensors. In: Proceedings of the 19th International Conference on World Wide Web, pp. 851–860. ACM (2010)
13. Zhao, W.X., Jiang, J., He, J., Song, Y., Achananuparp, P., Lim, E.-P., Li, X.: Topical keyphrase extraction from Twitter. In: ACL: Human Language Technologies, pp. 379–388 (2011)
14. Zong, B., Wu, Y., Song, J., Singh, A.K., Cam, H., Han, J., Yan, X.: Towards scalable critical alert mining. In: 20th ACM SIGKDD International Conference on Knowledge Discovery and Data Mining, pp. 1057–1066 (2014)

A Cluster-Based Epidemic Model for Retweeting Trend Prediction on Micro-blog

Zhuonan Feng[✉], Yiping Li, Li Jin, and Ling Feng

Department of Computer Science and Technology, Tsinghua University,
Beijing, China
fzn0302@163.com
{liyp09,l-jin12}@mails.tsinghua.edu.cn, fengling@tsinghua.edu.cn

Abstract. Tweets spread on social micro-blog bears some similarity to epidemic spread. Based on the findings from a user study on tweets' short-term retweeting characteristics, we extend the classic Susceptible-Infected-Susceptible (SIS) epidemic model for tweet's retweeting trend prediction, featured by the multiple retweeting peaks, retweeting lifetime, and total retweeting amount. We cluster micro-blog users with similar retweeting influence together, and train the model using the least square method on the historic retweeting data to obtain different groups' retweeting rates. We demonstrate its effectiveness on a real micro-blog platform.

Keywords: Micro-blog · Tweet · Retweet · Cluster-based epidemic model

1 Introduction

The emergence of social networks are changing people's communication styles. At the same time, the style of information propagation is undergoing profound changes. Some new media such as Twitter, Facebook, and WhatsApp are widely used and become new information carriers and communication tools of social networks. Compared with the traditional media like TV and newspaper, these new media have several significant characteristics.

1. *Different information propagation modes.* Traditional media adopt "broadcast" mode to propagate information, i.e., "one information center, thousands of listeners" in short. New media such as micro-blog is "we-media" whose information propagation is "mouth to mouth"-like, that is, everyone can become a radio station, transmitting information to his/her listeners.
2. *Different information capacity and user groups.* Traditional media can transmit massive information with a large capacity each time. But ordinary people have no chance to release news on these media. New media is lightweight and easy to publish messages. A micro-blog only requires 140 words or less. Someone who wants to announce something only needs a mobile phone and a client, typing some words and pressing "ok", a message could be sent all over the world.

Q. Chen et al. (Eds.): DEXA 2015, Part I, LNCS 9261, pp. 558–573, 2015.
DOI: 10.1007/978-3-319-22849-5_39

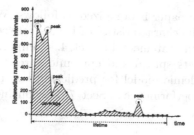

Fig. 1. Retweeting trend for a tweet posted by a real estate celebrity Ren, Z.Q

3. *Different audiences.* Traditional media look like "official voice", and are often influenced by politics. New media are often regarded as the voice of public and are very popular to young generation.

Like traditional media's (such as TV programs) rating surveys, the trend of information diffusion on social networks is also people's concern, and benefits many applications. One important field is micro-blog based advertising and marketing. With the help of microblog's trend analysis and prediction, merchants could know when, where, and who to promote their commodities to. Another important field is fake messages prevention. Network supervisors could prevent rumors diffusion just like disease prevention. For social network tool micro-blogs, the important quota to reflect information diffusion is tweets retweeting. If someone A published a tweet, his/her follower B could see the tweet and retweet it, and B's follower C could also do the same thing. So for micro-blog, information's diffusion trend could be simplified to its retweeting number's variation trend.

In the literature, [7] used a triad (Intensity, Coverage, Duration) to describe and predict information trend via a dynamic activeness (DA) model on the DBLP academic exchange platform. The work reported in this paper differs from this closely related work in the following two aspects. First, the DA model fits the relatively long-term trend of scholars' research interests, whose time granularity is about one year. But in micro-blog, the trend of information dissemination is a short-term trend. Most messages on micro-blog only last several days from their birth to the death. The time granularity is around an hour or minute. Compared with long-term trends, a tweet's diffusion trend fluctuates much drastically, and does not exhibit exponential distribution, which is the theoretical base of the DA model. Our data analysis result shows that DA model is not suitable for short-term trends. Figure 1 shows multiple retweeting peaks during the spread of a tweet posted by a real estate celebrity Ren, Z.Q. Second, based on the characteristics of short-term tweets spread obtained through a user study on real micro-blog data, we compare the similarity and difference between tweets diffusion and epidemic disease spread, and propose a cluster-based SIS (Susceptible-Infected-Susceptible) method for multiple retweeting peaks, lifetime, and coverage prediction. We compare the performance of the method with the existing ones on a micro-blog platform, and the result shows our method is effective and could achieve better performance on each prediction item.

The remainder of the paper is organized as follows. We review related work in Sect. 2. We analyze the characteristics of tweets retweeting trends in Sect. 3, and provide the problem statement in Sect. 4. Comparing the similarity and differences between tweets spread and epidemic spread in Sect. 5, we present a cluster-based SIS epidemic model for predicting tweets retweeting trends in Sect. 6, and evaluate its performance in Sect. 7. We conclude the paper in Sect. 8.

2   Related Work

2.1   Analysis and Prediction of Retweeting Behaviors

[1] examined conversional practices of retweeting in Twitter, such as how people retweet, why they retweet, and what they retweet. A number of features which may influence tweets retweetability are studied. [11] pointed out that two content features (URLs and hashtags) and two context features (number of followers/followees and the account age) have strong relationship with the retweetability of tweets, but the number of past tweets shows little correlation with the retweeting rate. [15] found that the most important features for retweeting prediction are the identity between the tweet and the retweeter. It trained a probabilistic collaborative filtering model called Matchbox to predict the probability whether a follower would retweet a tweet or not. [14] found that almost 25.5 % of tweets were retweeted from friends' blog spaces, and presented a factor graph model to predict users' retweeting behaviors. [9] used a passive-aggressive algorithm in machine learning to predict retweeting. The performance of the algorithm is dominated by such social features as number of followers and users interests, as well as some tweet-related features like hashtag, URL, trending words, and so on. Besides the features mentioned above, some researchers noticed the "celebrity effect" in social networks and tried to find influential nodes, which contribute to information's diffusion. [4] developed a decentralized version of the influential maximization problem by influencing k neighbors rather than arbitrary users in an entire network. It presented several reasonable neighbor selection schemes to find influential spreaders on twitter. [6] proved in general directed graphs finding k-effectors is a NP-hard problem. By transforming the graph to the most probable active tree, the problem could be solved optimally in polynomial time.

2.2   Retweeting Trend Prediction

[7] defined the information's diffusion trend in social networks as a triad (*intensity, coverage* and *duration*). where trend intensity is the volume of actions in general during a fixed length of time, trend coverage is the number of people taking the given action during a fixed length of time, and the trend duration is the time span that coverage is above a given threshold. Based on the three elements the author designed a Dynamic Activeness (DA) model for information's future trend prediction. Different from [7,13] predicted information diffusion from three

major aspects *speed* (how quickly a tweet will produce an offspring tweet), *scale* (number of child nodes the tweet will produce), and *range* (number of hops in the diffusion chain). Its Cox proportional hazards regression result showed that the rate with which a user is mentioned historically is a strong predictor of information diffusion in Twitter. [2] proposed two novel trend definitions called coordinated and uncoordinated trends to detect popular topics. The author also introduced a novel information diffusion model called ITFM to distinguish viral diffusion of information from diffusion through external entities such as news media. Besides above studies, there are some researches mainly about an aspect of information's diffusion trend. [5] predicted a tweet's lifespan by generating a time series based on tweet's first-hour retweeting number, and comparing it with those of historic tweets of the similar author and post time. Then top-k historic similar tweets were identified, whose mean lifespan was computed as the predicted value of the new tweet.

3   Characteristics of Tweets' Retweeting on Micro-Blog

We focus on three types of tweets which may usually incur large-scale retweeting. (a) Headline news tweets, whose contents are about breaking news around us, can be reached through various media portals. (b) Celebrities' tweets, which have great influence on social networks. The tweets of the estate agent Ren, Z.Q. belong to this category. They are often controversial and give rise to discussion. (c) Entertainment stars' tweets, like the tweets of the popular Hunan TV host Xie, N., who has over 50 million direct followers (also called fans).

To examine the characteristics of these tweets' spread on micro-blog, we study 4000 tweets and their 5 million retweeting activities on sina micro-blog (the biggest social networking platform in China with billions of users and up to 60 million active users per day[1]). Among them, 2300 are headline news tweets, 1000 are celebrities' tweets, and 500 are entertainment stars' tweets. We trace the users who retweet these tweets, and crawl about 370 thousand tweets posted by the retweeting users from November 2013 to December 2014. From the millions of retweeting records, we have some interesting findings.

(1) *Tweets spread on micro-blog through three media - direct followers, indirect followers, and external visitors.* Due to the microblog's transmission mechanism, a tweet is always firstly seen and possibly retweeted by the direct followers of its author. The followers of these retweeting users (also called indirect followers of the original author) can then see and may also retweet the same tweet to their respective followers. Through such a direct/indirect following relationship, the tweet spreads on micro-blog. Besides, as some hot tweets may be reprinted by certain portal sites and media channels, users who are not in the direct/indirect following chains could also come across the tweet and become a part of disseminators. We call them external visitors.

(2) *Direct followers constitute the main force of tweet's propagation at an early stage, while indirect followers and external visitors contribute more at the*

---

[1] http://blog.sina.com.

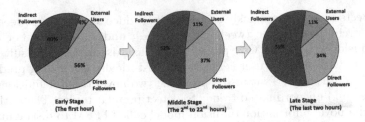

Fig. 2. Proportion of the three retweeting disseminators at different stages after a news tweet is posted

late stage. Figure 2 plots the proportion of the three retweeting disseminators at different stages since the post of a news tweet on average. The majority of retweeting happens within the next 24 h. At the beginning, direct followers are the main disseminators, indirect followers and external visitors' proportions are small. As time goes on, more and more indirect followers and external visitors take part in the propagation. Their influence becomes greater, and direct followers' participation gradually drops off. For most tweets, the proportion of external visitors fluctuates from 4 % to 11 %, except for some very hot news tweets.

(3) *Different types of tweets have different main propagators.* For headline news tweets which are externally visible, For news tweets, indirect followers and external visitors contribute a lot in the retweeting compared to other types (Fig. 3 (a)). In comparison, the spread of entertainment stars tweets relies on the direct followers of the author (Fig. 3 (b)). For celebrities who may also have important celebrity followers, both direct and indirect followers play important roles in the tweets spread (Fig. 3(c)).

(4) *If retweeting a tweet is mainly caused by direct followers, the tweet's diffusion trend exhibits an exponential distribution. Otherwise, the tweet's diffusion trend is a superposed result by direct/indirect followers and external visitors.* We make a statistics for about three thousand unnoticed tweets whose retweeting numbers are less than one thousand. We found the messages are mainly retweeted by direct followers. Proportion of external visitors is small enough that could be igored. These tweets' diffusion trends obey exponential distribution as we could see from Fig. 4 (a). The reason might be that direct followers'

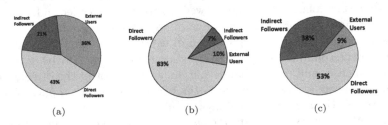

Fig. 3. Proportion of the three retweeting disseminators for (a) a news tweet on MH370 (b) an entertainment star Xie, N.'s tweet (c) a celebrity Ren, Z.Q.'s tweet

participation decays with time. However, we also observe that some notable tweets, such as MH370 whose total retweeting amount is over ten thousand, have an irregular trend rather than an exponential distribution, as many external factors may influence the propagation, not just because of time, as shown in Fig. 4 (b) and (c).

(5) *Retweeting by celebrity followers who have millions of followers will always bring a burst in the process of tweet's retweeting, as their participation could again attract a lot of respective followers to retweet.* The example tweet by Ren, Z.Q. has such a celebrity follower Pan, S.Y., whose retweeting causes another burst in the trend curve in Fig. 4 (c).

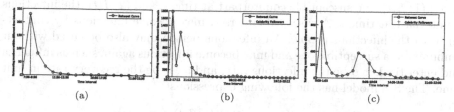

(a)   (b)   (c)

Fig. 4. Retweeting trend for (a) an unnoticed news tweet (b) a hot news an MH370 (c) celebrity Ren, Z.Q.'s tweet

4   Problem Definition

A tweet's retweeting trend can be characterized by multiple *peaks*, *coverage*, and *lifetime*, defined as follows. Let $T=[T.s, T.e]$ denote a time interval T, whose start time is $T.s$ and end time is $T.e$. The temporal length of T is $|T|=T.e-T.s$, which can be an hour, two hours, etc. Let $Ret(T)$ denote the retweeting number of a tweet within time interval T.

Definition 1. *Given a list of equal-length consecutive time intervals $T_1, T_2, \cdots, T_n$, where $\forall i \ (1 \leq i \leq n-1) \ (T_i.e=T_{i+1}.s)$ and $(|T_1|=\cdots=|T_n|)$. $T_1, T_2, \cdots, T_n$ constitute a **complete retweeting period of a tweet**, if and only if $T_1.s$ is the post time of the tweet, and for the last l consecutive time intervals, the retweeting number remains consistently lower than the minimal retweeting amount times a coefficient, i.e., $Ret(T_{n-l}), Ret(T_{n-l+1}), \cdots, Ret(T_n) < g \cdot \min_{1 \leq i \leq n-l-1} Ret(T_i)$, where $(1 \leq l \leq n-1) \wedge (0 < g \leq 1)$. In this study, $l = 4$ and $g = 0.1$.* ☐

Definition 2. *Let $T_1, T_2, \cdots, T_n$ be a complete retweeting period of a tweet.*
*(1) The retweeting **peaks** are a set of pairs $\{(T_i, Ret(T_i) \mid (1 < i < n) \wedge (Ret(T_{i-1}), Ret(T_{i+1}) < Ret(T_i)) \wedge (Ret(T_i) > p \cdot max_{1 \leq j \leq n} Ret(T_j))\}$, where $(0 < p \leq 1)$. In this study, $p=0.5$.*
*(2) The retweeting **lifetime** of the tweet is $n \cdot |T_1|$.*
*(3) The retweeting **coverage** of the tweet is $\sum_{i=1}^{n} Ret(T_i)$.* ☐

Figure 1 illustrates the multiple peaks, lifetime, and coverage in a tweet's retweeting trend.

Given a tweet, the problem of its retweeting trend prediction is to predict the tweet's multiple peaks, lifetime, and coverage after its first launch on micro-blog.

5 Analogy Between Tweets Spread and Epidemic Spread

5.1 Subjects

A classic epidemic model for the spread of infectious diseases is the Susceptible-Infected-Susceptible (SIS) model [3]. It divides the population into two counterparts: (1) $S(t)$: the susceptible counterpart at time t, and (2) $I(t)$: the infectious counterpart at time t. A susceptible person may become infectious by contacting with the infectious people. An infections person may also be cured without immunity as a susceptible one and may become infectious again, such as influenza and enteritis. Let β be the infectious rate, and let α be the recovery rate in unit time. The SIS model has the following expressions.

$$
\begin{aligned}
S(t+1) - S(t) &= -\beta \cdot S(t) \cdot I(t) + \alpha \cdot I(t); \\
I(t+1) - I(t) &= \beta \cdot S(t) \cdot I(t) - \alpha \cdot I(t); \\
N(t) &= S(t) + I(t) \equiv K.
\end{aligned}
\tag{1}
$$

Here K is the total population remaining unchanged.

If we view a tweet message as an infectious disease, action *"retweet"* as "infect", then *retweeters* are like infectors. In the SIS model, becoming a member of susceptible crowd $S(t)$ is subject to the following conditions: (1) A susceptible person has the chance of contacting infectors; (2) A susceptible person has poor immunity to this kind of disease; (3) An infectious person can recover later and become a susceptible one, possibly being infected again.

Accordingly, we could determine the susceptible crowd of tweets propagation on micro-blog: (1) A susceptible person should be a direct follower of the tweet's author or its retweeter, and should be interested in the tweet; (2) An infectious person who has retweeted the tweet may possibly retweet it again, just like a patient's repeated infection. So some infectors may become susceptible again.

5.2 Influence Factors

The reason why a disease becomes an epidemic has two factors: the gene of disease and the infectivity of infectors. If a disease contains a highly pathogenic gene such as H5N1, the disease will be aggressive and easy to spread. If a virus carrier is active and has a lot of contact with others, s/he will increase the disease's spread. Tweets spread in social network obeys the same principle. Two aspepcts contribute to a tweet's propagation: *tweet's gene* and *infector's gene*.

Tweet's Gene. After data analysis, we observe some important features closely relate to the tweet's heat.

- *Hashtag.* Hashtag is the symbol following "#" in a tweet. The content between two "#"s is a topic. If the topic is popular and hot, the tweet will most probably get more attentions.
- *URL.* URL is the link in a tweet. The URL may arouse followers' interest to the unknown content.
- *Multimedia.* Compared with shriveled words, pictures and videos are visual and easy to attract followers' attentions.
- *Post-time.* Tweets published at daytime will be retweeted more than those at wee hours.
- *Content-keywords.* The most important thing determining a tweet's influence is its content. Our experiments show if a tweet contains hot keywords, it will most probably get popular.

Retweeter's Gene. Retweeter's gene includes tweet's author and tweet's retweeters who are retweeting infectors. The tweet propagation from its author to the author's direct followers is viewed as a kind of retweeting action by the author. Different types of users have different influence power, and celebrity's influence is obviously larger than ordinary users'. Below is a list of features about an infector's influence.

- *Number of direct followers.* A retweeter with a large amount of direct followers is always popular in social network.
- *Number of being mentioned.* If retweeter is frequently mentioned, s/he appears to be more popular and active than those less mentioned.
- *Account verification.* A retweeter verified by microblog authentication mechanism is trustworthy and is well received by the public.
- *Account age.* A senior (re)tweeter, who is on micro-blog for a long time, gains lots of popularity and is thus easy to be focused.
- *Retweeting stimulation ratio.* Assume a direct follower A retweets n number of tweets by the same author. Via A, m number of tweets are retweeted by A's direct followers. The retweeting stimulation ratio is the ratio between m and n, i.e., m/n.

5.3 Spread Mechanisms

Differences exist between tweet's retweeting and disease spread mechanisms.

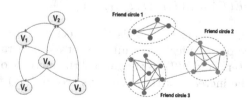

Fig. 5. Strongly-connected graph vs. social network

(1) *The classic SIS epidemic model is based on a strongly connected graph in theory, but social network micro-blog is a weakly connected graph.* In SIS, every infectious person has the chance of infecting others. While on microblog, everyone has his/her own friend circle (Fig. 5). People in different circles may not contact each other, and thus have no chance to infect the other sides. Moreover, in social network, information's transmission is often unidirectional and irreversible. This does not meet the strongly connected graph condition.

(2) *The SIS model assumes the size of the population is a constant, but in social network this is almost impossible.* SIS requires a closed counterpart $N(t)$. However, due to the social network's openness (external visitors), finding a completely closed environment is impossible. Direct application of the SIS model to the social network field is not appropriate.

(3) *In SIS, the infectious rate λ is the same for all infectors. This is not the case for tweet's retweeting: some users have more direct/indirect followers, and thus higher infectious rates than those with less followers.*

6   A Cluster-Based SIS Model for Retweeting Trend Prediction

6.1   The Model

Based on the above analysis, we extend the SIS model for tweets retweeting trend prediction. Each counterpart in the epidemic model is divided into several friend circles. Although the entire social contact graph is not fully connected, each friend circle still meets the connection condition. To highlight celebrity effect of social network, people with different influence power have different infectious rates. We cluster infectious crowd $I(T)$ into small groups according to infectors' infectious rates, i.e., influence power. We also consider the influence of external retweeting visitors. Given a tweet message at time t, we use $I(T)$, $S(T)$, and $E(T)$ to denote the number of retweeters, number of direct followers of the retweeters during time interval T, and number of external retweeting visitors, respectively. Once an external visitor retweets the tweet, s/he becomes a member of the retweeting crowd $I(t)$. Initially, $I(T_1)$ is 1, signifying the tweet's author, $S(T_1)$ is the number of author's direct followers, and $E(T_1)$ is 0.

$$\begin{aligned}
S(T_{i+1}) - S(T_i) &= -\sum_{i=1}^{\#c} \beta_i \cdot S_i(T_i) + \alpha \cdot I(T_i) + \gamma \cdot I(T_i); \\
I(T_{i+1}) - I(T_i) &= \sum_{i=1}^{\#c} \beta_i \cdot S_i(T_i) + E(T_{i+1}) - E(T_i); \\
E(T_{i+1}) - E(T_i) &= \omega(T_i) \cdot (I(T_{i+1}) - I(T_i)).
\end{aligned} \tag{2}$$

Here, $\#c$ is the number of clusters, β_i is the infection rate of the i-th cluster crowd, α is the infectors' reinfection (multiple retweeting) rate, γ is the direct fan-out of the retweeters, and $\omega(T_i)$ is the ratio of the increase of external retweeting visitors versus the increase of the infectious crowd $I(T_i)$ from time interval T_i to T_{i+1}.

6.2   Clustering of Infectious and Susceptible Crowds

Clustering infectious and susceptible crowds into #c groups is based on micro-blog users' influence power (i.e., infectious rates), determined by tweets' and retweeters' genes, described in Sect. 5. It proceeds in the following three steps.

Step-1:Tweets Clustering. A tweet's gene has two types of attributes: textual attribute (*content-keywords*) and numerical attributes (*hashtag, URL, multimedia, post-time*). As tweets on different topics attract different users, for example, some may be interested in politics, while some in entertainment news, and the two infectious/susceptible crowds are independent and have minor intersection. Hence, we first cluster tweets into groups based on their textual contents. We divide tweeter's total tweets by day. On each day's tweets, we extract keywords from their textual contents, and gather similar tweets together by the similarity of their keywords. We merge and sort these similar keywords by their $tf * idf$ values, and take the top-k keywords as the cluster's topic keywords. We gather different days' tweets together by the similarity of their top-k topic keywords, and combine the keywords of similar topics. The process repeats until the clustering result is stable. The advantage of this textual cluster approach is its simplicity and high efficiency compared with classic topic model LDA. It can bring the tweets with the same themes together, such as the air crashes in 2014: MH370's missing, MH17's being shot down, and QZ8501's fatal accident.

After content-based clustering, we further split tweets within one group into several sub-groups using a multi-dimensional KNN (K-Nearest Neighbor) method based on their numerical attributes values. Here, we assign value 1 or 0 to the attribute *hashtag, URL*, and *multimedia* according to their existence in the tweet. Attribute *post-time* is mapped to 0 if the tweet was posted at midnight when most micro-blog users are inactive, and 1 otherwise.

Step-2: Retweeters (Infectors) Clustering. From each tweets cluster obtained after Step-1, we can obtain all the tweets' retweeters (infectors). Based on these retweeters' genes (*number of direct followers, number of being mentioned, account verification, account age,* and *retweeting stimulation ratio*), we cluster these retweeters into several groups via the multi-dimensional KNN method.

Step-3: Susceptible Crowd Clustering. Each retweeters (infectious) cluster leads to a corresponding susceptible crowd cluster, whose members are the direct followers of at lease one retweeter in the former cluster, and are interested in the tweets by either posting/retweeting similar tweets before, or having similar self-description as the tweeter.

6.3   Predicting a Tweet's Retweeting Trend

After splitting the susceptible population into #c clusters for a set of tweets in each category (headline news, celebrity, or entertainment stars), we can apply the least square method to learn parameters $\beta_1, \cdots, \beta_{\#c}, \alpha, \gamma$ in Formula 2. Let Γ be the set of tweets belonging to the same tweet category. Let $T_1, T_2, \cdots, T_n$

be the complete retweeting period of the tweets in Γ, starting from the birth to the end. For a tweet $\tau \in \Gamma$, the prediction error between the real values $S(T_{i+1})$, $I(T_{i+1})$, $E(T_{i+1})$ and the predicted values $S^*(T_{i+1})$, $I^*(T_{i+1})$, $E^*(T_{i+1})$ at time interval T_{i+1} derived from Formula 2 can be expressed through function

$$f = \sum_{\tau \in \Gamma} \sum_{i=1}^{n} [\ (S(T_{i+1}) - S^*(T_{i+1}))^2 + (I(T_{i+1}) - I^*(T_{i+1}))^2 + (E(T_{i+1}) - E^*(T_{i+1}))^2]$$

where

$$S^*(T_{i+1}) = S(T_i) - \sum_{i=1}^{\#c} \beta_i \cdot S_i(T_i) + \alpha \cdot I(T_i) + \gamma \cdot I(T_i);$$
$$I^*(T_{i+1}) = I(T_i) + \sum_{i=1}^{\#c} \beta_i \cdot S_i(T_i) + E(T_{i+1}) - E(T_i);$$
$$E^*(T_{i+1}) = E(T_i) + \omega(T_i) \cdot (I(T_{i+1}) - I(T_i)).$$

To minimize the prediction error $min\ f$, we compute f's partial derivatives $\partial f/\partial \beta_i = 0$ ($i = 1, \cdots, \#c$), $\partial f/\partial \alpha = 0$, $\partial f/\partial \gamma = 0$, $\partial f/\partial \omega = 0$, and obtain the values of parameters $\beta_1 \beta_2 \cdots \beta_{\#c}$, α, γ in Formula 2. We can then estimate the retweeting amounts of a coming tweet at different future time intervals using the materialized Formula 2 obtained in the corresponding category, and finally derive the retweeting peaks, lifetime, and coverage of the tweet according to definition 2. The detail procedure is shown as follows:

1. Initialize the variables of $peaks, coverage$ and $lifetime$. Here $coverage = 0$, $peaks$ is an empty list and $lifetime$ is the tweets start time.
2. Set threshold $\theta_1 = I(T_1), \theta_2 = I(T_1).counter = 0$.
3. While(true)
   i get retweeting number $I(T_i)$ at T_i, $coverage = coverage + I(Ti)$.
   ii Compute $I(T_{i+1})$ with Formula 2, and get retweeting number $I(T_{i-1})$, at T_{i-1}. If $I(T_i) > I(T_{i+1})$ and $I(T_i) < I(T_{i-1})$ and $I(T_i) > 0.5 * \theta_1$, add$(T_i, I(T_i))$ into peaks, update $\theta_1 = I(T_i)$.
   iii If $I(T_i) < 0.1 * \theta_2$, Set $counter = counter + 1$.Else if $I(T_i) < \theta_2$, Update $\theta_2 = I(T_i)$.
   iv when $counter = 4; lifetime = T_i - T_0$ break.
4. End while
5. Get current $peaks, coverage, lifetime$ as the final result.

7   Evaluation

7.1   Set-Up

We crawl three kinds of tweets from sina micro-blog, including 2300 headline news tweets, 1000 celebrities' tweets, and 500 entertainment stars' tweets. We randomly pick up 100 tweets from each category as the test data, use the rest to train and obtain three cluster-based SIS models, each corresponding to one tweet category. The prediction performance is measured by MAPE (Mean Absolute Percentage Error), defined as $MAPE = \sum_{\tau \in \Gamma} \frac{|RealVal - PredictVal|}{RealVal} / |\Gamma|$. Besides, we use recall, precision, and F1-measure to measure multiple peaks prediction results. We compare our cluster-based SIS prediction model (SISC) with another

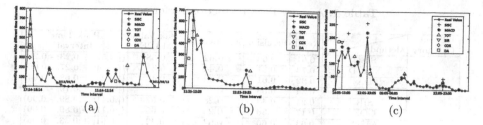

Fig. 6. Retweeting peaks prediction for (a) a Headline news (b) a celebrity Ren, Z.Q.'s tweet (c) an entertainment star Xie, N.'s tweet

four information diffusion models, which can be used or extended to resolve the retweeting trend prediction problem.

- DA (Dynamic Activeness) model [7] predicts information propagation trends (intensity, coverage, and duration) on a DBLP platform based on the concept of node activeness. It uses the law that the decrease of activeness roughly obeys exponential distribution to predict a tweet's future trend.
- Cox proportional hazards regression model [13] is used to predict whether and when a tweet produces its first offspring node based on the features of users and tweets. With this model, people could predict the speed, scale, and range of information's diffusion trends.
- MACD (Moving Average Convergence-Divergence) model [8], which predicts topic trends on Twitter based on the deviation between short-term moving average and long-term moving average. When the short term average is greater than the long term average, the trend will be upward. Conversely, the trend will be downward.
- TOT (Topics over Time) model [12] is a time dependent topic model based on LDA. It models topic distribution conditioned on time stamps. Then the model could determine the topic's future trend according to the distribution.
- SIR model [10] determines threshold conditions for arbitrary cascading models on arbitrary networks. It utilizes the balance point of infectors and rehabilitees to get the threshold of a tweet's diffusion, which is the sign of peak's coming. But the condition of the threshold's existence requires a closed environment, where the total amount of susceptibles and infectors is a constant.

7.2 Performance of Multiple Retweeting Peaks Prediction

Test 1: We randomly pick a tweet from each category, and examine their retweeting peaks (*peak time interval, peak value*). In Fig. 6(a), among the five peaks occuring on a news tweet's retweeting curve, SISC can predict four peaks of them. However, COX and SIR only can predict the first peak at 2014/3/13 17:24 and miss the other four. Although MACD predicts the five peaks, some non-peak points are also involved in the final results. DA and TOT just predict

Table 1. Average Peak Prediction Performance

Tweet Category	Metric / Method	Recall	Precision	F1-Measure	Peak Value MAPE	Peak Time Interval
News	SISC	0.61	0.63	0.62	0.38	(|:0.57; +:0.23; -:0.20)
	MACD	0.73	0.31	0.44	0.35	(|:0.64; +:0.27; -:0.09)
	DA	0.32	0.61	0.42	0.46	(|:0.28; +:0.68; -:0.04)
	TOT	0.34	0.58	0.43	-	(|:0.23; +:0.70; -:0.07)
	SIR	0.21	0.43	0.28	0.52	(|:0.11; +:0.89; -:0.00)
	COX	0.17	0.36	0.23	0.65	(|:0.11; +:0.89; -:0.00)
Celebrity	SISC	0.63	0.66	0.64	0.42	(|:0.61; +:0.33; -:0.06)
	MACD	0.68	0.27	0.39	0.31	(|:0.63; +:0.23; -:0.14)
	DA	0.36	0.54	0.43	0.63	(|:0.22; +:0.68; -:0.10)
	TOT	0.38	0.58	0.46	-	(|:0.20; +:0.69; -:0.11)
	SIR	0.31	0.51	0.39	0.72	(|:0.16; +:0.84; -:0.00)
	COX	0.22	0.32	0.26	0.68	(|:0.14; +:0.86; -:0.00)
Entertainment Star	SISC	0.71	0.78	0.74	0.33	(|:0.68; +:0.20; -:0.12)
	MACD	0.75	0.41	0.53	0.35	(|:0.64; +:0.25; -:0.11)
	DA	0.62	0.67	0.64	0.41	(|:0.59; +:0.20; -:0.21)
	TOT	0.64	0.68	0.66	-	(|:0.56; +:0.32; -:0.12)
	SIR	0.71	0.78	0.74	0.36	(|:0.64; +:0.36; -:0.00)
	COX	0.67	0.55	0.60	0.55	(|:0.61; +:0.39; -:0.00)

\|,+,-: the predicted peak time is equal to/ahead of/behind the real value respectively; - inapplicable

the first and the last one. As the time-dependent TOT model can not predict the precise peak values, we position its prediction results randomly in the figure. In general, SISC has a better performance than other algorithms on a tweet's peaks prediction. The results in Fig. 6(b) and (c) are similar to the one in Fig. 6(a).

Test 2: While Fig. 6 only shows the retweeting trend for an individual tweet, we examine the average performance of the methods on 100 tweets per category. We have two observations from the results in Table 1.

First, for the tweets of news and celebrity, the recall, precision, F1-measure, and MAPE values of our SISC method are (61 %, 63 %, 62 %, 38 %), each of which is better than (32 %, 61 %, 42 %, 46 %) of the closely related DA method. COX's performance is even worse than DA. The reason is that DA and COX are both based on simple mathematical assumptions (DA is based on exponential distribution and COX is based on normal distribution). They fit for long-term trend prediction, but ignore the influence of external factors. The extended MACD exhibits the best recall (73 %) and MAPE (35 %) in peaks prediction with the lowest precision (31 %) among the six methods. MACD's difficulty is the choice of time interval. Shorter time interval may make the deviation between the short and long term averages fluctuate frequently and produce many redundant points, thus reducing the prediction precision of MACD. The performance of the extended TOT and SIR in peaks prediction is not good. Because SIR requires a closed compartmental environment, its threshold determining algorithm can only predict the propagation mainly caused by direct followers, and ignore the burst caused by external factors. TOT suits topic-level trend prediction according to a certain distribution, but for fine-grained tweet-level trend prediction, it is insufficient. Our SISC method's recall and MAPE are close to those of MACD, but with a much better F1-measures.

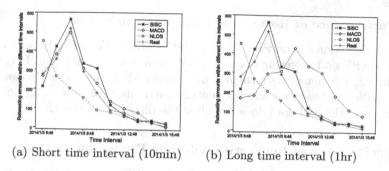

(a) Short time interval (10min)   (b) Long time interval (1hr)

Fig. 7. Peak values prediction with different time interval lengths

Second, for the tweets of entertainment star, the recall, precision, F1-measure, and MAPE values between SISC (71 %, 78 %, 74 %, 33 %) and DA (62 %, 67 %, 64 %, 41 %) are significantly reduced. The differences between SISC and other algorithms show the similar phenomena. The reason is the entertainment star's tweets are mainly retweeted by his/her direct fans, indirect fans and exteranl fans seldom participate. The environment of information's diffusion is nearly closed and the tweet's retweeting trend approximates to exponential distribution, which is fit for the application condition of SIR and DA. The situation is similar to COX and TOT. So for this type of tweets, SISC's performance approaches to other algorithms.

Test 3: To illustrate the limitation of MACD, we set different time intervals for MACD model. Figure 7(a) is the predicted result of short time interval (10 min), and Fig. 7(b) is the predicted result of long time interval (1 h). In Fig. 7(a), we can see the peak values of MACD approximate to real values. While in Fig. 7(b), the performance of MACD is not so good, where the predicted result of peak values is far from the real one. Yet other algorithms' predicted results remain unchanged. This test proves that selecting a appropriate time interval is sensitive to the prediction quality of MACD.

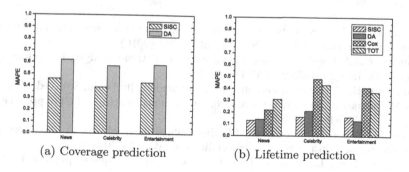

(a) Coverage prediction   (b) Lifetime prediction

Fig. 8. Average performance on three kinds of tweets

7.3   Performance of Retweeting Coverage Prediction

Coverage prediction can only be done by SISC and DA. Figure 8(a) compares the average MAPE on retweeting coverage prediction between SISC and DA, where the former has a better performance on average than the latter. Because DA makes prediction based on an exponential distribution of retweeting amounts. In most cases, the real data does not obey the distribution strictly.

7.4   Performance of Retweeting Lifetime Prediction

Among the six methods, only SISC, DA, COX, and TOT can predict the retweeting lifetime of a tweet. From Fig. 8(b), we can see that SISC and DA perform similarly in predicting tweet's retweeting lifetime on average. The reason is that the trend's termination conditions are similar for both DA and SISC. Compared to the data-driven methods like SISC and DA, the rough distribution (TOT:beta distribution, COX:log-normal distribution) based COX and TOT methods perform worse than SISC and DA, as real-time result is always more reliable than estimated one.

8   Conclusion

In this paper, we draw inspirations from the epidemic dynamic models developed in the medical field to predict retweeting trends on micro-blog. We extend the classic epidemic model (SIS) by clustering micro-blog users with similar retweeting rates together. Our performance study showed that the extended epidemic model is quite effective compared with the existing methods. This model provides a new way for information's diffusion trend analysis on social networks.

Acknowledgement. The work is supported by National Natural Science Foundation of China (61373022, 61073004), and Chinese Major State Basic Research Development 973 Program (2011CB302203-2).

References

1. Boyd, D., Golder, S., Lotan, G.: Tweet, tweet, retweet: conversational aspects of retweeting on twitter. In: Proceedings of HICSS (2010)
2. Budak, C., Agrawal, D., Abbadi, A.: Structural trend analysis for online social networks. In: Proceedings of VLDB (2011)
3. Hethcote, H.: A thousand and one epidemic models. In: Levin, S.A. (ed.) Frontiers in Mathematical Biology. Lecture notes in Biomathematics, vol. 100, pp. 504–515. Springer, Heidelberg (1984)
4. Kim, H., Wang, K., Yoneki, E.: Finding influential neighbors to maximize informationdiffusion in twitter. In: Proceedings of IW3C2 (2014)
5. Kong, S., Feng, L., Sun, G., Luo, K.: Predicting lifespans of popular tweets in microblog. In: Proceedings of SIGIR (2012)
6. Lappas, T., Terzi, E.: Finding effectors in social networks. In: Proceedings of KDD (2010)

7. Lin, S., Kong, X., Yu, P.: Predicting trends in social networks via dynamic activeness model. In: Proceedings of CIKM (2013)
8. Lu, R., Yang, Q.: Trend analysis of news topics on twitter. In: Proceedings of Machine Learning and Computing (2012)
9. Petrović, S., Osborne, M., Lavrenko, V.: RT to win! predicting message propagation in twitter. In: Proceedings of AAAI (2010)
10. Prakash, B.A., Chakrabarti, D., Faloutsos, M., Valler, N., Faloutsos, C.: Threshold conditions for arbitrary cascade models on arbitrary networks. In: Proceedings of ICDM (2011)
11. Suh, B., Hong, L., Pirolli, P., Chi, E.H.: Want to be retweeted? large scale analytics on factors impacting retweet in twitter network. In: Proceedings of SocialCom (2010)
12. Wang, X., McCallum, A.: Topics over time: a non-Markov continuous-time model of topical trends. In: Proceedings of KDD (2006)
13. Yang, J., Counts, S.: Predicting the speed, scale, and range of information diffusion in twitter. In: Proceedings of AAAI (2010)
14. Yang, Z., Guo, J., Tang, J., Li, J., Zhang, L., Su, Z.: Understanding retweeting behaviors in social networks. In: Proceedings of CIKM (2010)
15. Zaman, T., Herbrich, R., van Gael, J., Stern, D.: Predicting information spreading in twitter. In: Proceedings of NIPS Workshop (2010)

Author Index

Aberer, Karl II-215
Adachi, Jun II-162
Agrawal, Divyakant II-417
Akbarinia, Reza I-303, I-509, II-417
Alexandrov, Mikhail II-323
Alibrandi, Stephanie II-117
Almendros-Jiménez, Jesús M. II-268
Al-Sharji, Safiya II-205
Amagasa, Toshiyuki I-334, I-384
Amandi, Analía A. I-111
Arioua, Abdallah I-203
Armentano, Marcelo G. I-111

Bagheri, Ebrahim I-128, I-144
Bai, Lin I-119
Basu, Debabrota I-253
Batko, Michal II-520
Bebel, Bartosz I-416
Becerra-Terón, Antonio II-268
Beer, Martin II-205
Behkamal, Behshid I-144
Bellatreche, Ladjel II-430
Ben Hassen, Mariam II-36
Bennani, Nadia II-145
Berro, Alain II-383
Bianchini, Devis II-369
Bin-Thalab, Rasha II-235
Biondi, Salvatore Michele I-460
Biswas, Sudip I-13
Blanco, Xavier II-323
Bondiombouy, Carlyna I-170
Boukorca, Ahcène II-430
Bozanis, Panayiotis II-285
Brahmi, Hanen II-109
Brahmi, Imen II-109
Brandão, Michele A. I-527
Bravo, Sofia II-117
Bressan, Stéphane I-253, II-250
Bu, Kaili I-119

Caroprese, Luciano II-3
Carswell, James D. II-57

Carvalho, Daniel A.S. II-145
Catania, Vincenzo I-460
Cazalens, Sylvie II-507
Chang, Jae-Woo II-440
Charalambidis, Angelos I-285
Chebel, Mohamed I-517
Chen, Fanghui I-401
Chen, Guihai I-481, II-304
Chen, Weidong I-253
Chiky, Raja II-491
Chong, Eugene Inseok I-351
Cichowicz, Tomasz I-416
Combi, C. II-258
Conca, Piero I-460
Corral, Antonio I-43
Croitoru, Madalina I-203
Cruz, Mateus S.H. I-384
Cullen, Charlie II-57
Cuzzola, John I-128

da Silva, Alexandre Sawczuk II-134
Dang, Tran Khanh II-82
Danilova, Vera II-323
Dave, Vachik S. I-471
Davis Jr., Clodoveu A. I-186
De Antonellis, Valeria II-369
De Salve, Andrea II-479
Di Natale, Raffaele I-460
Dikaiakos, Marios D. I-432
Dou, Dejing I-94, II-339
Dufromentel, Sébastien II-507

Eadon, George I-351
El-Tazi, Neamat II-235

Faiz, Rim II-430
Fang, Yue II-339
Fazzinga, Bettina I-220
Feng, Jinbo II-195
Feng, Ling I-558
Feng, Zhuonan I-558
Fevgas, Athanasios II-285

Flesca, Sergio   I-220
Fung, Benjamin C.M.   I-401

Galante, Renata   II-331
Ganguly, Arnab   I-13
Gao, Libo   II-304
Gao, Xiaofeng   I-481, II-304
Gardiner, Keith   II-57
Garg, Vishesh   I-269
Gargouri, Faïez   II-36
Gaussier, Eric   I-517
Ghedira, Chirine   II-145
Giacomel, Felipe   II-331
Goncalves, Marlene   II-117
Granitzer, Michael   I-76
Gressier-Soudan, Eric   II-491
Gueye, Ibrahima   II-295

Hara, Takahiro   II-93
Haritsa, Jayant R.   I-269
Hasan, Mohammad Al   I-471
Hayashi, Fumitaka   I-334
Hikida, Satoshi   II-153
Hong, Seungtae   II-440
Hou, Wen-Chi   I-155
Hung, Nguyen Quoc Viet   II-215

Ikeda, Kosetsu   II-399
Intilisano, Antonio Rosario   I-460
Iosifidis, Vasileios   II-49
Ispoglou, Kyriakos   II-49

Jajodia, Sushil   II-455
Jaśkowiec, Krzysztof   II-13
Jemal, Dhouha   II-430
Jiang, Shangpu   I-94
Jin, Li   I-558
Jovanovic, Jelena   I-128

Kahani, Mohsen   I-144
Kawakatsu, Takaya   II-162
Kim, Hyung Jin   II-440
Kinoshita, Akira   II-162
Kiritoshi, Keisuke   II-73
Kitagawa, Hiroyuki   I-334, I-384
Kitayama, Daisuke   II-359
Kluska-Nawarecka, Stanisława   II-13
Kolev, Boyan   I-170

Koncilia, Christian   I-416
Konstantopoulos, Stasinos   I-285
Kotidis, Yannis   I-449
Kozawa, Yusuke   I-334, I-384
Krishna Reddy, P.   I-499
Kumar, Sathiya Prabhu   II-491
Kumara Swamy, M.   I-499
Küng, Josef   II-82
Kwon, Yeondae   II-399

Lamarre, Philippe   II-507
Latiri, Chiraz   I-517
Le, Hieu Hanh   II-153
Lee, HyunJo   II-440
Lee, Sanghoon   II-351
Lefebvre, Sylvain   II-491
Legień, Grzegorz   II-13
Lesueur, François   II-507
Levchenko, Oleksandra   I-170
Li, Huiying   I-61
Li, Yiping   I-558
Li, Yuanyuan   I-61
Lin, Qian   I-253
Ling, Tok Wang   II-250
Liroz-Gistau, Miguel   II-417
Litwin, Witold   II-455
Liu, Junqiang   I-401
Liu, Yuang   I-481
Loukopoulos, Thanasis   I-43
Lowd, Daniel   I-94
Lu, Amy Nan   I-367
Lykoura, Anna   I-285

Ma, Hui   II-134
Ma, Qiang   II-73
Makris, Christos   II-49
Manolopoulos, Yannis   I-43
Martinez-Gil, Jorge   II-21
Masini, A.   II-258
Masseglia, Florent   I-303, I-509
Matsumoto, Takuma   II-359
Mavroudi, Effrosyni   I-285
Megdiche, Imen   II-383
Meinel, Christoph   I-319
Melchiori, Michele   II-369
Monteleone, Giuseppe   I-460
Mori, Paolo   II-479
Moro, Mirella M.   I-186, I-527
Morzy, Tadeusz   I-416

Naacke, Hubert   II-295
Nalepa, Filip   II-520
Nawarecki, Edward   II-13
Ndong, Joseph   II-295
Neto, Plácido A. Souza   II-145
Nishio, Shojiro   II-93

Oliboni, B.   II-258

Pallis, George   I-432
Panangadan, Anand   I-543
Pang, Yin   I-119
Panno, Daniela   I-460
Paoletti, Alejandra Lorena   II-21
Papadopoulos, Andreas   I-432
Parisi, Francesco   I-220
Park, Dongwook   II-351
Pereira, Adriano C.M.   II-331
Perez-Cortes, Juan-Carlos   I-3
Phan, Trong Nhan   II-82
Pietramala, Adriana   I-220
Prasanna, Viktor K.   I-543

Rafailidis, Dimitrios   I-432
Raghavan, Ananth   I-351
Ricci, Laura   II-479
Roumelis, George   I-43
Rytwiński, Filip   I-416

Salah, Saber   I-303
Santos, Pedro O.   I-186
Sarr, Idrissa   II-295
Schewe, Klaus-Dieter   II-21
Schwarz, Thomas   II-455
Seifert, Christin   I-76
Senellart, Pierre   I-253
Shah, Rahul   I-13
Shang, Chao   I-543
Shao, Dongxu   II-250
Singh, Abhimanyu   I-269
Śnieżyński, Bartłomiej   II-13
Spyropoulos, Vasilis   I-449
Stamatiou, Yannis C.   II-49
Stavropoulos, Elias C.   II-49
Suzuki, Nobutaka   II-399

Takasu, Atsuhiro   I-28, II-162
Tam, Nguyen Thanh   II-215

Tamani, Nouredine   I-203
Tang, Ruiming   II-250
Tay, Y.C.   I-367
Teste, Olivier   II-383
Thang, Duong Chi   II-215
Tian, Chris Xing   I-367
Tsakalidis, Athanasios K.   II-49
Turki, Mohamed   II-36

Udomlamlert, Kamalas   II-93
Uruchurtu, Elizabeth   II-205

Valduriez, Patrick   I-170, II-417
van der Vet, Paul   I-236
van Keulen, Maurice   I-236
Van, Le Hong   I-28
Vargas-Solar, Genoveva   II-145
Vassilakopoulos, Michael   I-43
Vidal, Maria-Esther   II-117
Vo, Hoang Tam   I-253

Wanders, Brend   I-236
Wang, Cheng   I-319
Wang, Hao   II-339
Wiese, Lena   II-177
Wilk-Kołodziejczyk, Dorota   II-13
Wrembel, Robert   I-416
Wu, Huayu   II-250
Wu, Shengli   II-195
Wu, Yongsheng   I-401

Xia, Lu   I-367
Xu, Feifei   I-61

Yahia, Sadok Ben   I-509, II-109
Yang, Haojin   I-319
Yokota, Haruo   II-153
Yoon, Soungwoong   II-351
Yu, Binxiao   I-401
Yu, Feng   I-155
Yuan, Zihong   I-253

Zeng, Yong   I-367
Zezula, Pavel   II-520
Zhang, Mengjie   II-134
Zhang, Yatao   II-304
Zhang, Yongli   II-339
Zhong, Xinyu   I-61

Zhou, Qingfeng   I-401
Zitouni, Mehdi   I-509
Zorzi, M.   II-258

Zoulis, Nickolas   I-285
Zumpano, Ester   II-3
Zwicklbauer, Stefan   I-76

Printed in the United States
By Bookmasters